Complex Analysis

Second Edition

Complex Analysis

Second Edition

V. Karunakaran

Alpha Science International Ltd.
Harrow, U.K.

V. Karunakaran
School of Mathematics
Madurai Kamaraj University
Madurai, India

Copyright © 2002, 2005
Second Edition 2005

Alpha Science International Ltd.
Hygeia Building, 66 College Road,
Harrow, Middlesex HAI IBE, U.K.

ISBN 1-84265-171-4

Printed in India

Preface to the Second Edition

In this second edition I have divided the book into three parts (see contents). Part I is intended as a one-semester course in Complex Analysis for Undergraduate students. Part II is intended as a one-semester introductory course in Complex Analysis for postgraduate students and Part III is intended as an optional course in Complex Analysis at P.G/M.Phil level. In this edition I have completely reorganized the first chapter of the previous edition so as to be more useful to the Undergraduate students. In Part III, I have added one full section dealing with Analytic automorphisms of plane regions. I have also made the required corrections in respect of misprints and errors found in the first edition. I hope that this edition will be useful to the community of students and teachers of Mathematics.

I thank the many people who have read my first edition, pointed out misprints, errors and gave constructive criticism for the improvement of its contents.

I will be grateful to the readers if they come forward to point out mistakes/errors in this edition and share their comments/suggestions about this book with me.

V. KARUNAKARAN

Preface to the First Edition

The publication of this book fulfils one of my cherished desires. After two decades of teaching complex analysis at various levels, I am of the view that the rudiments of this beautiful topic should be split into two parts—one that should be taught at undergraduate and the other at postgraduate level. Thus, first three chapters (except for one section on Riemann Surfaces in Chapter 3) have been designed as part of an undergraduate course and Chapters 4 and 5 (except for two sections on Univalent Functions) can form a single semester course for postgraduate students.

In writing this book I have consulted several books on Complex Analysis (all of these have been included as references) and the fact that I have been influenced by these books is reflected in my presentation of this topic. This book can also be used for self-study by students of non-formal educational systems.

I have tried to make this book as self-contained as possible. However, certain amount of exposure to basic real analysis, basic topology, familiarity with concepts such as vector spaces and linear mappings between them are assumed on the part of the reader. Several exercises (some simple and others not so simple) are solved and a few others listed for each chapter. In the "notes" at the end of each chapter I have provided informations (some times historical too) for further studies with regard to a few topics in the relevant chapter with proper references. I hope that the student community will be benefited by reading this book.

I sincerely thank my former research students Dr. S. Subbulakshmi, Dr. A. Vijayan, Dr. T. Venugopal, Dr. N.V. Kalpakam and Mrs. V. Baby Thilaga for listening to my "evening lectures" on complex analysis and also for having discussions with me on these lectures. Indeed these activities have been the main source of inspiration for writing this book. My special thanks are due to Dr. T. Venugopal and Dr. N.V. Kalpakam for their valuable help in typesetting the manuscript in a computer using "ChiWriter". I must also record my sense of appreciation due to my wife K. Padma, son K. Srinivasan and daughter K. Sukanya for their encouragement, cooperation and support during the long period I have taken to write this book. My thanks are also due to Mr. R. Roopkumar, Mr. R. Vembu and Ms. K Bhuvaneswari for their help at the "proofreading" stage. I am also indebted to "M/s Macrosoft Computer Solutions", Tuticorin, in converting my illustrations into computer graphics. My thanks are due in no small measure to M/s Narosa Publishing House for publishing this book and for the neat execution of this job.

In spite of my sincere and best efforts, it is likely that there are errors in this book and I will be highly obliged to the readers if they come forward and bring these to my notice. I will be happy to incorporate the required corrections in the next edition of this book.

V. KARUNAKARAN

Contents

PART II: COMPLEX INTEGRATION

PART III: SPECIAL TOPICS IN COMPLEX ANALYSIS

A Bird's Eye View of the Complex Plane

1.1 Introduction

The earliest traces of the concept of square root of a negative number can be seen in Stereo Metrica of Heron of Alexandria, Arithmetica of Diophantus and Beeja Ganitha of Bhaskara. As early as 1494, Pacioli states in his book Suma that the quadratic equation $x^2 + c = bx$ cannot be solved unless $b^2 \geq 4c$ recognizing the impossibility of finding the value of $\sqrt{-a}$ $(a > 0)$. Wallis (1673) seems to be the first to have a graphic idea of the square root of negative numbers. He states that the notion of the square root of a negative number may seem impossible but the same can also be said of negative numbers themselves although one is able to explain the latter with the aid of physical applications. Further, he suggests that one can draw a line perpendicular to the real axis which can be called the imaginary axis. In 1702, J. Bernoulli introduced the complex numbers to higher mathematics by exhibiting the relationship between $\tan^{-1} y/x$ and the logarithm of $x + iy$. Again a British mathematician Cotes proved that

$$\log(\cos\phi + i\sin\phi) = i\phi$$

establishing the famous Euler's formula $e^{i\phi} = \cos\phi + i\sin\phi$ and the de Moivre's relation

$$(\cos\phi + i\sin\phi)^n = \cos n\phi + i\sin n\phi.$$

The use of i instead of $\sqrt{-1}$ is due to Euler in 1748. Cauchy suggested the name conjugates for $a + ib$ and $a - ib$ and the name modulus of $a + ib$ for $\sqrt{a^2 + b^2}$. Weierstrass calls modulus of $a + ib$ as the absolute value of $a + ib$ and writes it as $|a + ib|$.

In this chapter we shall divide the elementary study of complex numbers into five sections, viz.,

(i) Development of the complex numbers system and its algebra,

(ii) Arithmetic of the complex numbers,
(iii) Geometry of the complex plane,
(iv) Topology of the complex plane,
(v) Analysis of the complex domain,

and study them in some detail. However any study of pure mathematics requires a certain amount of mathematical logic, a brief account of which is given below.

1.2 Mathematical Logic

In mathematics we have to use precise language to communicate ideas. Since natural languages are not always precise we shall develop a special object language containing a set of declarative sentences. These we will also call as primary sentences or variables. Only those declarative sentences will be admitted in the object language which have one and only one of the two possible values called truth values. The two truth values are TRUE and FALSE and are denoted by the symbols T and F respectively. We shall use negation (complementation with connective symbol ~), conjunction ("and" with connective symbol $\wedge$) disjunction ("or" with connective symbol $\vee$) conditional ("implication" with connective symbol $\Rightarrow$), biconditional ("implies and is implied by" with connective symbol $\Leftrightarrow$) etc., to develop new compound statements from the given declarative statements A and B. A truth table indicating the validity of the new statements depending on the validities of A and B is given below.

A	B	$\sim A$	$A \vee B$	$A \wedge B$	$A \Rightarrow B$
T	T	F	T	T	T
T	F	F	T	F	F
F	T	T	T	F	T
F	F	T	F	F	T

A statement formula (or a mathematical statement) is an expression which is a string consisting of variables, parentheses and connective symbols. If A and B are two statement formulas and $P_1, P_2, \ldots, P_n$ denote all the variables (declarative statements) contained in both A and B we say A and B are equivalent if the truth values of A equals to the truth values of B for every one of the 2^n possible sets of truth values assigned to $P_1, P_2, \ldots, P_n$. For example $\sim\sim A$ is equivalent to A, $A \vee (\sim A)$ is equivalent to $B \vee (\sim B)$, $A \Rightarrow B$ is equivalent to $(\sim A) \vee B$ and that $A \Rightarrow B$ is equivalent to $(\sim B) \Rightarrow (\sim A)$ (also called contra positive of $A \Rightarrow B$). We shall not go into this in more detail and we refer the interested reader to [5] for further reading. However we shall emphasize that formulating a negation of a given mathematical statement is very important. We shall illustrate this in the following examples.

1. Let $f : \mathbb{R} \to \mathbb{R}$ be a real valued function of a real variable $x \in \mathbb{R}$ and $x_0 \in \mathbb{R}$. Let A be the following statement. Given any $\varepsilon > 0$ we can find

a $\delta > 0$ such that whenever $|x - x_0| < \delta$, we have $|f(x) - f(x_0)| < \varepsilon$. The negation, $\sim A$ will read as follows. We can find at least one $\varepsilon > 0$ (say ε_1) such that given any $\delta > 0$ there exists a corresponding x (at least one) with $|x - x_0| < \delta$ and $|f(x) - f(x_0)| \geq \varepsilon_1$.

2. Let $f : \mathbb{R} \to \mathbb{R}$ be a real-valued function of a real variable $x \in \mathbb{R}$. There exists $x_0 \in \mathbb{R}$ with the following property which we call as A. Given any $y \in \mathbb{R}$, any $\varepsilon > 0$ and $\delta > 0$ we can find $x_1 \in \mathbb{R}$ such that $|x_1 - x_0| < \delta$ and $|f(x_1) - y| < \varepsilon$. If this property were not true at the given point x_0, we have to formulate $\sim A$ which is given below. There exists at least one y say y_1, one $\varepsilon_1 > 0$ and one $\delta_1 > 0$ such that for all x with $|x - x_0| < \delta_1$ we have $|f(x) - y_1| \geq \varepsilon_1$.

3. Let $f : \mathbb{R} \to \mathbb{R}$ be a real valued function of a real variable $x \in \mathbb{R}$. The mathematical sentence A is the following. There is an $\varepsilon > 0$ and a real number N such that $|f(x)| \geq \varepsilon$ for $x \geq N$. The negation $\sim A$ of A, can be formulated as follows. Given any $\varepsilon > 0$ and any real number N we can find a point $x \geq N$ with $|f(x)| < \varepsilon$.

The reader should be able to learn from these examples how negations of mathematical sentences can be formulated.

An important difference between the reasoning used in any general discussion and that used in mathematics is that the premises (assumptions) used in general discussions are believed to be true either from experience or from faith, (these however can be questioned) and if proper rules are followed, then one expects the conclusion to be true. In mathematics, one is solely concerned with the conclusion which is obtained by following the rules of logic. This conclusion, called a theorem, can be inferred from a set of premises, called the axioms of the theory (and these cannot be questioned). The actual truth values of these axioms play no part in the theory.

A set of mathematical sentences $H_1, H_2, \ldots, H_m$ is said to be consistent if their conjunction has the truth value T for some assignment of the truth values to the variables appearing in $H_1, H_2, \ldots, H_m$. If, for every assignment to the truth values of the variables, at least one of the mathematical sentences $H_1, H_2, \ldots, H_m$ is false then the set of mathematical sentences $H_1, H_2, \ldots, H_m$ is said to be inconsistent.

Alternatively, a set of mathematical sentences $H_1, H_2, \ldots, H_m$ is inconsistent if their conjunction implies a contradiction (a statement which is always false), that is,

$$H_1 \wedge H_2 \wedge \ldots \wedge H_m \Rightarrow R \wedge \sim R$$

where R is any mathematical sentence. Note that $R \wedge \sim R$ is a contradiction, and thus the above implication is necessary and sufficient for $H_1 \wedge H_2 \wedge \ldots \wedge H_m$ to be a contradiction.

The notion of inconsistency is used in a procedure called *proof by contradiction or reductio ad absurdum or indirect method of proof.* In order to show that a conclusion C follows logically from the premises H_1, H_2, ..., H_m, we assume that C is false and consider $\sim C$ as an additional premise. If the new set of premises is inconsistent, so that they imply a contradiction, then the assumption that $\sim C$ is true does not hold simultaneously with $H_1 \wedge H_2 \wedge ... \wedge H_m$ being true. Therefore, C is true whenever $H_1 \wedge H_2 \wedge ... \wedge H_m$ is true. Thus, C follows logically from the premises H_1, H_2, ..., H_m.

In every theorem we will have a set of hypothesis whose validity is assumed to be true and we will be required to prove a set of conclusions using mathematical logic. Thus every theorem looks like $A \Rightarrow B$. However we can also prove this by proving $\sim B \Rightarrow \sim A$ as both are equivalent statements.

Any mathematical theory starts with certain axioms (statements whose validity is not questioned) and proceeds using mathematical logical to deduce conclusions and results. For example the Euclidean geometry assumes the parallel axiom viz. parallel lines do not meet. In pure mathematics we have several axioms one of which we shall state here and see how it is usefull in applications. This is the 'well-ordering principle' which states that "Any non-empty subset of natural numbers has a least element". We shall show how this axiom enables us to prove that $\sqrt{2}$ is irrational. We invite the reader to compare this with the usual proof.

If $\sqrt{2}$ were rational then we can find an integer $q \neq 0$ such that $q\sqrt{2}$ is also an integer. (Recall that a rational number is of the form p/q where p and q are integers with $q \neq 0$). Let S be the set of all positive integers which when multiplied by $\sqrt{2}$ gives an integer. It is sufficient to prove that $S = \phi$, the empty set. If S were non-empty then using well-ordering principle choose the least element say l of S. We know that $l\sqrt{2}$ is an integer and hence $l(\sqrt{2} - 1)$ is also an integer. But $l(\sqrt{2} - 1)$ when multiplied by $\sqrt{2}$ gives $l(\sqrt{2} - 1)\sqrt{2} = 2l - \sqrt{2}\, l$ which is also an integer. Hence $l(\sqrt{2} - 1)$ also belongs to S contradiction the minimality of l in S (Note that $l(\sqrt{2} - 1) < l$).

Another application of well-ordering principle is towards the proofs of the following principles.

(i) First principle of mathematical induction.

If a mathematical statement concerning natural numbers is true for $n = 1$ and whenever it is true for $n = m$ it is true for $n = m + 1$ then it is true for all natural numbers.

(ii) Second principle of mathematical induction.

If a mathematical statement concerning natural numbers is true for $n = 1$ and whenever it is true for all natural numbers $n < m$ it is true for m then the statement is true for all natural numbers.

The reader is encouraged to prove both the above principles using the well-ordering principle.

1.3 Development of the Complex Numbers System and Its Algebra

We shall assume that the reader is familiar with set theory, elementary algebraic structures like groups, rings, fields, vector spaces, ordered fields, and some elementary real analysis at the graduate level.

The numbers 1, 2, 3... which we use for counting purposes are called natural numbers. The set of natural numbers is denoted by $\mathbb{N}$. It was Peano who in 1899 made a successful attempt in describing the algebraic structure of the set of natural numbers. He was able to derive all the properties of the natural numbers system using the frame work of a set of five axioms.

Assuming the addition, multiplication and the order structure available for the set of natural numbers we consider the equation $x + m = n$ where n, m are natural numbers. When m is less than n we can always solve the equation within the natural number system. However, if m is greater than or equal to n, we have no solution in $\mathbb{N}$. This is because $\mathbb{N}$ has neither additive inverse for its elements nor an identity. Thus it is desirable to have a number system having all the properties of $\mathbb{N}$ intact and which possesses additive identity and additive inverse for all its elements. This extended system is called the system of integers and is denoted by $\mathbb{Z}$. This system is obtained from $\mathbb{N}$ as follows:

Consider $\mathbb{N} \times \mathbb{N}$. Define the following equivalence relation $\sim$ on $\mathbb{N} \times \mathbb{N}$: $(m, n) \sim (p, q)$ if and only if $m + q = n + p$. This equivalence relation $\sim$ partitions $\mathbb{N} \times \mathbb{N}$ into various equivalence classes each of which we call as an integers and denote the set of integers by $\mathbb{Z}$. Denoting the equivalence class containing (m, n) as $[(m, n)]$ we can introduce addition and multiplication in $\mathbb{Z}$ as follows:

$$[(m_1, n_1)] + [(m_2, n_2)] = [(m_1 + m_2, n_1 + n_2)]$$

and

$$[(m_1, n_1)] \cdot [(m_2, n_2)] = [(m_1 \cdot m_2 + n_1 \cdot n_2, m_1 n_2 + m_2 \cdot n_1)]$$

and verify that $\mathbb{Z}$ becomes a ring under these operations.

We can also imbed $\mathbb{N}$ in $\mathbb{Z}$ by the map $f : \mathbb{N} \to \mathbb{Z}$ defined by

$$f(n) = [(n + m, m)] \text{ for any } m \in \mathbb{N}$$

and verify that f is a bijection of $\mathbb{N}$ into (not onto) $\mathbb{Z}$ preserving 'addition' and 'multiplication'. Now we can solve the equation

$$x + m = n \qquad (m, n \in \mathbb{N} \text{ with } m > n)$$

which was not solvable in $\mathbb{N}$ in this system $\mathbb{Z}$ by interpreting the equation

as $\qquad x + [(m + p, p)] = [(n + q, q)] \ (p, q \in \mathbb{N})$

and verifying that $x = [(n, m)]$ solves this equation in $\mathbb{Z}$.

Conventionally we write $[(m, m)]$ as 0, $[(n + m, m)]$ as n and $[(m, n + m)]$ as $-n$ so that the solution we got by solving the equation $x + m = n$ can now be written as $x = n - m$. This also helps us to identify $\mathbb{N}$ in the usual way in $\mathbb{Z}$.

We again observe that an equation of the form $ax = b$ with $a \neq 0$ where a and b belong to $\mathbb{Z}$ may or may not admit solutions in $\mathbb{Z}$ depending on the integers a and b. Hence we develop the rational numbers system $\mathbb{Q}$ which is formally defined as equivalence classes in $\mathbb{Z} \times (\mathbb{Z} \setminus \{0\})$ with reference to the equivalence relation $\sim$ defined as follows: $(a, b) \sim (c, d)$ if and only if $ad = bc$. Again we can introduce addition and multiplication in $\mathbb{Q}$ as follows. As usual denoting the equivalence class containing (p, q) as $[(p, q)]$

$$[(p_1, q_1)] + [(p_2, q_2)] = [(p_1 \cdot q_2 + p_2 \cdot q_1, q_1 \cdot q_2)]$$

and

$$[(p_1, q_1)] \cdot [(p_2, q_2)] = [(p_1 \cdot p_2, q_1 \cdot q_2)].$$

In fact $\mathbb{Q}$ becomes a field under this addition and multiplication. On the other hand the natural order relation in $\mathbb{Z}$ can be extended to $\mathbb{Q}$ and hence $\mathbb{Q}$ becomes an ordered field. As usual we can identify $\mathbb{Z}$ in $\mathbb{Q}$ by the mapping $f : \mathbb{Z} \to \mathbb{Q}$ defined by

$$f(p) = [(p, 1)].$$

As before we interpret the equation $a \cdot x = b$ which was not solvable in $\mathbb{Z}$ if b is not a multiple of a by writing the equation as

$$[(a, 1)] \cdot x = [(b, 1)]$$

and giving the solution $x = [(b, a)]$. Again to keep the notations of elements of $\mathbb{Z}$ in $\mathbb{Q}$ intact we write $[(a, 1)]$ as a and $[(1, a)]$ as $\dfrac{1}{a}$ so that the solution x can be written as $\dfrac{b}{a}$ $\left(= b \cdot \dfrac{1}{a}\right)$.

The Archimedean property (for any $p, q \in \mathbb{Q}$ with $0 < p < q$ we can find $n \in \mathbb{N}$ such that $n \cdot p \geq q$) can also be verified in $\mathbb{Q}$ and hence the rational system becomes an Archimedean ordered field.

So far we have seen that the rational number system is closed for addition, subtraction, multiplication and division by non-zero numbers. From all these it may appear that $\mathbb{Q}$ is an adequate system of numbers for all practical purposes. But a careful look shows that this is not so. For instance, the equation $x^2 = 2$ has no solution in $\mathbb{Q}$. Hence we develop the real number system which contains the rational numbers and the so-called irrational numbers. There are two ways of constructing the real number system $\mathbb{R}$ from $\mathbb{Q}$. One method is due to Cantor and another due to Dedekind. Cantor's theory uses fundamental sequences of rational numbers and Dedekind's theory uses subsets of rational numbers which are called Dedekind's cuts. We shall assume that the reader is familiar with this development.

Recall that the real number system denoted by $\mathbb{R}$ is a field under the usual addition and multiplication. The numbers 0 and 1 are identity elements with respect to addition and multiplication respectively. Further the field $\mathbb{R}$ has an order relation (the usual order) denoted by '<' or '>' under which $\mathbb{R}$ becomes an Archimedean ordered field.

The various stages in the construction of the real number system starting from the natural number system can be viewed from the stand point of solutions of polynomial equations. Extending this argument further we see that the construction of the real number system is only partially satisfactory in the sense that any equation of the form $x^n = a$ has a solution for some n and a, but fails to have a solution when n is even and a is negative. In fact, using the properties of the order relation in $\mathbb{R}$, it is easy to prove that square of any real number is either zero or positive. Since $1 = 1^2$, 1 is always positive. Hence the equation $x^2 + 1 = 0$ cannot be solved in $\mathbb{R}$ because $x^2 + 1$ is always positive. To improve the situation we construct an extension of the real number system to a field in which the equation $x^2 = -1$ has a solution.

Let us now suppose that we can find a field $\mathcal{F}$ in which $\mathbb{R}$ is a subfield and wherein the equation $x^2 + 1 = 0$ can be solved (we have not so far shown the existence of such a field). Now consider a solution of this equation in $\mathcal{F}$. Let it be denoted by i so that $x^2 + 1 = (x + i)(x - i)$ and hence the equation has exactly two solutions in $\mathcal{F}$ namely i and $-i$. Consider the subset $\mathbb{C}$ of $\mathcal{F}$ consisting of all elements of the form $a + ib$ where a and b are reals. Here the + and . are the additions and multiplication in the field $\mathcal{F}$ and $\mathbb{R}$ is a subfield of $\mathcal{F}$ and a, b are elements of $\mathbb{R} \subset \mathcal{F}$. We first note that there cannot be two such representations for an element of $\mathcal{F}$ because if $a + ib = c + id$ then $a - c = i(b - d)$ and hence $(a - c)^2 = - (b - d)^2$. Since a, b, c, d are real this means to say that $(a - b)^2 + (b - d)^2 = 0$ or that $a = c$ and $b = d$. Now the subset $\mathbb{C}$ is indeed a subfield of $\mathcal{F}$. This is easily verified. On the other hand the structure of $\mathbb{C}$ is independent of the super field $\mathcal{F}$. This means that if $\mathcal{F}'$ is another field containing $\mathbb{R}$ and a root i' of the equation $x^2 + 1 = 0$ and the corresponding subset $\mathbb{C}'$ consists of $a + i'b$ with $a, b \in \mathbb{R}$ then $\mathbb{C}$ and $\mathbb{C}'$ are isomorphic as fields. In fact the isomorphism is given by $a + ib \rightarrow a + i'b$. Thus the field of complex numbers can be identified as a subfield of any field $\mathcal{F}$ which contains $\mathbb{R}$ as a subfield and in which $x^2 + 1 = 0$ has a solution. To construct such a field we consider the set $\mathbb{C}$ of ordered pairs (a, b) where a and b are real numbers and define addition and multiplication (using those in $\mathbb{R}$) as follows:

$$(a, b) + (c, d) = (a + c, b + d) \text{ and}$$
$$(a, b) \cdot (c, d) = (ac - bd, ad + bc).$$

It is easy to verify that $\mathbb{C}$ is a field under this addition and multiplication. The identity with respect to addition is $(0, 0)$, and the identity with respect to multiplication is $(1, 0)$. The additive inverse of (a, b) is $(-a, -b)$ and for $(a, b) \neq (0, 0)$ the multiplicative inverse is $\left(\dfrac{a}{a^2 + b^2}, \dfrac{-b}{a^2 + b^2} \right)$. The elements of the form $(a, 0)$ constitute a subfield isomorphic to $\mathbb{R}$ the isomorphism being $a \rightarrow (a, 0)$ and the element $(0, 1)$ solves the equation $x^2 + 1 = 0$. This is because $(0, 1) . (0, 1) = (-1, 0)$ and $(-1, 0) + (1, 0) = (0, 0)$. Hence the field $\mathbb{C}$ has the required properties and by our definition $\mathbb{C} = \{(a, 0) + i(0, b) : a, b \in \mathbb{R}\}$ or if we identify the elements of $\mathbb{C}$ which are images of real numbers (under the stated isomorphism) with $\mathbb{R}$, we can write a typical element of $\mathbb{C}$ as $a + ib$ where a and b are in $\mathbb{R}$. We can also think of the complex number system ($\mathbb{C}$) as the quotient field of the ring of polynomials with real coefficients modulo the irreducible polynomial $x^2 + 1$. In fact, every element of the quotient field will be of the form $ax + b + \langle x^2 + 1 \rangle$ where $\langle x^2 + 1 \rangle$ is the ideal generated by $x^2 + 1$. This element can be identified with a complex number $ai + b$. It is easy to see that this identification f is a field isomorphism because of the following identities:

$$f((ax + b) + \langle x^2 + 1 \rangle + (cx + d) + \langle x^2 + 1 \rangle)$$
$$= f((a + c)x + (b + d) + \langle x^2 + 1 \rangle)$$
$$= (a + c)i + (b + d)$$
$$= (ai + b) + (ci + d)$$
$$= f((ax + b) + \langle x^2 + 1 \rangle) + f((cx + d) + \langle x^2 + 1 \rangle).$$

Similarly,

$$f[((ax + b) + \langle x^2 + 1 \rangle) . ((cx + d) + \langle x^2 + 1 \rangle)]$$
$$= f(acx^2 + bd + (bc + ad)x + \langle x^2 + 1 \rangle)$$
$$= f(ac(x^2 + 1) - ac + bd + (bc + ad)x + \langle x^2 + 1 \rangle)$$
$$= f((bc + ad)x + bd - ac + \langle x^2 + 1 \rangle)$$
$$= (bc + ad)i + (bd - ac)$$
$$= (ai + b) \cdot (ci + d)$$
$$= f((ax + b) + \langle x^2 + 1 \rangle) \cdot f((cx + d) + \langle x^2 + 1 \rangle).$$

Again, the field of complex numbers is also isomorphic to the field M of all 2×2 real matrices of the special form $\begin{pmatrix} a & b \\ -b & a \end{pmatrix}$ wherein matrix addition and matrix multiplication are the operations. In fact the required isomorphism is

$$F : M \rightarrow \mathbb{C} \text{ defined by } f\left(\begin{pmatrix} a & b \\ -b & a \end{pmatrix}\right) = a + ib,$$

as can be easily verified.

While $\mathbb{R}$ is an ordered field, the field of complex numbers is not an ordered field (recall that an order relation '<' should satisfy the following properties (i) For any two elements x and y, either $x < y$ (also written as $y > x$) or $x = y$ or $y < x$. (ii) '<' is transitive. Also in an ordered field, if $x > 0$ and $y > 0$ they $xy > 0$ and if $x < 0$ and $y < 0$ then $xy > 0$ and if $x > 0$ implies $-x < 0$ and vice versa). Using these properties we can easily see that the field of complex numbers cannot be ordered. In fact, since $i \neq 0$, either i should be greater than 0 or i should be less than 0. But since $1^2 = 1$, $1 > 0$. If $i > 0$ then $i^2 > 0$ which implies that $-1 > 0$ or that $1 < 0$ which is a contradiction. Similarly $i < 0$ also leads to a contradiction and hence $\mathbb{C}$ cannot be ordered.

1.4 Arithmetic of the Complex Numbers

If a complex number z is the form $x + iy$, x, y real then x is called the real part of z and y the imaginary part of z and we write $x = Re\, z$ and $y = Im\, z$. The complex conjugate $\bar{z}$ of $z = x + iy$ is defined by $\bar{z} = x - iy$. If $z = x + iy$ and $w = u + iv$, then $z = w$ if and only if $x = u$ and $y = v$ and in particular $z = 0$ if and only if both x and y are zero. If $z = x + iy$ then $-z = (-x) + i(-y)$. We define $|z|$ as the non-negative square root of $x^2 + y^2$ and also call it the absolute value of z or modulus of z. We list the following properties of the complex numbers

(i) $|z|^2 = z\,\bar{z}$

(ii) $Re\, z = (z + \bar{z})/2$ and $Im\, z = (z + \bar{z})/2i$

(iii) $\overline{z + w} = \bar{z} + \bar{w}$

(iv) $\overline{z \cdot w} = \bar{z} \cdot \bar{w}$

(v) $|zw| = |z|\,|w|$

(vi) $|z| = |\bar{z}|$

(vii) If $z = x + iy$ with $z \neq 0$ then $Re\,(1/z) = x/(x^2 + y^2)$ and $Im\,(1/z) = -y/(x^2 + y^2)$.

These properties are easily verified using the definitions.

Theorem 1.4.1

If z and w are two complex numbers then we have

(i) $|z + w|^2 = |z|^2 + |w|^2 + 2Re\, z\bar{w}$

(ii) $|z - w|^2 = |z|^2 + |w|^2 - 2Re\, z\bar{w}$

(iii) $|z + w| \leq |z| + |w|$ *(triangle inequality)*

(iv) $|z + w|^2 + |z - w|^2 = 2(|z|^2 + |w|^2)$

(v) $|z - w| \geq ||z| - |w||$

(vi) *If z_i $(1 \leq i \leq n)$ are complex numbers then* $|z_1 + z_2 + \ldots + z_n| \leq |z_1| + |z_2| + \ldots + |z_n|$ *with equality if and only if the ratio of any two non-zero terms is real and positive.*

(vii) *(Cauchy inequality)* $\left|\sum_{i=1}^{n} z_i \bar{w}_i\right|^2 \leq \sum_{i=1}^{n} |z_i|^2 \sum_{i=1}^{n} |w_i|^2$ *where z_i and w_i are*

complex numbers.

Proof

$$|z + w|^2 = (z + w) \left(\overline{z + w}\right) = (z + w) (\overline{z} + \overline{w}) = z\overline{z} + w\overline{w} + z\overline{w} + w\overline{w}$$

$$= |z|^2 + |w|^2 + 2 \, Re \, z\overline{w}$$

This proves (i).

Similarly $|z - w|^2 = |z|^2 + |w|^2 - 2 \, Re \, z\overline{w}$ proving (ii).

Adding (i) and (ii) we get (iv).

To prove (iii) we observe that if $z = x + iy$ then

$$Re \, z = x \leq \sqrt{x^2 + y^2} = |z|, \; - Re \, z = -x \leq \sqrt{x^2 + y^2} = |z|$$

and hence $-|z| \leq Re \, z \leq |z|$ and similarly $-|z| \leq Im \, z \leq |z|$. Now

$$|z + w|^2 = |z|^2 + |w|^2 + 2 \, Re \, z\overline{w} \leq |z|^2 + |w|^2 + 2 |z\overline{w}|$$

$$\leq |z|^2 + |w|^2 + 2|z| \, |w| = (|z| + |w|)^2.$$

Taking positive square roots on both sides we get $|z + w| \leq |z| + |w|$.

$$|z - w|^2 = |z|^2 + |w|^2 - 2 \, Re \, z\overline{w} \geq |z|^2 + |w|^2 - 2 |z\overline{w}|$$

$$\geq |z|^2 + |w|^2 - 2|z| \, |w| = (|z| - |w|)^2$$

Taking the positive square roots on both sides we get $|z + w| \geq ||z| - |w||$.

To prove (vi) we observe that the inequality for $n = 2$ is the same as (iii). We prove the general case by induction. Assume that the inequality is true for $n = m$. Then

$$|z_1 + z_2 + \ldots + z_{m+1}| \leq |z_1 + z_2 + \ldots + z_m| + |z_{m+1}| \qquad \text{(by (i))}$$

$$\leq |z_1| + |z_2| + \ldots + |z_m| + |z_{m+1}|$$

by induction hypothesis.

Let us consider the case of equality. If all these numbers are zero there is nothing to prove. So assume $z_1 \neq 0$. We first observe the equality in (iii) with $z \neq 0$ and $w \neq 0$ holds if and only if z/w is real and positive. This is easy to verify because the required equality is same as $Re \, z\overline{w} = |z\overline{w}|$ which happens if and only if $z\overline{w}$ is real and positive which is equivalent to $|w|^2(z/w)$ is real and positive or that z/w is real and positive. Now suppose that equality holds in (vi) with $z_1 \neq 0$ and $z_2 \neq 0$. Then

$$|z_1| + |z_2| + \ldots + |z_n| = |z_1 + z_2 + \ldots + z_n| = |(z_1 + z_2) + \ldots z_n|$$

$$\leq |z_1 + z_2| + |z_3| + \ldots + |z_n|$$

$$\leq |z_1| + |z_2| + \ldots |z_n|$$

Hence $|z_1 + z_2| = |z_1| + |z_2|$ and hence z_2/z_1 is real and positive. Since the numbering is arbitrary it also follows that the ratio of any two non-zero terms must be positive. Conversely, let this condition be satisfied. If $z_1 \neq 0$ we obtain

$$|z_1 + z_2 + \ldots + z_n| = |z_1| \, (|1 + z_2/z_1 + \ldots + z_n/z_1|)$$

$$= |z_1| \, (1 + z_2/z_1 + \ldots + z_n/z_1)$$

$$= |z_1| (1 + |z_2/z_1| + \ldots + |z_n/z_1|)$$
$$= |z_1| + |z_2| + \ldots + |z_n|$$

This proves (vi).

To prove the Cauchy's inequality we let λ to be any arbitrary complex number and get

$$\sum_{i=1}^{n} |z_i - \lambda w_i|^2 = \sum_{i=1}^{n} |z_i|^2 + |\lambda|^2 \sum_{i=1}^{n} |w_i|^2 - 2\operatorname{Re} \bar{\lambda} \sum_{i=1}^{n} z_i \bar{w_i}$$

The above expression being the sum of non-negative terms is always greater than or equal to zero for any λ. If all the w_i's are zero then the inequality is trivial. Hence we can assume that not all w_i's are zero.

Take $\quad \lambda = \sum_{i=1}^{n} z_i \bar{w_i} \Big/ \sum_{i=1}^{n} |w_i|^2 \quad$ and put

$$A = \sum_{i=1}^{n} |z_i|^2, B = \sum_{i=1}^{n} |w_i|^2 \neq 0 \text{ and } C = \sum_{i=1}^{n} z_i \bar{w_i}.$$

Therefore, we have

$$A + |\lambda|^2 B - 2\operatorname{Re} \bar{\lambda} C \geq 0.$$

Substituting C/B for λ we get $AB + |C|^2 - 2|C|^2 \geq 0$ or that $|C|^2 \leq AB$, which is the required inequality.

1.5 Geometry of the Complex Plane

With respect to a given rectangular coordinate system in the plane $\mathbb{C}$, a complex number $z = x + iy$ can always be identified with the point whose Cartesian coordinates are x, y.

The numbers corresponding to the points on the x-axis are called real numbers and those on y-axis are called imaginary numbers (purely imaginary to be specific). Since the point $i^2 = -1$, on the negative real axis, is obtained from the point representing 1 on the positive real axis by a rotation anti-clockwise through $180°$ or $(2\pi/2)$ radians, it is reasonable to expect that the point i is obtained by rotating 1 through $90°$ or $\pi/2$ radians in the anti-clockwise sense. Thus the operation of rotation of 1 through a right angle in the anti-clockwise sense is symbolized by i.

Another way of representing a non-zero complex number (also called the polar representation of z) is as follows. Let $z = x + iy \neq 0$ be a complex number. Determine $r > 0$ and θ by the identities

$$x = r \cos\theta$$
$$y = r \sin\theta$$

so that $r^2 = x^2 + y^2$ or that $r = |z|$. Geometrically r represents the distance OP where O is the origin and P represents z in the coordinate plane. Since

$\sin\theta = y/r$ and $\cos\theta = x/r$, θ represents (in radians) the angle through which the x-axis must be rotated to get into the direction of $\overline{OP}$. (This angle is considered positive if the rotation is anti-clockwise and negative if the rotation is clockwise.) Thus this θ (also called argument of z) is determined uniquely only up to a multiple of 2π and the set of all such angles is denoted by *arg z*. However all the values in this set represent the same direction in the complex plane. Further for each $z \neq 0$ there is one and only one value of *arg z* say ϕ satisfying $-\pi < \phi \leq \pi$. This value will henceforth be denoted by *Arg z* and is called the principal branch of *arg z*. Its properties can be described as follows:

(i) $-\pi < Arg\ z \leq \pi\ \forall z \neq 0$

(ii) $Arg\ (z) = \begin{cases} 0 \text{ for positive reals} \\[4pt] 0 < Arg\ z < \dfrac{\pi}{2} \text{ for points in the I quadrant} \\[4pt] \dfrac{\pi}{2} \text{ for the positive imaginary axis} \\[4pt] \dfrac{\pi}{2} < Arg\ z < \pi \text{ for the points in the II quadrant} \\[4pt] \pi \text{ for negative reals} \\[4pt] -\pi < Arg\ z < -\dfrac{\pi}{2} \text{ for points in the III quadrant} \\[4pt] -\dfrac{\pi}{2} \text{ for negative imaginary axis} \\[4pt] -\dfrac{\pi}{2} < Arg\ z < 0 \text{ for points in the IV quadrant} \end{cases}$

(iii) $\theta \in arg\ z \Leftrightarrow \theta = \phi + 2k\pi$ where $\phi = Arg\ z$ and $k \in \mathbb{Z}$.

Using (iii) above one can easily see that given any semi-open interval of length 2π one can determine a unique value of *arg z* (for each $z \neq 0$) called a branch of *arg z*. For example there is also a branch of *arg z* say $\theta(z)$ (written as *ARG z*) satisfying $0 \leq \theta(z) < 2\pi$.
   If

$$a_1 = r_1 e^{i\theta_1} = r_1(\cos\theta_1 + i\sin\theta_1),$$
$$a_2 = r_2 e^{i\theta_2} = r_2(\cos\theta_2 + i\sin\theta_2),$$

are the polar representations of two non-zero complex numbers, a_1 and a_2 then using addition theorems of sine and cosine we find that

$$a_1 a_2 = r_1 r_2(\cos(\theta_1 + \theta_2) + i\sin(\theta_1 + \theta_2)) = r_1 r_2 e^{i(\theta_1+\theta_2)}.$$

Thus

$$|a_1 a_2| = r_1 r_2 = |a_1|\,|a_2|$$

and

$$arg\ a_1 a_2 = arg\ a_1 + arg\ a_2. \qquad (1.1)$$

The meaning of the last equality is that of the equality of sets where for two sets A and B we write

$$A + B = \{s + t : s \in A, t \in B\}.$$

Using induction we can now assert that for non-zero complex numbers $a_i (1 \leq i \leq n)$,

$$arg\ (a_1 a_2 \ldots a_n) = arg\ a_1 + arg\ a_2 + \ldots + arg\ a_n. \qquad (1.2)$$

Similarly

$$arg\ (\frac{a_1}{a_2}) = arg\ a_1 - arg\ a_2. \qquad (1.3)$$

One should not fail to notice the geometric significance of (1.1) and (1.3), (1.1) tells us that the direction of $a_1 a_2$ can be got from that of a_1 by rotating through an angle belonging $arg\ a_2$ (anti-clockwise for positive sign and clockwise for negative sign). Similarly (1.3) tells us that the direction of a_1/a_2 can be got by rotating the direction of a_1 by an angle belonging to $-\ arg\ a_2$ with the same convention for sign. (Occasionally we will also use $arg\ z$ to represent one particular value in the set $arg\ z$).

Yet another way of representing a complex number would be to use a

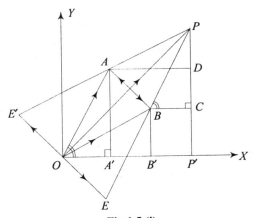

Fig 1.5 (i)

vector joining origin and the given point (which represents the complex number in the Argand diagram). If we identify all vectors which can be obtained from each other by parallel displacements we can denote a given complex number by the vector associated to it. Using this vector notation we can easily illustrate the arithmetic operations like addition, subtraction, multiplication and division. Let $a = a_1 + ia_2$ and $b = b_1 + ib_2$ be two non-zero complex numbers represented by the points A and B and by the vectors $\overline{OA}$ and $\overline{OB}$ respectively. Construct the parallelogram $O\,APB$. If OX and OY are the axes, draw $BB' \perp OX$, $AA' \perp OX$, $PP' \perp OX$, $BC \perp PP'$, $AD \perp PP'$, $OE \parallel^l AB$, $OE = AB$ and $OE' \parallel^l BA$, $OE' = BA$. Figure 1.5(i) illustrates the above construction.

From the above figure the triangles OAA' and BPC are easily seen to be congruent ($OA = BP$, $A\hat{O}A' = P\hat{B}C$, $O\hat{A}'A = B\hat{C}P$). Thus, $OP' = OB' + B'P' = OB' + OA' = b_1 + a_1$ and $PP' = PC + CP' = AA' + BB' = a_2 + b_2$. (Note that $BCP'B'$ is a parallelogram). In other words, P represents $a + b$ and the corresponding vector is $\overline{OP}$ which is the diagonal of the parallelogram whose adjacent sides represent a and b respectively. Thus the parallelogram law for addition holds good for complex numbers. Further from our constructions $AOEB$ and $BAE'O$ are also parallelograms. Hence if E and E' represent the complex numbers c and d then by what we have seen already $\overline{OB} = \overline{OA} + \overline{OE}$ or $\overline{OE} = \overline{OB} - \overline{OA}$. Similarly $\overline{OE'} + \overline{OB} = \overline{OA}$. Thus $c = b - a$ and $d = a - b$. These considerations help us to fix the positions of $a + b$, $a - b$ and $b - a$ in the plane as soon as the points a and b are given. We shall now give a geometrical construction for representing the points corresponding to the complex numbers ab and a/b ($b \neq 0$) as soon as a and b are given.

In Figure 1.5(ii) we rotate the direction of B (say $\phi \in arg\ b$) by an angle $\theta \in arg\ a$ (any member of $arg\ a$ will do provided we use anti-clockwise

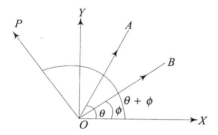

Fig 1.5 (ii) Construction of the point P corresponding to ab.

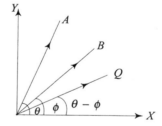

Fig 1.5 (iii) Construction of the point Q corresponding to a/b.

rotation for a positive member and clockwise for negative member) and pick P in this direction such that $OP = OA \cdot OB$. Thus P represents ab. Indeed $\left|\overline{OP}\right| = OA \cdot OB = |a| \cdot |b|$ and the direction of $\overline{OP}$ is given by $\phi + \theta \in arg\ a + arg\ b = arg\ ab$. Thus for the vector $\overline{OP}$, modulus is $|a|\,|b|$ and its direction is that of ab. Thus $\overline{OP}$ represents ab or P represents ab.

In Figure 1.5(iii) above we rotate the direction A (say $\theta \in arg\ a$) by an angle $\phi \in - arg\ b$ (using the same convention for sign as before) and pick Q on this direction such that $OQ = OA/OB$. Then as before $\left|\overline{OQ}\right| = |a/b|$ and the direction of $\overline{OQ}$ is given by $\theta - \phi \in arg\ a - arg\ b = arg\ a/b$. Thus $\overline{OQ}$ represents a/b or Q represents a/b.

Now let 'a' be any non-zero complex number. To find the n-th roots of a, we have to solve the equation $z^n = a$. We write $a = \rho e^{i\phi}$ ($0 \leq \phi < 2\pi$), $z = re^{i\theta}$ ($0 \leq \theta < 2\pi$). It follows by equating the modulii and arguments on both sides of $z^n = a$, that $r^n = \rho$, $n\theta = \phi + 2k\pi$ ($k \in \mathbb{Z}$). Thus $r = \rho^{1/n}$, the unique positive n-th root of ρ and $\theta = (\phi + 2k\pi)/n$. Putting $k = 0, 1, 2,$ ... $n - 1$ successively we obtain all the n-th roots of a. Thus all the n-th roots of a are situated on the same circle centre origin and radius r but are separated from each other by an angle which is an integral multiple of $2\pi/n$. Geometrically the n-th roots of a form the vertices of a regular polygon with n sides. In particular if $a = 1$ the roots are called the nth roots of unity and if $\omega = \cos\dfrac{2\pi}{n} + i \sin \dfrac{2\pi}{n}$ then all the nth roots of unity are given by $1, \omega, \omega^2, \dots \omega^{n-1}$.

Using the vector notation for complex numbers it is easy to see that the equation of a straight line in the parametric form is given by $z = a + bt$ where $b \neq 0$ and a are complex numbers and t real. This represents the line through a and parallel to the vector represented by b. Two such equations $z = a + bt$ and $z = c + dt$ represent the same line if and only if $c - a$ and d are real multiples of b. In fact if they represent the same line then the vectors represented by b and d are parallel and hence d is a real multiple of b. Again, since both the lines are same, for some real t, $c = a + bt$. Thus $c - a$ is a real multiple of b. Conversely if $c = a + qb$ and $d = sb$ where q, s are reals, then $z = c + dt$ is also equivalent to $z = a + bq + (bs)t = a + b (q + st)$. As t varies over $\mathbb{R}$, $t' = q + st$ also varies over $\mathbb{R}$, and so $z = c + dt$ and $z = a + bt$ represent the same line. Also if $z = a + bt$ and $z = c + dt$ are the equations of two different lines then the angle between those two lines is the same as the angle between and b and d which is represented by $arg\,(b/d)$. Note that the angle depends on the order in which the lines are named. If the order is reversed then the angle will change its sign because $arg\,(b/d) = -arg\,(d/b)$. Of course the lines are orthogonal if and only if $arg\,(b/d) = \pm\pi/2$ or b/d is purely imaginary.

Any line with equation $z = a + bt$ is such that for all points z on this line, $(z - a)/b$ will be real. i.e., $Im\,[(z - a)/b] = 0$. On the other hand, points not belonging to this line can be divided into two disjoint sets, one given by the set of all z such that $Im\,[(z - a)/b] < 0$ and another defined by the set of all z such that $Im\,[(z - a)/b] > 0$.

These sets are called the half planes determined by the line and the following argument shows that this distinction is independent of the parametric representation. Let $z = a + bt$ and $z = a' + b't$ represent the same line. Then we know that $a' - a$ and b' are multiples of b. Hence

$$Im\,[(z - a')/b'] = Im\,[(z - (a + sb))/rb] = \frac{1}{r}Im\,[(z - a)/b]$$

for some reals s and r.

Hence the pair of sets one determined by $Im\,[(z - a')/b'] < 0$ and another by $Im\,[(z - a')/b'] > 0$ are the same as the pair of the sets one determined by $Im\,[(z - a)/b] < 0$ and another by $Im\,[(z - a)/b] > 0$, that is, the half planes determined by $z = a + bt$ are the same as the half planes determined by $z = a' + b't$. Note that if the two lines are equally directed then b' will be a positive multiple of b or that $r > 0$ and hence the set determined by $Im[(z - a)/b] < 0$ will be identical with the set determined by $Im[(z - a')/b'] < 0$ and the set determined by $Im[(z - a)/b] > 0$ will be identical with the set determined by $Im[(z - a')/b'] > 0$. If $r < 0$ the planes will get interchanged.

Another way of writing down the equation of a line is as follows:

$$z = a + bt \Rightarrow \bar{z} = \bar{a} + \bar{b}t \Rightarrow \bar{z} = \bar{a} + \bar{b}\left(\frac{z - a}{b}\right)$$

$$\Rightarrow b\bar{z} = \bar{a}b + \bar{b}z - a\bar{b}$$

$$\Rightarrow \bar{b}z - b\bar{z} + \bar{a}b - a\bar{b} = 0$$

$$\Rightarrow \bar{b}z - b\bar{z} + 2i\delta = 0 \quad (\delta = Im\ \bar{a}b \text{ which is real})$$

$$\Rightarrow \frac{\bar{b}}{i}z - \frac{b}{i}\bar{z} + \gamma = 0 \quad (\gamma = 2\delta,\ \gamma \text{ real})$$

$$\Rightarrow \beta z + \bar{\beta}\bar{z} + \gamma = 0 \quad \text{with } \beta = \frac{\bar{b}}{i}.$$

Thus its equation can also be written as $\beta z + \bar{\beta}\bar{z} + \gamma = 0$. ($\gamma$ real).

Conversely it is easy to see that the locus of z satisfying $\beta z + \bar{\beta}\bar{z} + \gamma = 0$, ($\beta \neq 0$ and γ real) is a straight line. (Indeed if $\beta = \beta_1 + i\beta_2 \neq 0$ and $z = x + iy$ then this equation is equivalent to $2(\beta_1 x - \beta_2 y) + \gamma = 0$ which is evidently the equation of a line in the xy plane.)

The equation of a circle in the complex plane is very simple to describe since the distance between two points in the plane is equal to the modulus of their difference. The circle with centre a and radius r which is the locus of the point z whose distance from a is equal to r, is described by $|z - a| = r$ which is equivalent to $(z - a)(\bar{z} - \bar{a}) = r^2$; i.e., $z\bar{z} - a\bar{z} - \bar{a}z + a\bar{a} = r^2$. Its parametric form is $z = a + re^{i\theta}$ ($0 \leq \theta \leq 2\pi$).

Just as a line determines two half planes, a circle center a and radius $r > 0$ determines two components in the plane called the inside and the outside described respectively by $|z - a| < r$ and $|z - a| > r$. In general the equation of a line and a circle can be combined in a single equation of the form $\alpha z\bar{z} + \beta z + \bar{\beta}\bar{z} + \gamma = 0$ where α and γ are real contants and β is a complex constant with $\beta\bar{\beta} > \alpha\gamma$ (using $z = x + iy$ and $\beta = u + iv$ we can verify that this condition is same as saying that the radius of the circle is positive provided $\alpha \neq 0$). However if $\alpha = 0$, this is the equation of a straight line as we already saw. Thus the equation $\alpha z\bar{z} + \beta z + \bar{\beta}\bar{z} + \gamma = 0$ where $\alpha \neq 0$ real, γ real and $\beta\bar{\beta} > \alpha\gamma$ represents a circle with centre $-\bar{\beta}/\alpha$ and radius $\sqrt{(\beta\bar{\beta} - \alpha\gamma)}\big/\alpha$. Thus the combined equation of a circle or a straight line can be written in the form

$$\alpha z \bar{z} + \beta z + \bar{\beta} \bar{z} + \gamma = 0$$

with α, γ real and $\beta \bar{\beta} > \alpha \gamma$. $\alpha = 0$ gives a line and $\alpha \neq 0$ gives a circle.

1.6 Extended Complex Plane and the Stereographic Projection

Recall that the extended real number system can be obtained from the real number system by adding two symbols $+\infty$ and $-\infty$ with certain arithmetical laws. Similarly we can construct the extended complex number system by introducing a single symbol ∞ and adding it to the finite complex number system $\mathbb{C}$. We shall establish the following arithmetical laws for the extended complex numbers.

(i) $a + \infty = \infty + a = \infty$ for all complex numbers a.
(ii) $b \cdot \infty = \infty \cdot b = \infty$ for all extended complex numbers other than zero.
(iii) $a/0 = \infty$ for $a \neq 0$ and $b/\infty = 0$ for $b \neq \infty$.

To obtain a geometric model for the extended complex number system we consider the unit sphere S in three dimensional space which is described by the equation $x^2 + y^2 + z^2 = 1$. We also identify the xy-plane with the finite complex plane in which the x-axis and y-axis correspond respectively to the real and imaginary axis. The point $(0, 0, 1)$ will be called N. We shall now construct a one-to-one correspondence between the sphere S (hence forth called Riemann sphere) and the extended complex number system (the xy-plane together with the point ∞). Let $z = x + iy$ be a finite complex number. Join z to N and let the line zN meet the unit sphere at a point Z. Conversely take any point $z \neq N$ on the unit sphere. Join zN by a straight line and let this line meet the complex plane at Z. Let N correspond to ∞. This is the correspondence between the extended complex number system on the one hand and Riemann sphere on the other. (see Figure 1.5 (iv)).

To obtain analytical expressions for this correspondence, we proceed as follows. We first consider the correspondence between the extended plane and the sphere S. Since the points $z = (x, y, 0)$, $Z = (x_1, x_2, x_3)$ and $N = (0, 0, 1)$ are in a straight line, it follows that

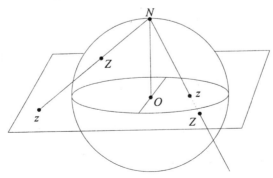

Fig 1.5. (iv) Stereographic projection.

$$Z = t(0, 0, 1) + (1 - t)(x, y, 0)$$

for some real t. But Z also lies on the sphere. Hence $x_1^2 + x_2^2 + x_3^2 = 1$ or

$$[(1 - t)x]^2 + [(1 - t)y]^2 + t^2 = 1 \text{ or } t = \frac{|z|^2 - 1}{|z|^2 + 1}.$$

Hence

$$x_1 = (1 - t)x = 2x/(1 + |z|^2) = (z + \bar{z})/(1 + |z|^2)$$
$$x_2 = (1 - t)y = 2y/(1 + |z|^2) = (z - \bar{z})/i(1 + |z|^2)$$

$$x_3 = t = \frac{|z|^2 - 1}{|z|^2 + 1}$$

Note that $t \neq 1$ if $z \neq \infty$. For the inverse correspondence between the sphere S and the extended plane we proceed as follows. If $z \neq N$ were given and $z = (x_1, x_2, x_3)$ then we can find $Z = x + iy \in \mathbb{C}$ by writing $t(0, 0, 1) + (1 - t)(x_1, x_2, x_3) = (x, y, 0) = Z$, that is $(1 - t)x_3 + t = 0$ or $t = -x_3/(1 - x_3)$. Thus $x = (1 - t)x_1 = x_1/(1 - x_3)$, $y = (1 - t)x_2 = x_2/(1 - x_3)$ so that $Z = x + iy = (x_1 + ix_2)/(1 - x_3)$. The advantage in this spherical representation is that the point at ∞ (namely the point N) is no longer a distinguished point. Further it enables us to introduce a notion of distance between two points in the extended complex number system in the following way:

If z and z' are two extended complex numbers and $Z = (x_1, x_2, x_3)$ and $Z' = (x_1', x_2', x_3')$ are their stereographic projections then we define the distance between z and z' by the formula

$$d(z, z') = [(x_1 - x_1')^2 + (x_2 - x_2')^2 + (x_3 - x_3')^2]^{1/2}$$

With the relationships that we have already obtained between z and x_i and z' and x_i' we can easily compute the following:

$$[d(z, z')]^2 = (x_1 - x_1')^2 + (x_2 - x_2')^2 + (x_3 - x_3')^2$$
$$= 2 - 2(x_1 x_1' + x_2 x_2' + x_3 x_3').$$

But

$$(x_1 x_1' + x_2 x_2' + x_3 x_3')$$

$$= \frac{(z + \bar{z})(z' + \bar{z}') - (z - \bar{z})(z' - \bar{z}') + (|z|^2 - 1)(|z'|^2 - 1)}{(1 + |z|^2)(1 + |z'|^2)}$$

$$= \frac{(1 + |z|^2)(1 + |z'|^2) - 2|z - z'|^2}{(1 + |z|^2)(1 + |z'|^2)}.$$

Therefore $d(z, z') = \dfrac{2|z - z'|}{\sqrt{(1 + |z|^2)(1 + |z'|^2)}}$ $\quad (z \neq \infty \neq z')$.

If $z \neq \infty$ then $d(z, \infty) = \dfrac{2}{\sqrt{(1+|z|^2)}}$.

The word stereographic projection can be used for any one of the maps between the Riemann sphere and the extended complex plane. It is geometrically evident that the stereographic projection transforms every straight line in the z-plane onto a circle on the sphere through $N = (0, 0, 1)$ and vice-versa. Indeed any circle Γ on the sphere through N is the intersection of S with a non-tangential plane say L through N. Thus the projection of points on Γ (these are points on the line joining N and the points of Γ lying in L) must be on L and also on the xy-plane. But the intersection of the two planes must then be a line. Conversely, if l is a line in the xy-plane their projections must lie on the plane containing l and N and also on the sphere S. This latter set is the intersection of a plane and S and so is a circle through N. But any circle on the sphere corresponds to a circle or a straight line in the z plane. This can be proved as follows.

A circle on the sphere lies in a plane of the form $\alpha_1 x_1 + \alpha_2 x_2 + \alpha_3 x_3 = \alpha_0$, where $\alpha_1, \alpha_2, \alpha_3$ are the direction cosines of the normal to the plane (and hence $\alpha_1^2 + \alpha_2^2 + \alpha_3^2 = 1$) and α_0 is the distance of the plane from the origin (and hence $0 \leq \alpha_0 < 1$ if the plane cuts the unit sphere). Substituting the values of x_1, x_2, x_3 in terms of z, this equation takes the form

$$\alpha_1 (z+\bar{z}) - i\alpha_2 (z-\bar{z}) + \alpha_3 (|z|^2 - 1) = \alpha_0(|z|^2 + 1)$$

(i.e.) $\quad (\alpha_0 - \alpha_3) z\bar{z} - (\alpha_1 - i\alpha_2) z - (\alpha_1 + i\alpha_2) \bar{z} + (\alpha_0 + \alpha_3) = 0.$

This equation, as we have already seen, is a combined equation of a circle or a straight line depending on whether $\alpha_0 \neq \alpha_3$ or $\alpha_0 = \alpha_3$.

1.7 Topology of the Complex Plane

Let us recall that by a topological space (X, τ) we mean a non-empty set X together with a collection τ of subsets of X satisfying the following properties: (i) $X \in \tau$, $\phi \in \tau$ and (ii) τ is closed under arbitrary unions and finite intersections. (i.e., any arbitrary union of members of τ is a member of τ and any finite intersection of members of τ is a member of τ). Members of τ are called open sets and a set whose complement is open is said to be closed. A neighbourhood N of a point $x \in X$ is an open set containing x. A base for a topological space (X, τ) is a sub collection τ' of τ with the following property: Every member of τ is a certain union of members of τ'. Let (X, τ) be a topological space and E a subset of X. The interior of E denoted by E° (points of E° are also called interior points of E) is the largest open set contained in E and can be characterized as the union of all open sets contained in E or the set of all points for which E is neighbourhood (the proof of this last statement depends on the easily verifiable fact that the set G of all points for which E is a neighbourhood is itself open).

The closure of E denoted by $\overline{E}$ is the smallest closed set containing E or the intersection of all closed sets containing E or the set of all points whose arbitrary neighbourhoods intersect E. (The proof of this last statement depends on the easily verifiable fact that the complement of the set F of all points whose arbitrary neighbourhoods intersect E is itself open and hence F is closed.) The boundary of E denoted by ∂E can be defined as the set of all points whose arbitrary neighbourhoods intersect both E and E^c. The exterior of E is the interior of E^c which is also equal to $(\overline{E})^c$. Always $E^\circ \subseteq E \subseteq \overline{E}$. E is open if and only if $E^\circ = E$ and that E is closed if and only if $\overline{E} = E$. A point $x \in X$ is an accumulation point or a limit point of $E \subseteq X$ if every neighbourhood of x contains infinitely many points from E. A point $x \in E$ is an isolated point of E if we can find a neighbourhood of x whose intersection with E is the singleton $\{x\}$. A subset E of a topological space X is said to be dense in X if $\overline{E} = X$ (Of course this is equivalent to the easily verifiable fact that every non-empty open subset of X intersects E). A topological space is Hausdorff if any two distinct points are contained in disjoint open sets. Let $f: (X, \tau) \rightarrow (Y, \tau')$ be a map between two topological spaces X and Y. We say f is continuous if $f^{-1}(V)$ is open in X for each open V in Y. We say f is open if $f(U)$ is open in Y for each open set U in X. Two topological spaces (X, τ) and (Y, τ') are said to be homeomorphic if there exists a bijection $f : X \rightarrow Y$ such that f and f^{-1} are continuous in the sense that $f^{-1}(V) \in \tau$, $\forall V \in \tau'$ and $f(U) \in \tau'$, $\forall U \in \tau$.

The ordered pair (X, d) is said to be a metric space if we can define a non-negative real function $d : X \times X \rightarrow \mathbb{R}$ such that

(i) $d(x, y) = d(y, x)$
(ii) $d(x, y) = 0 \Leftrightarrow x = y$
(iii) $d(x, y) \leq d(x, z) + d(z, y)$ for all x, y, z in X.

Let (X, d) be a metric space and $x \in X$. An open ball or an open disc of centre x and radius $\delta > 0$ denoted by $B(x, \delta)$ is defined by $B(x, \delta) = \{y \in X : d(x, y) < \delta\}$. We call a set S in X open if around each point x of S we can find a $\delta > 0$ such that $B(x, \delta) \subset S$. For example every open ball is an open set as can be verified easily. It is easy to verify that the collection of all open sets defined in this way gives a topology on X called the topology induced by the metric d. For example in $\mathbb{R}$ or $\mathbb{C}$ we can define a metric (called the usual metric) by $d(x, y) = |x - y|$.

More generally in $\mathbb{R}^n$ or $\mathbb{C}^n$ we can define $[d(x, y)]^2 = \sum_{i=1}^{n} |x_i - y_i|^2$ where $x = (x_1, x_2, \ldots, x_n)$ and $y = (y_1, y_2, \ldots, y_n)$. On other hand, in the extended complex plane we can define the following metric

$$d(z, z') = \frac{2|z - z'|^2}{(1 + |z|^2)(1 + |z'|^2)} \quad \text{if } z, z' \neq \infty$$

$$d(z, \infty) = \frac{2}{\sqrt{(1 + |z|^2)}} \quad \text{if } z \neq \infty.$$

(Of course, $d(\infty, \infty) = 0$)

Another example of a metric space is the space $B([a, b])$ of all bounded real-valued functions defined on the interval $a \leq x \leq b$. The metric is $d(f, g) = \max_{a \leq x \leq b} |f(x) - g(x)|$.

Two metric spaces (X, d) and (Y, ρ) are said to be isometric to each other if there exists a bijection (called isometry) $f : X \to Y$ such that $\rho(f(x), f(y)) = d(x, y) \; \forall x, y \in X$. Note that an isometry is also a homeomorphism.

We shall now recall certain important properties of the real line. Every non-empty subset of $\mathbb{R}$ has a greatest lower bound and a least upper bound (by convention a set which is unbounded above has ∞ as the least upper bound and a set which is unbounded below has $-\infty$ as the greatest lower bound). An interval in $\mathbb{R}$ is defined by the set of all x for which any one of the following four inequalities hold: (i) $a < x < b$ (ii) $a \leq x < b$ (iii) $a < x \leq b$ (iv) $a \leq x \leq b$. Note that in (i) and (iii) a can be $-\infty$ and in (i) and (ii) b can be $+\infty$. These intervals respectively are denoted by (a, b); $[a, b)$; $(a, b]$; $[a, b]$.

Using the definition of open sets in the metric space $\mathbb{C}$, it is easy to see that for the usual topology induced by the usual metric any open set V in the complex plane can be written as an arbitrary union of open discs or as an arbitrary union of open rectangles (i.e., cartesian product of two open intervals in $\mathbb{R}$). For this reason, the collection of all open discs or the collection of all open rectangles form a base for the usual topology of the plane. Further the complex plane is a Hausdorff space because any two distinct points z and z' are contained in disjoint open sets namely the open discs centered at z and z' and radius ρ which is less than $|z - z'|/2$.

Let S be a subset of a metric space (X, d). Then S is also a metric space under the induced metric of X. (The metric $d : X \times X \to \mathbb{R}$ can be restricted to $S \times S$ and this gives raise to a metric on S). It is now easy to see that a subset A of S is open relative to S if and only if $A = S \cap U$ where U is open in X. Indeed if $A \subseteq S$ is relatively open in S then to each point $x \in A$ there exists $\delta_x > 0$ such that

$$N_x = \{y \in S : d(x, y) < \delta_x\} \subset A \text{ and hence } A = \bigcup_{x \in A} N_x.$$

But $N_x = \{y \in S : d(x, y) < \delta_x\} = \{y \in X : d(x, y) < \delta_x\} \cap S = M_x \cap S$ (say). Since $A = \bigcup_{x \in A} N_x = \bigcup_{x \in A} M_x \cap S = S \cap U$ where $U = \bigcup_{x \in A} M_x$ is open in X because it is a certain union of open discs in X. Conversely, if U is open in X and $A = S \cap U$ then $x \in A$ implies $x \in U$ and as U is open in X there exists $\delta_x > 0$ such that $M_x = \{y \in X : d(x, y) < \delta_x\} \subset U$. But then $N_x = S \cap M_x = \{y \in S : d(x, y) < \delta_x\} \subset A$ is a relative neighbourhood of x in S and so each point of A is an interior point of A relative to S. Thus A is relatively open in S. Similarly, a subset B of S is relatively closed in S if and only if $B = S \cap F$ where F is closed in X. Also S is said to be bounded if $d(x, y) \leq M$ for all $x, y \in S$ and for some fixed constant M.

The above discussion motivates us to define relative open subsets and relative closed subset of a subset Y of an arbitrary topological space (X, τ) as follows.

Definition 1.7.1
$S \subseteq Y \subset X$ is said to be relatively open in Y if $S = Y \cap U$ where $U \in \tau$ and $F \subseteq Y \subset X$ is relatively closed in Y if $F = Y \cap K$ where K is closed in X.

In what follows we shall examine in detail the properties of subsets of the complex plane with reference to notions like connectedness and compactness.

Definition 1.7.2
A subset S of a metric space X is disconnected if $S = A \cup B$ where A and B are non-empty sets with $\overline{A} \cap B = \phi = A \cap \overline{B}$. (Such a pair of sets A and B is said to form a separation for S.) Thus a subset of a metric space is connected if and only if it is not disconnected.

Example 1.7.3
The empty set ϕ, singleton sets, open or closed discs are all connected subsets of the plane whereas a finite set consisting of more than one element, disjoint union of open discs are all disconnected in the plane.

Theorem 1.7.4

  (i) The only connected subsets of the real line are the intervals (open, closed, semi-open or semi-closed).

 (ii) A non-empty open subset of the complex plane is connected if and only if any two of its points can be joined by a polygon lying in the set where by a polygon we mean a finite union of successive line segments such that the end point of each line segment is the initial point of the next line segment.

(iii) A non-empty open subset of the complex plane is connected if and only if any two of its points can be joined by a polygon lying in the set having its segments parallel to the coordinate axes. The last segment can be chosen at will.

Proof
(i) Let E be a non-empty connected subset of $\mathbb{R}$ and $a = \inf E$ and $b = \sup E$ (here a can be $-\infty$ and b can be $+\infty$). We claim that E is one of the

four intervals with left end point a and right end point b. Let us take the case in which the lub and glb of E are members of E. Then for all $x \in E$, $a \leq x \leq b$ and so $E \subset [a, b]$. On the other hand if $a < y < b$ and $y \notin E$ then the sets defined by $A = \{x \in E : x < y\}$ and $B = \{x \in E : x > y\}$ are non- empty and form a separation for E. (Note that $\overline{A} = \{x \in E : x \leq y\}$ and $\overline{B} = \{x \in E : x \geq y\}$. This contradiction shows that $a < y < b \Rightarrow$ $y \in E$, and hence $E = [a, b]$. Similar arguments prove that $E = (a, b)$ or $[a, b)$ or $(a, b]$ respectively if both a, b do not belong to E, or $a \in E$ and $b \notin E$, or $a \notin E$ and $b \in E$.

Conversely, let E be an interval. If it is disconnected we can write $E = A \cup B$ where A and B form a separation for E. Take $a_1 \in A$ and $b_1 \in B$. Without loss of generality assume that $a_1 < b_1$. Now bisect the interval (a_1, b_1) and observe that one of the two subintervals has its left end point in A and right end point in B. Call this interval as (a_2, b_2) and continue the process infinitely to get a decreasing sequence of intervals (a_n, b_n) with $a_n \in A$ and $b_n \in B$. Since the lengths of the intervals tend to zero as n tends to infinity, the sequences $\{a_n\}$ and $\{b_n\}$ have a common limit c (By Cantor intersection theorem of real analysis). Further since a_n's are increasing and b_n's are decreasing the common limit satisfies $a_n \leq c \leq b_n$. Thus $c \in E$ (as E is an interval and $a_n, b_n \in E$). But then $c \in A$ or B. In as much as $c \in \overline{A}$ $\cap \overline{B}$ (Limit points of A and B are members of $\overline{A}$ and $\overline{B}$ respectively) either $\overline{A} \cap B \neq \phi$ or $A \cap \overline{B} \neq \phi$ a contradiction to the separation of E. Thus E is connected.

(ii) Let A be an open connected set in the complex plane. Let $a \in A$. Let A_1 be the subset of A whose points can be joined to 'a' by a polygon in A and let $A_2 = A \backslash A_1$ so that $A_1 \cap A_2 = \phi$. We claim that A_1 and A_2 are both open. If $a_1 \in A_1 \subset A$ then there exists a neighbourhood $B(a_1, \varepsilon)$ of a_1 contained in A. In this open disc of centre a_1 and radius ε all points can be joined to a_1 by simple line segments. Since $a_1 \in A_1$ it follows that all points in this neighbourhood are in A_1. Thus A_1 is open. For the same reason if $a_2 \in A_2$ and $B(a_2, \delta)$ is contained in A then no point in this neighbourhood can be joined to 'a' by another polygon. Hence $B(a_2, \delta)$ is completely contained in A_2 and hence A_2 is open. Since A_1 and A_2 are disjoint it follows that $A_1 \cap \overline{A_2} = \phi = \overline{A_1} \cap A_2$. (For example if $x \in A_1$ $\cap \overline{A_2}$ then we can find a neighbourhood of x which is completely contained in A_1 and intersecting A_2 also). But since A is connected, either A_1 $= \phi$ or $A_2 = \phi$. But A_1 cannot be empty because $a \in A_1$. Thus $A_2 = \phi$ and $A = A_1$. Conversely, we assume that A satisfies this condition and prove that A cannot be disconnected. If A were to be disconnected then $A = A_1 \cup A_2$ where A_1 and A_2 form a separation for A. Choose $a_1 \in A_1$ and $a_2 \in A_2$ and take a polygon connecting these two points. One of the line segments of this polygon must then join a point of A_1 to a point of A_2. Call them as b_1 and b_2. This closed line segment joining b_1 and b_2 can be written as $E = \{b_1 + t(b_2 - b_1) : 0 \le t \le 1\}$. Consider the subsets of the interval $0 \le t \le 1$ which correspond to points in A_1 or in A_2. Call these as B_1 and B_2. Then $0 \in B_1$, $1 \in B_2$ and B_1, B_2 form a separation for interval $[0, 1]$. Indeed if $t_0 \in \overline{B_1} \cap B_2$ then there exists a sequence $t_n \in B_1$ such that $t_n \to t_0$, i.e., $(1 - t_n) b_1 + t_n b_2 \in A_1$ and tends to $(1 - t_0) b_1 + t_0 b_2 \in A_2$. Thus $(1 - t_0) b_1 + t_0 b_2 \in A_2 \cap \overline{A_1}$, a contradiction. Therefore $\overline{B_1} \cap B_2 = \phi$ and similarly $B_1 \cap \overline{B_2} = \phi$. Thus the closed interval $[0, 1]$ is disconnected contradicting the fact that intervals in $\mathbb{R}$ are connected.

(iii) The proof is exactly same if we let A_1 to be the subset of A whose points can be joined to 'a' by a desired polygon in A with its line segments parallel to the coordinate axes.

Theorem 1.7.5

Every non-empty subset of the complex plane has a unique decomposition into maximal connected sets called components.

Proof

Let S be a non-empty subset of the complex plane. Let $a \in S$. Denote by $C(a)$ the union of all connected subsets of S containing a. $C(a)$ is non-empty since the singleton set $\{a\}$ is connected and it contains a. We assert that $C(a)$ is connected. Suppose it were not connected, then $C(a) = A \cup B$ where A and B are non-empty, disjoint subsets of $C(a)$ with $\overline{A} \cap B = A \cap \overline{B} = \phi$. Since $a \in C(a)$ we can assume that $a \in A$ and choose $b \in B$ (since B is non-empty). Now since $b \in C(a)$ there exists a connected set $S' \subset S$ such that $a, b \in S'$. Now $S' \subset C(a)$ and we can write $S' = (S' \cap A) \cup (S' \cap B)$ where $(S' \cap A)$ and $(S' \cap B)$ are non-empty, disjoint subsets of S' with $(S' \cap A) \cap \overline{(S' \cap B)} \subset A \cap \overline{B} = \phi$, $\overline{(S' \cap A)} \cap (S' \cap B) \subset \overline{A} \cap B = \phi$. Hence S' is disconnected contradicting our assumption. Thus $C(a)$ is connected. However since $C(a)$ is the union of all connected sets containing 'a' it should be a maximal connected set. In other words, there is no bigger connected set containing $C(a)$. This maximal connected set is called a component and it is also clear that any two components are either identical or disjoint. For if $C(a) \cap C(b) \ne \phi$ take $c \in C(a) \cap C(b)$. Then $C(a) \subset C(c)$ by definition of $C(c)$ and the connectedness of $C(a)$. Hence $a \in C(c)$ and by the same logic $C(c) \subset C(a)$ and hence $C(a) = C(c)$ Similarly $C(b) = C(c)$ and so $C(a) = C(b)$.

Theorem 1.7.6

In $\mathbb{C}$, components of open sets are open.

Proof

Let S be open in $\mathbb{C}$ and $a \in S$. Let $C(a)$ be the component of S containing 'a'. Since S is open there exists an open disc around 'a' of radius r denoted by $N \subset S$. But in as much as N is connected (by Theorem 1.7.4) and contains a, $N \subset C(a)$. This means that a is an interior point of $C(a)$. But if b is any other point of $C(a)$ then b will be interior point of $C(b)$ but then $C(a) = C(b)$. Hence $C(a)$ is open and the theorem follows.

Theorem 1.7.7
Every open subset S of $\mathbb{C}$ is at most a countable union of pairwise disjoint open connected sets.

Proof
By the previous theorem the open set S is a disjoint union of non-empty open connected subsets (namely its components). The fact that $\mathbb{Q}$ is dense in $\mathbb{R}$ enables us to infer that the points with rational coordinates are dense in $\mathbb{C}$. (Prove that $\mathbb{Q} \times \mathbb{Q}$ is dense in $\mathbb{C}$). So in every component there is a point with rational coordinates. In as much as the components are disjoint, we can associate distinct points (with rational coordinates) with distinct components. In particular the various components of S can be put into one-to-one correspondence with a subset of a countable set. Hence the union is atmost countable.

Definition 1.7.8
A non-empty open connected subset of $\mathbb{C}$ is called a region.

Definition 1.7.9
Let X be a topological space. We say X is compact if every open cover of X admits a finite subcover. (A collection of sets is said to be a cover for X if X is contained in the union of the members in the collection.)

Let X be a topological space and S a subset of X. According to the definition, S is compact if and only if every relative open cover of S admits a finite subcover. Since any relative open set in S is of the form $U \cap S$ where U is open in X, it follows that if every open cover of S by open subsets of X admits a finite subcover then S is compact. From this reasoning it follows that if $K \subset S \subset X$, then K is compact relative to X if and only if K is compact relative to S.

We shall now recall the concepts of a sequence and a subsequence in a metric space.

Definition 1.7.10
Let (X, d) be a metric space. By a sequence in X we mean a map $f : \mathbb{N} \to X$ and denote it by (x_n) where $f(n) = x_n \in X$. Further if $n_1 < n_2 < n_3 < \dots$ is an increasing sequence of natural numbers then (x_{n_i}) is called a sub sequence of (x_n).

Definition 1.7.11
Let (X, d) be a metric space and (x_n) a sequence in X. We say (x_n) tends to x or converges to $x \in X$ as n tends to infinity if $d(x_n, x) \to 0$ as $n \to \infty$ and y is a limit point of (x_n) if there exists a subsequence (x_{n_i}) converging to y. We say (x_n) is Cauchy if $d(x_n, x_m) \to 0$ as $m, n \to \infty$.

Note 1.7.12

If A is a subset of a metric space (X, d) then $x \in \bar{A}$ if and only if there exists a sequence (x_n) in A such that $x_n \to x$ as $n \to \infty$. Further every convergent sequence is Cauchy.

Definition 1.7.13

A metric space X is said to be complete if every Cauchy sequence in X is convergent.

Recall that $\mathbb{R}$ is a complete metric space but an open interval (a, b) in $\mathbb{R}$ is not a complete metric space.

Theorem 1.7.14

The complex plane $\mathbb{C}$ is a complete metric space.

Proof

Let (x_n) be a Cauchy sequence in $\mathbb{C}$. Using the inequalities $|Re\, x_n - Re\, x_m| \leq |x_n - x_m|$ and $|Im\, x_n - Im\, x_m| \leq |x_n - x_m|$ we get that $(Re\, x_n)$ and $(Im\, x_n)$ are Cauchy sequences in $\mathbb{R}$ which is a complete metric space. Hence there exist $\alpha, \beta \in \mathbb{R}$ such that $Re\, x_n \to \alpha$ as $n \to \infty$ and $Im\, x_n \to \beta$ as $n \to \infty$ in $\mathbb{R}$. Now it is easy to show (using the triangle inequality again) that $x_n \to \alpha + i\beta$ as $n \to \infty$ in $\mathbb{C}$. Hence $\mathbb{C}$ is complete.

Theorem 1.7.15

Every compact metric space is complete.

Let X be a compact metric space and (x_n) a Cauchy sequence in X. We claim that this sequence is convergent in X. We first observe that if this sequence has a limit point then the sequence is convergent to it. (Let (x_{n_j}) be a subsequence of (x_n) converging to $x \in X$. Now $d(x_n, x) \leq d(x_n, x_{n_j}) + d(x_{n_j}, x)$ implies $d(x_n, x) \to 0$ as $n \to \infty$.) Thus our claim will be proved once we prove that if (x_n) has no limit point then X is not compact. Assume that each point y in X is not a limit point of (x_n). Now we can find an $\varepsilon_y > 0$ such that $B(y, \varepsilon_y)$ contains x_n only for finitely many n. This collection $\{B(y, \varepsilon_y) : y \in X\}$ is an open cover for X which cannot admit a finite subcover for X. (A finite cover would mean that X contains x_n's only for finitely many n's and thus the sequence is a finite sequence). This proves that every compact metric space is complete.

Definition 1.7.16

Let (X, d) be a metric space. A subset S of X is said to be totally bounded if for every $\varepsilon > 0$, S can be covered by finitely many open balls (in X) of radius ε with centres in S.

Note that any compact set is totally bounded because the collection of all open balls of radius ε is an open cover of X and hence should admit a finite sub cover. On the other hand any totally bounded set is bounded because if X is covered by n open balls centered at $x_1, x_2, \ldots x_n$ with radius

ε, then the distance between any two points of X is less than or equal to $2\varepsilon + \max d(x_i, x_j)$ $(1 \le i, j \le n)$.

From these observations it follows that any compact subset of a metric space is complete and totally bounded (and hence bounded also). It is interesting to note that the converse is also true as per the following theorem.

Theorem 1.7.17
A subset S of a metric space X is compact if and only if S is complete and totally bounded.

Proof
We have already proved one part and hence we can assume that $S \subseteq X$ is complete and totally bounded and prove that S is compact. Suppose we have an open cover for S which does not admit a finite subcover. Let $\varepsilon_n = 2^{-n}$ and cover S by finitely many open balls of radius ε_1 say $B_{\varepsilon_1}(x_1)$, $B_{\varepsilon_1}(x_2), \dots B_{\varepsilon_1}(x_k)$. If each of the $B_{\varepsilon_1}(x_i)$ admitted a finite subcover from the given cover then the same will be true of S. Thus we can assume $\{B_{\varepsilon_1}(y_1) : y_1 = x_i$ for some $i\}$ does not admit a finite sub cover. In a sense, starting from a totally bounded set S we are able to get a ball $B_{\varepsilon_1}(y_1)$ which does not admit a finite subcover. Now since $B_{\varepsilon_1}(y_1)$ is also totally bounded (Prove that a subset of a totally bounded set is totally bounded.) we can get a ball $B_{\varepsilon_2}(y_2)$ with $y_2 \in B_{\varepsilon_1}(y_1)$ (note the change in the radius) such that $B_{\varepsilon_2}(y_2)$ has no finite subcover. We now proceed by induction and get a sequence (y_n) with the property that $y_{n+1} \in B_{\varepsilon_n}(y_n)$ and $B_{\varepsilon_n}(y_n)$ does not admit a finite sub cover. Consider the sequence (y_n). It is clear that $d(y_n, y_{n+1}) < \varepsilon_n$ and so for $m > n$,

$$d(y_n, y_m) < \varepsilon_n + \varepsilon_{n+1} + \dots \varepsilon_m < 2^{-n+1}$$

(use triangle inequality and the fact that $\sum_{n}^{m} 2^{-k} < 2^{-n+1}$). Hence (y_n) is Cauchy. Since we assume the completeness of S, there exists $y \in S$ such that $d(y_n, y) \to 0$. But this y belongs to at least one open set in the given cover say U. Now it follows that there exists a $\delta > 0$ such that the open ball $B_\delta(y) \subseteq U$. Choose n large so that $d(y_n, y) < \delta/2$ and $\varepsilon_n < \delta/2$. Then $B_{\varepsilon_n}(y_n) \subseteq B_\delta(y)$ (if $t \in B_{\varepsilon_n}(y_n)$ then

$$d(t, y) \le d(t, y_n) + d(y_n, y) < \delta/2 + \delta/2 = \delta$$

which implies that $t \in B_\delta(y)$). But we assume each $B_{\varepsilon_n}(y_n)$ does not admit a finite subcover contradicting the fact that $B_{\varepsilon_n}(y_n) \subseteq B_\delta(y) \subset U$ (a single set in the open cover). Thus every open cover of S admits a finite subcover and so S is compact.

Corollary 1.7.18
A subset of $\mathbb{C}$ is compact if and only if it is closed and bounded.

Proof

It is easy to see that if $S \subset \mathbb{C}$ is compact, then by the previous theorem it is complete and totally bounded. But total boundedness implies boundedness and completeness of S in $\mathbb{C}$ (which is a complete metric space by Theorem 1.7.14) implies closedness. This is because if x is a limit point of S then there is a sequence (x_n) in S converging to x. But then (x_n) is Cauchy in S and every Cauchy sequence must be convergent in S (S is complete) and hence $x \in S$.

Conversely we know that any closed subset of a complete space is complete (easy to check, using Note 1.7.12) and all that we need to verify is that the given bounded set is also totally bounded. Let S be bounded in $\mathbb{C}$ and hence contained in a closed square. This square can be subdivided into a finite number of squares whose sides are arbitrarily small, say squares whose sides are less than $\varepsilon/\sqrt{2}$ for any $\varepsilon > 0$. Then each of the square is covered by an open disc with the same centre as that of the square and with radius $\varepsilon/2$. Thus we have covered S by a finite number of open balls of radius $\varepsilon/2$. However the centres may not be in S. It is now possible for us to cover S by finite number of open balls of radius ε with centres also in S. For this we first omit those open balls which does not intersect S. However if an open ball of radius $\varepsilon/2$ contains a point x of S we replace this $\varepsilon/2$-ball by an ε-ball around x. We now obtain a finite cover for S by ε-balls with centres in S and this proves that S is totally bounded. Hence S is both totally bounded and complete and hence compact by Theorem 1.7.17.

Corollary 1.7.19

A subset of $\mathbb{R}^n$ is compact if and only if it is closed and bounded.

The proof is similar to the previous corollary. The only difference is that we have to replace squares and circles by n-cells of the form $\{(x_1, x_2, \ldots x_n)/a_i \le x_i \le a_i + \delta, \text{ with } \delta > 0\}$ and spheres in $\mathbb{R}^n$.

Theorem 1.7.20

A metric space is compact if and only if every infinite sequence has a limit point.

Proof

Let (x_n) be an infinite sequence in the given compact metric space. If y is not a limit point of (x_n) it has a neighbourhood which contains (x_n) only for finitely many n. If there were no limit points, every point has a neighbourhood containing x_n only for finitely many n. These open sets containing only finitely many x_n's would then form a cover for the given compact space and if we can extract a finite subcover, it only means that the sequence is finite. Thus in a compact space every infinite sequence has a limit point.

Conversely, if every infinite sequence had a limit point then every Cauchy sequence will also have a limit point and in that case it has to converge and hence the space will have to be complete. We claim that it is also totally bounded. If not, there exists an $\varepsilon > 0$ such that the space cannot be covered by finitely many ε-balls. Choose x_1 arbitrary and take x_2 not in $B_\varepsilon(x_1)$ and x_3 not in $B_\varepsilon(x_2) \cup B_\varepsilon(x_1)$ etc. This is always possible because any finite number of these balls do not cover the whole space. It is easy to see that $d(x_n, x_m)$ is always greater than or equal to ε for all m and n and so the sequence (x_n) cannot have a convergent subsequence contradicting the hypothesis that every infinite sequence has a limit point. Thus the space is totally bounded and now it follows that the space is compact.

Definition 1.7.21

Let $f: X \to Y$ be a function from one metric space (X, d) into another space (Y, ρ). We say that f is continuous at $x_0 \in X$ if given $\varepsilon > 0$ we can find $\delta > 0$ such that $\rho(f(x), f(x_0)) < \varepsilon$ whenever $d(x, x_0) < \delta$.

We say that f is continuous on $S \subseteq X$ if it is continuous at every point of S.

Theorem 1.7.22

Let $f : X \to Y$ be a function between metric spaces (X, d) and (Y, ρ). f is continuous at $x_0 \in X$ if and only if $x_n \to x_0 \Rightarrow f(x_n) \to f(x_0)$.

Proof

Let f be continuous at $x_0 \in X$ and $x_n \to x_0$ as $n \to \infty$ in X. By continuity given $\varepsilon > 0$ there exists $\delta > 0$ such that $d(x, x_0) < \delta \Rightarrow \rho(f(x), f(x_0)) < \varepsilon$. Choose N so large that for $n \geq N$, $d(x_n, x_0) < \delta$. Hence $\rho(f(x_n), f(x_0)) < \varepsilon$ for $n \geq N$. Since $\varepsilon > 0$ is arbitrary, it follows that $f(x_n) \to f(x_0)$. To prove the converse let us assume that f is not continuous at x_0. Thus there exists $\varepsilon_1 > 0$ such that for every $\delta > 0$ there exists a corresponding point x_δ such that $d(x_\delta, x_0) < \delta$ but $\rho(f(x_\delta), f(x_0)) \geq \varepsilon_1$. Choosing $\delta = 1$, $1/2, 1/3, \ldots$ successively we have a sequence (w_n) in X such that $d(w_n, x_0) \to 0$ but $\rho(f(w_n), f(x_0)) \geq \varepsilon_1$ (implying $f(w_n) \nrightarrow f(x_0)$ contradicting the given hypothesis). Hence the theorem.

Theorem 1.7.23

Let $f: X \to Y$ be a function between metric spaces (X, d) and (Y, ρ). Then f is continuous on X if and only if $f^{-1}(V)$ is open in X for every open V in Y.

Proof

Let $f : X \to Y$ be continuous and $V \subseteq Y$ be open. We have to prove that every point of $f^{-1}(V)$ is interior to $f^{-1}(V)$. Let $x_0 \in f^{-1}(V)$ so that $f(x_0) \in V$. Since V is open there exists a ball $B_\varepsilon(f(x_0)) \subset V$. For this ε (by continuity) we can find a $\delta > 0$ such that $f(B_\delta(x_0)) \subset B_\varepsilon(f(x_0))$. Hence $B_\delta(x_0) \subset f^{-1}(V)$. Hence x_0 is interior to $f^{-1}(V)$. Since x_0 is arbitrary it follows that $f^{-1}(V)$ is open.

Conversely, assume that $f^{-1}(V)$ is open for every open set V in Y. We shall now prove that f is continuous at an arbitrary point $x_0 \in X$. Let $\varepsilon >$

0 be given. Take $V = B_\varepsilon(f(x_0))$ which is clearly open in Y. By hypothesis $f^{-1}(V)$ is open and $x_0 \in f^{-1}(V)$. Thus there exists a $\delta > 0$ such that $B_\delta(x_0) \subset f^{-1}(V)$. Thus whenever $d(x, x_0) < \delta$, we have $\rho(f(x), f(x_0)) < \varepsilon$. This proves that f is continuous at x_0.

Corollary 1.7.24
Let $f : X \to Y$ be a function between metric spaces (X, d) and (Y, ρ). Then f is continuous on X if and only if the inverse image of every closed set in Y is closed in X.
The proof follows from Theorem 1.7.23 and the fact that $F \subset Y$ is closed if and only if $Y \backslash F$ is open in Y.

Example 1.7.25
If $f : \mathbb{C} \to \mathbb{C}$ is defined by $f(z) = z^m$ (m is a positive integer) then f is continuous on $\mathbb{C}$. But $f(z) = 1/z$ is not continuous at the origin whatever value we assign for $f(0)$.

Theorem 1.7.26
Let f and g be complex continuous functions on a metric space X. Then $f + g$, fg and f/g (provided $g(x) \neq 0$ for all $x \in X$) are all continuous on X.

Proof
Exercise.

Theorem 1.7.27
(a) Let $f_1, f_2, \dots, f_n$ be real functions on a metric space X and let $f : X \to \mathbb{R}^n$ be defined by
$$f(x) = (f_1(x), f_2(x), \dots, f_n(x)) \quad x \in X.$$
Then f is continuous if and only if each of the functions f_i, $(1 \leq i \leq n)$ is continuous.
(b) If f and g are continuous mappings of X into $\mathbb{R}^n$, then $f + g$ and $f \cdot g$ are continuous on X. (Here $(f \cdot g)(x)$ is the inner product of $f(x)$ and $g(x)$ in $\mathbb{R}^n$.)

Proof
Part (a) follows from the inequalities
$$|f_j(x) - f_j(y)| \leq |f(x) - f(y)| = \left\{ \sum_{i=1}^{n} |f_i(x) - f_i(y)|^2 \right\}^{\frac{1}{2}},$$
for $j = 1, 2, \dots n$. Part (b) follows from (a) and Theorem 1.7.26.

Definition 1.7.28
$f : (X, d) \to (Y, \rho)$ is uniformly continuous if given $\varepsilon > 0$ there is a $\delta > 0$ such that $\rho(f(x), f(y)) < \varepsilon$ for all $x, y \in Y$ with $d(x, y) < \delta$. The

importance of the choice of δ here is that it does not depend on the pair of points x, y in X.

Theorem 1.7.29
 (i) Continuous image of a connected set is connected.
 (ii) Continuous image of a compact set is compact.
 (iii) A continuous function on a compact space is uniformly continuous.

Proof
(i) Let $f : X \to Y$ be continuous and S be a connected subset of X. If $f(S)$ is not connected, write $f(S) = A \cup B$ where A and B form a separation for $f(S)$. Then

$$S = (f^{-1}(A) \cap S) \cup (f^{-1}(B) \cap S).$$

Observe that A and B are non-empty implies that there exists $a_1 = f(s_1)$ $\in A$ and $b_1 = f(s_2) \in B$ with s_1, $s_2 \in S$. Thus, $s_1 \in (f^{-1}(A) \cap S)$ and $s_2 \in (f^{-1}(B)) \cap S$ and so $(f^{-1}(A) \cap S)$ and $(f^{-1}(B) \cap S)$ are also non-empty. If $s \in (f^{-1}(A) \cap S) \cap (f^{-1}(B) \cap S)$ then we can find a sequence of points $s_n \in (f^{-1}(B) \cap S)$ such that $s_n \to s$ so that by continuity $f(s_n)$ $\to f(s)$ and hence $f(s) \in \overline{B} \cap A$, (note that $f(s_n) \in B$ and $f(s) \in A$), a contradiction. Similarly $(f^{-1}(A) \cap S) \cap (f^{-1}(B) \cap S) = \phi$ and thus S is disconnected, a contradiction. Thus we have proved that the image of a connected set is connected.

   (ii) Let $f : X \to Y$ be continuous and S be a compact subset of X. Consider an open cover for $f(S)$ say (U_α). Then $(f^{-1}(U_\alpha))$ will be an open cover for S. Since S is compact $S \subseteq \underset{1 \leq i \leq n}{\cup} f^{-1}(U_{\alpha_i}) = f^{-1}\left(\underset{1 \leq i \leq n}{\cup} U_{\alpha_i} \right)$ and hence $f(S) \subset \underset{1 \leq i \leq n}{\cup} U_{\alpha_i}$ and hence $f(S)$ is compact.

   (iii) Suppose f is continuous on a compact set X. By continuity, given $\varepsilon > 0$ and $y \in X$ there exists a $\delta_y > 0$ depending on y such that $d(x, y) <$ δ_y implies $\rho(f(x), f(y)) < \varepsilon/2$ where d and ρ are metrics on X and Y respectively. Cover X by open balls of radius $\delta_y/2$ around each point y of X. Since X is compact we can extract a finite sub cover. So

$$X \subset B(y_1, \delta_{y_1}/2) \cup B(y_2, \delta_{y_2}/2) \cup \ldots \cup B(y_n, \delta_{y_n}/2)$$

Let $\delta > 0$ be the minimum of $\delta_{y_1}/2, \delta_{y_2}/2, \ldots \delta_{y_n}/2$. Let x_1, x_2 be two points in X with $d(x_1, x_2) < \delta$. Then there exists y_i with $d(x_1, y_i) < \delta_y/2$ and so $d(x_2, y_i) \leq d(x_2, x_1) + d(x_1, y_i) < \delta_{y_i}$. Thus $\rho(f(x_1), f(y_i)) < \varepsilon/2$ and $\rho(f(x_2), f(y_i)) < \varepsilon/2$. Thus $\rho(f(x_1), f(x_2)) < \varepsilon$ (by triangle inequality).

1.8 One-Point Compactification and the Riemann Sphere
In this section we shall define the one-point compactification of a locally compact Hausdorff space and prove that the stereographic projection effects a homeomorphism between the one-point compactification of $\mathbb{C}$ and

the Riemann sphere with its Euclidean metric (denoted by ρ). We recall that in § 1.6 we have exhibited a one-to-one correspondence between the Riemann sphere and the extended complex plane. It is easy to see (from the definition) that this correspondence is actually an isometry between the Riemann sphere with the Euclidean metric and the extended complex plane with the metric $d(z, z')$ defined as follows.

$$d(z, z') = \frac{2|z - z'|}{\sqrt{(1+|z|^2)(1+|z'|^2)}} \quad \text{if } z, z' \neq \infty$$

$$d(z, \infty) = \frac{2}{\sqrt{(1+|z|^2)}} \quad \text{if } z \neq \infty.$$

In the sequel we shall prove that $\mathbb{C}^* = \mathbb{C} \cup \{\infty\}$ can be given a new topology τ (by retaining the usual topology of $\mathbb{C}$ and defining neighbourhoods of ∞) called the one-point compactification of $\mathbb{C}$. We shall further prove that the stereographic projection effects a homeomorphism between $(\mathbb{C}^*, \tau)$ and (S, ρ). This result enables us to show that the one-point compactification of $\mathbb{C}$ is metrizable (compatible with the above metric d on $\mathbb{C}^*$). We shall now define the one-point compactification of a non-compact locally compact Hausdorff space in general and apply it to $\mathbb{C}$. At this point we need some topological preliminaries which we shall proceed to obtain.

Lemma 1.8.1
(X, τ) be a compact topological space and F be a closed subset of X. Then F is compact.

Proof
Let $\{U_\alpha\}$ be an open cover for F. Then U_α's together with F^c form an open cover for X. This open cover admits a finite subcover, say $U_{\alpha_1}, U_{\alpha_2}, \ldots, U_{\alpha_m}$ and perhaps F^c. Since no point of F can belong to F^c it follows that $\left\{U_{\alpha_i}\right\}_{i=1}^m$ form a finite subcover for F. This completes the proof.

Lemma 1.8.2
(X, τ) be a Hausdorff topological space and F be a compact subset of X. Then F is closed.

Proof
Fix $x \in F^c$. For each $y \in F$ there exist open sets U_y and V_y such that $x \in U_y$, $y \in V_y$ and $U_y \cap V_y = \phi$. (Note that $x \neq y$ and X is Hausdorff). Now the collection V_y as y varies in F is an open cover for F and hence we can extract a finite subcover for F say $V_{y_1}, V_{y_2}, \ldots, V_{y_m}$. If we take

$$U = \bigcap_{i=1}^m U_{y_i} \quad \text{and} \quad V = \bigcup_{i=1}^m V_{y_i} \quad \text{then clearly } U \text{ and } V \text{ are open, } x \in U \text{ and}$$

$U \cap V = \phi$, $F \subset V$. It follows that $U \subset V^c \subset F^c$. Thus x is interior to F^c, and as x is arbitrary it follows that F^c is open or that F is closed.

Definition 1.8.3

A topological space X is said to be locally compact if given any point x of X there exist an open neighbourhood of x say V such that $\overline{V}$ is compact.

Definition 1.8.4

Let (X, τ) be a locally compact, non-compact Hausdorff space. Let ∞ denote an object that is not in X and put $Y = X \cup \{\infty\}$. In Y the following collection of sets are declared to be open: (i) U where U is open in X (ii) $Y \setminus K$ where K is any compact subset of X. Denote this collection as σ.

Lemma 1.8.5

σ defines a topology on Y.

Proof

$\phi \in \tau \subset \sigma$. Since $Y = Y \setminus \phi$ and ϕ is compact in X, it also follows that $Y \in \sigma$. We now observe that if $U \in \tau$ and $V = Y \setminus K$ where K is compact in X, then $U \cup V \in \sigma$. Indeed $U \cup V = Y \setminus (K \cap U^c)$. Since K is a compact subset of the Hausdorff space X, K is closed and hence $K \cap U^c$ (being a closed subset of the compact set K) is compact. Thus $U \cup V \in \sigma$. In a similar way we can prove that if $K_\alpha \subset K$ are compact then $\underset{\alpha}{\cup} (Y \setminus K_\alpha) = Y \setminus (\underset{\alpha}{\cap} K_\alpha)$ is also of the form $Y \setminus K$ where K is compact in X (note that intersection of any collection of compact sets in a Hausdorff space is compact). Now we shall prove that any arbitrary union of members of σ is a member of σ. Let $\{W_\alpha\}$ be a collection of members of σ. Let U be the union of all those W_α's which are members of τ. Let V be the union of all other W_α's. It is now clear that $U \in \tau$ and V is of the form $Y \setminus K$ where $K \in X$ is compact. As already observed $\underset{\alpha}{\cup} W_\alpha = U \cup V \in \sigma$. Similarly using

(i) $\overset{n}{\underset{i=1}{\cap}} (Y \setminus K_i) = Y \setminus (\overset{n}{\underset{i=1}{\cup}} K_i)$,

(ii) finite union of compact sets is compact,

(iii) If $U \in \tau$ and $K \subset X$ is compact then $U \cap (Y \setminus K) = U \cap (X \setminus K) \in \tau$

we can show that σ is closed under finite intersections.

Theorem 1.8.6

The one point compactification Y of a locally compact, Hausdorff, non-compact space X is a compact Hausdorff space. Moreover the relative topology of X in Y is the original topology on X and $\overline{X} = Y$.

Proof

To prove that Y is compact, let $\{V_\alpha\}$ be an open cover of Y. In as much as ∞ belongs to Y, $\infty \in V_{\alpha_0} = Y \setminus K$ for some α_0. Now take $\{V_\alpha \cap X\}$ $(\alpha \neq \alpha_0)$. This is an open cover for K. Because K is compact, finitely many of them cover K. If we add $V_{\alpha_0} = Y \setminus K$ to this finite collection, it becomes a finite subcover for Y. Hence Y is compact.

To show that Y is Hausdorff, let x and y be two points in Y with $x \neq y$. If both x and y belong to X, then since X is Hausdorff there exist open sets U and V in X containing x and y respectively such that $U \cap V = \phi$. Since U and V are also open in Y, x and y are contained in disjoint open sets in Y. On the other hand if $x \in X$ and $y = \infty$, choose V a neighbourhood of x such that $\overline{V}$ is compact (use the local compactness of X). Since $x \in V$, $y \in Y \backslash \overline{V}$ and $V \cap (Y \backslash \overline{V}) = \phi$ (V and $Y \backslash \overline{V}$ are both open sets in Y), again we have disjoint open sets of Y containing x and y.

We shall now show that any relative open set U of X in Y is open in the original topology of X. Let $U = X \cap V$ where V is open in Y. By definition either V is open in X or $V = Y \backslash K$ where K is compact in X. If V is open in X then $U = V$ is open in X. (i.e. in the original topology of X). However if $V = Y \backslash K$ where K is compact in X then $U = X \backslash K$ which is also open in X (in the original topology of X) since K is closed (being a compact subset of a Hausdorff space X). Now we shall prove that any open set U in X (in the original topology of X) is also open in the relative topology of X in Y. Clearly $U = U \cap X$ where U is also open in Y. Thus U is open in the relative topology of X in Y. Hence the original topology of X is the same as the relative topology of X in Y.

Now we shall prove that $\overline{X} = Y$. Since $\overline{X} \subset Y$ and $Y \backslash X = \{\infty\}$ it suffices to show that $\infty \in \overline{X}$. Since X is not compact each open set of the form $Y \backslash K$ containing the point ∞ intersects X (if not $(Y \backslash K) \cap X = (X \backslash K) = \phi$ and $X = K$). Thus ∞ is a limit point of X and hence $\overline{X} = Y$. This completes the proof.

Theorem 1.8.7
The stereographic projection is a homeomorphism between the one-point compactification of $\mathbb{C}$ and the Riemann sphere S with its Euclidean metric ρ.

Proof
First observe that the extended complex plane $\mathbb{C}^* = \mathbb{C} \cup \{\infty\}$ can be given the topology of one-point compactification of $\mathbb{C}$ in accordance with the definition 1.8.4 where we replace X by $\mathbb{C}$ and denote this by $(\mathbb{C}^*, \tau)$. We have already seen that the stereographic projections $f : \mathbb{C}^* \to S$ and $f^{-1} = g : S \to \mathbb{C}^*$ are bijections with $f(\infty) = N$ and $g(N) = \infty$ where $N = (0, 0, 1)$ is the north pole. Further their restrictions $f_1 : \mathbb{C} \to S/\{N\}$ and $g_1 : S/\{N\} \to \mathbb{C}$ are also bijections given by

$$f_1(z) = \left(\frac{z + \bar{z}}{1 + |z|^2}, \frac{z - \bar{z}}{i(1 + |z|^2)}, \frac{|z|^2 - 1}{1 + |z|^2} \right) \quad (z \in \mathbb{C})$$

$$g_1(x_1, x_2, x_3) = \frac{x_1 + i x_2}{1 - x_3} \quad [(0, 0, 1) = N \neq (x_1, x_2, x_3) \in S].$$

From these expressions for f_1 and g_1 it is clear that both f_1 and g_1 are continuous in their respective domains. (See Theorem 1.7.27, which incidentally implies that coordinate projections are continuous on $\mathbb{R}^n$). Hence open subsets of $\mathbb{C}$ are mapped onto open subset of $S\backslash\{N\}$ and vice-versa by these maps. To show $f : \mathbb{C}^* \to S$ and $g : S \to \mathbb{C}^*$ are continuous it suffices to show that open sets containing ∞ in $\mathbb{C}^*$ are mapped onto open subsets of S containing N and vice-versa. Let $U \subset \tau$ and $\infty \in U$. Then $U = \mathbb{C}^*\backslash K$, where $K \subset \mathbb{C}$ is compact. Since f is bijective $f(U) = f(\mathbb{C}^*)\backslash f(K) = S\backslash f(K)$. As f_1 is continuous on $\mathbb{C}$, $f_1(K) = f(K)$ is compact and so closed in S or that $S\backslash f(K)$ is open in S proving $f(U)$ is open in (S, ρ). Conversely if V is open in (S, ρ) and $N \in V$ then $V = S\backslash(S\backslash V)$ and $S\backslash V \subset S\backslash\{N\}$ and is compact by Corollary 1.7.19 (as $S\backslash V$ is closed and bounded). Again as g is bijective $g(V) = g(S)\backslash g(S\backslash V) = \mathbb{C}^*\backslash g(S\backslash V)$ and $g_1(S\backslash V) = g(S\backslash V)$ is compact (Since g_1 is continuous on $S\backslash\{N\}$). Thus $g(V)$ is open in$(\mathbb{C}^*, \tau)$ and the proof is complete.

Corollary 1.8.8
The topology of one-point compactification of $\mathbb{C}$ is metrizable and a compatible metric is given by

$$d(z, z') = \frac{2|z - z'|}{\sqrt{(1+|z|^2)(1+|z'|^2)}} \quad \text{if } z, z' \neq \infty$$

$$d(z, \infty) = \frac{2}{\sqrt{(1+|z|^2)}} \quad \text{if } z \neq \infty$$

Proof
From Theorem 1.8.7, $(\mathbb{C}^*, \tau)$ is homeomorphic to (S, ρ) under the stereographic projection. Under the same map $(\mathbb{C}^*, d)$ and (S, ρ) are homeomorphic (in fact even isometric as we already saw). Thus $(\mathbb{C}^*, \tau)$ and $(\mathbb{C}^*, d)$ are homeomorphic and this is the content of the corollary.

1.9 Analysis in the Complex Domain
Since real analysis is fundamental to the analysis of complex domain we shall recall the following concepts. Let A be a nonempty subset of $\mathbb{R}$. l is called an upper bound for A if $a \leq l$ for all $a \in A$ and g is called a lower bound for A if $g \leq a$ for all $a \in A$. It is clear that if l is an upper bound and g a lower bound, then $l' > l$ is also an upper bound and $g' < g$ is also a lower bound. The least upper bound of A is denoted by sup A or *lub A* and is defined to be the number α with the following properties: (i) α is an upper bound for A and (ii) if l is any other upper bound of A then $\alpha \leq l$. Similarly the greatest lower bound β of A is denoted by *inf A* or *glb A* and is defined to be the number β satisfying the following properties: (i) β is a lower bound for A and (ii) if g is any other lower bound of A then $g \leq \beta$.

Assuming the notion of $a_n \to l$ where $\{a_n\}$ is a real sequence and $l \in [-\infty, \infty]$, we recall the following from real analysis. The limit supremum of $\{a_n\}$ denoted by $\limsup\limits_{n \to \infty} a_n$, or $\overline{\lim\limits_{a \to \infty}} a_n$ is defined as $L = \sup A'$ where A' is the set of all limit points of $\{a_n\}$ (the point $l \in [-\infty, \infty]$ is a limit point of $\{a_n\}$ if there is a subsequence of $\{a_n\}$ tending to l). Similarly $\liminf\limits_{n \to \infty} a_n$ or $\underline{\lim\limits_{n \to \infty}} a_n$ is defined by $G = \inf A'$. Note that A' can contain ∞ or $-\infty$ and hence L and G are, in general, extended real numbers. Note that $L \in A'$ and if $l > L$ there is an integer N such that $a_n \leq l$ for all $n \geq N$ and L is the only number with these properties. A similar description applies to G. Note that these characterizations are valid if L and G are finite. However if ∞ is a limit point of $\{a_n\}$ (this happens if $\{a_n\}$ is not bounded above) then $\infty = \limsup\limits_{n \to \infty} \{a_n\}$. Similarly if $\{a_n\}$ is unbounded below then $-\infty$ is a limit point and $G = -\infty$. The concept of the limit of a function as is available in the real case can be easily extended to the complex case as follows. Let f be a complex function of a complex variable defined in a deleted neighbourhood of a point w. (By this we mean the set of all points z in the complex plane with $0 < |z - w| < \delta$ for some $\delta > 0$). Then we say $f(z)$ has a limit L as $z \to w$ (written as $\lim\limits_{z \to w} f(z) = L$) if given $\varepsilon > 0$ there exists $\delta > 0$ such that whenever $0 < |z - w| < \delta$ we have $|f(z) - L| < \varepsilon$. The well known results concerning the limit of a sum, product and quotient hold good as in the real case. In addition we also have the following:

If $\lim\limits_{z \to w} f(z) = L$ then (i) $\lim\limits_{z \to w} \overline{f(z)} = \overline{L}$, (ii) $\lim\limits_{z \to w} Re\, f(z) = Re\, L$,

(iii) $\lim\limits_{z \to w} Im\, f(z) = Im\, L$, (iv) $\lim\limits_{z \to w} |f(z)| = |L|$.

These results can be proved using the following facts:

$$|\overline{L}| = |L|, \; |Re\, L| \leq |L|, \; |Im\, L| \leq |L|, \text{ etc.}$$

Suppose $f(z)$ is a complex valued function defined in an open disc around w then we say $f(z)$ is a continuous at w if $\lim\limits_{z \to w} f(z) = f(w)$. It is clear that (as in the real case) the sum and product of two continuous functions at w and the quotient of two continuous functions at w is also continuous at w provided the denominator is non-zero at w. Further if $f(z)$ is continuous at w then $Re\, f(z)$, $Im\, f(z)$ and $|f(z)|$ are also continuous at w. Note that the concept of continuity is local in character. i.e., if f is a map from some domain $\mathcal{D}$ of the complex plane (with complex values) then f is defined to be continuous in $\mathcal{D}$ if f is continuous at every point of $\mathcal{D}$. Since for both the concepts of limit and continuity, we need the values of the function in a disc around that point, the definition of continuity in a domain $\mathcal{D}$ is non-trivial only if around each point there is an open disc around that point, which is completely contained in $\mathcal{D}$. This is what is usually described by saying that 'the domain is open'.

In the case of a real function of a real variable f we define the derivative of f at a in its domain of definition by

$$f'(a) = \lim_{h \to 0} \frac{f(a+h) - f(a)}{h}. \qquad \dots (1)$$

The notion of derivative of a real function of a real variable can be extended to the case of a complex-valued function of a complex variable in many ways. One is the notion of the partial derivatives. In this, we treat $f(x, y)$ as a function of one variable at a time, holding the other fixed. The partial derivative of $f(x, y)$ with respect to x (or y) at (x_0, y_0) is the derivative of $g(x) = f(x, y_0)$ at x_0 (or $h(y) = f(x_0, y)$ at y_0). The partial derivative can be viewed as the rate of change of the function along either of the coordinate axes. The directional derivative generalizes the partial derivatives and is the rate of change of the function in an arbitrary direction. We now proceed to motivate yet another way of generalizing the derivative of a real function of a real variable. This notion is called the real differentiability of a function from sub domains of $\mathbb{C}$ to $\mathbb{C}$. Let $f : \mathbb{R} \to \mathbb{R}$ be differentiable at x. We have

$$f'(x) = \lim_{h \to 0} \frac{f(x+h) - f(x)}{h}$$

Hence

$$f(x+h) - f(x) = f'(x)h + r(h) \quad (h \text{ small}) \qquad (1.4)$$

where the remainder $r(h)$ satisfies $\lim_{h \to 0} \dfrac{r(h)}{h} = 0$. We now interpret this as follows: The difference $f(x+h) - f(x) = A(h) + r(h)$ where $A(h)$ is a linear continuous function from $\mathbb{R}$ to $\mathbb{R}$ defined by $A(h) = f'(x)h$ and $r(h)$ is a small remainder. We can therefore regard the derivative of f not as a real number but as a linear function on $\mathbb{R}$ given by $A(h)$. On the other hand, if A is a linear map form $\mathbb{R}$ to $\mathbb{R}$ that takes h to $A(h)$ and

$$f(x+h) - f(x) = A(h) + r(h) \quad (h \text{ small})$$

where $\lim_{h \to 0} \dfrac{r(h)}{h} = 0$, then also $f'(x)$ exists as a real number. This is because every linear function from $\mathbb{R}$ to $\mathbb{R}$ is always given by $x \to \alpha x$ for some fixed $\alpha \in \mathbb{R}$. (Note that if A is a real linear map from $\mathbb{R}$ to $\mathbb{R}$ then $A(h) = hA(1) = \alpha h$ where $\alpha = A(1)$). Therefore $A(h) = \alpha h$ for some real α and

$$f'(x) = \lim_{h \to 0} \frac{f(x+h) - f(x)}{h} = \alpha .$$

At this stage we observe that $\mathbb{R}^2$ can be regarded as a vector space over $\mathbb{R}$ or as a vector space over $\mathbb{C}$ (identifying $\mathbb{R}^2$ with $\mathbb{C}$). Thus there are two possibilities of generalizing (1) for maps from $\mathbb{R}^2$ to $\mathbb{R}^2$ by requiring A to be either real linear or complex linear.

Let us now define these concepts formally and study the interconnection between them. Let us note that a complex-valued function of a complex variable $z = x + iy$ can be viewed as a function $f(x, y)$ of two real variables x and y.

Definition 1.9.1

Let $f(x, y)$ be a complex-valued function of a complex variable $z = x + iy$ (or as a function of two real variables x and y) defined in an open subset Ω of $\mathbb{C}$. Then

(a) The partial derivatives of f with respect to x and y at a general point $(x, y) \in \Omega$ are defined as follows:

$$\frac{\partial f}{\partial x} = \lim_{h \to 0} \frac{f(x+h, y) - f(x, y)}{h} \quad (h \in \mathbb{R})$$

$$\frac{\partial f}{\partial y} = \lim_{k \to 0} \frac{f(x, y+k) - f(x, y)}{k} \quad (k \in \mathbb{R})$$

(b) The directional derivative of f at $z \in \Omega$ in the direction $\alpha \, (0 \le \alpha < 2\pi)$ denoted by $Df(z, \alpha)$ is defined as

$$Df(z, \alpha) = \lim_{h \to 0} \frac{f(z + he^{i\alpha}) - f(z)}{h} \quad (h \in \mathbb{R})$$

(c) f is said to be real differentiable at $(x, y) \in \Omega$ if there exists a neighbourhood N of $(0, 0)$ and a real linear map of $\mathbb{R}^2$ to $\mathbb{R}^2$ say L such that

$$f(x + h, y + k) = f(x, y) + L(h, k) + R(h, k) \quad ((h, k) \in N)$$

where $R(h, k)$ is a remainder term satisfying $\dfrac{|R(h,k)|}{\sqrt{(h^2 + k^2)}} \to 0$ as $(h, k) \to (0, 0)$ in $\mathbb{R}^2$.

(d) f is said to be complex differentiable at z if there exists a neighbourhood N of $0 \in \mathbb{C}$ and a complex linear map L of $\mathbb{C}$ to $\mathbb{C}$ such that

$$f(z + h + ik) = f(z) + L(h + ik) + R(h + ik) \quad (h + ik \in N)$$

where $R(h + ik)$ is a remainder term satisfying $\dfrac{|R(h+ik)|}{|h+ik|} \to 0$ as $(h + ik) \to 0$ in $\mathbb{C}$.

Theorem 1.9.2

Let $f(z)$ be a complex function defined on an open set Ω of $\mathbb{C}$. If f is complex differentiable at $z \in \Omega$ then f is real differentiable at $z = (x, y) \in \Omega$.

Proof

Writing $f(z) = f(x + iy) = f(x, y)$ we see that the only difference between real and complex differentiability of f at $z \in \Omega$ is that the linear map $L : \mathbb{C} \to \mathbb{C}$ is real linear in the former case and is complex linear in the later. But since any complex linear map from $\mathbb{C} \to \mathbb{C}$ is automatically real linear $(L(\alpha z) = \alpha L(z) \; \forall \; \alpha \in \mathbb{R})$ the theorem follows.

Theorem 1.9.3

Let $f(z)$ be a complex function defined on an open set Ω of $\mathbb{C}$. Then f is real differentiable if and only if there exists a 2×2 real matrix A and a neighbourhood N of $(0, 0)$ such that

$$f(x + h, y + k) = f(x, y) + \left[A \begin{pmatrix} h \\ k \end{pmatrix} \right]^T + R(h, k) \quad ((h, k) \in N)$$

where $\left[A \begin{pmatrix} h \\ k \end{pmatrix} \right]^T$ is the transpose of the matrix multiplication of the 2×2

matrix A and $\begin{pmatrix} h \\ k \end{pmatrix}$ representing $(h, k) \in N$ and $R(h, k)$ is a remainder term

satisfying $\dfrac{|R(h, k)|}{\sqrt{(h^2 + k^2)}} \to 0$ as $(h, k) \to (0, 0)$ in $\mathbb{R}^2$.

Proof

Let f be real differentiable at $(x, y) \in \Omega$. We have a real linear map L from $\mathbb{R}^2 \to \mathbb{R}^2$ such that

$$f(x + h, y + k) = f(x, y) + L(h, k) + R(h, k) \quad ((h, k) \in N) \qquad (1.5)$$

with $\dfrac{|R(h, k)|}{\sqrt{(h^2 + k^2)}} \to 0$ as $(h, k) \to (0, 0)$ in $\mathbb{R}^2$. To complete one part of

the proof one need to verify that $L(h, k) = \left[A \begin{pmatrix} h \\ k \end{pmatrix} \right]^T$ for some real 2×2

matrix A. Using the standard basis $(1, 0)$ and $(0, 1)$ of the vector space $\mathbb{R}^2$ over $\mathbb{R}$ we can easily see that the real linear map satisfies

$$L(h, k) = L(h(1, 0) + k(0,1)) = hL(1, 0) + kL(0, 1)$$

for all $(h, k) \in \mathbb{R}^2$. If we let $L(1, 0) = (\alpha, \beta)$ and $L(0, 1) = (\gamma, \delta)$ then

$$L(h, k) = (h\alpha + k\gamma, h\beta + k\delta) = \left[\begin{pmatrix} \alpha & \gamma \\ \beta & \delta \end{pmatrix} \begin{pmatrix} h \\ k \end{pmatrix} \right]^T .$$

Thus the required A can be taken as $\begin{pmatrix} \alpha & \gamma \\ \beta & \delta \end{pmatrix}$ and one part of our theorem is proved. To prove the converse we shall assume

$$f(x + h, y + k) = f(x, y) + L(h, k) + R(h, k) \quad ((h, k) \in N)$$

and define $L(h, k) = \left[A \begin{pmatrix} h \\ k \end{pmatrix} \right]^{T}$ from $\mathbb{R}^2 \to \mathbb{R}^2$. We shall verify that this L is real linear so that by definition f will be real differentiable. Now

$$L((h, k) + (u, v)) = L(h + u, k + v) = \left[A \begin{pmatrix} h + u \\ k + v \end{pmatrix} \right]^{T}$$

$$= \left[A \left[\begin{pmatrix} h \\ k \end{pmatrix} + \begin{pmatrix} u \\ v \end{pmatrix} \right] \right]^{T}$$

$$= \left[A \left[\begin{pmatrix} h \\ k \end{pmatrix} + A \begin{pmatrix} u \\ v \end{pmatrix} \right] \right]^{T}$$

$$= L(h, k) + L(u, v)$$

by the distributive property of matrix multiplication over matrix addition. Similarly, for α real

$$L(\alpha(h, k)) = L(\alpha h, \alpha k) = \left[A \begin{pmatrix} \alpha h \\ \alpha k \end{pmatrix} \right]^{T} = \left[A \alpha \begin{pmatrix} h \\ k \end{pmatrix} \right]^{T} = \left[\alpha A \begin{pmatrix} h \\ k \end{pmatrix} \right]^{T} = \alpha L$$

(h, k). Thus L is real linear and this completes the proof.

Note 1.9.4

By previous theorem, every real differentiable function at $(x, y) \in \Omega$ gives rise to a 2×2 real matrix A. We call this matrix as the matrix associated to the real differentiable function at (x, y). Occasionally we shall call the real linear map L associated with a real differentiable map f (as in Theorem 1.9.1(c)) as the total differential of f at (x, y) and denote it by $df \,|\, (x, y)$.

Theorem 1.9.5

Let $f(z)$ be a complex function of a complex variable defined in an open subset Ω of $\mathbb{C}$. Then

(a) f is complex differentiable at $z \in \Omega$ if and only if there exists a complex number w and a neighbourhood N of $0 \in \mathbb{C}$ such that

$$f(z + h + ik) = f(z) + w(h + ik) + R(h + ik) \quad (h + ik \in N) \quad (1.6)$$

where $\dfrac{|R(h + ik)|}{|h + ik|} \to 0$ as $(h + ik) \to 0$ in $\mathbb{C}$.

(b) f is complex differentiable at $z \in \Omega$ if and only if

$$f'(z) = \lim_{\Delta z \to 0} \frac{f(z + \Delta z) - f(z)}{\Delta z} \qquad (1.7)$$

exists as a complex number.

Proof

(a) Assume f is complex differentiable at $z \in \Omega$ and write

$$f(z + h + ik) = f(z) + L(h + ik) + R(h + ik) \quad (h + ik \in N)$$

where L is complex linear map from $\mathbb{C}$ to $\mathbb{C}$ and $\dfrac{|R(h + ik)|}{|h + ik|} \to 0$ as

$(h + ik) \to 0$ in $\mathbb{C}$. The complex linearity of L implies that $L(h + ik) = L((h + ik) . 1) = (h + ik)L(1) = (h + ik)w$ (say) Thus (1.6) holds with $w = L(1)$. Conversely the map $L(z) = wz$ is a complex linear map from $\mathbb{C} \to \mathbb{C}$ and so (1.6) $\Rightarrow f$ is complex differentiable at $z \in \Omega$. This proves (a).

(b) If f is complex differentiable at $z \in \Omega$ then by (a) (1.6) holds for some complex number w and (1.6) $\Rightarrow$ (1.7) with $f'(z) = w$. Conversely if (1.7) holds, writing $\Delta z = h + ik$ we have in a neighbourhood N of $0 \in \mathbb{C}$

$$\frac{f(z + h + ik) - f(z)}{h + ik} - f'(z) = R_1(h + ik)$$

where $R_1(h + ik) \to 0$ as $(h + ik) \to 0$ in $\mathbb{C}$. Thus

$$f(z + h + ik) = f(z) + f'(z) (h + ik) + (h + ik)R_1(h + ik)$$

for $h + ik \in N$. Hence (1.6) holds with $w = f'(z)$ and $R(h + ik) = (h + ik)R_1(h + ik)$. Thus f is complex differentiable at z by (a).

Note 1.9.6

Condition (1.7) of Theorem 1.9.5(b) is described by saying that f has a complex derivative at z denoted by $f'(z)$. Thus we have proved that the complex differentiability of f at z is equivalent to the existence of the complex derivative of f at z.

Theorem 1.9.7

Let f be a complex function defined on an open set $\Omega \subset \mathbb{C}$. If f is complex differentiable at z (or equivalently f has complex derivative at $z \in \Omega$) then

at $z = (x, y)$, $\dfrac{\partial f}{\partial x}$ and $\dfrac{\partial f}{\partial y}$ exist and satisfy $f'(z) = \dfrac{\partial f}{\partial x} = -i\dfrac{\partial f}{\partial y}$. (The

equation $\dfrac{\partial f}{\partial x} = -i\dfrac{\partial f}{\partial y}$ is called the Cauchy-Riemann equation or C-R

equation at z).

Proof

From Theorem 1.9.5 we can find a complex number $w = f'(z)$ such that for $h + ik$ in a neighbourhood of $0 \in \mathbb{C}$

$$f(z + h + ik) - f(z) = w(h + ik) + R(h + ik) \tag{1.8}$$

with $\dfrac{|R(h + ik)|}{|h + ik|} \to 0$ as $(h + ik) \to 0$. Putting $k = 0$ and $h = 0$ successively in (1.8) above and writing $f(z) = f(x, y)$ we have

$$\lim_{h \to 0} \frac{f(x + h, y) - f(x, y)}{h} = w = f'(z) \quad (h \in \mathbb{R})$$

$$\lim_{k \to 0} \frac{f(x, y + k) - f(x, y)}{k} = iw \quad (k \in \mathbb{R})$$

Hence $\dfrac{\partial f}{\partial x} = w$; $\dfrac{\partial f}{\partial y} = iw = i\dfrac{\partial f}{\partial x}$ or $\dfrac{\partial f}{\partial x} = -i\dfrac{\partial f}{\partial y}$ at $z = (x, y) \in \Omega$.

This proves the theorem.

Theorem 1.9.8

Let f be a complex function defined on an open set Ω of the complex plane. Let f be real differentiable at $(x, y) \in \Omega$ with associated 2×2 real matrix being $A = \begin{pmatrix} \alpha & \gamma \\ \beta & \delta \end{pmatrix}$. Then

(i) f is continuous

(ii) $Df(z, \theta)$ exists for all $\theta \in [0, 2\pi)$

(iii) $\dfrac{\partial f}{\partial x}, \dfrac{\partial f}{\partial y}$ both exist at $(x, y) \in \Omega$ and $\dfrac{\partial f}{\partial x} = \alpha + i\beta = (\alpha, \beta)$; $\dfrac{\partial f}{\partial y} = \gamma + i\delta = (\gamma, \delta)$ (i.e., the column vectors of A represent $\dfrac{\partial f}{\partial x}$ and $\dfrac{\partial f}{\partial y}$).

Proof

(i) By Theorem 1.9.3, we have

$$f(x + h, y + k) - f(x, y) = \left[\begin{pmatrix} \alpha & \gamma \\ \beta & \delta \end{pmatrix} \begin{pmatrix} h \\ k \end{pmatrix} \right]^T + R(h, k) \tag{1.9}$$

where (h, k) belongs to a suitable neighbourhood (say N) of $(0, 0)$ and

$\dfrac{|R(h, k)|}{\sqrt{h^2 + k^2}} \to 0$ as $(h, k) \to (0, 0)$. Thus for $(h, k) \in N$

$$|f(x + h, y + k) - f(x, y)| \leq |\alpha h + \gamma k + i(\beta h + \delta k)| + |R(h, k)|$$
$$\leq M(|h| + |k|) + |R(h, k)|$$

where we can choose $M = \max\{|\alpha| + |\beta|, |\gamma| + |\delta|\}$. Allowing $(h, k) \to$ $(0, 0)$ we see that (Note that $|R(h, k)| \to 0$ as $(h, k) \to (0, 0)$)
$$f(x + h, y + k) \to f(x, y).$$
Thus f is continuous at (x, y) and this proves (i).
(ii) Now using (1.9),

$$Df(z, \theta) = \lim_{h \to 0} \frac{f(z + he^{i\theta}) - f(z)}{h}$$
$$= \lim_{h \to 0} \frac{f(x + h\cos\theta, y + h\sin\theta) - f(x, y)}{h}$$
$$= \lim_{h \to 0} \frac{h[(\alpha\cos\theta + \gamma\sin\theta) + i(\beta\cos\theta + \delta\sin\theta)] + R(h\cos\theta, h\sin\theta)}{h}$$
$$= (\alpha\cos\theta + \gamma\sin\theta) + i(\beta\cos\theta + \delta\sin\theta)$$
$$= (\alpha + i\beta)\cos\theta + (\gamma + i\delta)\sin\theta.$$

Thus (ii) is proved.

(iii) Now, note that $\dfrac{\partial f}{\partial x} = Df(z, 0)$ and $\dfrac{\partial f}{\partial y} = Df\left(z, \dfrac{\pi}{2}\right)$. Hence both $\dfrac{\partial f}{\partial x}$ and $\dfrac{\partial f}{\partial y}$ exist at $z = (x, y)$ and further $\dfrac{\partial f}{\partial x} = Df(z, 0) = \alpha + i\beta = (\alpha, \beta)$ and $\dfrac{\partial f}{\partial y} = Df\left(z, \dfrac{\pi}{2}\right) = \gamma + i\delta = (\gamma, \delta)$. Thus (iii) is proved. This completes the proof of the theorem.

Note 1.9.9
In general the converses of Theorems 1.9.2, 1.9.7 and 1.9.8 are not true. The following counter examples testify to this.

Examples 1.9.10
(i) $f(z) = \bar{z}$ is a real linear map of $\mathbb{C} \to \mathbb{C}$ and so $f(x + h, y + k) - f(x, y)$ $= f(h, k) + R(h, k)$ with $R(h, k) = 0$. Thus the real differentiability condition holds for f with $L = f$. However since

$$\frac{f(z + \Delta z) - f(z)}{\Delta z} = \frac{\overline{\Delta z}}{\Delta z}$$

which is $+1$ or -1 depending on whether Δz is real or purely imaginary we see that f does not posses a complex derivative and so (by Note 1.9.6) f cannot be complex differentiable at any $z \in \mathbb{C}$.
(ii) Consider

$$f(z) = \begin{cases} \dfrac{(\bar{z})^2}{z} & \text{when } z \neq 0 \\ 0 & \text{when } z = 0 \end{cases}$$

$$\left(\frac{\partial f}{\partial x}\right)_{(0,0)} = \lim_{\substack{h\to 0 \\ h\in R}} \frac{f(h,0) - f(0,0)}{h} = \lim_{\substack{h\to 0 \\ h\in R}} \left(\frac{\overline{h}}{h}\right)^2 = 1$$

$$\left(\frac{\partial f}{\partial y}\right)_{(0,0)} = \lim_{\substack{k\to 0 \\ k\in R}} \frac{f(0,k) - f(0,0)}{k} = \lim_{\substack{k\to 0 \\ k\in R}} \frac{(-ik)^2}{ik^2} = i$$

Thus $\dfrac{\partial f}{\partial x}$, $\dfrac{\partial f}{\partial y}$ both exist at $(0, 0)$ and satisfy the Cauchy-Riemann

equation $\dfrac{\partial f}{\partial x} = -i\dfrac{\partial f}{\partial y}$, However, $f'(0)$ does not exist. Indeed, if

$f'(0)$ exists then $f'(0) = \lim\limits_{(h+ik)\to 0} \dfrac{f(h+ik) - f(0)}{h+ik}$ must be indepen-

dent of the approach of $h + ik$ to 0. However, allowing $h + ik \to 0$

through reals (i.e. $k = 0$) we have $\lim\limits_{\substack{h\to 0 \\ h\in R}} \dfrac{f(h) - f(0)}{h} = 1$. But as

$h + ik \to 0$ through the line $x = y$ (i.e., $h = k$) we have

$\lim\limits_{\substack{h\to 0 \\ h\in R}} \dfrac{f(h) - f(0)}{h} = \lim\limits_{\substack{h\to 0 \\ h\in R}} \dfrac{h^2((1+i)^2)}{h^2(1+i)^2} = \dfrac{-2i}{2i} = -1$. Hence $f'(0)$ does

not exists and by Note 1.9.6, f cannot be complex differentiable.

(iii) Consider

$$f(x, y) = \begin{cases} \dfrac{x^3}{x^2 + y^2} & (x, y) \neq (0, 0) \\ 0 & (x, y) = (0, 0) \end{cases}$$

$$|f(x, y) - f(0, 0)| = \left|\dfrac{x^3}{x^2 + y^2}\right| \leq |x| \to 0 \text{ as } (x, y) \to (0, 0). \text{ Thus}$$

f is continuous at $(0, 0)$. Further

$$Df(0, \alpha) = \lim_{\substack{h\to 0 \\ h\in R}} \frac{f(he^{i\alpha}) - f(0)}{h} = \lim_{\substack{h\to 0 \\ h\in R}} \frac{h^3\cos^3\alpha}{h^3} = \cos^3\alpha$$

exists for all $\alpha \in [0, 2\pi)$ and in particular $\dfrac{\partial f}{\partial x} = 1$ and $\dfrac{\partial f}{\partial y} = 0$ at

$(0, 0)$. However $f(x, y)$ is not real differentiable at $(0, 0)$. This can
be proved by "reductio ad absurdum". Assume the contrary. In a
neighbourhood of $(0, 0)$ we must have

$$f(h, k) - f(0, 0) = \left[\begin{pmatrix} \alpha & \gamma \\ \beta & \delta \end{pmatrix}\begin{pmatrix} h \\ k \end{pmatrix}\right]^T + R(h, k)$$

where

$$\alpha + i\beta = \left(\frac{\partial f}{\partial x}\right)_{(0,0)} = 1 \text{ and } \gamma + i\delta = \left(\frac{\partial f}{\partial y}\right)_{(0,0)} = 0$$

and $\dfrac{|R(h+ik)|}{|(h+ik)|} \to 0$ as $(h+ik) \to 0$ using Theorem 1.9.8 and equation (1.9) thereof. Thus,

$$\frac{h^3}{h^2+k^2} = \left[\begin{pmatrix} 1 & 0 \\ 0 & 0 \end{pmatrix}\begin{pmatrix} h \\ k \end{pmatrix}\right]^T + R(h,\ k)$$

or $R(h,k) = \dfrac{h^3}{h^2+k^2} - h = \dfrac{-k^2 h}{h^2+k^2}$. Then $\dfrac{R(h,k)}{\sqrt{h^2+k^2}} = \dfrac{-k^2 h}{(h^2+k^2)^{3/2}}$

$\to \dfrac{-m^2}{(1+m^2)^{3/2}}$ as $(h,\ k) \to (0,\ 0)$ along the line $k = mh$ since

$\dfrac{-m^2}{(1+m^2)^{3/2}}$ gives different non-zero values for different $m \neq 0$ we see

that $\dfrac{R(h,k)}{\sqrt{h^2+k^2}}$ does not tend to zero as $(h,\ k) \to (0,\ 0)$ proving that

$f(x, y)$ is not real differentiable at $(0, 0)$.

(iv) Consider

$$f(z) = \begin{cases} e^{-1/z^4} & z \neq 0 \\ 0 & z = 0 \end{cases}$$

$f(z)$ is defined in the entire finite complex plane. We shall consider its behaviour at $(0, 0)$

$$\left(\frac{\partial f}{\partial x}\right)_{(0,0)} = \lim_{\substack{h\to 0 \\ h\in R}} \frac{f(h,0)-f(0,0)}{h} = \lim_{\substack{h\to 0 \\ h\in R}} \frac{e^{-1/h^4}}{h} = 0$$

$$\left(\frac{\partial f}{\partial y}\right)_{(0,0)} = \lim_{\substack{k\to 0 \\ k\in R}} \frac{f(0,k)-f(0,0)}{k} = \lim_{\substack{k\to 0 \\ k\in R}} \frac{e^{-1/k^4}}{k} = 0$$

(Note that by *L'* Hospital's rule $\lim_{t\to\infty} te^{-t^4} = \lim_{t\to\infty} (t/e^{t^4}) = 0$)

$$Df\left(0,\frac{\pi}{4}\right) = \lim_{\substack{h\to 0 \\ h\in R}} \frac{f(he^{i\pi/4})-f(0)}{h} = \lim_{\substack{h\to 0 \\ h\in R}} \frac{e^{1/h^4}}{h} = \infty.$$

Further, limit of $f(z)$ as $z \to 0$ through the real axis is given by

$$\lim_{x \to 0} f(x) = \lim_{x \to 0} e^{-1/x^4} = 0$$

But its limit as $z \to 0$ through the line $he^{i\pi/4}$ is

$$\lim_{h \to 0} f(he^{i\pi/4}) = \lim_{x \to 0} e^{1/h^4} = \infty.$$

Thus $f(z)$ is not even continuous at $z = 0$.

Note 1.9.11

(1) Examples 1.9.10 (i), (ii) and (iii) provide counter examples for the reverse implications in Theorems 1.9.2, 1.9.7, 1.9.8 respectively. Example (iv) illustrates that for some functions $\dfrac{\partial f}{\partial x}$ and $\dfrac{\partial f}{\partial y}$ may exist at a point but some other directional derivative may not. This example also illustrates the simple fact that mere existence of partial derivatives for a function does not even guarantee its continuity. Thus the existence of partial derivatives for a complex-valued function of a complex variable is the least satisfying way of generalizing the concept of a real derivative for functions from $\mathbb{R} \to \mathbb{R}$. Since complex differentiability (which is equivalent to the existence of complex derivative) implies real differentiability and the existence of all directional derivatives, it is the most satisfying generalization of the concept of real derivative for functions from $\mathbb{R} \to \mathbb{R}$.

(2) In view of the counter examples provided above it is interesting to ask what additional hypothesis on the function will ensure that the converses of Theorems 1.9.2, 1.9.7 and 1.9.8 hold. These questions are answered by the following theorems.

Theorem 1.9.12

Let $f(z)$ be a complex function defined on an open set Ω of $\mathbb{C}$. If f is real differentiable at $z = (x, y) \in \Omega$ and at (x, y) the Cauchy-Riemann equation $\dfrac{\partial f}{\partial x} = -i\dfrac{\partial f}{\partial y}$ holds then f is a complex differentiable at z with

$$f_{(z)} = \frac{\partial f}{\partial x} = -i\frac{\partial f}{\partial y}.$$

Proof

By Theorem 1.9.8 and equation 1.9 there of, we can find a neighbourhood N of $(0, 0)$ and a real 2×2 matrix $A = \begin{pmatrix} \alpha & \gamma \\ \beta & \delta \end{pmatrix}$ with $\dfrac{\partial f}{\partial x} = \alpha + i\beta$ and

$\dfrac{\partial f}{\partial y} = \gamma + i\delta$ such that for $(h, k) \in N$

$$f(x + h, y + k) - f(x, y) = \left[\begin{pmatrix} \alpha & \gamma \\ \beta & \delta \end{pmatrix} \begin{pmatrix} h \\ k \end{pmatrix} \right]^T + R(h, k) \qquad (1.10)$$

with $\dfrac{|R(h,k)|}{\sqrt{h^2 + k^2}} \to 0$ as $(h, k) \to (0, 0)$. But

$$\frac{\partial f}{\partial x} = -i \frac{\partial f}{\partial y} \Rightarrow \alpha + i\beta = -i\gamma + \delta \Rightarrow \alpha = \delta; \; \beta = -\gamma$$

and in the complex notation we can write (1.10) as

$$\begin{aligned} f(z + h + ik) &= f(z) + \alpha h + \gamma k + i(\beta h + \delta k) + R(h, k) \\ &= f(z) + \alpha(h + ik) + i\beta(h + ik) + R(h, k) \\ &= f(z) + (\alpha + i\beta)(h + ik) + R(h, k). \end{aligned}$$

It follows from Theorem 1.9.5(a) with $w = \alpha + i\beta$ that f is complex differentiable.

Theorem 1.9.13
Let $f(z)$ be a complex function defined on an open set Ω of the plane. If

at $z = (x, y) \in \Omega$, f is continuous, $\dfrac{\partial f}{\partial x}, \dfrac{\partial f}{\partial y}$ exist in a neighbourhood of

z and are also continuous at z then f is real differentiable at z.

Proof
For (h, k) belonging to a suitable neighbourhood of $(0, 0)$ we write
$f(x + h, y + k) - f(x, y) = f(x + h, y + k) - f(x, y + k) + f(x, y + k)$
$- f(x, y)$. Using one variable mean-value theorem of differential calculus we can write the right hand side of the above equality as

$$h f_x(x + h_1, y + k) + k f_y(x + h, y + k_1)$$

for some h_1 between 0 and h and k_1 between 0 and k where f_x, f_y denote the partial derivatives of f. By continuity of $f_x = f_x(x, y)$ and $f_y = f_y(x, y)$ at (x, y) we can write

$$f_x(x + h_1, y + k) = f_x + \varepsilon_1 \quad \text{and} \quad f_y(x + h, y + k_1) = f_y + \varepsilon_2.$$

If we denote $f_x = \alpha + i\beta$ and $f_y = \gamma + i\delta$, by their column vectors then

$$f(x + h, y + k) - f(x, y) = \left[\begin{pmatrix} \alpha & \gamma \\ \beta & \delta \end{pmatrix} \begin{pmatrix} h \\ k \end{pmatrix} \right]^T + \varepsilon_1 h + \varepsilon_2 k$$

where $R(h, k) = \varepsilon_1 h + \varepsilon_2 k$ is such that $\dfrac{|R(h,k)|}{\sqrt{h^2 + k^2}} \leq \varepsilon_1 + \varepsilon_2 \to 0$ as $(h,$

$k) \to (0, 0)$.
Thus by Theorem 1.9.3, f is real differentiable at $(x, y) \in \Omega$.

Theorem 1.9.14   Let f be a complex function defined on an open set Ω of the plane. If at $z = (x, y) \in \Omega$, f is continuous, $\dfrac{\partial f}{\partial x}$, $\dfrac{\partial f}{\partial y}$ exist in a neighbourhood of z and are continuous at z and satisfy the Cauchy-Riemann equation $\dfrac{\partial f}{\partial x} = -i \dfrac{\partial f}{\partial y}$ then f is complex differentiable at z.

Proof

By Theorem 1.9.13, f is real differentiable. Further $\dfrac{\partial f}{\partial x}$ and $\dfrac{\partial f}{\partial y}$ satisfy the Cauchy-Riemann equation. An application of Theorem 1.9.12 now completes the proof.

Note 1.9.15
From the above theorems it is clear that Cauchy-Riemann equation plays a very crucial role in the complex differentiability of a complex valued function of a complex variable. However, the Cauchy-Riemann equation takes different forms depending on the nature of the coordinate systems that are used in the domain z-plane and the image w-plane. We would like to bring all these equations together under a single setup and derive them.

Theorem 1.9.16   (Cauchy-Riemann Equations)
Let $f(z)$ be a complex-valued function of a complex variable defined on an open subset Ω of the plane. Let f be complex differentiable at $z \in \Omega$ (or that the complex derivative $f'(z)$ exists). Let $z = x + iy = re^{i\theta}$ and $f(z) = f(x, y) = f(r, \theta)$. Further let $\vec{t} = z + he^{i\alpha}$ and $\vec{n} = z + he^{i\beta}$ $\left(\beta = \alpha + \dfrac{\pi}{2} \right)$ be a pair of mutually perpendicular directions through z and define

$$\frac{\partial f}{\partial \vec{t}} = Df(z, \alpha) \text{ and } \frac{\partial f}{\partial \vec{n}} = Df(z, \beta)$$

Then

$$(i) \ \frac{\partial f}{\partial x} = -i \frac{\partial f}{\partial y} \quad (ii) \ \frac{\partial f}{\partial r} = -i \frac{1}{r} \frac{\partial f}{\partial \theta} \quad (iii) \ \frac{\partial f}{\partial \vec{t}} = -i \frac{\partial f}{\partial \vec{n}}$$

Proof
In view of Theorem 1.9.5(b)

$$f'(z) = \lim_{\Delta z \to 0} \frac{f(z + \Delta z) - f(z)}{\Delta z} \tag{1.11}$$

exists and is independent of the mode of approach of Δz to 0. Taking Δz real and purely imaginary successively in (1.11) we get $f'(z) = \dfrac{\partial f}{\partial x} = \dfrac{1}{i} \dfrac{\partial f}{\partial y} = -i \dfrac{\partial f}{\partial y}$. This gives (i). Using the notation $f(z) = f(re^{i\theta}) = f(r, \theta)$ we also have

$$f'(z) = \lim_{\Delta r \to 0} \frac{f((r + \Delta r)e^{i\theta}) - f(re^{i\theta})}{\Delta r e^{i\theta}} \quad \text{for any fixed } r = |z|.$$

But from the definitions, we have

$$\frac{\partial f}{\partial r} = \lim_{\Delta r \to 0} \frac{f(r + \Delta r, \theta) - f(r, \theta)}{\Delta r}$$

$$= \lim_{\Delta r \to 0} \frac{f((r + \Delta r)e^{i\theta}) - f(re^{i\theta})}{\Delta r}$$

$$= f'(z)e^{i\theta}$$

$$\frac{\partial f}{\partial \theta} = \lim_{\Delta \theta \to 0} \frac{f(r, \theta + \Delta \theta) - f(r, \theta)}{\Delta \theta}$$

$$= \lim_{\Delta \theta \to 0} \frac{f(re^{i(\theta + \Delta \theta)}) - f(re^{i\theta})}{\Delta \theta}$$

$$= \lim_{\Delta \theta \to 0} \frac{f(re^{i(\theta + \Delta \theta)}) - f(re^{i\theta})}{re^{i(\theta + \Delta \theta)} - re^{i\theta}} \lim_{\Delta \theta \to 0} \frac{re^{i(\theta + \Delta \theta)} - re^{i\theta}}{\Delta \theta}$$

$$= if'(z)re^{i\theta}.$$

(Note that from the definition, $ie^{i\theta} = \dfrac{d}{d\theta}(e^{i\theta}) = \lim_{\Delta \theta \to 0} \dfrac{e^{i(\theta + \Delta \theta)} - e^{i\theta}}{\Delta \theta}$).

Thus

$$\frac{\partial f}{\partial r} = f'(z)e^{i\theta} = \frac{ire^{i\theta} f'(z)}{ir} = -i\frac{1}{r}\frac{\partial f}{\partial \theta}$$

proving (ii).

$$\frac{\partial f}{\partial \vec{t}} = Df(z, \alpha) = \lim_{h \to 0} \frac{f(z + he^{i\alpha}) - f(z)}{h}$$

$$= \lim_{h \to 0} \frac{f(z + he^{i\alpha}) - f(z)}{he^{i\alpha}} e^{i\alpha}$$

$$= e^{i\alpha} f'(z)$$

$$\frac{\partial f}{\partial \vec{n}} = Df(z, \beta) = e^{i\beta} f'(z) = e^{i(\alpha + \pi/2)} f'(z) = i\frac{\partial f}{\partial \vec{t}}.$$

Thus $\qquad \dfrac{\partial f}{\partial \vec{t}} = -i\dfrac{\partial f}{\partial \vec{n}}$ proving (iii).

Corollary 1.9.17 (Cauchy-Riemann Equations)

Let $f(z)$ be a complex-valued function of a complex variable defined on an open subset Ω of the plane. Let f be complex differentiable at $z \in \Omega$ or that $f'(z)$ exists.

(i) If $z = x + iy$; $f(z) = f(x, y) = u(x, y) + iv(x, y)$ then

$$\frac{\partial u}{\partial x} = \frac{\partial v}{\partial y}; \frac{\partial u}{\partial y} = -\frac{\partial v}{\partial x}$$

If $z = x + iy$; $f(z) = Re^{i\phi}$ where R and ϕ are function of x and y then

$$\frac{\partial R}{\partial x} = R\frac{\partial \phi}{\partial y}; \quad R\frac{\partial \phi}{\partial x} = -\frac{\partial R}{\partial y}$$

(ii) If $z = re^{i\theta}$; $f(z) = f(r, \theta) = u(r, \theta) + iv(r, \theta)$ then

$$\frac{\partial u}{\partial r} = \frac{1}{r}\frac{\partial v}{\partial \theta}; \quad \frac{\partial v}{\partial r} = -\frac{1}{r}\frac{\partial u}{\partial \theta};$$

If $z = re^{i\theta}$; $f(z) = Re^{i\phi}$ where R and ϕ are function of r and θ then

$$\frac{\partial R}{\partial r} = \frac{R}{r}\frac{\partial \phi}{\partial \theta}; \quad R\frac{\partial \phi}{\partial r} = -\frac{1}{r}\frac{\partial R}{\partial \theta}$$

Proof

From Theorem 1.9.16(i) we have $\dfrac{\partial f}{\partial x} = -i\dfrac{\partial f}{\partial y}$. Write $f = u + iv$ and compare real and imaginary parts to get

$$\frac{\partial u}{\partial x} = \frac{\partial v}{\partial y}; \quad \frac{\partial u}{\partial y} = -\frac{\partial v}{\partial x}.$$

Write $f = Re^{i\phi}$ and get

$$\frac{\partial f}{\partial x} = Re^{i\phi} i\frac{\partial \phi}{\partial x} + e^{i\phi}\frac{\partial R}{\partial x}$$

$$\frac{\partial f}{\partial y} = Re^{i\phi} i\frac{\partial \phi}{\partial y} + e^{i\phi}\frac{\partial R}{\partial y}.$$

Again $\dfrac{\partial f}{\partial x} = -i\dfrac{\partial f}{\partial y}$ gives, after cancelling $e^{i\phi}$ and comparing real and imaginary parts, $\dfrac{\partial R}{\partial x} = R\dfrac{\partial \phi}{\partial y}$; $R\dfrac{\partial \phi}{\partial x} = -\dfrac{\partial R}{\partial y}$. These prove (i). Using theorem 1.9.16(ii) we have $\dfrac{\partial f}{\partial r} = -i\dfrac{1}{r}\dfrac{\partial f}{\partial \theta}$.

Writing $f = u + iv$ we have

$$\left[\frac{\partial u}{\partial r} + i\frac{\partial v}{\partial r}\right] = -\frac{i}{r}\left[\frac{\partial u}{\partial \theta} + i\frac{\partial v}{\partial \theta}\right] \Rightarrow \frac{\partial u}{\partial r} = \frac{1}{r}\frac{\partial v}{\partial \theta}; \quad \frac{\partial v}{\partial r} = -\frac{1}{r}\frac{\partial u}{\partial \theta}.$$

Similarly writing $f = Re^{i\phi}$, we have

$$Re^{i\phi} i\frac{\partial \phi}{\partial r} + e^{i\phi}\frac{\partial R}{\partial r} = -\frac{i}{r}\left[Re^{i\phi} i\frac{\partial \phi}{\partial \theta} + e^{i\phi}\frac{\partial R}{\partial \theta}\right]$$

or that

$$\frac{\partial R}{\partial r} = \frac{R}{r}\frac{\partial \phi}{\partial \theta} ; \quad R\frac{\partial \phi}{\partial r} = -\frac{1}{r}\frac{\partial R}{\partial \theta}$$

This completes the proof of (ii).

Note 1.9.18
Using Theorem 1.9.16 (iii), writing $f = u + iv$ or $f = Re^{i\phi}$ one gets

$$\frac{\partial u}{\partial \vec{t}} = \frac{\partial v}{\partial \vec{n}} ; \quad \frac{\partial v}{\partial \vec{t}} = -\frac{\partial u}{\partial \vec{n}} \quad \text{and} \quad \frac{\partial R}{\partial \vec{t}} = R\frac{\partial \phi}{\partial \vec{n}} ; \quad R\frac{\partial \phi}{\partial \vec{t}} = \frac{\partial R}{\partial \vec{n}}$$

which are also Cauchy-Riemann equations involving directional derivatives. Since these equations are seldom used we have chosen to just mention them here.

1.10 Sequences and Series

Recall that by definition (See Definition 1.7.10 where X is replaced by $\mathbb{C}$) a complex sequence is a map from $\mathbb{N}$ into $\mathbb{C}$. It is usually denoted by $\{a_n\}$ where each a_n is the image of $n \in \mathbb{N}$ under the given map. It converges to $A \in \mathbb{C}$ if given $\varepsilon > 0$ we can find $N \in \mathbb{N}$ such that $|a_n - A| < \varepsilon$ for all $n \geq N$. A complex sequence diverges if it does not converge. If $\{a_n\}$ converges to A we also write $a_n \to A$ as $n \to \infty$ or $\lim_{n \to \infty} a_n = A$. $\{a_n\}$ is Cauchy if given $\varepsilon > 0$ we can find $N \in \mathbb{N}$ such that $|a_n - a_m| < \varepsilon$ for all $m, n \geq N$.

On the other hand a series of complex numbers is a formal infinite sum

$\sum_{n=1}^{\infty} a_n$ ($a_n \in \mathbb{C}$ for all n). The series converges by definition if its associated

sequence $\{s_n\}$ of partial sums defined by $s_n = a_1 + a_2 + \dots a_n$ ($n = 1, 2, \dots$)

converges and its limit is denoted by $\sum_{n=1}^{\infty} a_n$. $\sum_{n=1}^{\infty} a_n$ is said to be convergent

absolutely if $\sum_{n=1}^{\infty} |a_n|$ converges. We shall obtain certain properties of con-

vergent sequences and series.

Theorem 1.10.1

If $\{s_n\}$ and $\{t_n\}$ are complex sequences with $\lim_{n \to \infty} s_n = s$ and $\lim_{n \to \infty} t_n = t$ then

(i) $\lim_{n \to \infty} (s_n \pm t_n) = s \pm t$

(ii) $\lim_{n \to \infty} s_n t_n = st$

(iii) If $t_n \neq 0$, $t \neq 0$, $\lim\limits_{n \to \infty} \dfrac{s_n}{t_n} = \dfrac{s}{t}$

(iv) If $|s_n| \leq |t_n|$ for $n \geq N$ then $|s| \leq |t|$.

(v) $\lim\limits_{n \to \infty} (Re\ s_n) = Re\ s$, $\lim\limits_{n \to \infty} (Im\ s_n) = Im\ s$ and $\lim\limits_{n \to \infty} |s_n| = |s|$.

Theorem 1.10.2

If $z_n = a_n + ib_n$ $(n = 1, 2, \ldots)$ then $\sum\limits_{1}^{\infty} z_n = \sum\limits_{1}^{\infty} (a_n + ib_n) = a + ib$ if and

only if $\sum\limits_{1}^{\infty} a_n = a$ and $\sum\limits_{1}^{\infty} b_n = b$

Theorem 1.10.3   (Comparison test)

Let $\sum\limits_{1}^{\infty} a_n$ and $\sum\limits_{1}^{\infty} b_n$ are two series of non-negative (real) terms with a_n

$\leq b_n$ $\forall n \geq N$ (for some $N \in \mathbb{N}$). If $\sum\limits_{1}^{\infty} b_n$ converges then $\sum\limits_{1}^{\infty} a_n$ converges

and if $\sum\limits_{1}^{\infty} a_n$ diverges then $\sum\limits_{1}^{\infty} b_n$ diverges.

Theorem 1.10.4   (Ratio test)

Let $\sum\limits_{1}^{\infty} a_n$ be a series of non-zero complex numbers. Let $\lim\limits_{n \to \infty} \left| \dfrac{a_{n+1}}{a_n} \right| =$

ρ. Then

(i) $\sum\limits_{1}^{\infty} a_n$ converges absolutely if $\rho < 1$

(ii) $\sum\limits_{1}^{\infty} a_n$ diverges if $\rho > 1$

(iii) If $\rho = 1$ the series may or may not be convergent.

Theorem 1.10.5   (Root test)

Let $\sum\limits_{1}^{\infty} a_n$ be a complex series. Put $\alpha = \lim\limits_{n \to \infty} \sup \sqrt[n]{|a_n|}$. Then

(i) $\sum\limits_{1}^{\infty} a_n$ converges absolutely if $\alpha < 1$

(ii) $\sum\limits_{1}^{\infty} a_n$ diverges if $\alpha > 1$

(iii) If $\alpha = 1$ the series may or may not be convergent.

The proofs of theorems 1.10.1, 1.10.3, 1.10.4, 1.10.5 are so similar to their analogues in real analysis that we prefer to omit them while the proof of theorem 1.10.2 is trivial.

Theorem 1.10.6 (Cauchy Criterion)

A complex sequence $\{z_n\}$ converges if and only if it is Cauchy.

Proof

Follows from Note 1.7.12 and the proof of theorem 1.7.14.

Corollary 1.10.7

A complex series $\sum_{n=1}^{\infty} a_n$ is convergent if and only if given $\varepsilon > 0$ there exists $N = N(\varepsilon) \in \mathbb{N}$ depending on ε such that

$$\left| \sum_{k=n}^{m} a_k \right| < \varepsilon \quad \text{for all } n, m \geq N$$

Proof

Apply Theorem 1.10.6 to the sequence of partial sums of $\sum_{n=1}^{\infty} a_n$.

Note 1.10.8

From Corollary 1.10.7, it is clear that if $\sum_{n=1}^{\infty} a_n$ is convergent then $a_n \to 0$ as $n \to \infty$. (Given $\varepsilon > 0$ take $m = n \geq N(\varepsilon)$). However the converse is not true as is seen by the example $\sum_{n=1}^{\infty} \frac{1}{n}$.

Definition 1.10.9

Suppose $\{f_n\}$, $n = 1, 2, \ldots$ is a sequence of complex-valued functions defined on a metric space X, and suppose that the sequence of complex number $\{f_n(x)\}$ converges for every $x \in X$. We can then define a function f by

$$f(x) = \lim_{n \to \infty} f_n(x) \quad (x \in X). \tag{1.14}$$

Under these circumstances we say that $\{f_n\}$ converges to f on X and that f is the limit, or the limit function of $\{f_n\}$. Sometimes we shall use a more descriptive terminology and shall say that "$\{f_n\}$ converges to f point-wise on X" if (1.14) holds. Similarly if $\sum_{n=1}^{\infty} f_n(x)$ converges for every $x \in X$, and if we define

$$f(x) = \sum_{n=1}^{\infty} f_n(x) \quad (x \in X)$$

the function f is called the sum of the series $\sum_{n=1}^{\infty} f_n$.

Definition 1.10.10
We say that a sequence of complex functions $\{f_n\}$ converges uniformly on X to a function f if for every $\varepsilon > 0$ there is an integer N such that $n \geq N$ implies that

$$|f_n(x) - f(x)| < \varepsilon \quad \text{for all } x \in X.$$

It is clear that every uniformly convergent sequence is point-wise convergent. Quite explicitly, the difference between the two concepts is this: If $\{f_n\}$ converges pointwise on X, then there exists a function f such that, for every $\varepsilon > 0$, and for every $x \in X$, there is an integer N, depending on ε and x, such that $|f_n(x) - f(x)| < \varepsilon$ if $n \geq N$; if $\{f_n\}$ converges uniformly on X, it is possible, for each $\varepsilon > 0$, to find one integer N which will do for all $x \in X$.

We say that a series $\sum_{n=1}^{\infty} f_n(x)$ converges uniformly on X if the sequence $\{s_n(x)\}$ of partial sums defined by

$$\sum_{i=1}^{n} f_i(x)$$

converges uniformly on X.

The Cauchy criterion for uniform convergence is as follows;

Theorem 1.10.11
The sequence of complex functions $\{f_n\}$ defined on a metric space X, converges uniformly on X if and only if for every $\varepsilon > 0$ there exists an integer N such that $n \geq N$, $m \geq N$, $x \in X$ implies

$$|f_n(x) - f_m(x)| < \varepsilon.$$

Proof
Suppose $\{f_n\}$ converges uniformly on X, and let f be the limit function. Then there is an integer N such that $n \geq N$, $x \in X$ implies

$$|f_n(x) - f(x)| < \frac{\varepsilon}{2},$$

so that

$$|f_n(x) - f_m(x)| \leq |f_n(x) - f(x)| + |f(x) - f_m(x)| < \varepsilon$$

if $n \geq N$, $m \geq N$, $x \in X$.

Conversely, suppose that the Cauchy condition holds. By Theorem 1.10.6, the sequence $\{f_n(x)\}$ converges, for every $x \in X$, to a limit which we may call as $f(x)$. Thus the sequence $\{f_n\}$ converges on X, pointwise to f. We have to prove that this convergence is uniform.

Let $\varepsilon > 0$ be given, and choose N such that $|f_n(x) - f_m(x)| < \dfrac{\varepsilon}{2}$ $\forall m, n \geq N$. Fix n, and let $m \to \infty$ in the above inequality. Since $f_m(x) \to f(x)$ as $m \to \infty$, this gives

$$|f_n(x) - f(x)| < \varepsilon$$

for every $n \geq N$ and for every $x \in X$, which completes the proof.

The following criterion is sometimes useful and can be easily proved (we omit the details).

Theorem 1.10.12

Suppose $\lim_{n \to \infty} f_n(x) = f(x)$ $(x \in X)$. Put $M_n = \sup_{x \in X} |f_n(x) - f(x)|$. Then $f_n \to f$ uniformly on X if and only if $M_n \to 0$ as $n \to \infty$.

For series, there is very convenient test for uniform convergence due to Weierstrass.

Theorem 1.10.13 (Weierstrass M-test)

Suppose $\{f_n\}$ is a sequence of functions defined on X, and suppose

$$|f_n(x)| \leq M_n \quad (x \in X, n = 1, 2, \ldots).$$

Then $\sum f_n$ converges uniformly on X if $\sum M_n$ converges.

Note that the converse is not asserted (and is, in fact, not true).

Proof

If $\sum M_n$ converges, then, for arbitrary $\varepsilon > 0$,

$$\left| \sum_{i=n}^{m} f_i(x) \right| \leq \sum_{i=n}^{m} M_i \leq \varepsilon \quad (x \in X),$$

provided m and n are large enough. Uniform convergence now follows from Cauchy criterion for sequence of functions.

Theorem 1.10.14

The limit of a uniformly convergent sequence of complex continuous functions on a metric space is continuous.

Proof

Let $f_n \to f$ as $n \to \infty$ uniformly on a metric space (X, d), where f_n's are continuous complex functions on X. Given $\varepsilon > 0$, we can find $N \in \mathbb{N}$ such that $|f_n(x) - f(x)| < \dfrac{\varepsilon}{3}$ $\forall n \geq N$ and $\forall x \in X$. Let x_0 be a point in X. Since f_N is continuous at x_0 we can find a $\delta > 0$ such that $|f_N(x) - f_N(x_0)| < \dfrac{\varepsilon}{3}$ if $d(x, x_0) < \delta$. Under the same condition on x it follows that $|f(x) - f(x_0)| \leq |f(x) - f_N(x)| + |f_N(x) - f_N(x_0)| + |f_N(x_0) - f(x_0)| < \varepsilon$, and we have proved that f is continuous at x_0. Since x_0 is arbitrary the proof follows.

SOLVED EXERCISES

Problem 1 Prove $\dfrac{|a - b|}{|1 - \bar{a}b|} < 1$ if $|a| < 1$ and $|b| < 1$.

Solution Suppose that $|a| < 1$ and $|b| < 1$.

$$|\bar{a}b| = |a|\,|b| < 1 \Rightarrow 1 - \bar{a}b \neq 0$$

and

$$(1 - |a|^2)(1 - |b|^2) > 0$$

$$\Rightarrow \qquad |a|^2 + |b|^2 < 1 + |a|^2\,|b|^2 = 1 + |\bar{a}b|^2$$

Hence

$$\left|\frac{a - b}{1 - \bar{a}b}\right|^2 = \frac{|a|^2 + |b|^2 - 2\,\mathrm{Re}\,(a\bar{b})}{1 + |\bar{a}b|^2 - 2\,\mathrm{Re}\,(a\bar{b})} < 1.$$

This proves that $\dfrac{|a - b|}{|1 - \bar{a}b|} < 1$.

Problem 2 (i) Prove that $\dfrac{|a - b|}{|1 - \bar{a}b|} = 1$ if either $|a| = 1$ or $|b| = 1$ but not both. (ii) If $|a| = |b| = 1$ what exception must be made for its validity?

Solution (i) Let $|a| = 1$, $|b| \neq 1$.

$$\left|\frac{a - b}{1 - \bar{a}b}\right|^2 = \frac{|a|^2 + |b|^2 - 2\,\mathrm{Re}\,(a\bar{b})}{1 + |\bar{a}b|^2 - 2\,\mathrm{Re}\,(a\bar{b})} = \frac{1 + |b|^2 - 2\,\mathrm{Re}\,(a\bar{b})}{1 + |b|^2 - 2\,\mathrm{Re}\,(a\bar{b})} = 1.$$

Thus $\left|\dfrac{a - b}{1 - \bar{a}b}\right| = 1$. At this stage $1 + |b|^2 - 2\mathrm{Re}\,(a\bar{b}) \neq 0$ as otherwise we have $1 + |\bar{a}b|^2 - 2\mathrm{Re}\,a\bar{b} = 0 \Rightarrow |1 - a\bar{b}|^2 = 0 \Rightarrow a\bar{b} = 1 \Rightarrow |b| = 1$. The case $|b| = 1$, $|a| \neq 1$ is dealt with in the same manner.

(ii) Let $|a| = |b| = 1$. Then $\left|\dfrac{a - b}{1 - \bar{a}b}\right|^2 = \dfrac{2 - 2\,\mathrm{Re}\,\bar{a}b}{2 - 2\,\mathrm{Re}\,a\bar{b}}$. This represents 1 when $2 - 2$ Re $a\bar{b} \neq 0$ i.e., when Re $a\bar{b} \neq 1$. But Re $a\bar{b} = 1$ if and only if $a = b$. [For if $a = e^{i\theta}$ and $b = e^{i\phi}$, Re $a\bar{b} = 1 \Rightarrow$ Re $a\bar{b} = \cos(\theta - \phi) = 1 \Rightarrow \theta = \phi + 2k\pi$ ($k \in \mathbb{Z}$) $\Rightarrow a = b$. Conversely, $a = b \Rightarrow a\bar{b} = |b|^2 = 1 \Rightarrow$ Re $a\bar{b} = 1$]. Hence Re $a\bar{b} \neq 1 \Leftrightarrow a \neq b$. Thus if $|a| = 1 = |b|$, the equality is not true when $a = b$. This is the required exceptional case.

Problem 3 Prove Lagrange identity in the complex form

$$\left| \sum_{i=1}^{n} a_i b_i \right|^2 = \sum_{i=1}^{n} |a_i|^2 \sum_{i=1}^{n} |b_i|^2 - \sum_{1 \leq i < j \leq n} |a_i \bar{b}_j - a_j \bar{b}_i|^2$$

Solution We shall prove the result by induction on n. Let $n = 1 \cdot |a_1 b_1|^2 = |a_1|^2 |b_1|^2$. Hence the result is true for $n = 1$. Assume the result to be true for $(n - 1)$, i.e.,

$$\left| \sum_{i=1}^{n-1} a_i b_i \right|^2 = \sum_{i=1}^{n-1} |a_i|^2 \sum_{i=1}^{n-1} |b_i|^2 - \sum_{1 \leq i < j \leq n-1} |a_i \bar{b}_j - a_j \bar{b}_i|^2 \qquad (1)$$

Now

$$\left| \sum_{i=1}^{n} a_i b_i \right|^2 = \left| \sum_{i=1}^{n-1} a_i b_i + a_n b_n \right|^2$$

$$= \left| \sum_{i=1}^{n-1} a_i b_i \right|^2 + |a_n b_n|^2 + \left(\sum_{i=1}^{n-1} a_i b_i \right) \bar{a}_n \bar{b}_n + \left(\sum_{i=1}^{n-1} \bar{a}_i \bar{b}_i \right) a_n b_n$$

$$= \left[\sum_{i=1}^{n-1} |a_i|^2 \sum_{i=1}^{n-1} |b_i|^2 - \sum_{1 \leq i < j \leq n-1} |a_i \bar{b}_j - a_j \bar{b}_i|^2 \right] + |a_n b_n|^2$$

$$+ \left(\sum_{i=1}^{n-1} a_i b_i \right) \bar{a}_n \bar{b}_n + \left(\sum_{i=1}^{n-1} \bar{a}_i \bar{b}_i \right) a_n b_n \text{ by (1)}$$

$$= \sum_{i=1}^{n} |a_i|^2 \sum_{i=1}^{n} |b_i|^2 - |a_n|^2 \sum_{i=1}^{n-1} |b_i|^2 - |b_n|^2 \sum_{i=1}^{n-1} |a_i|^2$$

$$- \sum_{1 \leq i < j \leq n-1} |a_i \bar{b}_j - a_j \bar{b}_i|^2 + \left(\sum_{i=1}^{n-1} a_i b_i \right) \bar{a}_n \bar{b}_n + \left(\sum_{i=1}^{n-1} \bar{a}_i \bar{b}_i \right) a_n b_n \qquad (2)$$

We need to prove that

$$\left| \sum_{i=1}^{n} a_i b_i \right|^2 = \sum_{i=1}^{n} |a_i|^2 \sum_{i=1}^{n} |b_i|^2 - \sum_{1 \leq i < j \leq n} |a_i \bar{b}_j - a_j \bar{b}_i|^2 .$$

From (2) it suffices to prove that

$$\sum_{1 \leq i < j \leq n} |a_i \bar{b}_j - a_j \bar{b}_i|^2 = |a_n|^2 \sum_{i=1}^{n-1} |b_i|^2 + |b_n|^2 \sum_{i=1}^{n-1} |a_i|^2$$

$$+ \sum_{1 \le i < j \le n-1} |a_i \overline{b}_j - a_j \overline{b}_i|^2 - \left(\sum_{i=1}^{n-1} a_i b_i \right) \overline{a}_n \overline{b}_n - \left(\sum_{i=1}^{n-1} \overline{a}_i \overline{b}_i \right) a_n b_n \qquad (3)$$

But

$$\sum_{1 \le i < j \le n} |a_i \overline{b}_j - a_j \overline{b}_i|^2 = \sum_{1 \le i < j \le n-1} |a_i \overline{b}_j - a_j \overline{b}_i|^2$$

$$+ \sum_{1 \le i \le n-1} |a_i \overline{b}_n - a_n \overline{b}_i|^2 \qquad (4)$$

From (3) and (4) it is enough to prove that

$$\sum_{1 \le i \le n-1} |a_i \overline{b}_n - a_n \overline{b}_i|^2 = |a_n|^2 \sum_{i=1}^{n-1} |b_i|^2 + |b_n|^2 \sum_{i=1}^{n-1} |a_i|^2$$

$$- \left(\sum_{i=1}^{n-1} a_i b_i \right) \overline{a}_n \overline{b}_n - \left(\sum_{i=1}^{n-1} \overline{a}_i \overline{b}_i \right) a_n b_n \qquad (5)$$

Now in (5), LHS $= \sum_{1 \le i \le n-1} (a_i \overline{b}_n - a_n \overline{b}_i)(\overline{a}_i b_n - \overline{a}_n b_i)$

$$= |a_n|^2 \sum_{i=1}^{n-1} |b_i|^2 + |b_n|^2 \sum_{i=1}^{n-1} |a_i|^2 - \overline{a}_n \overline{b}_n \sum_{i=1}^{n-1} a_i b_i - a_n b_n \sum_{i=1}^{n-1} \overline{a}_i \overline{b}_i$$

$$= \text{RHS of (5)}$$

Thus the result is ture for n and the proof by induction is complete.

Problem 4 Find the condition under which the equation $az + b\overline{z} + c = 0$ in one complex variable z has exactly one solution and compute the solution. When does this equation represent a line?

Solution We have

$$az + b\overline{z} + c = 0 \qquad (1)$$

Hence

$$\overline{a}\overline{z} + \overline{b}z + \overline{c} = 0 \qquad (2)$$

Solving the two equations we have $(a\overline{a} - b\overline{b})z + (c\overline{a} - b\overline{c}) = 0$. Hence

$$z = \frac{b\overline{c} - c\overline{a}}{|a|^2 - |b|^2} \qquad (3)$$

From (3) the equations have exactly one complex solution when $|a|^2 - |b|^2 \ne 0$, i.e, when $|a| \ne |b|$. In this case, the expression given by (3) is the solution. Equation (1) represents a single line when z is indeterminate. i.e., when $b\overline{c} - c\overline{a} = 0$ and $|a| = |b|$.

Problem 5 Show that there exists complex numbers z satisfying $|z + a| + |z - a| = 2|c|$ if and only if $|a| \le |c|$. Under this condition compute the smallest and largest values of $|z|$.

Solution Suppose that there exists z with $|z + a| + |z - a| = 2|c|$. Then $2|a| = |(z + a) - (z - a)| \le |z + a| + |z - a| = 2|c| \Rightarrow |a| \le |c|$. Conversely, let $|a| \le |c|$.

Case (i) $a = 0$.

In this case all points on the circle with centre 0 and radius $|c|$ satisfy the required condition. Further max $|z|$ and min $|z|$ in this locus are equal to $|c|$.

Case (ii) $a \neq 0$, $|a| = |c|$.

All points in the line segment joining "$-a$" and "a" satisfy $|z + a| + |z - a| = 2|a| = 2|c|$ by assumption. These are the only points satisfying this property, by triangle law. Here, max $|z| = |a|$ and min $|z| = 0$.

Case (iii) $a \neq 0$, $|a| < |c|$.

We consider all points in the ellipse whose focii are $\pm a$ and eccentricity

$e = \left|\dfrac{a}{c}\right| < 1$. By geometric properties of the ellipse if z is represented by P,

"a" by S and "$-a$" by S', we have

$$SP + S'P = \text{major axis} = \frac{2|a|}{e} = 2\left|\frac{a}{a}\right||c| = 2|c|$$

i.e., $|z + a| + |z - a| = 2|c|$. In this case, max $|z|$ and min $|z|$ are semi-major and semi-minor axes, i.e., $|c|$ and $\sqrt{|c|^2 - |a|^2}$.

Problem 6 Prove that the points a_1, a_2, a_3 are the vertices of an equilateral triangle if and only if

$$a_1^2 + a_2^2 + a_3^2 = a_1a_2 + a_2a_3 + a_3a_1$$

Solution If $a_1 = 0$, a_2 is real and positive and a_3 is an arbitrary complex number then a_1, a_2, a_3 are the vertices of an equilateral triangle if and only

if $a_3 = a_2\, e^{\pm i\pi/3}$ or if and only if $a_3^2 = a_2^2\left(\dfrac{-1}{2} \pm i\dfrac{\sqrt{3}}{2}\right)$ or if and only if

$a_2^2 + a_3^2 = a_2a_3$.

$$\left[\text{Since } a_2^2 + a_3^2 = a_2^2\left(1 - \frac{1}{2} \pm i\frac{\sqrt{3}}{2}\right) = a_2^2\left(\frac{1}{2} \pm i\frac{\sqrt{3}}{2}\right) = a_2a_2\left(\frac{1}{2} \pm i\frac{\sqrt{3}}{2}\right)\right.$$

$\left. = a_2a_2e^{\pm i\pi/3} = a_2a_3\right]$. Hence the result is true in this case. On the other hand if a_1, a_2, a_3 are arbitrary complex numbers (by translation), they form the vertices of an equilateral triangle if and only if $0 = a_1 - a_1$, $a_2 - a_1$, $a_3 - a_1$ form the vertices of the equilateral triangle. Again this happens if and only if (by rotation) $A_1 = 0$, $A_2 = (a_2 - a_1)e^{i\theta} = |a_2 - a_1|$, $A_3 = (a_3 - a_1)e^{i\theta}$ form the vertices of an equilateral triangle, where $\theta = -\arg(a_2 - a_1)$.

Now since $A_1 = 0$, A_2 is real positive, A_3 is arbitrary, by the previous case A_1, A_2, A_3 form the vertices of an equilateral triangle if and only if $A_2^2 + A_3^2 = A_2A_3$ or equivalently

$$(a_2 - a_1)^2e^{2i\theta} + (a_3 - a_1)^2e^{2i\theta} = (a_2 - a_1)e^{i\theta}(a_3 - a_1)e^{i\theta}$$

i.e., if and only if $a_1^2 + a_2^2 + a_3^2 = a_1a_2 + a_2a_3 + a_3a_1$. Hence the result follows.

Problem 7 Show that all circles passing through a and $1/\bar{a}$ cut $|z| = 1$ at right angles.

Solution Let A, A' represent the points a, $1/\bar{a}$ (note that O, A, A' lie on the same line) and C' the circle passing through them. Let C represent the circle $|z| = 1$ and P be one of the points of intersection of C and C'. Now $OP^2 = 1 = |a| \cdot 1/|a| = OA \cdot OA'$, $OP^2 = OA \cdot OA' \Rightarrow OP$ is the tangent at P to the circle through A and A'. But OP is the radius of the unit circle at the point of intersection. Hence C and C' cut orthogonally.

Problem 8 Show that z and z' correspond to diametrically opposite points on the Riemann sphere if and only if $z\bar{z}' = -1$.

Solution z and z' represent diametrically opposite points on the Riemann sphere if and only if $d(z, z') = 2$.

$$d(z, z') = 2 \Leftrightarrow 1 = \frac{|z - z'|}{\left((1 + |z|^2)^{1/2} \left(1 + |z'|^2\right)^{1/2}\right)}$$

$\Leftrightarrow \quad |z - z'|^2 = (1 + |z|^2)(1 + |z'|^2).$

$\Leftrightarrow \quad |z|^2 + |z'|^2 - 2\text{Re } z\bar{z}' = 1 + |z|^2 + |z'|^2 + |zz'|^2$

$\Leftrightarrow \quad 1 + |zz'|^2 + 2\text{Re } z\bar{z}' = 0.$

$\Leftrightarrow \quad |1 + z\bar{z}'|^2 = 0$

$\Leftrightarrow \quad z\bar{z}' = -1.$

Problem 9 Let the complex function $f(z)$ defined in the plane have a derivative at z_0 and at $\bar{z}_0$. Define $g(z) = f(\bar{z})$. If $f'(z_0) = \alpha$ and $f'(\bar{z}_0) = \beta$ then which of the following is true? Why?

(i) $g'(z_0)$ does not exist.
(ii) $g'(z_0)$ exists and is equal to α.
(iii) $g'(z_0)$ exists and is equal to $\bar{\alpha}$.
(iv) $g'(z_0)$ exists and is equal to β.
(v) $g'(z_0)$ exists and is equal to $\bar{\beta}$.

Solution $g'(z_0) = \underset{z \to z_0}{\text{Lt}} \frac{g(z) - g(z_0)}{z - z_0} = \underset{z \to z_0}{\text{Lt}} \frac{f(\bar{z}) - f(\bar{z}_0)}{z - z_0}$

$= \underset{z \to z_0}{\text{Lt}} \frac{f(\bar{z}) - f(\bar{z}_0)}{\bar{z} - \bar{z}_0} \frac{\bar{z} - \bar{z}_0}{z - z_0}$

The right side limit is a product, of which one limit exists and is equal

to $f'(\bar{z}_0) = \beta$ and the second one does not exist as $\dfrac{\bar{z} - \bar{z}_0}{z - z_0} \to \pm 1$

depending on whether $z \to z_0$ through a horizontal line at z_0 or through a vertical line at z_0.

Hence the correct answer is (i).

Problem 10 If Σc_n is convergent and $|\arg c_n| \le \alpha < \pi/2$ then prove that it converges absolutely.

Solution Let $c_n = a_n + ib_n$. Let $\tan^{-1}x$ denote the unique inverse of the bijection h: $\left(-\dfrac{\pi}{2}, \dfrac{\pi}{2}\right) \to \mathbb{R}$ defined by $x = h(y) = \tan y$. Then

$$|\tan^{-1}(b_n/a_n)| = |\arg c_n| \le \alpha < \frac{\pi}{2} \Rightarrow -\alpha \le \tan^{-1}(b_n/a_n) \le \alpha$$

$$\Rightarrow \quad -A \le (b_n/a_n) \le A \quad \text{where} \quad A = \tan \alpha < \infty.$$

$$\Rightarrow \quad |b_n/a_n| \le A.$$

Σc_n is convergent $\Rightarrow \Sigma a_n$ is convergent.

$$|\arg c_n| \le \alpha < \frac{\pi}{2} \Rightarrow \mathrm{Re}\ c_n > 0. \Rightarrow a_n > 0.$$

Thus Σa_n is convergent $\Rightarrow \Sigma|a_n|$ is convergent. Now

$$\Sigma|c_n| \le \Sigma|a_n| + \Sigma|b_n| \le \Sigma|a_n| + A\ \Sigma|a_n| \le (1 + A)\ \Sigma|a_n| < \infty.$$

Hence $\Sigma|c_n|$ is convergent. i.e., Σc_n converges absolutely.

Problem 11 If Σc_n and Σc_n^2 are convergent and $\mathrm{Re}\ (c_n) \ge 0$ then prove that $\Sigma|c_n|^2$ is convergent.

Solution Let $c_n = a_n + ib_n$. Then Σc_n is convergent $\Rightarrow \Sigma a_n$ is convergent $\Rightarrow a_n \to 0$ as $n \to \infty$, i.e., for sufficiently large n, $a_n < \varepsilon$ (since $|a_n| = a_n$ as $\mathrm{Re}\ c_n = a_n \ge 0$). Σa_n^2 is dominated by $\varepsilon \Sigma a_n$ after some stage. Hence Σa_n^2 is convergent. But Σc_n^2 is convergent implies that $\Sigma(a_n^2 - b_n^2)$ is convergent. Now Σa_n^2 is also convergent. Hence Σb_n^2 is convergent. Therefore $\Sigma a_n^2 + \Sigma b_n^2$ is convergent, i.e., $\Sigma(a_n^2 + b_n^2) = \Sigma|c_n|^2$ is convergent.

Problem 12 Construct a homeomorphism between the open unit disc and the finite complex plane.

Solution Define a mapping f from the finite complex plane to the open unit disc U by $f(z) = z/(1 + |z|)$, $\forall z \in \mathbb{C}$. $|f(z)| = |z|/(1 + |z|) < 1$. Therefore the mapping f maps $\mathbb{C}$ into U. We shall now prove that f is one-to-one.

$$f(z_1) = f(z_2) \Rightarrow z_1/(1 + |z_1|) = z_2/(1 + |z_2|) \tag{1}$$
$$\Rightarrow |z_1| (1 + |z_2|) = |z_2|(1 + |z_1|)$$
$$\Rightarrow |z_1| = |z_2|.$$

Substituting this in (1) we get $z_1 = z_2$. Hence f is one-to-one. For every w in open unit disc, there exists $w/(1 - |w|)$ in $\mathbb{C}$ such that $w = f(w/(1 - |w|))$. Hence f is onto. Now we can define f^{-1} as $f^{-1}(w) = w/(1 - |w|)$, for every w in the open unit disc. As the mappings $z \to z$, $z \to |z|$, $z \to 1 + |z| \neq 0$ are all continuous, the mapping $f(z) = z/(1 + |z|)$ is continuous. For similar reasons, f^{-1} is continuous. Thus we have proved that f is bijective and bi-continuous from the finite complex plane to the open unit disc. Hence it is a homeomorphism.

Problem 13 Without any reference to the stereographic projection prove that $(\mathbb{C}^*, d)$ and $(\mathbb{C}^*, \tau)$ are homeomorphic where

$$d(z, z') = \frac{2|z - z'|}{\sqrt{[(1 + |z|^2)(1 + |z'|^2)]}} \qquad \text{if } z, z' \neq \infty$$

$$d(z, \infty) = \frac{2}{\sqrt{(1 + |z|^2)}}$$

and τ is the topology of one-point compactification of $\mathbb{C}$.

Solution Consider the identity map $I : (\mathbb{C}^*, d) \to (\mathbb{C}^*, \tau)$ which is evidently bijective. We shall prove that I is continuous and open (I open implies I^{-1} is continuous), i.e., U is open in $(\mathbb{C}^*, d)$ if and only if U is open in $(\mathbb{C}^*, \tau)$. Let U be open in $(\mathbb{C}^*, d)$. We shall prove that any $z \in U$ is interior to U in $(\mathbb{C}^*, \tau)$. $z \in U$, U open in $(\mathbb{C}^*, d)$ implies there exists $0 < \delta < 2$ such that

$$N = \{w \in \mathbb{C}^* : d(z, w) < \delta\} \subset U \tag{1}$$

Case 1 $z = \infty \in U$. Put

$$V = \{w \in \mathbb{C}^* : w = \infty \quad \text{or} \quad w \in \mathbb{C} \text{ with } |w| > \sqrt{(4/\delta^2) - 1}\}.$$

Clearly $\infty \in V \in \tau$. Further $w \in V \Rightarrow 1 + |w|^2 > 4/\delta^2 \Rightarrow d(z, w) = \frac{2}{\sqrt{1 + |w|^2}} < \delta \Rightarrow w \in N \subset U \Rightarrow V \subset U$ by (1). Thus $z = \infty$ is an interior point of U in $(\mathbb{C}^*, \tau)$.

Case 2 $z \neq \infty$, $z \in U$. Put $V = \{w \in \mathbb{C} : |w - z| < \delta/2\}$. Clearly $z \in V \in \tau$. Further $w \in V \Rightarrow d(z, w) = \frac{2|z - w|}{\sqrt{(1 + |z|^2)(1 + |w|^2)}} < 2|z - w| < \delta \Rightarrow w$

$\in N \subset U \Rightarrow V \subset U$ by (1). Thus $z \neq \infty$ is also an interior point of U in $(\mathbb{C}^*, \tau)$.

Conversely assume U is open in $(\mathbb{C}^*, \tau)$. We shall prove that any $z \in U$ is interior to U in $(\mathbb{C}^*, d)$. Again there are two cases to consider.

Case 1 $U \subset \mathbb{C}$ is open in the usual topology of $\mathbb{C}$. $z \in U \Rightarrow$ there exists $0 < \delta < \dfrac{1}{2}$ such that $N = \{w \in \mathbb{C} : |w - z| < \delta\} \subset U$.

We first observe that

$$\sqrt{(1+|z|^2)} \leq 1 + |z| \quad \text{and} \quad \sqrt{(1+|w|^2)} \leq 1 + |w| \leq 1 + |z| + |w - z|.$$

Hence $\qquad \dfrac{2|z - w|}{(1+|z|)(1+|z|+|w-z|)} \leq \dfrac{2|z-w|}{\sqrt{(1+|z|)^2(1+|w|)^2}} = d(z, w) \qquad (2)$

Now choose $\delta' = \dfrac{2\delta}{(1+|z|)(1+|z|+\delta)}$ and $V = \{w \in \mathbb{C} \mid d(z, w) < \delta'\}$.

Clearly $z \in V$, V is open in $(\mathbb{C}^*, d)$ and

$w \in V \Rightarrow \dfrac{2|z - w|}{(1+|z|)(1+|z|+|w-z|)} < \delta' = \dfrac{2\delta}{(1+|z|)(1+|z|+\delta)}$ by (2).

This inequality further implies that

$$\frac{|z - w|}{1+|z|+|w-z|} < \frac{\delta}{1+|z|+\delta}$$

$\Leftrightarrow \qquad |z - w|\,(1 + |z| + \delta - \delta) < (1 + |z|)\delta$

$\Leftrightarrow \qquad |z - w| < \delta \Leftrightarrow w \in N \subset U$ from the definition of N.

Thus $z \in V \subset U$ and V is open in $(\mathbb{C}^*, d)$, that is z is interior to U in $(\mathbb{C}^*, d)$. As z is arbitrary it follows that U is open in $(\mathbb{C}^*, d)$.

Case 2 $U = \mathbb{C}^* \backslash K = \{\infty\} \cup \mathbb{C} \backslash K$ where K is a compact subset of $\mathbb{C}$. By Case 1 all points of $\mathbb{C} \backslash K$ are interior to $\mathbb{C} \backslash K$ and hence interior to U in $(\mathbb{C}^*, d)$. Thus it suffices to prove that ∞ is also an interior point of U. Choose r large enough so that $|w| \leq r$ holds for all $w \in K$ (since K is compact K is bounded in $\mathbb{C}$). For $z = \infty \in U$ put

$$V = \left\{ w \in \mathbb{C}^* : d(w, \infty) < \frac{2}{\sqrt{(1+r^2)}} \right\}$$

Clearly $\infty \in V$ and V is open in $(\mathbb{C}^*, d)$. Further $w \in V$ and $w \neq \infty \Rightarrow$

$w \in \mathbb{C}$ and $\dfrac{2}{\sqrt{(1+|w|^2)}} < \dfrac{2}{\sqrt{(1+r^2)}} \Rightarrow |w| > r \Rightarrow w \notin K \Rightarrow w \in U$. Thus

$V \subset U$ and ∞ is interior to U in $(\mathbb{C}^*, d)$. This completes the proof.

EXERCISES

1. Which of the following is true? (i) Every real number is a complex number (ii) Every complex number is a real number.
2. How will you find all integral powers of "i"?
3. Find the value of $(1 + i)^n + (1 - i)^n$ for n = 1, 2, 3, ...,.
4. If $\omega \neq 1$ is a primitive nth root of unity, show that $1 + \omega^h + \omega^{2h} + \ldots$
   $+ \omega^{(n-1)h} = 0$ whenever h is an integer that is not a multiple of n.
5. Prove that a convergent complex sequence is bounded.
6. If $\lim_{n \to \infty} z_n = a$ prove that $\lim_{n \to \infty} (z_1 + z_2 + \ldots + z_n)/n = a$.
7. Show that the union of two regions is a region if and only if they have a common point.
8. Prove that the components of closed sets in the plane are closed.
9. Prove that in a metric space a set is compact if and only if every collection of closed sets with empty intersection contains a finite sub-collection with empty intersection.
10. Prove that any closed and bounded subset of $\mathbb{R}$ has a maximum and a minimum.
11. If $E_1 \supset E_2 \supset \ldots$ is a decreasing sequence of non-empty compact sets in a metric space then show that $\cap E_n$ is non-empty.
12. Prove that for real sequences (α_n) and (β_n)

    (i) $\underline{\lim}_{n \to \infty} (\alpha_n) + \underline{\lim}_{n \to \infty} (\beta_n) \leq \underline{\lim}_{n \to \infty} (\alpha_n + \beta_n) \leq \underline{\lim}_{n \to \infty} (\alpha_n) + \overline{\lim}_{n \to \infty} (\beta_n)$
    $\leq \overline{\lim}_{n \to \infty} (\alpha_n + \beta_n) \leq \overline{\lim}_{n \to \infty} (\alpha_n) + \overline{\lim}_{n \to \infty} (\beta_n)$.

    (ii) If $\alpha_n > 0$, $\beta > 0$ and if $\overline{\lim}_{n \to \infty} (\alpha_n) = \alpha < \infty$ and $\overline{\lim}_{n \to \infty} (\beta_n) = \beta < \infty$
    then $\overline{\lim}_{n \to \infty} (\alpha_n \beta_n) \leq \alpha\beta$.

13. If A and B are disjoint sets in a metric space with B closed and A compact show that the distance $d(A, B) > 0$ where
    $$d(A, B) = \inf \{d(a, b)/a \in A \text{ and } b \in B\}.$$
14. Let X and Ω be metric spaces and $f : X \to \Omega$ be uniformly continuous. Show that if (x_n) is Cauchy in X then the sequence $(f(x_n))$ is Cauchy in Ω. What happens if f is merely continuous?
15. Show that the open unit disc in the complex plane is not compact.
16. Show that for any non-real complex a, the point $1/a$ is obtained by intersecting the straight line that joins origin to $\bar{a}$ with the circle that passes through 0 and e and is tangent to the line $\overline{ea}$ where e is the point $(1, 0)$ on the real axis. Describe a similar construction that produces $\pm\sqrt{a}$.
17. Let $f(x) = x^2$ and $g(x) = \sqrt{x}$. Which of the following is true?

     (i) Both f and g are uniformly continuous in $(1, \infty)$.

     (ii) Neither f nor g is uniformly continuous in $(1, \infty)$.

     (iii) f is uniformly continuous but g is not in $(1, \infty)$.

     (iv) g is uniformly continuous but f is not in $(1, \infty)$.

18. If $f : \mathbb{R} \to \mathbb{R}$ defined by $f(x) = |x|$ we know that f is differentiable at all points other than origin. Let $f : \mathbb{C} \to \mathbb{C}$ be defined by $f(z) = |z|$. Which of the following is true?

     (i) $f'(z)$ exists at no point (ii) $f'(z)$ exists at all points except origin.

     (iii) $f'(z)$ exists at all points (iv) $f'(z)$ exists only at origin.

19. Let $\mathcal{D} = \{z/|z| < 1\}$ and $\Omega = \{z/0 < |z| < 1\}$. Then which of the following is true?

     (i) $\mathcal{D}$ and Ω are both connected.

     (ii) $\mathcal{D}$ is connected but Ω is not.

     (iii) Ω is connected but $\mathcal{D}$ is not.

     (iv) Both $\mathcal{D}$ and Ω are disconnected.

NOTES

For more informations regarding the history of mathematics in general and the history of complex numbers in particular, refer to [4]. The subject of mathematical logic is very vast and an insight into it can be had from [5]. For full details about the logical development of the various number systems mentioned in this chapter one can refer to [2]. In 1.5 we have given geometric constructions to interpret complex addition, subtraction, multiplication and division by non-zero number. Certain other arithmetical operations (just as construction of $\sqrt{a}$ where a is a complex number) can also be illustrated using geometric constructions, (see [1]). Under certain conditions a topological space which is not compact can be imbedded in a compact space. These are called compactifications. An example is the one-point compactification given in 1.10. There is one other important compactification called "Stone-Čech compactification for a non-compact topological space. For details refer [3].

References

1. C. Caratheodory, Theory of Functions, Vol. I, Chelsea Publishing Company, New York, 1978, p. 17-20.
2. L. W. Cohen and G. Ehrlich, The Structure of the Real Number System, Affiliated East-West Press Pvt. Ltd., New Delhi, Second East-West Reprint, 1969.
3. J.R. Munkres, Topology—A first course, Prentice-Hall of India Pvt. Ltd., 1991.
4. D.E. Smith, History of Mathematics, Vol I and II, Dover Publication Inc., New York, 1958.
5. J.P. Tremblay and R. Manohar, Discrete Mathematical Structures with Application to Computer Science, McGraw-Hill Book Company, Singapore, 1987.

✦ 2

Elementary Properties of Analytic Functions

2.1 Introduction to the Concept of an Analytic Function

When we deal with complex-valued functions of a complex variable we have to come across real-valued functions of a complex variable, complex valued functions of a real variable, real-valued functions of a real variable and functions whose domains and ranges contain complex numbers. Among these classes of functions the real-valued functions of a real variable are the concerns of real analysis and so we no longer consider them here. Similarly a complex valued function of a real variable can be considered as a pair (its real and imaginary parts) of real functions of a real variable and again we prefer to omit them. Now the immediate question for us will be "what should be the domain of definition for a function of a complex variable?". First we observe that the properties of functions which we are going to study relate to concepts such as continuity, differentiability, analyticity etc. All these notions require that the function be defined in a neighbourhood of 'a' as soon as it is defined at 'a'. Thus a necessary condition for the domains of definition of our functions seems to be at least open sets. Further, most of the properties that we are going to study in complex analysis are local in character (These properties are the ones which hold throughout a domain $\mathcal{D}$ whenever it holds at every point of $\mathcal{D}$). Thus we are free to choose more special subsets of open sets as our domain of definition. Since components of open sets continue to be open (in addition connected) and any open set can be decomposed into its components, it is but natural that we take open connected sets as our domains of definitions. We give a special name for these sets by defining a region to be a non-empty open connected set.

Let f be a complex-valued function of a complex variable defined in a region Ω. We say that $f(z)$ has a derivative at $z \in \Omega$ if $\lim_{h \to 0} \dfrac{f(z+h) - f(z)}{h}$ exists as a complex number. We call it as $f'(z)$. We say f is holomorphic in Ω or f is analytic in Ω if $f'(z)$ exists at every $z \in \Omega$. (Note that this condition is equivalent to the complex differentiability of f at each z in Ω. See Theorem 1.8.5 (b)). Since we have defined analyticity only in regions we agree that whenever we say f is analytic or holomorphic in an arbitrary subset A of the plane we mean f is analytic in a region containing A. In particular f is analytic or holomorphic at 'a' if and only if f is analytic or holomorphic in an open disc containing 'a'.

Theorem 2.1.1
Let f and g be analytic in a region Ω. Then
   (i) f is continuous in Ω.
   (ii) $f + g$ is analytic in Ω and $(f + g)'(z) = f'(z) + g'(z)$
   (iii) fg is analytic in Ω and $(fg)'(z) = f(z)g'(z) + f'(z)\, g(z)$
   (iv) f/g is analytic at all points $z \in \Omega$ where $g(z) \neq 0$ and

$$(f/g)'\,(z) = \left[\frac{g(z)f'(z) - f(z)g'(z)}{(g(z))^2} \right]$$

Proof
(i) This is already known to us from the results of § 1.9. But the following argument is much simpler. If h is small enough

$$f(z + h) - f(z) = h\frac{f(z+h) - f(z)}{h}$$

and so, $\lim\limits_{h \to 0}(f(z+h) - f(z)) = 0.f'(z) = 0$. Thus $f(z + h) \to f(z)$ as $h \to 0$ proving continuity of f at z.

    The proofs of (ii), (iii) and (iv) are similar to their analogues in the real variable case and so we prefer to omit them.

Theorem 2.1.2 (Chain rule)
Let f be analytic in a region Ω and g analytic in a region G and suppose that $f(\Omega) \subset G$. Then $g \circ f$ is analytic in Ω and $(g \circ f)'\,(z) = g'(f(z))f'\,(z)$ for every $z \in \Omega$.

Proof
Let $z_0 \in \Omega$ be fixed and let $w_0 = f(z_0)$. Then there exists an ε-neighbourhood $U = \{w: |w - w_0| < \varepsilon\}$ of w_0 such that for $w \in U$ we can define

$$h(w) = \frac{g(w) - g(w_0)}{w - w_0} - g'(w_0)$$

when $w \neq w_0$ and $h(w_0) = 0$. By the definition of the derivative, h is continuous at w_0. Now the above expression for $h(w)$ can be used to write

$$g(w) - g(w_0) = [g'(w_0) + h(w)]\,(w - w_0)\quad(w \in U)$$

(Note that the above expression is valid even for $w = w_0$.)

As f is continuous we can find a $\delta > 0$ such that $f(z) \in U$ whenever $|z - z_0| < \delta$. Therefore we can replace w by $f(z)$ whenever $|z - z_0| < \delta$ and w_0 by $f(z_0)$ and get that for $0 < |z - z_0| < \delta$

$$\frac{g[f(z)] - g[f(z_0)]}{z - z_0} = \{g'[f(z_0)] + h[f(z)]\}\left\{\frac{f(z) - f(z_0)}{z - z_0}\right\}.$$

Since f and h are continuous at z_0 and w_0 respectively $h \circ f$ is also continuous at z_0 and hence as $z \to z_0$, $(g \circ f)'(z_0) = g'(f(z_0)) f'(z_0)$. Since $z_0 \in \Omega$ is arbitrary the result follows.

Theorem 2.1.3

Suppose f is a complex-function defined in a region Ω such that f is real differentiable in Ω. Let

$$\partial = \frac{1}{2}\left(\frac{\partial}{\partial x} - i\frac{\partial}{\partial y}\right) \quad \text{and} \quad \bar{\partial} = \frac{1}{2}\left(\frac{\partial}{\partial x} + i\frac{\partial}{\partial y}\right)$$

be the Cauchy-Riemann differential operators. Then f is analytic in Ω if and only if the Cauchy-Riemann equation $(\bar{\partial}f)(z) = 0$ holds for every $z \in \Omega$. In this case we also have $f'(z) = (\partial f)(z)$.

Proof

See Theorem 1.9.7 and Theorem 1.9.12.

Remark 2.1.4

If f is written in the form $u + iv$ where u and v are real-valued functions of two real variables x and y, then the Cauchy-Riemann equation $\bar{\partial}f(z) = 0$ splits into

$$\frac{\partial u}{\partial x} = \frac{\partial v}{\partial y} \quad \text{and} \quad \frac{\partial u}{\partial y} = -\frac{\partial v}{\partial x}.$$

Theorem 2.1.5

An analytic function f defined in a region Ω whose derivative vanishes identically should reduce to a constant. Similarly if Re f or Im f or $|f|$ or a branch of arg f is constant throughout the region then also f reduces to a constant.

Proof

Let $f = u + iv$ be such that $f'(z) \equiv 0$. Then using the expression for derivatives and Cauchy-Riemann equations it follows that $\dfrac{\partial u}{\partial x} = \dfrac{\partial v}{\partial y} = \dfrac{\partial u}{\partial y} = \dfrac{\partial v}{\partial x} \equiv 0$.

(Note that $\dfrac{\partial u}{\partial x} \equiv 0$ implies that u is a constant on each horizontal line, i.e. on lines parallel to the x-axis). It follows that u and v are constants on any line segment in Ω which is parallel to one of the coordinate axes. But in a region Ω, any two points can be joined by a polygon whose sides are horizontal and vertical lines (Recall the characterization of connectedness of sets in the plane given by Theorem 1.7.4 (iii)). Thus if $z_0 \in \Omega$ is fixed and $z \in \Omega$ is arbitrary $f(z_0) = u_0 + iv_0 \equiv u + iv = f(z)$. Thus $u + iv$ is constant throughout the region. On the other hand, if u or v is constant then $f'(z)$

$$= \frac{\partial u}{\partial x} - i\frac{\partial u}{\partial y} = \frac{\partial v}{\partial y} + i\frac{\partial v}{\partial x} = 0$$ and so f reduces to a constant. If $|f|^2$ or

$u^2 + v^2$ is constant then differentiation yields

$$u\frac{\partial u}{\partial x} + v\frac{\partial v}{\partial x} = 0 \quad \text{and} \quad u\frac{\partial u}{\partial y} + v\frac{\partial v}{\partial y} = -u\frac{\partial v}{\partial x} + v\frac{\partial u}{\partial x} = 0.$$

Solving these equations we conclude that $\dfrac{\partial u}{\partial x} = \dfrac{\partial v}{\partial x} \equiv 0$ unless the deter-

minant of the system $u^2 + v^2$ is zero. But if $u^2 + v^2 = 0$ at a single point it is constantly zero (our assumption is that $u^2 + v^2$ is constant) and hence

$u = v \equiv 0$ or $f(z) = 0$. But the conclusion $\dfrac{\partial u}{\partial x} = \dfrac{\partial v}{\partial x} = 0$ also implies

$\dfrac{\partial u}{\partial y} = \dfrac{\partial v}{\partial y} = 0$ and as before $f'(z) = 0$ and so $f(z)$ is a constant. If a branch of arg $f(z)$ ($f(z) = u + iv$) is constant, we can set $v/u = k$ (Unless $u \equiv 0$ in which case f is constant) and get $v - ku = 0$ which is the same as saying that Im $((1 - ik)f) = 0$. But then $(1 - ik)f$ is constant and hence f is also a constant. Note that in this theorem the connectedness of Ω is used. If this is not assumed the only thing we can assert is that in each component of the domain of definition the function reduces to a constant.

Definition 2.1.6
Let u be a real-valued continuous function of a complex variable z or two real variables x any y defined in region Ω. If u has continuous partial derivatives up to two orders and satisfies the Laplace equation

$$\Delta u = \frac{\partial^2 u}{\partial x^2} + \frac{\partial^2 u}{\partial y^2} = 0$$

then u is called a harmonic function. Further if u and v are two harmonic

functions with the property that $\dfrac{\partial u}{\partial x} = \dfrac{\partial v}{\partial y}$ and $\dfrac{\partial u}{\partial y} = -\dfrac{\partial v}{\partial x}$ then v is said to

be the harmonic conjugate of u. Under the same situation $-u$ will be the harmonic conjugate of v.

Generally the harmonic conjugate of a given harmonic function can be found by integration and in certain simple cases the expressions can be explicitly evaluated. We give the following examples:

Example 2.1.7

(i) if $u = x^2 - y^2$, then it is easy to see that u is harmonic in the plane and $\dfrac{\partial u}{\partial x} = 2x$ and $\dfrac{\partial u}{\partial y} = -2y$. Thus if we assume v is the harmonic conjugate

of u then by definition $\dfrac{\partial v}{\partial x} = 2y$ and $\dfrac{\partial v}{\partial y} = 2x$. From the first of these

equations we get, $v = 2xy + \phi(y)$ where $\phi(y)$ is a function of y alone. Substituting this in the second of these equations we get $\phi'(y) = 0$ and hence $\phi(y)$ is a constant. Hence the most general conjugate harmonic function of u is given by $2xy + c$ where c is a constant. It follows that $f(z) = u + iv = (x^2 - y^2) + i(2xy + c) = z^2 + ic$.

(ii)   As a second example we would like to find out the most general harmonic polynomial of the form $u = ax^3 + bx^2y + cxy^2 + dy^3$ and determine its harmonic conjugate v and the corresponding analytic function $f = u + iv$. Since the regularity assumptions are easy to check we find that the Laplace equation for u gives $(6a + 2c) x + (2b + 6d) y = 0$. Since this is an identity for all x and y we get $6a + 2c = 0$ and $2b + 6d = 0$. i.e, $c = -3a$ and $b = -3d$. Thus the most general harmonic polynomial should be $ax^3 - 3dx^2y - 3axy^2 + dy^3$. To determine its harmonic conjugate v we use

$$\frac{\partial v}{\partial y} = 3ax^2 - 6dxy - 3ay^2 \qquad (1)$$

$$\frac{\partial v}{\partial x} = 3dx^2 + 6axy - 3dy^2 \qquad (2)$$

Integrating (2) with respect to x we get

$$v = dx^3 + 3ax^2y - 3dxy^2 + \phi(y).$$

Substituting in (1) we see that

$$\phi'(y) = -3ay^2 \text{ or } \phi(y) = -ay^3 + k.$$

Thus $v = dx^3 + 3ax^2 y - 3dy^2 x - ay^3 + k$. It is easily verified that v is harmonic. The corresponding analytic function is given by $f = u + iv = (a + id)x^3 + (-3d + i3a) x^2y + (-3a - i3d) xy^2 + (d - ia) y^3 + ik$
$= (a + id) x^3 + i3x^2y (a + id) + 3xy^2i^2 (a + id) + i^3y^3 (a + id) + ik$
$= (a + id) (x^3 + i3x^2y + 3xy^2 i^2 + i^3y^3) + ik$
$= (a + id) (x + iy)^3 + ik = (a + id)z^3 + ik.$

Note 2.1.8

We now observe that if $u(z)$ is harmonic in Ω and $\overline{\Omega} = \{z : \overline{z} \in \Omega\}$ then $u(\overline{z})$ is harmonic in $\overline{\Omega}$. This is because if we let $v(z) = u(\overline{z})$ $(z \in \overline{\Omega})$ then

$$\left(\frac{\partial v}{\partial x}\right)_{(x_0, y_0)} = \left(\frac{\partial u}{\partial x}\right)_{(x_0, -y_0)} \text{ and } \left(\frac{\partial v}{\partial y}\right)_{(x_0, y_0)} = -\left(\frac{\partial u}{\partial y}\right)_{(x_0, -y_0)}$$

$$\left(\frac{\partial^2 v}{\partial x^2}\right)_{(x_0, y_0)} = \left(\frac{\partial^2 u}{\partial x^2}\right)_{(x_0, -y_0)} \quad \text{and} \quad \left(\frac{\partial^2 v}{\partial y^2}\right)_{(x_0, y_0)} = \left(\frac{\partial^2 u}{\partial y^2}\right)_{(x_0, -y_0)}$$

and hence $\Delta v = 0$ at all points of $\overline{\Omega}$ as $\Delta u = 0$ at every corresponding point in Ω.

Remark 2.1.9

We shall now describe a very simple method of calculating the analytic function whose real part is a given harmonic function $u(x, y)$. However, it should be cautioned that this method is purely formal and is limited to functions u which are rational functions of x and y (the functions must have meaning for complex values of the argument when x and y are replaced by $(z + \bar{z})/2$ and $(z - \bar{z})/2i$ respectively). We first remark that if $f(z)$ is the analytic function whose real part is u then formally we consider $f(x, y)$ as a function of z and $\bar{z}$ as follows. First $f(x, y)$ is a function of x and y and x and y are functions of z and $\bar{z}$ given by the relation.

$$x = \frac{z + \bar{z}}{2}, y = \frac{z - \bar{z}}{2i}.$$

Thus using several variable calculus in a formal sense we see that

$$\frac{\partial f}{\partial z} = \frac{\partial f}{\partial x}\frac{\partial x}{\partial z} + \frac{\partial f}{\partial y}\frac{\partial y}{\partial z}$$

and

$$\frac{\partial f}{\partial \bar{z}} = \frac{\partial f}{\partial x}\frac{\partial x}{\partial \bar{z}} + \frac{\partial f}{\partial y}\frac{\partial y}{\partial \bar{z}}.$$

i.e.,

$$\frac{\partial f}{\partial z} = \frac{1}{2}\left(\frac{\partial f}{\partial x} - i\frac{\partial f}{\partial y}\right) \quad \text{and} \quad \frac{\partial f}{\partial \bar{z}} = \frac{1}{2}\left(\frac{\partial f}{\partial x} + i\frac{\partial f}{\partial y}\right).$$

But using Cauchy-Riemann equations as in Theorem 2.1.3 we get $\frac{\partial f}{\partial \bar{z}} = \bar{\partial}f(z) = 0$. We now remark that the function $\bar{f}(z) = \overline{f(z)}$ should have zero partial derivative with respect to z. This is because writing $f = u + iv$, $\bar{f} = u - iv$, it is first of all clear that

$$\frac{\partial \bar{f}}{\partial x} = \overline{\frac{\partial f}{\partial x}} \quad \text{and} \quad \frac{\partial \bar{f}}{\partial y} = \overline{\frac{\partial f}{\partial y}}. \tag{1}$$

On the other hand differentiating formally and using

$$x = \frac{z + \bar{z}}{2}, y = \frac{z - \bar{z}}{2i},$$

$$\frac{\partial \bar{f}}{\partial z} = \frac{\partial \bar{f}}{\partial x}\frac{\partial x}{\partial z} + \frac{\partial \bar{f}}{\partial y}\frac{\partial y}{\partial z} = \frac{1}{2}\frac{\partial \bar{f}}{\partial x} + \frac{1}{2i}\frac{\partial \bar{f}}{\partial y}$$

$$\frac{\partial f}{\partial \bar{z}} = \frac{\partial f}{\partial x}\frac{\partial x}{\partial \bar{z}} + \frac{\partial f}{\partial y}\frac{\partial y}{\partial \bar{z}} = \frac{1}{2}\frac{\partial f}{\partial x} - \frac{1}{2i}\frac{\partial f}{\partial y}$$

A comparison using (1) shows that $\dfrac{\partial \bar{f}}{\partial z} = \overline{\left(\dfrac{\partial f}{\partial \bar{z}}\right)} = 0$. Hence in the vari-

ables z and $\bar{z}, \overline{f(z)}$ can be considered purely as a function of $\bar{z}$ and we

denote this function by $\tilde{f}(\bar{z}) = \overline{f(z)}$. With this notation we can now write

$$u(x, y) = [f(x + iy) + \tilde{f}(x - iy)]/2.$$

In a formal way we treat this as an identity and substitute $x = z/2$ and $y = z/2i$ so that

$$u(z/2, z/2i) = [f(z) + f(0)]/2.$$

Since $f(z)$ needs to be determined only up to a purely imaginary constant we can as well assume that $f(0)$ is real (if $f(0)$ is not real replace $f(z)$ by $f(z) - i\,\mathrm{Im}\,f(0)$) which implies $f(0) = \overline{f(0)} = \tilde{f}(0) = u(0, 0)$. The function $f(z)$ has therefore an explicit formula

$$f(z) = 2u(z/2, z/2i) - u(0, 0) + ic.$$

Note that a purely imaginary constant can be added at will without violating our requirements that $u = \mathrm{Re}\,f$. One can check this with our Examples 2.1.7 (i) and (ii).

Definition 2.1.10

A polynomial in one complex variable z is an expression of the form $p(z) = a_0 + a_1 z + a_2 z^2 + \ldots a_n z^n$ where a_i's are complex numbers for $1 \leq i \leq n$.

We note that the identity function z is a non-constant analytic function in the whole plane with derivative 1. More generally the function z^n (n, a non-negetive integer) is analytic in the whole plane with derivative nz^{n-1}. Since the sums and the products of analytic functions are analytic we see that the above polynomial $p(z)$ is analytic in the whole plane with derivative $a_1 + 2a_2 z + \ldots + na_n z^n$. In the above notation we shall always assume that $a_n \neq 0$ and call $p(z)$ a polynomial of degree n. We shall call a complex number α a root of the polynomial or a zero for the polynomial $p(z)$ if $p(\alpha) = 0$. Assuming the fundamental theorem of algebra (See Corollary 4.4.9) and using the division algorithm we can write any polynomial $p(z)$ of degree n as $p(z) = a_n(z - \alpha_1)(z - \alpha_2) \ldots (z - \alpha_n)$ where α_i's are complex numbers not necessarily distinct. If just 'h' of the α_j's coinside the zero α_j is called a zero of order h for $p(z)$. The order of a zero α for $p(z)$ can also be characterized by the following properties. (verify this).

$$p(\alpha) = p'(\alpha) = \ldots = p^{(h-1)}(\alpha) = 0 \text{ and } p^{(h)}(\alpha) \neq 0.$$

In other words the order of the zero α for $p(z)$ is exactly equal to the order of the first non-vanishing derivative for $p(z)$ at α. A zero of order one is also called a simple zero. We shall now state and prove the following theorem of Lucas concerning the zeros of a polynomial (We shall assume the validity of the fundamental theorem of algebra for the present).

Theorem 2.1.11 (Lucas Theorem)
Let $z = a + bt$ be a line in the complex plane. If all the zeros of a polynomial $P(z)$ lie in the half plane given by Im $[(z - a)/b] < 0$ then all the zeros of the derived polynomial $P'(z)$ also lie in the same half plane.

Proof
Let the factorization of the polynomial P be given by $P(z) = c(z - \alpha_1) \ldots (z - \alpha_n)$ where $\alpha_i (1 \le i \le n)$ are not necessarily distinct. The hypothesis implies that Im $[\alpha_i - a)/b]$ is negative for $1 \le i \le n$. If z is not in the half plane in which α_j's lie then Im $\dfrac{z-a}{b} \ge 0$. For any such z we also have

$$\frac{P'(z)}{P(z)} = \sum_{i=1}^{n} \frac{1}{z - \alpha_i}$$

(note here that $P(z) \ne 0$) and hence

$$\text{Im}\left(\frac{z - \alpha_i}{b}\right) = \text{Im}\left(\frac{z - a}{b}\right) - \text{Im}\left(\frac{\alpha_i - a}{b}\right) > 0.$$

But the imaginary parts of reciprocal numbers have opposite sign and we conclude that

$$\text{Im}\left(\frac{b}{z - \alpha_i}\right) < 0 \quad (1 \le i \le n).$$

Hence

$$\text{Im}\left(b\frac{P'(z)}{P(z)}\right) = \sum_{i=1}^{n} \text{Im}\left(\frac{b}{z - \alpha_i}\right) < 0 \text{ and hence } P'(z) \ne 0. \text{ We have}$$

proved that if z is not in the same half plane as α_i's then $P'(z) \ne 0$. Thus all the zeros of $P'(z)$ lie in the same half plane given by $\text{Im}\left(\dfrac{z - a}{b}\right) < 0.$

Corollary 2.1.12
The smallest convex polygon that contains all the zeros of $P(z)$ also contains all the zeros of $P'(z)$.

Proof
Apply the Lucas theorem with reference to the half planes determined by each side of the polygon.

Definition 2.1.13
A rational function is a function of the form $R(z) = P(z)/Q(z)$ where $P(z)$ and $Q(z)$ are two polynomials having no common factors.

By definition, $R(z) = \infty$ at zeros of $Q(z)$ which are also called the poles of
$R(z)$. In general any rational function has finite zeros which are the zeros
of P and finite poles which are the zeros of Q. If 'a' is a finite zero of
R then the order of the zero for $R(z)$ at 'a' is the same as the order of
zero for P at 'a'. Similarly if b is a finite pole for R then the order of the
pole for $R(z)$ at b is the same as the order of the zero for Q at b.
To determine the nature of R at ∞ we shall define a new rational function
$R_1(z) = R(1/z)$ and set $R(\infty) = R_1(0)$. If $R_1(0) = 0$ then R has a zero at ∞
and the order of zero at ∞ for R is the same as the order of zero for R_1
at the origin. Similarly if $R_1(0) = \infty$ then ∞ is a pole for R and the order
of the pole at ∞ for R is the same as the order of the pole for R_1 at origin.
If $R_1(0) \neq 0$ or ∞, R has neither a zero nor a pole at ∞.

Theorem 2.1.14
A rational function $R(z)$ has the same number of zeros and poles including
those at ∞ and is equal to p, the maximum of m and n where m and n are
the degrees of the denominator polynomial and the numerator polynomial
respectively. This number p is called the order of the rational function and
for any complex number 'a' the number of solutions of the equation $R(z)$
$= a$ is also equal to p.

Proof
We write $R(z) = P(z)/Q(z)$ where $P(z) = a_0 + a_1 z + \dots + a_n z^n$ ($a_n \neq 0$)
and $Q(z) = b_0 + b_1 z + \dots + b_m z^m$ ($b_m \neq 0$). By the fundamental theorem of
algebra $R(z)$ has n finite zeros and m finite poles. To calculate the number
of zeros or poles at ∞ for R we consider

$$R_1(z) = R(1/z) = \frac{a_0 z^n + a_1 z^{n-1} + \dots + a_n}{b_0 z^m + b_1 z^{m-1} + \dots + b_m} z^{m-n}$$

We now consider three cases.

Case (i) $m > n$. Now, $R(z)$ has a zero of order $m - n$ at ∞ (and no
pole at ∞) and the number of finite zeros equals n and hence the total
number of zeros is $m - n + n = m = p$. The number of poles for R is m
$= p$.

Case (ii) $m < n$. In this case ∞ is a pole of order $n - m$ for R (and
no zeros at ∞) and the number of finite poles is m and hence the total
number of poles is $n - m + m = n = p$. On the other hand there are only
finite zeros whose number is $n = p$.

Case (iii) $m = n$. Now $R(\infty) = R_1(0) = a_n/b_m \neq 0, \neq \infty$ and hence the
number of zeros of R is $n = p$ and the number of poles for R is $m = p$.
Thus the count shows that the total number of zeros or poles including
those at ∞ always equals the order of R.

If 'a' is any complex number the rational functions $R(z) - a$ and $R(z)$
have the same order as is easily verified. Thus the zeros of $R(z) - a$ which

are nothing but the solutions of the equation $R(z) = a$ should be equal in number to p.

Theorem 2.1.15
Every rational function has a partial fraction expansion.

Proof
Let $R(z)$ be a rational function given by $R(z) = P(z)/Q(z)$. If the degree of P is greater than that of Q ((i.e) $R(z)$ has a pole at ∞), by division algorithm, we can write $R(z) = G(z) + H(z)$ where $G(z)$ is a polynomial without constant term and $H(z)$ is a rational function with the property that the degree of its denominator is greater than or equal to the degree of its numerator. If however the degree of $P(z)$ is less than or equal to that of Q this process is not necessary and $H(z) = R(z)$. The above condition on $H(z)$ is the same thing as saying that $H(z)$ is finite at ∞. The degree of the polynomial $G(z)$ is the same as the order of the pole at ∞ for $R(z)$ and the polynomial $G(z)$ is also called the singular part of $R(z)$ at ∞. Summarizing we get that if $R(z)$ has a pole at ∞ then $R(z) = G(z) + H(z)$ where $G(z)$ is a polynomial without constant term and $H(z)$ is finite at ∞. Let us now denote the finite distinct poles of $R(z)$ by β_1, β_2, ..., β_l. Consider the new function $R(\beta_j + 1/\zeta)$ (for each fixed j, $1 \le j \le l$) as a rational function of ζ. Clearly it has a pole at $\zeta = \infty$. Applying the previous case we can write $R(\beta_j + 1/\zeta)$ as $G_j(\zeta) + H_j(\zeta)$ where $G_j(\zeta)$ is a polynomial in ζ without constant term and $H_j(\zeta)$ is finite at $\zeta = \infty$. Reverting to the original sub-

stitution $z = \beta_j + 1/\zeta$ we get $R(z) = G_j\left(\dfrac{1}{z - \beta_j}\right) + H_j\left(\dfrac{1}{z - \beta_j}\right)$. Now

$G_j\left(\dfrac{1}{z - \beta_j}\right)$ is a polynomial in $\left(\dfrac{1}{z - \beta_j}\right)$ without constant term (the

singular part of $R(z)$ at β_j, and $H_j\left(\dfrac{1}{z - \beta_j}\right)$ is finite at $z = \beta_j$). This can

be done for each β_j $(1 \le j \le l)$.

Now consider the expression $S(z) = R(z) - G(z) - \sum\limits_{j=1}^{l} G_j\left(\dfrac{1}{z - \beta_j}\right)$. This

is evidently a rational function in z. The possible poles for $R(z)$ (and therefore for $S(z)$) are β_1, β_2, ..., β_l and ∞. $G(z)$ has no pole except at ∞ and

similarly $G_j\left(\dfrac{1}{z - \beta_j}\right)$'s have no poles expect at β_j's. Thus the only

possible poles for $S(z)$ are β_1, β_2, . . . β_l and ∞. At $z = \infty$ both $R(z)$ and $G(z)$ become ∞ but their difference $H(z)$ is finite at ∞. Similarly at each β_j,

$R(z)$ and the corresponding $G_j\left(\dfrac{1}{z-\beta_j}\right)$ become ∞, but their difference

$H_j\left(\dfrac{1}{z-\beta_j}\right)$ is finite at $z = \beta_j$. Thus $S(z)$ has neither finite poles nor a pole

at ∞. This means that the order of the rational function $S(z)$ is zero and hence $S(z)$ reduces to a constant and absorbing this constant in $G(z)$ we get

$$R(z) = G(z) + \sum_j G_j\left(\frac{1}{z-\beta_j}\right).$$

This is the well-known partial fraction expansion for $R(z)$ which is used in integration theory. It is interesting to note that only with the theory of complex analysis the reduction of a rational function into its partial fraction becomes rigorous.

2.2 Power Series

We have already seen that polynomials are the simplest examples of analytic functions in the plane. The next important examples of analytic functions come from the notion of a power series with complex coefficients. By definition a power series is an expression of the form

$$a_0 + a_1 z + a_2 z^2 + \ldots + a_n z^n + \ldots..$$

where the coefficients are complex numbers and z is a complex variable. More specifically, these are all power series centered at origin in contrast

to more general series such as $\sum_0^\infty a_n(z - z_0)^n$ which are centered at z_0. A

trivial example is the geometric series $\sum_0^\infty z^n$ whose partial sums are $\dfrac{1-z^n}{1-z}$.

Since $z^n \to 0$ for $|z| < 1$ and $|z|^n \geq 1$ for $|z| \geq 1$ (and therefore z^n does

not tend to zero for $|z| \geq 1$) we conclude that the geometric series $\sum_0^\infty z^n$

converges to $\dfrac{1}{1-z}$ for $|z| < 1$ and does not converge for $|z| \geq 1$. It is

interesting to note that the behavior of a general power series is very similar to that of the geometric series in as much as every power series has a circle inside of which it is convergent and outside of which it is not convergent. In extreme cases however the circle may reduce either to a point-circle or to the whole plane (considered as a circle with radius 0 or ∞ respectively). The following theorem makes this idea explicit.

Theorem 2.2.1
(Abel's theorem on radius of convergence of a power series)

Let $\sum_{0}^{\infty} a_n z^n$ be a power series. Then there exists a real number R, $0 \le R \le \infty$, called the radius of convergence with the following properties.
  (i) The series converges absolutely for each z with $|z| < R$.
 (ii) If $0 < r < R$, the convergence is uniform in $|z| \le r$.
(iii) If $|z| > R$ the series is not convergent at z.
(iv) In $|z| < R$ the sum of the series represents an analytic function whose derivative is obtained by term-wise differentiation. The derived series also has the same radius of convergence.
[Note that the circle $|z| = R$ is called the circle of convergence for the power series.]

Proof
We shall define a number R by the formula
$$1/R = \limsup_{n \to \infty} |a_n|^{1/n}$$
and verify that the conclusions of the theorem are true for this R. Note that in the trivial case $R = 0$, (i), (ii) and (iv) does not make any sense. Let z be such that $|z| < R$. Choose a positive real number ρ such that $|z| < \rho < R$. Then $(1/\rho) > (1/R)$ and by the definition of limit supremum, there exists a stage N such that $|a_n|^{1/n} < 1/\rho$ for $n \ge N$. But this implies that $|a_n z^n| < (|z|/\rho)^n$ for $n \ge N$. Thus the power series $\sum_{0}^{\infty} |a_n z^n|$ is dominated by a convergent geometric series (note that $|z|/\rho < 1$) and so is convergent. Hence the series $\sum_{0}^{\infty} a_n z^n$ converges absolutely for every z with $|z| < R$. This proves (i).

If $r < R$, choose a ρ with $r < \rho < R$. As before there exists a stage N such that $|a_n z^n| < (|z|/\rho)^n$ for $n \ge N$. However if z varies in the compact set $|z| \le r$, then $|a_n z^n| < (r/\rho)^n$ for $n \ge N$. Thus the series $\sum |a_n z^n|$ is again dominated by a convergent series of non-negative terms (in fact, by a geometric series of common ratio less than one). We now apply Weierstrass M-test and conclude that the power series $\sum a_n z^n$ is uniformly convergent in $|z| \le r < R$. This proves (ii). If z is such that $|z| > R$ choose ρ so that $R < \rho < |z|$. Since $(1/\rho) < (1/R)$ by the definition of limit supremum there are infinitely many values of n for which $|a_n|^{1/n} > 1/\rho$ and consequently $|a_n| > 1/\rho^n$. Thus $|a_n z^n| > (|z|/\rho)^n$ holds for infinitely many n and as $|z|/\rho > 1$, $|a_n z^n| > 1$ for infinitely many n and consequently $a_n z^n$ will not tend to zero as $n \to \infty$. Since this contradicts a necessary condition for the convergence of $\sum a_n z^n$ (the nth term of a convergent series tends to zero as $n \to \infty$) we conclude that $\sum a_n z^n$ is not convergent. This proves (iii).

Before proceeding with the proof of the remaining part of the theorem, we first note that the series, $\Sigma n a_n z^{n-1}$ obtained by differentiating $\Sigma a_n z^n$ term-by-term, has the same radius of convergence R. For this we first prove that $n^{1/n} \to 1$ as $n \to \infty$. Put $n^{1/n} = 1 + \delta_n$ $(\delta_n > 0)$. Then $n = (1 + \delta_n)^n > 1 + [n(n-1)\delta_n^2]/2$ by the binomial theorem. This inequality shows that $\delta_n^2 < 2/n$ and hence $\delta_n \to 0$ as $n \to \infty$. We also observe that the radius of convergence R' of the derived series is given by

$$1/R' = \limsup_{n\to\infty}| n a_n |^{1/n} = \limsup_{n\to\infty}| a_n |^{1/n} = 1/R.$$

Hence $R' = R$. (Here one has to prove that if $a_n \geq 0$, $b_n \geq 0$ and $b_n \to b$ then $\limsup_{n\to\infty} a_n b_n = b \limsup_{n\to\infty} a_n$.) From this observation it follows that the series $\Sigma n a_n z^{n-1}$ converges absolutely at every point z with $|z| < R$. We shall use this in proving (iv).

For $|z| < R$ we shall write $f(z) = \sum_0^\infty a_n z^n = S_n(z) + R_n(z)$ where $S_n(z) = a_0 + a_1 z + \ldots + a_{n-1} z^{n-1}$ and $R_n(z) = \sum_n^\infty a_k z^k$. We shall also put

$$F(z) = \sum_1^\infty n\, a_n z^{n-1} = \lim_{n\to\infty} S_n'(z).$$

We have to show that $f(z)$ is analytic in $|z| < R$ and $f'(z) = F(z)$. Now fix $|z_0| < R$ and consider the following identity.

$$\frac{f(z)-f(z_0)}{z-z_0} - F(z_0) = \frac{S_n(z)-S_n(z_0)}{z-z_0} - S_n'(z_0) + S_n'(z_0) - F(z_0)$$

$$+ \frac{R_n(z)-R_n(z_0)}{z-z_0}. \tag{1}$$

Here we assume $|z|$ and $|z_0|$ are less than R and $z \neq z_0$. Choose a $\rho > 0$ such that $|z| < \rho < R$ and $|z_0| < \rho < R$. (It is possible to choose such a ρ since we can assume z is very near z_0. For example choose $|z - z_0| <$

$$\delta = \frac{R-|z_0|}{2} \quad \text{and} \quad \rho = \frac{R+|z_0|}{2}).$$

$$\left| \frac{R_n(z)-R_n(z_0)}{z-z_0} \right| \leq \sum_n^\infty |a_k| \,[z^{k-1} + z^{k-2} z_0 + \ldots + z_0^{k-1}]|$$

$$\leq \sum_n^\infty k|a_k|\rho^{k-1}.$$

Let $\varepsilon > 0$ be given. Since the series $\sum_1^\infty k a_k z^{k-1}$ is absolutely convergent at $z = \rho$, $\sum_n^\infty k\,|a_k|\,\rho^{k-1}$ can be made sufficiently small, say less than $\varepsilon/3$ for $n \geq N$. Hence

$$\left|\frac{R_n(z) - R_n(z_0)}{z - z_0}\right| < \varepsilon/3 \tag{2}$$

for $n \geq N$. We can find another stage M such that $\forall\, n \geq M$

$$|S_n'(z_0) - F(z_0)| < \varepsilon/3. \tag{3}$$

We now fix an integer n greater than both N and M. For this n, $S_n(z)$ is analytic with derivative $S_n'(z_0)$ at z_0. Thus given $\varepsilon/3$ we can find a $\delta > 0$ such that $0 < |z - z_0| < \delta$ implies

$$\left|\frac{S_n(z) - S_n(z_0)}{z - z_0} - S_n'(z_0)\right| < \varepsilon/3. \tag{4}$$

Using (2), (3) and (4) and applying the triangle inequality in (1) we get that for $0 < |z - z_0| < \delta$ (for a suitable $\delta > 0$)

$$\left|\frac{f(z) - f(z_0)}{z - z_0} - F(z_0)\right| < \varepsilon.$$

Since $\varepsilon > 0$ is arbitrary, this proves that $f'(z_0)$ exists and is equal to $F(z_0)$.

Remark 2.2.2

From the above theorem it follows that if a power series $\sum_0^\infty a_n z^n$ converges at ζ and diverges at η then ζ cannot lie outside the circle of convergence nor can η lie inside. Hence the radius of convergence R satisfies $|\zeta| \leq R \leq |\eta|$. If it so happens that $|\zeta| = |\eta|$ then the common value has to be the radius of convergence. For example consider the power series $\sum_1^\infty z^n/n$ which converges at -1 but diverges at 1. As both -1 and 1 have the same modulus 1 it easily follows that the radius of convergence of the power series is 1. Further the above theorem does not say anything about the convergence or otherwise of the power series on the circle of convergence.

In fact even if a power series $\sum_0^\infty a_n z^n$ is convergent at a point on the circle of convergence no information regarding the continuity or analyticity of the function at that point can be obtained in general. But what best can be said regarding the continuity of f at such a point? Yet another theorem again due to Abel discusses this problem. In this context we loose no generality by the assumptions that $R = 1$ and that the convergence takes place at $z = 1$. But before proceeding with this theorem, we would like to characterize a certain limit taking place in a so-called "Stolz angle".

Definition 2.2.3

z is said to stay in a Stolz angle at 1 if z lies in an angular sector with total angle less than $180°$ with vertex 1, symmetric to the part of $(-\infty, 1)$ of the real axis.

Proposition 2.2.4

z tends to 1 in a Stolz angle if and only if z tends to 1 in such a way that $|1 - z|/(1 - |z|)$ remains bounded.

Proof

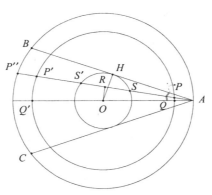

Fig 2.2 (i)

In the above figure O is the origin, A is 1 and P is z and is near 1. We denote half of the Stolz angle by α. Let B and C be the end points of the two chords of the unit circle enclosing the Stolz angle at A. Draw a circle centre O radius OP and draw another circle for which the pair of tangents are AB and AC with H as a point of contact. Let the line AP cut the latter circle at S and S'. Let P' and Q' be the points of intersection of the lines AS and AO with the circle with centre O and radius OP. Let Q be such that it lies on the line AO with $OQ = OQ'$. Let R be the foot of the perpendicular from O to AP'. From the figure it is clear that $RS^2 = OS^2 - OR^2$ and $RP'^2 = OP'^2 - OR^2$ and as $OS < 1$, $OP' > OS$ ($OP' = OP = |z|$ and OP' is greater than any fixed number less than 1 if $|z|$ is near 1) and consequently $RP' > RS = RS'$ and $AP' = AR + RP' > AS' = AR + RS' > AS$. But either $AS \geq AH$ or $AS' \geq AH$ (as otherwise $AS \cdot AS' < AH^2$ which is impossible). Thus $AS' \geq AH$ and $AP' \geq AH$. The same is true even if S and S' are interchanged. Further

$$AP \cdot AP' = AQ \cdot AQ'.$$

Thus $\qquad \dfrac{|1 - z|}{1 - |z|} = \dfrac{AP}{AQ} = \dfrac{AQ'}{AP'} \leq \dfrac{2}{AH} = 2 \sec\alpha.$

As $\alpha < \pi/2$, $\sec\alpha$ is finite and so as $z \to 1$ in this Stolz angle, $\dfrac{|1 - z|}{1 - |z|}$ $\leq 2\sec\alpha = M$ (say).

Conversely if P'' is the intersection of the unit circle with the line AP' and if it is given that $\dfrac{|1 - z|}{1 - |z|} \leq M$ as $z \to 1$ then with $\angle OAP = \theta$ we have,

$0 < \theta < \pi/2$ and

$$\frac{1}{2}\sec\theta = \frac{1}{2}\frac{2}{AP''} = \frac{1}{AP''} = \frac{AO}{AP''} \le \frac{AQ'}{AP''} \le \frac{AQ'}{AP''} = \frac{AP}{AP'} = \frac{AP}{AQ} = \frac{|1-z|}{1-|z|} \le M.$$

Hence $\sec\theta \le 2M$ or $\theta \le \sec^{-1}(2M) = \beta < \pi/2$ (M is finite). Thus z stays in the Stolz angle β. Let us now state and prove the following Abel's limit theorem.

Theorem 2.2.5 (Abel's Limit Theorem)

If $f(z) = \sum_0^\infty a_n z^n$ is a power series converging at $z = 1$, then $f(z)$ tends to $f(1)$ as z tends to 1 in any Stolz angle. ($f(z)$ has $f(1)$ as a non-tangential limit).

Note that the radius of convergence of this power series is at least 1 and so $f(z)$ is defined for each z with $|z| < 1$.

Proof

Without loss of generality we can assume that $f(1) = 0$ (otherwise $g(z) = a_0 - f(1) + \sum_1^\infty a_n z^n$ will be a power series with $g(1) = 0$ and $g(z) \to 0 = g(1)$ is the same as $f(z) \to f(1)$). We now write $s_n = a_0 + a_1 + \ldots + a_n$ and $s_n(z) = a_0 + a_1 z + \ldots + a_n z^n = s_0(1 - z) + s_1 z(1 - z) + \ldots + s_{n-1} z^{n-1}(1 - z) + s_n z^n = (1 - z)(s_0 + s_1 z + \ldots + s_{n-1} z^{n-1}) + s_n z^n$. By our assumption s_n tends to zero and so $s_n z^n$ tends to zero as $n \to \infty$ ($|z|^n < 1$). Hence $f(z) = (1 - z)\sum_0^\infty s_n z^n$. As z tends to 1 in a Stolz angle we may assume that $|1 - z| \le M(1 - |z|)$. Given $\varepsilon > 0$ choose N so large that $|s_k| < \varepsilon$ for every $k \ge N$. Now

$$|f(z)| \le |1 - z|\left|\sum_0^{N-1} s_k z^k\right| + |1 - z|\left|\sum_N^\infty s_k z^k\right|$$

$$\le |1 - z|\left(A + \varepsilon\sum_N^\infty |z|^k\right)$$

$$= |1 - z|A + \varepsilon\frac{|1 - z|}{1 - |z|}.$$

Here we have denoted $\sum_0^{N-1} |s_k|$ by A. Now as $z \to 1$ in the given Stolz angle we can choose $|1 - z| < \varepsilon/A$. It now follows that $|f(z)| < \varepsilon(1 + M)$. Thus $f(z) \to 0 = f(1)$ as $z \to 1$ in any Stolz angle.

2.3 Linear Fractional Transformations

Definition 2.3.1

A linear fractional transformation is a rational function of order 1.

Since the order of a rational function is the maximum of the degrees of the numerator and the denominator polynomials it follows that the most general linear fractional transformation is of the form $\dfrac{az+b}{cz+d}$ where a, b, c, d are complex constants. However if $ad - bc$ were zero, this transformation will reduce to a constant and so will be of order zero (prove this yourself). Thus a linear fractional transformation is of the form $S(z) = \dfrac{az+b}{cz+d}$ where $ad - bc \neq 0$. Geometrically these transformations effect a one-to-one correspondence from the extended complex plane onto itself. For this we need only define $S(\infty) = \dfrac{a}{c}$ and $S^{-1}(\infty) = -\dfrac{d}{c}$. Actually with this convention S effects a bijective bicontinuous map of the extended plane onto itself where the topology is defined by the metric

$$d(z, z') = \frac{2|z - z'|}{\sqrt{[(1+|z|^2)(1+|z'|^2)]}} \qquad \text{if } z, z' \neq \infty$$

$$d(z, \infty) = \frac{2}{\sqrt{[(1+|z|^2)]}}.$$

To see this we observe that

$$\lim_{z \to \infty} S(z) = S(\infty) = a/c, \quad \lim_{z \to -d/c} S(z) = S(-d/c) = \infty \qquad (1)$$

where the limits are in our usual sense. Further $z \to \infty$ in the usual sense if and only if $d(z, \infty) \to 0$ and $z \to l$ (l a finite complex number) in the usual sense if and only if $d(z, l) \to 0$. With these observations and the fact that the one-point compactification of $\mathbb{C}$ is homeomorphic to the extended complex plane with the above metric, it follows that S effects a bijective bicontinuous map of the extended complex plane $\mathbb{C}^*$ onto itself (with the topology for $\mathbb{C}^*$ being τ, the topology of one-point compactification).

The set of all linear fractional transformations form a group with composition playing the role of multiplication. The identity in this group is the identity transformation $w = z$ and the inverse of $w = \dfrac{az+b}{cz+d}$ is $z = \dfrac{dw-b}{-cw+a}$.

There are certain special elementary linear fractional transformations with geometric significance. These are
(i) $w = z + a$, ($a \in \mathbb{C}$) a translation.
(ii) $w = az$ with $|a| = 1$, a rotation.
(iii) $w = az$ with a real and positive, a homothetic transformation.
(iv) $w = 1/z$, an inversion.

It now follows that if $w = az$, where a is an arbitrary non-zero complex number, then such a transformation is a composition of a rotation $w_1 = e^{i(\arg a)}z$ and a homothetic transformation $w_2 = |a|z$.

If $c \neq 0$, we can write $az + b = \dfrac{a}{c}(cz + d) - \dfrac{ad}{c} + b$ and thus

$$\frac{az + b}{cz + d} = \left[\left(\frac{bc - ad}{c^2}\right)\left(\frac{1}{z + \dfrac{d}{c}}\right)\right] + \frac{a}{c}.$$

This way of writing the transformation shows that any linear fractional transformation is a composition of a translation, inversion, rotation, homothetic transformation followed by another translation. That is, if $c \neq 0$ it is a composition of

$$z \to z + \frac{d}{c} \to \frac{1}{z + \dfrac{d}{c}}\left(\frac{bc - ad}{c^2}\right)\left(\frac{1}{z + \dfrac{d}{c}}\right) \to \left[\left(\frac{bc - ad}{c^2}\right)\left(\frac{1}{z + \dfrac{d}{c}}\right)\right] + \frac{a}{c}$$

and when $c = 0$ it is a composition of $z \to az/d \to az/d + b/d$. (Note that $d \neq 0$ here).

Of the four elementary linear fractional transformations that we have identified, translations, rotations and homothetic transformations carry straight lines onto straight lines and circles onto circles. (Suppose we take a translation $w = z + a$. If z varies over the circle $|z - \alpha| = r$ then w varies over the circle $|w - (a + \alpha)| = r$ and if z varies over the straight line $z = \alpha + \beta t$ then w varies over the straight line $w = a + \alpha + \beta t$ and similarly for rotation and homothetic transformations). However inversion does not always take circles to circles and straight lines to straight lines. In fact, under the inversion a straight line can be mapped onto either a straight line or a circle and similarly a circle can be mapped onto either a circle or a straight line. To see this, we first write down the equation of either a circle or a straight line in the following form $\alpha z\bar{z} + \beta z + \bar{\beta}\bar{z} + \gamma = 0$, where α and γ are real, β complex with $\beta\bar{\beta} > \alpha\gamma$ (If $\alpha \neq 0$ this equation represents a circle and if $\alpha = 0$, it represents a line. The condition $\beta\bar{\beta} > \alpha\gamma$ is equivalent to the radius being positive in the case of a circle and nondegenerate in the case of a line). If now we find the locus of $w = 1/z$ from the above equation we get $\alpha + \beta\bar{w} + \bar{\beta}w + \gamma w\bar{w} = 0$ which is again the equation of either a circle or a straight line. Thus the inversion maps a circle through origin ($\gamma = 0$) onto a line and maps circles not through origin ($\gamma \neq 0$) onto circles. In a similar way the inversion takes straight lines through origin ($\alpha = \gamma = 0$) onto straight lines and straight lines not through origin ($\gamma \neq 0$) onto circles.

From the above observations it follows that any linear fractional transformation, being a composition of elementary transformations, takes the family of circles and straight lines onto itself. For this reason in the context of linear fractional transformations circles and straight lines will be referred to as "circles".

The expression for a linear fractional transformation shows that there are three effective constants to be determined for fixing a linear fractional transformation. In fact the condition $ad - bc \neq 0$ implies that at least one of them is not equal to zero (say a) and then the transformation reduces to $\left(z + \dfrac{b}{a}\right)\Big/\left(\dfrac{c}{a}z + \dfrac{d}{a}\right)$ wherein only $\dfrac{b}{a}, \dfrac{c}{a}, \dfrac{d}{a}$ need to be determined. This suggests that three independent conditions should determine a linear fractional transformation uniquely. For example if a, b, c are three distinct extended complex numbers we shall prove that there exists a unique linear fractional transformation that takes a, b, c to 1, 0, ∞ in that order. This is

given by $S(z) = \dfrac{(a-c)(z-b)}{(a-b)(z-c)}$. While it is easy to check that this transfor-

mation takes a, b. c to 1, 0, ∞ in that order, we shall prove that this is unique. If T is any other transformation that does the same job then $(z - b)$ should be a factor of the numerator for T and $(z - c)$ should be a factor in the denominator for T. So any such map T will be of the form $T(z) =$

$k\dfrac{(z-b)}{(z-c)}$. If further $T(a) = 1$, $k = \dfrac{(a-c)}{(a-b)}$. Thus $T = S$ and indeed we have

proved our result. From this we can also get that there exists a unique linear fractional transformations that takes a, b, c to a', b', c' where a, b, c and a', b', c' are two triplets of (distinct) extended complex numbers. In fact if S takes a, b, c to 1, 0, ∞ and T takes a', b', c' to 1, 0, ∞ the required transformation is $T^{-1} \circ S$. This final observation enables us to conclude that there exists a unique linear fractional transformation that takes C to C' where C and C' are members of the family consisting of circles and straight lines. This fact will be very useful later on, in exhibiting conformal maps between open discs or half planes.

Definition 2.3.2 (Cross Ratio)
The cross ratio of four points z, a, b, c in the extended complex plane where a, b, c are distinct is defined as the image of z under the unique linear fractional transformation S that takes a, b, c into 1, 0, ∞ in that order and is usually denoted by (z, a, b, c).

More explicitly if none of the points is ∞, then Sz is given by $\dfrac{(z-b)(a-c)}{(z-c)(a-b)}$ and if a, b or $c = \infty$ then Sz is respectively given by

$\dfrac{(z-b)}{(z-c)}, \dfrac{(a-c)}{(z-c)}, \dfrac{(z-b)}{(a-b)}$. Since S is uniquely determined by these condi-

tions, the cross ratio is well defined. The following theorem gives a few properties of the cross ratio.

Theorem 2.3.3
(i) If a, b, c are distinct extended complex numbers and T is any linear fractional transformation then $(z, a, b, c) = (Tz, Ta, Tb, Tc)$
(ii) $\overline{(z, a, b, c)} = (\bar{z}, \bar{a}, \bar{b}, \bar{c})$

(iii) The cross ratio of four distinct points z, a, b, c is real if and only if the four points lie on a circle or on a straight line.

Proof

(i) If S is the linear fractional transformation that takes a, b, c to 1, 0, ∞ then by definition $Sz = (z, a, b, c)$. But then ST^{-1} carries Ta, Tb, Tc into 1, 0, ∞ in that order. Hence by definition

$$(Tz, Ta, Tb, Tc) = ST^{-1}(Tz) = Sz = (z, a, b, c).$$

(ii) Under the notation of (i) $\overline{(z, a, b, c)} = \overline{Sz}, (\overline{z}, \overline{a}, \overline{b}, \overline{c}) = S_1 \overline{z}$ where S_1 is the linear fractional transformation that takes $\overline{a}, \overline{b}, \overline{c}$ to 1, 0, ∞ in that order. We shall show that if Sw is given by $\dfrac{Aw + B}{Cw + D}$ then S_1 is given by

$$S_1 w = \frac{\overline{A}w + \overline{B}}{\overline{C}w + \overline{D}}.$$ Once this is shown it will follow that $S_1(\overline{z}) = \dfrac{\overline{A}z + \overline{B}}{\overline{C}z + \overline{D}}$

$$= \overline{\left(\frac{Az + B}{Cz + D}\right)} = \overline{Sz}.$$ In fact since S_1 is uniquely determined by the conditions

that $S_1(\overline{a}) = 1$, $S_1(\overline{b}) = 0$ and $S_1(\overline{c}) = \infty$ and the transformation given by

$\dfrac{\overline{A}w + \overline{B}}{\overline{C}w + \overline{D}}$ does indeed satisfy these conditions

$$\left[\frac{\overline{A}\overline{a} + \overline{B}}{\overline{C}\overline{a} + \overline{D}} = \overline{\frac{Aa + B}{Ca + D}} = \overline{Sa} = 1 \text{ etc.} \right] \text{ we get } S_1 w = \frac{\overline{A}w + \overline{B}}{\overline{C}w + \overline{D}}.$$

(iii) We shall first show that (iii) follows if the image of the real axis under any linear fractional transformation is either a circle or a straight line and then show that this is indeed the case. If the cross ratio of four points (z, a, b, c) is real then Sz is real where $Sa = 1$, $Sb = 0$ and $Sc = \infty$. So z, a, b, c lie on the image of the real axis under S^{-1} which by our assumption is either a circle or a straight line. Thus if the cross ratio of four points is real then the points lie on a circle or a straight line. On the other hand, if the four points z, a, b, c lie on a circle or a straight line C, take S such that $Sa = 1$, $Sb = 0$ and $Sc = \infty$ then as S^{-1} takes the real line (by our assumption) onto a circle or a straight line and in as much as z, a, b, c belong to this image, $C = S^{-1}(\mathbb{R})$ or $S(C) = \mathbb{R}$. In particular Sz should be real. Now we prove that the image of the real axis under any linear fractional transformation is a circle or a straight line. Let T be any linear fractional transformation. Let $T^{-1}(w) = z = \dfrac{Aw + B}{Cw + D}$. We need the locus of w when z satisfies $z = \overline{z}$. The later condition is equivalent to

$$\frac{Aw + B}{Cw + D} = \frac{\overline{A}\overline{w} + \overline{B}}{\overline{C}\overline{w} + \overline{D}}.$$

Hence

$$(A\overline{C} - C\overline{A})w\overline{w} + (A\overline{D} - C\overline{B})w + (B\overline{C} - D\overline{A})\overline{w} + (B\overline{D} - D\overline{B}) = 0.$$

Note here that $(A\overline{C} - C\overline{A})$ and $(B\overline{D} - D\overline{B})$ are purely imaginary and so we shall multiply the above equation by i and observe that this latter equation conforms to the standard combined equation of a circle or a straight line. Hence this equation represents a circle if $A\overline{C} - C\overline{A} \neq 0$ and a straight line if $A\overline{C} - C\overline{A} = 0$ [$A\overline{C} - C\overline{A} = 0$ implies $A\overline{D} - C\overline{B} \neq 0$ as $AD - CB \neq 0$].

Corollary 2.3.4

A linear fractional transformation carries "circles" to "circles".

Proof

Even though we have already proved this, this is also a simple consequence of (iii) and (i) of Theorem 2.3.3.

Definition 2.3.5 (Symmetric Points)

Two points z and z^* are said to be symmetric with respect to a "circle" C passing through a, b, c, if $(z^*, a, b, c) = \overline{(z, a, b, c)}$.

Remark 2.3.6

It looks as though the definition is dependent on the points a, b, c that we choose. However this is not the case and the notion of symmetry depends only on the circle C and not on the points that we choose on C. To see this suppose a', b', c' were three other points on the same circle C, then using the linear fractional transformation S with $Sa = 1$, $Sb = 0$ and $Sc = \infty$ we have $Sz^* = \overline{Sz}$ and $\overline{(z^*, a', b', c')} = \overline{(Sz^*, Sa', Sb', Sc')} = (\overline{Sz^*}, \overline{Sa'}, \overline{Sb'}, \overline{Sc'}) = (Sz, Sa', Sb', Sc') = (z, a', b', c')$. Note that $Sa' = (a', a, b, c)$ is real as a', a, b, c lie on the same circle and hence $\overline{Sa'} = Sa'$. Similarly $\overline{Sb'} = Sb'$ and $\overline{Sc'} = Sc'$. The geometric significance of symmetry is as follows: Suppose first that C is a straight line, then we can choose $c = \infty$, a, $b \in C$ and the condition for symmetry becomes

$$\frac{z^* - b}{a - b} = \frac{\overline{z} - \overline{b}}{\overline{a} - \overline{b}} \tag{2}$$

Taking absolute values we get that $|z^* - b| = |z - b|$. But since b can be any arbitrary point on C it follows that z and z^* are equidistant from all points of C. Further (2) gives, $\text{Im} \dfrac{z^* - b}{a - b} = -\text{Im} \dfrac{z - b}{a - b}$ and so z and z^* are in different half planes determined by C. This is obviously a reflection with respect to C. In the case of a finite circle C with centre p and radius r, systematic use of the invariance of cross ratios under linear fractional transformation tells us that

$$(z^*, a, b, c) = \overline{(z, a, b, c)}$$
$$= \overline{(z - p, a - p, b - p, c - p)}$$

$$= (\bar{z} - \bar{p}, \bar{a} - \bar{p}, \bar{b} - \bar{p}, \bar{c} - \bar{p})$$

$$= \left(\bar{z} - \bar{p}, \frac{r^2}{a-p}, \frac{r^2}{b-p}, \frac{r^2}{c-p} \right)$$

$$= \left(\frac{r^2}{\bar{z} - \bar{p}}, a - p, b - p, c - p \right) = \left(\frac{r^2}{\bar{z} - \bar{p}} + p, a, b, c \right).$$

Thus $z^* = \dfrac{r^2}{\bar{z} - \bar{p}} + p$ or that $(z^* - p)(\bar{z} - \bar{p}) = r^2$. Taking modulus it

follows that the product of the distances of z and z^* from p equals r^2.

Further $\dfrac{z^* - p}{z - p} = \dfrac{r^2}{|z - p|^2} > 0$ which means that z and z^* are situated on the

same half line through p. Geometrically this means exactly that z^* is the inverse point of z with respect to C.

Theorem 2.3.7 (Symmetry Principle)
If a linear fractional transformation carries a "circle" C onto a "circle" C' then it transforms a pair of points symmetric with respect to C onto a pair of points symmetric with respect to C'.

Proof
Let z and z^* be symmetric with respect to C and T be the given linear fractional transformation taking C onto C'.

Fix three points a, b, c, on C. Then we have

$$(z^*, a, b, c) = \overline{(z, a, b, c)}$$

This implies that

$(Tz^*, Ta, Tb, Tc) = \overline{(Tz, Ta, Tb, Tc)}$. Thus Tz^* and Tz are symmetric points with respect to the circle through Ta, Tb, Tc, i.e., C'.

2.4 Exponential and Trigonometric Functions

Definition 2.4.1
The exponential function in the complex plane denoted by e^z is defined by

$$e^z = \exp (z) = \sum_0^\infty \frac{z^n}{n!}.$$

The series in the right side of the last equality has a radius of convergence ∞ ($1/R = \limsup (1/n!)^{1/n}$ and $\log n \to \infty$ as $n \to \infty$ implies

$$(\log 1 + \log 2 + \ldots + \log n)/n \to \infty \text{ as } n \to \infty$$

or that $\log (n!)^{1/n} \to \infty$ which in turn implies $(n!)^{1/n} \to \infty$ as $n \to \infty$ and hence $R = \infty$).

Theorem 2.4.2
The exponential function $f(z) = e^z$ has the following properties:

(i) $f(z)$ is analytic in the whole plane and $f'(z) = f(z)$ with $f(0) = 1$.

(ii) $e^{a+b} = e^a e^b$ and $e^z \neq 0$ for any z.

(iii) $f|\mathbb{R}$ is a monotonically increasing positive function and $e^x \to \infty$ as $x \to \infty$ and $e^x \to 0$ as $x \to -\infty$. Also its range equals $(0, \infty)$.

(iv) There exists a positive number π such that $e^{2i\pi} = 1$ and $e^z = 1$ if and only if z is an integral multiple of $2\pi i$.

(v) $f(z)$ is periodic with period $2\pi i$ and all its periods are integral multiples of $2\pi i$.

(vi) the mapping $t \to e^{it}$ maps the real axis onto the unit circle.

(vii) If w is a non-zero complex number then there exists z in the plane such that $w = e^z$.

Proof

(i) By Abel's radius of convergence theorem, the radius of convergence of the power series representing e^z is ∞. Hence the series converges absolutely for every finite complex z and further the convergence is uniform on every bounded subset of the complex plane. Also, it is analytic in the entire complex plane. Using the same theorem we also have

$$(d/dz)\ (e^z) = \sum_1^\infty \frac{z^{n-1}}{(n-1)!} = \sum_0^\infty \frac{z^n}{n!}, \text{ i.e. } f'(z) = f(z). \text{ Evidently } f(0) = 1.$$

(ii) If we let $g(z) = e^z e^{c-z}$ for any given complex number c, then $g(z)$ is analytic in the whole plane and $g'(z) = 0$ at all points. Thus $g(z)$ is a constant. Since the constant can be evaluated at any point we have $g(0) = e^c$. Thus $f(z) f(c - z) = f(c)$ for all z in the plane. Taking $c = a + b$, $z = a$ we get $f(a + b) = f(a) f(b)$ or that $e^{a+b} = e^a e^b$.

In particular $e^z e^{-z} = e^{z-z} = 1$. Hence e^z can never be equal to zero.

(iii) For z real and equal to x we have $e^x = \sum_0^\infty \frac{x^n}{n!}$ and hence $e^x > 1$ if $x > 0$ and by taking reciprocals $e^x < 1$ if $x < 0$. Thus $x > y$ implies $x - y > 0$ which in turn implies $e^{x-y} > 1$ or that $e^x > e^y$.

As $e^x > 1 + x$, for $x > 0$ we also get that e^x tends to ∞ as x tends to ∞ and e^{-x} tends to zero as x tends to ∞ (which is the same thing as saying that e^x tends to zero as $x \to -\infty$). Since e^x is continuous it has the intermediate value property and as e^x tends ∞ or 0 respectively as $x \to +\infty$ or as $x \to -\infty$ it follows that e^x is onto $(0, \infty)$.

(iv) Let t be real. Now $e^{it} = \sum_0^\infty \frac{(it)^n}{n!}$ and $e^{-it} = \sum_0^\infty \frac{(-it)^n}{n!} = \overline{e^{it}}$. Thus

$|e^{it}|^2 = e^{it}\overline{e^{it}} = e^{it}e^{-it} = 1$. Thus $|e^{it}| = 1$ and so e^{it} for all real t lies on the unit circle. Let us define $\cos t = \operatorname{Re} e^{it}$ and $\sin t = \operatorname{Im} e^{it}$ for real t. Thus $e^{it} = \cos t + i \sin t$. Differentiating with respect to t we get

$$\frac{d}{dt}(\cos t + i \sin t) = i e^{it} = i(\cos t + i \sin t) = -\sin t + i \cos t.$$

Equating the real and imaginary parts we get

$$\frac{d}{dt}(\cos t) = -\sin t \text{ and } \frac{d}{dt}(\sin t) = \cos t.$$

Using the power series representation for the exponential function we also get

$$\cos t = 1 - \frac{t^2}{2!} + \frac{t^4}{4!} - \cdots$$

and

$$\sin t = t - \frac{t^3}{3!} + \frac{t^5}{5!} - \cdots$$

Since all the above series are absolutely convergent the terms can be grouped in any way we like and a simple computation (prove $\cos t < 1 - \frac{t^2}{2!} + \frac{t^4}{4!}$ for $t^2 \leq 56$) shows that $\cos 0 = 1 > 0$ and $\cos 2 < -1/3 < 0$. Since $\cos t$ is continuous we get by intermediate value theorem that there exists a smallest positive number t_0, with $0 < t_0 < 2$ such that $\cos t_0 = 0$. Define $\pi = 2t_0$ and it follows that $\sin^2 t_0 = 1 - \cos^2 t_0 = 1$ and so $\sin t_0 = \pm 1$. Since $d/dt (\sin t) = \cos t > 0$ for $0 < t < t_0$ and $\sin 0 = 0$ we have $\sin t_0 > 0$ and hence $\sin t_0 = 1$. Thus $e^{i\pi/2} = e^{it_0} = \cos t_0 + i \sin t_0 = i$. It follows that $(e^{i\pi/2})^2 = i^2 = -1$ and $e^{2\pi i} = (e^{i\pi})^2 = (-1)^2 = 1$. Hence $e^{2\pi in} = 1$ for every integer n (positive, negative or zero). For the converse we observe that if $z = x + iy$ (x, y real) then $e^z = e^x e^{iy}$. Hence $|e^z| = e^x$. Thus if $e^z = 1$ then $e^x = 1$. But $e^x > 1$ if $x > 0$ and $e^x < 1$ if $x < 0$. Thus $e^x = 1$ implies $x = 0$. Thus $z = iy$ for some real y. If y is not an integral multiple of 2π then y lies between two consecutive integral multiples of 2π (or rather $y/2\pi$ lies between two consecutive integers) and hence there exists an integer m such that $2m\pi < y < 2 (m + 1)\pi$. Also $e^{iy} = e^{i(y - 2m\pi) + 2m\pi i} = e^{i(y - 2m\pi)} = e^{it}$ where $t = y - 2m\pi$ ($0 < t < 2\pi$). We now claim that $e^{it} \neq 1$ if $0 < t < 2\pi$. Let $e^{it/4} = u + iv$, u, v real so that $u^2 + v^2 = 1$. Since $0 < \frac{t}{4} < \frac{\pi}{2}$, we have $0 < u < 1$ and $0 < v < 1$ and

$$e^{it} = (u + iv)^4 = u^4 - 6u^2v^2 + v^4 + 4iuv(u^2 - v^2) \qquad (1)$$

Now the right side of (1) is real only if $u^2 = v^2$. Since $u^2 + v^2 = 1$ we get $u^2 = v^2 = 1/2$. Thus (1) again shows that $e^{it} = -1 \neq 1$. Hence $e^z = 1$ implies $z = 2m\pi i$ ($m \in \mathbb{Z}$)

(v) A function $f(z)$ is said to be periodic with period c if $f(z + c) = f(z)$ for all permissible z. As we have already noted in (ii) and (iv), $e^{z+2\pi i} = e^z$ and therefore $2\pi i$ is a period for $f(z) = e^z$. On the other hand, if c is any period then $e^{z+c} = e^z$ for all z and this implies $e^c = 1$. By (iv) this means that c should be an integral multiple of $2\pi i$.

(vi) We have already seen the Euler formula $e^{it} = \cos t + i \sin t$ and that $|e^{it}| = 1$ whenever t is real. Thus the mapping $t \to e^{it}$ takes the real axis into the unit circle. To prove that this mapping is onto we have to show that given any complex w, with $|w| = 1$ there exists t real such that $w = e^{it}$. To simplify the situation let us first take the case when $w = u + iv$ where u and v are both non negative. Since $u^2 + v^2 = 1$, $u \leq 1$ but the definition of π shows that there exists a 't' such that $0 \leq t \leq \pi/2$ and $\cos t = u$, then $\sin^2 t = 1 - u^2 = v^2$ and as $\sin t \geq 0$ in $0 \leq t \leq \pi/2$ we should have $\sin t = v$. Thus $w = u + iv = \cos t + i \sin t = e^{it}$.

The second possibility for w is that $w = u + iv$ where $u < 0$ and $v > 0$. But in this case $-iw$ is such that both its real and imaginary parts are nonnegative and $|-iw| = |w| = 1$. Using the previous case we can get $-iw = e^{it}$ for some real t and then $w = e^{i(t + \pi/2)}$. If $u > 0$ and $v < 0$ we can take $-w = e^{it}$ and $w = e^{i(t+\pi)}$. Finally if v is also negative along with u then using the first case we can get $-w = e^{it}$ for some real t and then $w = e^{i(\pi + t)}$.

(vii) If $w \neq 0$ take $\alpha = w/|w|$ so that $w = |w|\alpha$. However we can always find *a real* x such that $|w| = e^x$ (e^y is a continuous function of y in $\mathbb{R}$ taking all values between 0 and ∞ by (iii)). Also since $|\alpha| = 1$ (vi) shows that $\alpha = e^{iy}$ for some real y. Hence $w = |w|\alpha = e^x e^{iy} = e^{x+iy}$. This completes the proof of (vii) and also the proof of the theorem.

Note 2.4.3
In the course of the proof of the previous theorem we have defined $\sin t$ and $\cos t$ and obtained many of their properties. We shall recall them as follows:
(i) $\cos t = \text{Re }(e^{it})$ and $\sin t = \text{Im }(e^{it})$
(ii) $\cos^2 t + \sin^2 t = 1$
(iii) $\cos t + i \sin t = e^{it}$ and $\cos(-t) + i \sin(-t) = e^{-it}$ and hence $\cos(-t) = \cos t$ and $\sin(-t) = -\sin t$.
(iv) $\dfrac{d}{dt}(\cos t) = -\sin t$ and $\dfrac{d}{dt}(\sin t) = \cos t$.
(v) $\cos t = 1 - \dfrac{t^2}{2!} + \dfrac{t^4}{4!} - \ldots$ and $\sin t = t - \dfrac{t^3}{3!} + \dfrac{t^5}{5!} - \ldots$

We shall now define the following trigonometric functions of a complex variable.

Definition 2.4.4
$\cos z = (e^{iz} + e^{-iz})/2$ and $\sin z = (e^{iz} - e^{-iz})/2i$.
Observe the following properties of $\cos z$ and $\sin z$ which are immediate consequences of the definition and the properties of the exponential function.

(i) $\cos z = 1 - \dfrac{z^2}{2!} + \dfrac{z^4}{4!} - \ldots$ and $\sin z = z - \dfrac{z^3}{3!} + \dfrac{z^5}{5!} - \ldots$
(ii) $\cos z + i \sin z = e^{iz}$
(iii) $\cos^2 z + \sin^2 z = 1$

(iv) $d/dz\,(\cos z) = -\sin z$ and $d/dz\,(\sin z) = \cos z$

(v) $\cos(z + w) = \cos z\,\cos w - \sin z\,\sin w$

(vi) $\sin(z + w) = \sin z\,\cos w + \cos z\,\sin w$

The other trigonometric functions are defined canonically by $\tan z = \sin z/\cos z$, $\cot z = 1/\tan z$, $\operatorname{cosec} z = 1/\sin z$ and $\sec z = 1/\cos z$.

Definition 2.4.5

We shall define $z = \log w$ to be a solution of the equation $e^z = w$. Since $e^z \neq 0$, 0 has no logarithm and for $w \neq 0$ the equation $e^z = w$ ($z = x + iy$) is equivalent to $e^x = |w|$ and $e^{iy} = w/|w|$. The first of these equations has a unique solution $x = \log|w|$, the real logarithm of the positive number $|w|$. The second of these equations has one and only one solution in $0 \le y < 2\pi$. (Note that y is an argument of w.) Further, any integral multiple of 2π added to y is also a solution of the second equation. Hence every complex number other than 0 has infinitely many logarithms which differ from each other by integral multiples of $2\pi i$.

The imaginary part of $\log w$ is nothing but the set $\arg w$ and geometrically this is the angle in radians through which the positive real axis must be rotated to obtain the directed line segment from 0 to w. Thus each complex number has infinitely many values for its arguments differing by integral multiples of 2π. Now $\log w = \log|w| + i\,\arg w$. The addition theorem of exponential function translates into either of the equalities: $\log (zw) = \log z + \log w$ or $\arg (z \cdot w) = \arg z + \arg w$ in the sense that both sides represent the same infinite set of complex numbers. We also define for complex numbers a and b ($a \neq 0$) $a^b = \exp (b \log a)$. In general a^b has infinitely many values. We also define $z = \cos^{-1} w$ as solution of the equation $\cos z = (e^{iz} + e^{-iz})/2 = w$. Solving this quadratic equation in e^{iz} we get all the values of z as $\cos^{-1} w = \arccos w = -i \log (w \pm \sqrt{(w^2 - 1)})$. We also define $\sin^{-1} w$ by $\sin^{-1} w = \arcsin w = \pi/2 - \cos^{-1} w$. We can also define the hyperbolic trigonometric functions by $\cosh z = (e^z + e^{-z})/2$ and $\sinh z = (e^z - e^{-z})/2$ and get the identities like $\cosh^2 z - \sinh^2 z = 1$ etc.

We would like to remark that all these elementary transcendental functions are expressible in terms of the transcendental function e^z and its inverse $\log z$.

Remark 2.4.6

The definition of $\log z$ enables us to define z^α as above whenever z, α are non-zero complex numbers. This definition makes sense but for each $z \neq 0$ it gives infinitely many values corresponding to infinitely many values of $\log z$. We shall later see how we can define z^α as a single-valued function of z in certain regions of the complex plane.

SOLVED EXERCISES

Problem 1 If $f(z)$ is analytic in a region Ω, is $\overline{f(\bar{z})}$ analytic? If so where and how? What about the converse?

Solution Let $f(z)$ be analytic in Ω. Hence if $z_0 \in \Omega$,

$$\lim_{z \to z_0} \frac{f(z) - f(z_0)}{z' - z_0} \tag{1}$$

exists. We shall now prove that $g(z) = \overline{f(\bar{z})}$ is analytic in $\overline{\Omega}$, the reflection of Ω in the real axis. By definition

$$\overline{\Omega} = \{z \in \mathbb{C} : \bar{z} \in \Omega\}$$

Let $z_0 \in \overline{\Omega}$. Then $\bar{z}_0 \in \Omega$. From (1) we have

$$\lim_{\bar{z} \to \bar{z}_0} \frac{f(\bar{z}) - f(\bar{z}_0)}{\bar{z} - \bar{z}_0} = f'(\bar{z}_0) = \alpha \tag{2}$$

exists in Ω. As the mapping $z \to \bar{z}$ is continuous, we have from (2),

$\lim_{z \to z_0} \dfrac{g(z) - g(z_0)}{z - z_0} = \bar{\alpha}$ exists whenever $z_0 \in \overline{\Omega}$.

Hence $g(z) = \overline{f(\bar{z})}$ is analytic in $\overline{\Omega}$. Conversely let $g(z) = \overline{f(\bar{z})}$ be analytic in $\overline{\Omega}$. By the previous case we have $\overline{g(\bar{z})}$ is analytic in $\overline{\overline{\Omega}} = \Omega$, i.e., $f(z)$ is analytic in Ω. Hence the converse is also true.

Problem 2 Prove that the functions $u(z)$ and $u(\bar{z})$ are simultaneously harmonic in their respective domains of definition.

Solution Let $u(z)$ be harmonic in Ω. Then u is continuous along with its partial derivatives up to two orders and

$$\frac{\partial^2 u}{\partial x^2} + \frac{\partial^2 u}{\partial y^2} = 0 \ \forall \ (x, y) \in \Omega. \tag{1}$$

Let $v(z) = u(\bar{z})$. i.e. $v(x, y) = u(x, -y)$. We have to prove that $v(z)$ is harmonic in $\overline{\Omega}$. As the mapping $(x, y) \to (x, -y)$ $(z \to \bar{z})$ is infinitely real differentiable $v(x, y)$ is also continuous with its partial derviatives up to two orders. Let $(x_0, y_0) \in \overline{\Omega}$. Therefore $(x_0, -y_0) \in \Omega$ and hence (1) is true.

i.e.,
$$\frac{\partial^2 u}{\partial x^2} + \frac{\partial^2 u}{\partial y^2} = 0 \text{ at } (x_0, -y_0)$$

$$\left[\frac{\partial v}{\partial x}\right]_{(x_0, y_0)} = \lim_{h \to 0} \frac{v(x_0 + h, y_0) - v(x_0, y_0)}{h}$$

$$= \lim_{h \to 0} \frac{u(x_0 + h, -y_0) - u(x_0, -y_0)}{h}$$

$$= \left[\frac{\partial u}{\partial x} \right]_{(x_0, -y_0)}$$

$$\left[\frac{\partial^2 v}{\partial x^2} \right]_{(x_0, y_0)} = \lim_{h \to 0} \frac{\frac{\partial v}{\partial x}(x_0 + h, y_0) - \frac{\partial v}{\partial x}(x_0, y_0)}{h}$$

$$= \lim_{h \to 0} \frac{\frac{\partial u}{\partial x}(x_0 + h, -y_0) - \frac{\partial u}{\partial x}(x_0, -y_0)}{h}$$

$$= \left[\frac{\partial^2 u}{\partial x^2} \right]_{(x_0, -y_0)} \tag{2}$$

Similarly we have

$$\left[\frac{\partial v}{\partial y} \right]_{(x_0, y_0)} = -\left[\frac{\partial u}{\partial y} \right]_{(x_0, -y_0)}$$

$$\left[\frac{\partial^2 v}{\partial y^2} \right]_{(x_0, y_0)} = \left[\frac{\partial^2 u}{\partial y^2} \right]_{(x_0, -y_0)} \tag{3}$$

From (2) and (3) we have

$$\left[\frac{\partial^2 v}{\partial x^2} + \frac{\partial^2 v}{\partial y^2} \right]_{(x_0, y_0)} = \left[\frac{\partial^2 u}{\partial x^2} + \frac{\partial^2 u}{\partial y^2} \right]_{(x_0, -y_0)}$$

$$= 0 \text{ by (1).}$$

Hence $v(z) = u(\bar{z})$ is harmonic in $\overline{\Omega}$. Conversely let $u(\bar{z})$ be harmonic in $\overline{\Omega}$. If we take $v(z) = u(\bar{z})$ then $v(\bar{z})$ is harmonic in $\overline{\overline{\Omega}}$, i.e., $u(z)$ is harmonic in Ω. Hence the result.

Problem 3 Show that a harmonic function u satisfies the formal differential equation $\dfrac{\partial^2 u}{\partial z \partial \bar{z}} = 0$.

Solution Let $u(z)$ be a harmonic function in Ω. We have $u_{xx} + u_{yy} = 0$ at all points of Ω. We write $x = \dfrac{z + \bar{z}}{2}$ and $y = \dfrac{z - \bar{z}}{2i}$. Now formally

$$\frac{\partial u}{\partial \bar{z}} = \frac{\partial u}{\partial x}\frac{\partial x}{\partial \bar{z}} + \frac{\partial u}{\partial y}\frac{\partial y}{\partial \bar{z}} = \frac{\partial u}{\partial x}\frac{1}{2} + i\frac{\partial u}{\partial y}\frac{1}{2}.$$

Similarly $\dfrac{\partial u}{\partial z} = \dfrac{\partial u}{\partial x}\dfrac{1}{2} + \dfrac{\partial u}{\partial y}\dfrac{1}{2i}$. Hence

$$\frac{\partial^2 u}{\partial z\,\partial \bar{z}} = \frac{\partial}{\partial z}\left(\frac{\partial u}{\partial \bar{z}}\right) = \frac{1}{2}\frac{\partial}{\partial \bar{z}}\left(\frac{\partial u}{\partial x}\right) + \frac{i}{2}\frac{\partial}{\partial z}\left(\frac{\partial u}{\partial y}\right)$$

$$= \frac{1}{2}\left[\frac{1}{2}\frac{\partial^2 u}{\partial x^2} + \frac{1}{2i}\frac{\partial^2 u}{\partial y\,\partial x}\right] + \frac{i}{2}\left[\frac{1}{2}\frac{\partial^2 u}{\partial x\,\partial y} + \frac{1}{2i}\frac{\partial^2 u}{\partial y^2}\right]$$

$$= 0.$$

Note that by the continuity of the first order partial derivatives of u we have

$$\frac{\partial^2 u}{\partial y\,\partial x} = \frac{\partial^2 u}{\partial x\,\partial y}.$$

Problem 4 If Q is a polynomial of degree n with distinct roots α_1, α_2 . . ., α_n and if P is a polynomial of degree $< n$, show that

$$\frac{P(z)}{Q(z)} = \sum_{k=1}^{n}\frac{P(\alpha_k)}{Q'(\alpha_k)(z-\alpha_k)}$$

Solution Let $Q(z) = c\,(z - \alpha_1)\,(z - \alpha_2)\ \ldots\ (z - \alpha_n)$ with $c \neq 0$. Thus

$$\frac{P(z)}{Q(z)} = \frac{P(z)}{c(z-\alpha_1)(z-\alpha_2)\ldots(z-\alpha_n)}$$

$$c\frac{P(z)}{Q(z)} = \sum_{i=1}^{n}\frac{A_i}{(z-\alpha_i)} \qquad (1)$$

where the right side represents the partial fraction expansion of $c\dfrac{P(z)}{Q(z)}$. Thus

$$c\frac{P(z)}{Q(z)} = \frac{\displaystyle\sum_{j=1}^{n}A_j\prod_{i\neq j}(z-\alpha_i)}{Q(z)/c}$$

$$\therefore \qquad P(z) = \sum_{j=1}^{n}A_j\prod_{i\neq j}(z-\alpha_i)$$

Hence $P(\alpha_k) = A_k\prod_{i\neq k}(\alpha_k - \alpha_i)$ and therefore

$$A_k = \frac{P(\alpha_k)}{\prod\limits_{i \neq k}(\alpha_k - \alpha_i)}. \tag{2}$$

Now
$$Q'(\alpha_k) = c\prod\limits_{i \neq k}(\alpha_k - \alpha_i). \tag{3}$$

Using (3), equation (2) becomes $A_k = \dfrac{cP(\alpha_k)}{Q'(\alpha_k)}$. Hence from (1) we have

$$c\frac{P(z)}{Q(z)} = \sum\limits_{k=1}^{n}\frac{cP(\alpha_k)}{Q'(\alpha_k)(z - \alpha_k)}$$

i.e.,
$$\frac{P(z)}{Q(z)} = \sum\limits_{k=1}^{n}\frac{P(\alpha_k)}{Q'(\alpha_k)(z - \alpha_k)}.$$

Problem 5 Use the formula in the preceding problem to prove that there exists a unique polynomial P of degree less than n with given values c_k at the distinct points α_k ($k = 1, 2, \ldots, n$).

Solution We need a polynomial P of degree less than n with $P(\alpha_k) = c_k$. Let $Q(z) = (z - \alpha_1)(z - \alpha_2) \ldots (z - \alpha_n)$. If P is any polynomial of degree less than n then by the previous problem P and Q satisfy

$$\frac{P(z)}{Q(z)} = \sum\limits_{k=1}^{n}\frac{P(\alpha_k)}{Q'(\alpha_k)(z - \alpha_k)}.$$

Hence
$$P(z) = Q(z)\sum\limits_{k=1}^{n}\frac{P(\alpha_k)}{Q'(\alpha_k)(z - \alpha_k)}.$$

If we impose the additional condition that $P(\alpha_k) = c_k$ then

$$P(z) = \sum\limits_{k=1}^{n}\frac{c_k}{Q'(\alpha_k)}\prod\limits_{\substack{i=1 \\ i \neq k}}^{n}(z - \alpha_i).$$

Thus the required polynomial $P(z)$ is uniquely determined by this condition. On the other hand it is easy to check that this polynomial is of degree less than n and satisfies $P(\alpha_k) = c_k$.

Problem 6 Let $R(z)$ be a rational function. How will you find the order of $R'(z)$? How large and how small can this order be?

Solution Let $R(z) = \dfrac{P(z)}{Q(z)}$ where $(P, Q) = 1$, deg $P = l$, deg $Q = m$ and order of $R = n = \max(l, m)$. Consider the partial fraction expansion of $R(z)$. Assume $R(z)$ has $\beta_1, \beta_2, \ldots \beta_k$ as distinct finite poles of orders $n_1, n_2, \ldots$ n_k with $\sum\limits_{j=1}^{k}n_j = m$. We consider two cases.

Case 1 $l > m$. In this case $R(z) = G(z) + \sum\limits_{j=1}^{k} H_j \left(\dfrac{1}{z - \beta_j} \right)$ where $G(z)$ is

a polynomial of degree $(l - m)$, where $(l - m)$ represents the order of the

pole for $R(z)$ at ∞ and each H_j $(1 \leq j \leq k)$ is a polynomial in $\dfrac{1}{z - \beta_j}$ of

degree n_j. Evidently the order of $R(z)$ is obtained by adding the degrees of all the polynomials G and H_j $[(l - m) + \sum n_j = l = \text{order of } R(z)]$. We now have

$$R'(z) = G'(z) + \sum_{j=1}^{k} \left[H_j \left(\frac{1}{z - \beta_j} \right) \right]'$$

Thus the order of $R'(z)$ is equal to $l - m - 1 + \sum\limits_{j=1}^{k} (n_j + 1) = l - 1 + k$.

As k denotes the number of distinct finite poles of $R(z)$ the maximum and minimum of k are respectively m and 0. Thus the order $R'(z)$ equals $l + k - 1$ with the maximum being $l + m - 1$ and the minimum being $l - 1$.

Case 2 $l \leq m$. In this case the partial fraction expansion of $R(z)$ as in case 1 will be such that $G(z)$ is a constant which can be equal to 0 also. Thus

the order of $R'(z)$ is equal to $\sum\limits_{j=1}^{k} (n_j + 1) = m + k$. Again the maximum is

$2m$ and the minimum is m. This completes the solution of our problem.

Problem 7 Expand $\dfrac{2z + 3}{z + 1}$ in powers of $(z - 1)$. What is its radius of convergence?

Solution Let $f(z) = \dfrac{2z + 3}{z + 1}$. Put $h = z - 1$ or that $z = 1 + h$.

Now $\qquad f(z) = \dfrac{2(1 + h) + 3}{(1 + h) + 1}$

$\qquad\qquad = 2 + \dfrac{1}{h + 2}$

$\qquad\qquad = 2 + \dfrac{1}{2} \left(1 + \dfrac{h}{2} \right)^{-1}$

$\qquad\qquad = \dfrac{5}{2} + \dfrac{1}{2} \left[-\dfrac{h}{2} + \left(\dfrac{h}{2} \right)^2 - \ldots \right]$ (for h small with $h = z - 1$)

Since the radius of convergence is the largest disk around 1 in which the function is analytic and -1 is the only point where the function is not analytic and the distance of -1 from 1 is 2, the required radius of convergence $R = 2$.

Problem 8 Find the radius of convergence of $\Sigma z^{n!}$.

Soultion The coefficients of this series are 1 and 0. In fact $a_k = 1$ if $k = n!$ for some n and 0 otherwise. Therefore $\limsup\limits_{n \to \infty} |a_n|^{1/n} = 1$. Hence the radius of convergence $R = 1$.

Problem 9 If $\Sigma a_n z^n$ has radius of convergence R, what is the radius of convergence of $\Sigma a_n z^{2n}$ and of $\Sigma a_n^2 z^n$?

Solution (i) Consider $\Sigma a_n z^{2n} = \Sigma a_n w^n$ (say) where $w = z^2$. We know $\Sigma a_n w^n$ converges when $|w| < R$ and does not converge for $|w| > R \cdot |w| < R \Leftrightarrow |z^2| < R \Leftrightarrow |z| < \sqrt{R}$. Similarly $|w| > R \Leftrightarrow |z^2| > R \Leftrightarrow |z| > \sqrt{R}$. Hence the radius of convergence of $\Sigma a_n z^{2n}$ is $\sqrt{R}$.

(ii) Consider $\Sigma a_n^2 z^n$. Radius of convergence of $\Sigma a_n z^n$ is R, i.e.,

$$\frac{1}{R} = \limsup\limits_{n \to \infty} |a_n|^{1/n}.$$

Now
$$\limsup\limits_{n \to \infty} |a_n^2|^{1/n} = \limsup\limits_{n \to \infty} |a_n|^{2/n} = \frac{1}{R^2}$$

Hence the radius of convergence of $\Sigma a_n^2 z^n$ is R^2.

Problem 10 For what values of z is $\sum\limits_0^\infty \left(\dfrac{z}{1+z}\right)^n$ convergent?

Solution The given series is a geometric series with common ratio $\dfrac{z}{1+z}$.

Hence it is convergent if and only if $\left|\dfrac{z}{1+z}\right| < 1$, i.e.,

$$\Leftrightarrow |1 + z| > |z|$$
$$\Leftrightarrow |1 + z|^2 > |z|^2$$
$$\Leftrightarrow 1 + |z|^2 + 2\mathrm{Re}\ z > |z|^2$$
$$\Leftrightarrow \mathrm{Re}\ z > \frac{-1}{2}$$

i.e., $x > \dfrac{-1}{2}$ where $z = x + iy$.

Hence the given series is convergent in the half plane determined by $x > \dfrac{-1}{2}$.

Problem 11 Suppose that a linear fractional transformation carries one pair of concentric cricles onto another pair of concentric circles. Prove that the ratio of the radii must be the same.

Solution Let T be the given linear fractional transformation. Let C_1, C_2 be the two concentric circles centered at a with radii R_1, R_2 which are mapped onto the concentric circles C'_1, C'_2 centered at a' with radii R'_1, R'_2.

Now a and ∞ are the only pair of symmetric points with respect to both C_1, C_2. Recall that if z and z' are the symmetric points with respect to a circle centre a and radius R then $(\bar{z} - \bar{a})(z' - a) = R^2$. Hence by symmetry principle Ta and $T\infty$ are symmetric with respect to both C'_1, C'_2. But a' and ∞ are the only pair of points symmetric with respect to both C'_1 and C'_2. Thus $Ta = a'$ and $T\infty = \infty$ or $Ta = \infty$ and $T\infty = a'$. In the first case

$$Tz = \alpha z + a' - \alpha a. \ \left(\text{Write } Tz = \frac{Az + B}{Cz + D} \text{ and observe that } T\infty = \infty \Rightarrow C = 0\right.$$

and $Ta = a'$ implies $\dfrac{A}{D}a + \dfrac{B}{D} = a'$ or that $Tz = \alpha z + a' - \alpha a$ where $\alpha = $

$\left. \dfrac{A}{D}.\right)$ If we take $z_1 \in C_1$ and $z_2 \in C_2$ we have $Tz_1 \in C'_1$ and $Tz_2 \in C'_2$.

Now $\dfrac{R_1}{R_2} = \left|\dfrac{z_1 - a}{z_2 - a}\right|$ and

$$\frac{R'_1}{R'_2} = \left|\frac{Tz_1 - a'}{Tz_2 - a'}\right| = \left|\frac{\alpha z_1 - \alpha a}{\alpha z_2 - \alpha a}\right| = \frac{R_1}{R_2}$$

In the second case $Tz = \dfrac{a'z + b}{z - a}$. $\left(\text{Write } Tz = \dfrac{Az + B}{Cz + D} \text{ and observe that}\right.$

$Ta = \infty \Rightarrow Ca + D = 0$ and $T\infty = a'$ implies $\dfrac{A}{C} = a'$ or that $Tz = \dfrac{a'z + b}{z - a}$

where $b = \left.\dfrac{B}{C}.\right)$ Again as before $\dfrac{R_1}{R_2} = \left|\dfrac{z_1 - a}{z_2 - a}\right|$ and $\dfrac{R'_1}{R'_2} = \left|\dfrac{Tz_1 - a'}{Tz_2 - a'}\right| = $

$\dfrac{|b + aa'| \ |z_2 - a|}{|b + aa'| \ |z_1 - a|} = \dfrac{R_2}{R_1}$. Hence the ratio of the radii are the same.

Problem 12 Find a linear fractional transformation which carries $|z| = 1$ and $|z - (1/4)| = 1/4$ onto concentric circles. What is the ratio of the radii in the image plane?

Solution Let T map these circles onto circles with centre a and radii R_1 and R_2. Since a and ∞ are symmetric with respect to both the image circles their pre-images say α and β must be symmetric with respect to both these given circles. Thus $\overline{\alpha}\beta = 1$ and $\left(\overline{\alpha} - \frac{1}{4}\right)\left(\beta - \frac{1}{4}\right) = \frac{1}{16}$ i.e., $\overline{\alpha}\beta - \frac{1}{4}(\overline{\alpha} + \beta)$ $= 0$ which implies $\overline{\alpha} + \beta = 4$ and $\overline{\alpha} - \beta = \sqrt{(\overline{\alpha} + \beta)^2 - 4\overline{\alpha}\beta} = \pm 2\sqrt{3}$. Thus $\overline{\alpha} = 2 \pm \sqrt{3}, \beta = 2 \mp \sqrt{3}$. Hence T takes $2 + \sqrt{3}$ to a and $2 - \sqrt{3}$ to ∞ or $2 + \sqrt{3}$ to ∞ and $2 - \sqrt{3}$ to a. One such linear fractional transformation

is, $Tz = \dfrac{(z - (2 \pm \sqrt{3}))}{(z - (2 \mp \sqrt{3}))} + a$. Choose 1 on the unit circle and $\dfrac{1}{2}$ on the circle

centre $\dfrac{1}{4}$ and radius $\dfrac{1}{4}$. Then $T(1)$ is on the first of the concentric circles

and $T\left(\dfrac{1}{2}\right)$ is on the other concentric circle. Hence the ratio of the radii of

the image circles $\dfrac{R_1}{R_2}$ is given by $\dfrac{R_1}{R_2} = \left|\dfrac{T(1) - a}{T(1/2) - a}\right| = (2 - \sqrt{3})$, taking

$\alpha = (2 + \sqrt{3})$ and $\beta = 2 - \sqrt{3}$. The other assumption $\alpha = 2 - \sqrt{3}$ and

$\beta = 2 + \sqrt{3}$ gives $\dfrac{R_1}{R_2} = 2 + \sqrt{3}$.

Problem 13 Let f, g be analytic in a region G which contains 0. Suppose that

$$f(w + z) = f(z) \cdot f(w) - g(z) \cdot g(w) \tag{1}$$
$$g(w + z) = g(z) \cdot f(w) + g(w) \cdot f(z) \tag{2}$$

hold whenever $w, z, w + z$ belong to G. Show that if $f(0) = 1$ and $f'(0) = 0$ then there is a disc B centered at 0 and lying in G such that $f(z) = \cos bz$ and $g(z) = \sin bz$, $\forall z \in B$ where $b = g'(0)$.

Solution Define $h(z) = f(z) + ig(z)$. We want to show that $h(z) = e^{az}$ for z belonging to some disc B around 0 were $a = ib$. From the above equations (1) and (2) it is easy to check that

$$h(w + z) = h(z) \cdot h(w), \ \forall \ w, z, w + z \in G \tag{3}$$

Now $h(0) = 1$ by (3) (Since $h(0) = [h\ (0)]^2$ and Re $h(0) = f(0) \neq 0$). If z belongs to a small disc B around 0 then for sufficiently small w we have $z, w, w + z \in B$

$$\frac{h(z + w) - h(z)}{w} = \frac{h(z)[h(w) - 1]}{w} = \frac{h(w) - h(0)}{w} h(z).$$

As $h(z) = f(z) + ig(z)$, it is analytic in G and $h'(0)$ exists. Thus from the last identity we get by taking limit as $w \to 0$,

$$h'(z) = h'(0) \cdot h(z) = ib\ h(z) = a\ h(z)\ (z \in B). \tag{4}$$

From (3) it also follows that $h(z)$ is never zero (Since $h(z) h(-z) = h(0)$ $\neq 0$) but $\dfrac{h'(z)}{h(z)} = a$. Consider $s(z) = h(z) e^{-az}$. This is defined and analytic in G and its derviative is equal to $s'(z) = h'(z) e^{-az} - a h(z) e^{-az} = 0$ by (4). Since G is connected, $s(z)$ is identically a constant. Hence $s(z) = s(0) = h(0)$ $= 1$, i.e. $h(z) e^{-az} = 1 \Rightarrow h(z) = e^{az}$. This proves the result.

Problem 14 Suppose $f(z) = \sum_0^\infty c_n z^n$ is zero at all points of a non-zero sequence $\{z_k\}$ which converges to zero. Then prove that the power series is identically zero.

Solution Since $f(z_k) = 0$, $z_k \neq 0$ the series converges at some non-zero point and so it has a positive radius of convergence. Now inside this circle (say C) of convergence (which contains zero) $f(z)$ is continuous. Hence $z_k \to 0 \Rightarrow 0 = f(z_k) \to f(0)$. Therefore $f(0) = 0$ which implies that $c_0 = 0$. Now consider $g(z) = \dfrac{f(z)}{z}$. Inside the circle of convergence, $f(z)$ is continuous and z is also continuous. Hence $g(z)$ is continuous for $z \neq 0$. Now $g(z) = c_1 + c_2 z + \dots$ and this latter series is also convergent inside C as it represents $\dfrac{f(z)}{z}$ for $z \neq 0$ and c_1 at $z = 0$. $\left[\text{Note that } \sum_0^\infty c_n z^n \text{ converges}\right.$ and $z \neq 0$ implies that $\dfrac{1}{z} \sum_0^\infty c_n z^n$ converges. $\left.\right]$ Thus $g(z)$ is continuous inside C. Therefore

$$g(0) = \lim_{z_k \to 0} g(z_k) = \lim_{z_k \to 0} \frac{f(z_k)}{z_k} = 0.$$

If $c_j = 0$, $1 \leq j \leq n - 1$ then as before the power series representing $\dfrac{f(z)}{z^n}$ is convergent inside C and so represents a continuous function. Hence

$$\lim_{z_k \to 0} \frac{f(z_k)}{z_k^n} = 0 = \lim_{z_k \to 0} (c_n + c_{n+1} z_k + \dots) = c_n$$

Therefore by induction $c_n = 0$ for all n and $f(z) \equiv 0$.

Problem 15 Find the domains in which the function $f(z) = f(x, y) = |x^2 - y^2| + 2i|xy|$ is analytic.

Solution Refer Fig. 2.E.(i) which is self explanatory. Here Ω_i's are regions excluding their linear boundaries.

We consider the following three cases.

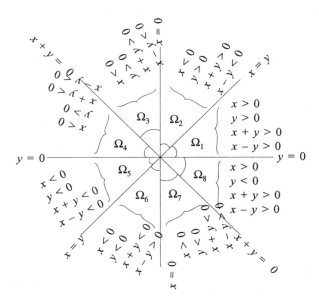

Fig. 2.E.(i)

Case (i) $xy(x^2 - y^2) > 0$. Geometrically $z = x + iy \in \Omega_1 \cup \Omega_3$ or $\Omega_5 \cup \Omega_7$. In this case

$$f(z) = \pm (x^2 - y^2 + 2ixy) = \pm (x + iy)^2 = \pm z^2$$

which is analytic.

Case (ii) $xy(x^2 - y^2) < 0$. Geometrically $z = x + iy \in \Omega_2 \cup \Omega_6$ or $\Omega_4 \cup \Omega_8$. In this case

$$f(z) = \pm (x^2 - y^2 - 2ixy) = \pm (x - iy)^2 = \pm \bar{z}^2$$

which is not analytic.

Case (iii) $xy(x^2 - y^2) = 0$. In this case $z \in$ x-axis or y-axis or the line $x = y$ or the line $x = -y$. In any neighbourhood of these points $f(z)$ is analytic at some point and not analytic at some other point. Hence $f(z)$ is not analytic in any neighbourhood of these points. Hence f is not analytic. [Recall that $f(z)$ is analytic at z_0 if and only if there exists a neighbourhood of z_0 in which f is analytic].

Problem 16 Let $f(z) = w = u + iv$ and assume that the functions u and v are real differentiable at z. Prove that the existence of either of the limits

$$\lim_{\Delta z \to 0} \left[\text{Re} \frac{\Delta w}{\Delta z} \right] \text{ or } \lim_{\Delta z \to 0} \left[\text{Im} \frac{\Delta w}{\Delta z} \right] \text{ at } z \text{ ensures the existence of the other limit}$$

and consequently the complex differentiability of the function $f(z)$ at z.

Solution Assume that u and v are real differentiable.

$$u(x + \Delta x, y + \Delta y) - u(x, y) = u_x \, \Delta x + u_y \, \Delta y + R(\Delta x, \Delta y)$$
$$v(x + \Delta x, y + \Delta y) - v(x, y) = v_x \, \Delta x + v_y \, \Delta y + S(\Delta x, \Delta y)$$

where $\dfrac{|R(\Delta x, \Delta y)|}{|\Delta x + i\Delta y|} \to 0$ $\dfrac{|S(\Delta x, \Delta y)|}{|\Delta x + i\Delta y|} \to 0$. Let $\lim\limits_{\Delta z \to 0}\left[\mathrm{Re}\,\dfrac{\Delta w}{\Delta z}\right]$ exist.

i.e., $\lim\limits_{h \to 0}\mathrm{Re}\left(\dfrac{f(z+h)-f(z)}{h}\right)$ exists independent of the direction along

which h tends to zero. Now take h to be real. This limit has to be equal to

$$\lim_{h \to 0}\left(\frac{u(x+h, y) - u(x, y)}{h}\right) = u_x$$

Similarly by taking $h = ik$ and allowing $k \to 0$ this limit has to be equal to v_y. Thus $u_x = v_y$. Now with $h = \Delta x + i\Delta y$,

$$\frac{f(z+h)-f(z)}{h} = \frac{\Delta x - i\Delta y}{(\Delta x)^2 + (\Delta y)^2}[(u_x + iv_x)\,\Delta x + (u_y + iv_y)\,\Delta y + R + iS].$$

Our hypothesis together with $u_x = v_y$ imply that (take real parts on both sides and apply limit).

$$u_x = \lim_{\substack{\Delta x \to 0 \\ \Delta y \to 0}}\left[u_x + \frac{\Delta x\,\Delta y}{(\Delta x)^2 + (\Delta y)^2}(u_y + v_x) + \frac{S\Delta y + R\Delta x}{(\Delta x)^2 + (\Delta y)^2}\right]$$

Therefore $\lim\limits_{\substack{\Delta x \to 0 \\ \Delta y \to 0}}\dfrac{\Delta x\,\Delta y}{(\Delta x)^2 + (\Delta y)^2}(u_y + v_x) = 0$. As $\dfrac{\Delta x\,\Delta y}{(\Delta x)^2 + (\Delta y)^2}$ does

not have a unique limit (along $\Delta y = m\,\Delta x$ this limit equals $\dfrac{m}{1 + m^2}$ giving

different values for different m's) the only possibility is that $u_y = -v_x$. But then taking imaginary parts on both sides of the relevant equality we get

$$\lim_{\Delta z \to 0}\left[\mathrm{Im}\,\frac{\Delta w}{\Delta z}\right] = \lim_{\substack{\Delta x \to 0 \\ \Delta y \to 0}}\left[v_x + \frac{S\Delta x - R\Delta y}{(\Delta x)^2 + (\Delta y)^2}\right] = v_x$$

The situation is similar when we assume $\lim\limits_{\Delta z \to 0}\left[\mathrm{Im}\,\dfrac{\Delta w}{\Delta z}\right]$ exists. However

when both these limits exist then clearly $\lim\limits_{\Delta z \to 0}\left[\dfrac{\Delta w}{\Delta z}\right]$ exists or that $f(z)$ has

a complex derivative at z which is equivalent to the complex differentiability of f at z. This completes the proof.

EXERCISES

1. Verify Cauchy-Riemann equations for the functions z^2, z^3.

2. If $\sum a_n z^n$ and $\sum b_n z^n$ have the radii of convergence R_1 and R_2 show that the radius of convergence of $\sum a_n b_n z^n$ is at least $R_1 R_2$.

3. If $\lim_{n \to \infty} \left| \dfrac{a_n}{a_{n+1}} \right| = R$ prove that $\sum a_n z^n$ has radius of convergence R.

4. For what values of z is $\sum_{0}^{\infty} \dfrac{z^n}{1 + z^{2n}}$ convergent?

5. Consider the solved problem number 6. Give a different method to find the order of $R'(z)$ (**Hint:** Assume $(P, Q) = 1$ and consider two cases: Case (i) $(Q, Q') = 1$; Case (ii) $(Q, Q') \neq 1$).

6. If $w(z) = \dfrac{az + b}{cz + d}$ is a rational function with $ad - bc = 0$ then prove that $w(z)$ is identically a constant.

7. Let a function $f(z)$ be analytic in a domain D. Prove that $f(z)$ must be constant in D if $\overline{f(z)}$ is analytic in D.

8. Show that $|\exp(z^2)| \leq \exp(|z^2|)$.

9. Prove that $|e^{-z}| < 1$ if and only if Re $z > 0$.

10. (a) Will u^2 be a harmonic function if u is harmonic.
    (b) Let u be a harmonic function. For what functions f, will function $f(u)$ also be harmonic?

11. Prove that the following linear fractional transformations $T_1 = z$,
$$T_2 = \frac{1}{z}, \; T_3 = 1 - z, \; T_4 = \frac{1}{1 - z}, \; T_5 = \frac{z - 1}{z}, \; T_6 = \frac{z}{z - 1} \text{ form a group}$$
(Group of harmonic ratios) with the composition of mappings as the group operation.

12. Let $k \in \mathbb{N}$ and Γ denote the boundary of a square in $\mathbb{C}$ which has the vertices $(\pm 1 \pm i) \pi k$. Show that $|\cos z| \geq 1$, $\forall z \in \Gamma$.

13. Assume that the power series $f(z) = \sum_{k=0}^{\infty} a_k z^k$ converges in a disk D around origin. Further assume that for all z in D such that $2z$ also is in D, f satisfies $f(2z) = (f(z))^2$. Show that if $f(0) \neq 0$ then $f(z) = e^{bz}$ with $b = f'(0) = a_1$.

14. Prove rigorously that any linear fractional transformation is bicontinuous from the extended complex plane onto itself where the topology of the extended plane is τ, the one-point compactification of $\mathbb{C}$ (see 2.3 for hints).

NOTES

A few historical remarks on the Cauchy-Riemann equations can be found in [4, page 49]. The theory of analytic functions and the theory of harmonic functions have a lot in common. We shall see later that several properties like integral formulas, maximum principle etc are shared by both analytic functions and harmonic functions. In some sense each theory enriches the other. The theory of linear fractional transformations enables us to obtain bijective bi-holomorphic maps (maps which together with their inverses are analytic or holomorphic) between specific regions in the complex plane. Moreover in some very interesting cases these are the only bijective biholomorphic maps between these regions. Further the theory of linear fractional transformations is related to the theory of general linear group $GL(2; \mathbb{C})$ which consists of non-singular 2×2-complex matrices with the group operation being matrix multiplication.

The formula for the radius of convergence of a power series was first given by Cauchy in 1821 (see [1]). It was rediscovered by J.S. Hadamard (see [2]).

There is an extensive literature dealing with the convergence behavior of a power series on the boundary of its circle of convergence. N. Lusin in 1911 constructed a power series $\sum c_n z^n$ with radius of convergence 1, satisfying $c_n \to 0$ and which diverges at every z with $|z| = 1$. W. Sierpinski in 1912 produced a power series with radius of convergence 1 which converges at 1 and nowhere else on the boundary. For more details see [3].

For algebraic properties of the collection of all power series centered at origin the reader is referred to [4, page 128]. For some interesting historical remarks on the logarithm function in the complex domain see [4, page 158-159].

References

1. A.L. Cauchy, Cours d'analyse de l'École Royale Polytechnique (Analyse algébrique). Paris, 1821. Reprinted by Wissenschaftliche Buchgesellschaft (1968).
2. J.S. Hadamard, Jour. Math. Pures et Appl. (4) 8, page 108.
3. E. Landau, Darstellung und Begründung einiger neuerer Ergebnisse der Funktionentheorie, Springer Verlag (1916), Berlin; 2nd ed. 1929; 3rd edition with supplements by D. Gaier, 1986.
4. R. Remmert. Theory of complex functions, Springer-Verlag, Graduate Texts in Mathematics, Vol. 122, 1991.

+ 3

Conformal Mappings

3.1 Definition and Properties of Conformal Mappings

Throughout this section we shall be dealing with geometric concepts involving curves, tangents, angles etc. All these ideas can be formulated mathematically and interpreted geometrically. For this purpose we shall define a few concepts and derive some auxiliary results.

Definitions 3.1.1

(a) A curve or an arc γ in the plane $\mathbb{C}$, is a continuous function $z(t)$ (also called a parametrization of γ) defined on a closed interval $[a, b] \subset \mathbb{R}$ with values in $\mathbb{C}$. The range $z([a, b])$ of the function $z(t)$ will be denoted by γ^* (also called the trace of the curve γ). Most often by abusing the notation we shall write γ for γ^*.

(b) A curve γ is said to be differentiable if there exists a parameterization $z(t)$ $(a \le t \le b)$ of γ such that $z'(t)$ is continuous on $[a, b]$ (i.e., $z(t) = x(t) + iy(t)$ with $x'(t)$ and $y'(t)$ continuous on $[a, b]$). If in addition $z'(t) \ne 0$ on $[a, b]$ we say γ is smooth or regular. For a differentiable or smooth curve we always use a parameterization with the required conditions. We shall say that γ is piece-wise differentiable (piece-wise smooth) if the parameterization $z(t)$ on $[a, b]$ is such that $z'(t)$ is continuous (continuous and non-zero) except for finitely many points on $[a, b]$.

(c) A curve γ is said to be a simple curve or a Jordan curve if it has a parameterization $z(t)$ $(a \le t \le b)$ such that $z(t)$ is one-to-one on $[a, b]$.

(d) A curve γ is said to be closed if its parameterization $z(t)$ $(a \le t \le b)$ satisfies $z(a) = z(b)$ (i.e. its end points coincide). A closed curve γ is said to be a simple closed curve or a closed Jordan curve if its parameterization $z(t)$ $(a \le t \le b)$ is one-to-one on (a, b). (Note that an exception is made here from the definition of a Jordan curve.)

Examples 3.1.2

(i) The unit circle in the complex plane is a smooth closed Jordan curve with the parameterization $z(t) = e^{2\pi it}$ $(0 \le t \le 1)$.

(ii) The ellipse $z(t) = a \cos t + ib \sin t$ $(0 \le t \le 2\pi)$ is also a smooth closed Jordan curve.

(iii) The parameterization $z(t) = t + i|t|$ $(-1 \le t \le 1)$ represents a curve which is not smooth (prove this yourself).

Theorem 3.1.3
If γ is a smooth curve with the parameterization $z = z(t)$ $(a \le t \le b)$ then arg $z'(t_0)$ represents the direction of the tangent to the curve γ at the point $z(t_0)$ $(a \le t_0 \le b)$.

Proof
Writing $z(t) = x(t) + iy(t)$ and $z'(t) = x'(t) + iy'(t)$ it is clear that

$$\tan (\arg z'(t_0)) = \frac{y'(t_0)}{x'(t_0)} = \left(\frac{dy}{dx}\right)_{(x(t_0), y(t_0))}.$$

By the geometric interpretation of the derivative $\left(\dfrac{dy}{dx}\right)$ when x and y are

given by parameteric representations we see that arg $z'(t_0)$ represents the angle made by the tangent to γ (at $z(t_0)$) with the positive direction of the x axis. Hence the result.

Theorem 3.1.4
Let $\Omega = \mathbb{C}\backslash(-\infty, 0]$. For each $z \in \Omega$, define its principal argument by Arg $z = \theta$ with $-\pi < \theta < \pi$. Then $z \to$ Arg z is continuous on Ω.

Proof

Let $z_0 \in \Omega$ and Arg $z_0 = \theta_0$ so that $-\pi < \theta_0 < \pi$. Let $\varepsilon > 0$ be given. Choose δ such that $0 < \delta < \varepsilon$, $-\pi < \theta_0 - \delta < \theta_0 + \delta < \pi$. Choose the infinite sector $S = \{z \in \mathbb{C}: |\mathrm{Arg}\ z - \theta_0| < \delta\} \subset \Omega$. If we choose $\delta' = |z_0| \sin \delta$ it is easy to see that

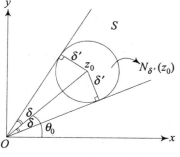

$$N_{\delta'}(z_0) = \{z \in \mathbb{C}: |z - z_0| < \delta'\} \subset S$$

Further for $z \in N_{\delta'}(z_0)$ we have $|\mathrm{Arg}\ z - \mathrm{Arg}\ z_0| < \delta < \varepsilon$ proving continuity of Arg z at $z = z_0$ (see Fig. 3.1(i)).

Fig 3.1 (i)

Theorem 3.1.5
Let $\Omega_1 = \mathbb{C}\backslash[0, \infty)$. For each $z \in \Omega_1$ define a new branch of arg z by ARG $z = \theta$ with $0 < \theta < 2\pi$. Then $z \to$ ARG z is continuous on Ω_1.

Proof
Similar to that of Theorem 3.1.4 and is left to the reader.

Corollary 3.1.6
Given any non-zero complex number z_0 we can define a continuous branch of arg z in a neighbourhood of z_0.

Proof

If $z_0 \notin (-\infty, 0]$ then z_0 belongs to the region Ω defined in Theorem 3.1.4. In this case we can take Arg z as the required branch around z_0. If however $z_0 \in (-\infty, 0)$ then z_0 belongs to the region Ω_1 defined in Theorem 3.1.5. In this case we can take ARG z as the required branch around z_0.

Definitions 3.1.7

Let $f(z)$ be a continuous mapping of a region $\Omega \subset \mathbb{C}$ into $\mathbb{C}$ with $z_0 \in \Omega$. Assume that there exists a $\delta > 0$ such that $N_\delta (z_0) = \{z \in \mathbb{C} / |z - z_0| < \delta\} \subset \Omega$ and that $f(z) \neq f(z_0)$ for $z_0 \neq z \in N_\delta (z_0)$.

(a) *Conformality of the first kind:* We say f is conformal of the first kind at z_0 if given any curve $z = z(t)$ in Ω with $z(t_0) = z_0$, $z(t) \neq z_0$ for $t \neq t_0$ and $w(t) = f(z(t))$ with $w(t_0) = f(z_0) = w_0$ we can define a single valued continuous branch for $\arg \left(\dfrac{w(t) - w_0}{z(t) - z_0} \right)$ as $t \to t_0$ (i.e., in a small neighbourhood of t_0) and that in this branch $\displaystyle\lim_{t \to t_0} \arg \left(\dfrac{w(t) - w_0}{z(t) - z_0} \right)$ exists and is independent of the curve $z(t)$ through z_0.

(b) *Conformality of the second kind:* We say f is conformal of the second kind at z_0 if given any curve $z = z(t)$ in Ω with $z(t_0) = z_0$, $z(t) \neq z_0$ for $t \neq t_0$ and $w(t) = f(z(t))$ with $w(t_0) = f(z_0) = w_0$, $\displaystyle\lim_{t \to t_0} \left| \dfrac{w(t) - w_0}{z(t) - z_0} \right|$ exists, not equal to zero and is independent of the curve $z(t)$ through z_0.

(c) *Conformality:* We say f is conformal at z_0 if f is conformal of both kinds at z_0.

Note 3.1.8

The geometric significance of the first and second kind of conformality can be explained as follows. If $z(t)$ and $f(z(t)) = w(t)$ are smooth curves then

$$\lim_{t \to t_0} \arg \left(\frac{w(t) - w_0}{z(t) - z_0} \right) = \arg \left(\frac{w'(t_0)}{z'(t_0)} \right)$$

measures the angular difference between the tangents to the curves $w(t)$ and $z(t)$ at w_0 and z_0 respectively and once this is independent of the curves $z(t)$ and its image $w(t)$ we see that the angle between any two curves at z_0 is equal in magnitude and sense to the angle between their images at w_0. Similarly if f is conformal of the second kind at z_0 then any chord of any curve $z(t)$ through z_0 lying in Ω is in the limit contracted or expanded by f in the same ratio. This limit can justifiably be called the magnification due to f at z_0 or the linear change of scale effected by f at z_0.

Example 3.1.9

(a) $w = e^z$ is conformal at all points in the plane.

(b) $w = \bar{z}$ is conformal of the second kind at all points but not conformal of the first kind at any point. In fact it preserves the magnitude of the angle between any two curves but reverses the sense. Such mappings are also called anti-conformal.

(c) $w = z|z|$ is conformal of the first kind at $z = 0$ but is not conformal of the second kind at the same point.

(d) $w = z^2$ is conformal of neither kind at $z = 0$.

The above observations are easy to verify and we prefer to omit the details.

Theorem 3.1.10

Let $\Omega \subset \mathbb{C}$ be a region and $f : \Omega \to \mathbb{C}$ be a continuous function. Assume $z_0 \in \Omega$.

(i) If $f'(z_0)$ exists and is non-zero then f is conformal at z_0.

(ii) If f is conformal (of both kinds) at z_0 then $f'(z_0)$ exists and is not equal to 0.

(iii) If f is real differentiable at z_0 and the differential is different from 0 at z_0 and if f is conformal of the first kind then $f'(z_0)$ exists and is different from zero.

(iv) If f has continuous first order partial derivatives at z_0 and is conformal of the first kind then $f'(z_0)$ exists.

(v) If f has continuous first order partial derivatives at z_0 and f is conformal of the second kind at z_0 then either f or $\bar{f}$ is analytic at z_0.

Proof

(i) Since $f'(z_0)$ exists and is not equal to zero, we first claim that there exists a deleted neighbourhood of z_0 in Ω such that $f(z) \neq f(z_0) = w_0$ there. If not we get a sequence of points $z_n \to z_0$, $z_n \neq z_0$ such $f(z_n) = f(z_0)$. For these z_n

$$\frac{f(z_n) - f(z_0)}{z_n - z_0} = 0$$

and so $\qquad f'(z_0) = \lim_{n \to \infty} \frac{f(z_n) - f(z_0)}{z_n - z_0} = 0$

which is a contradiction. Let $z(t)$ be any curve through z_0 with $z(t_0) = z_0$ and $w(t) = f(z(t))$ with $w(t_0) = w_0$. Then we have to show that for a suitable continuous branch of argument function

$$\lim_{t \to t_0} \arg \left(\frac{w(t) - w_0}{z(t) - z_0} \right)$$

is independent of $z(t)$. This can be proved as follows:

$$\left(\frac{w(t) - w_0}{z(t) - z_0} \right) \to f'(z_0) \neq 0$$

as $t \to t_0$. So by Theorem 3.1.6 a continuous branch for $\arg \left(\dfrac{w(t) - w_0}{z(t) - z_0} \right)$

can be defined as $t \to t_0$ and further in this branch

$$\lim_{t \to t_0} \arg\left(\frac{w(t) - w_0}{z(t) - z_0}\right) = \arg f'(z_0)$$

is independent of the curve $z(t)$ through z_0. Hence f is conformal of the first kind. Similarly

$$\lim_{t \to t_0}\left|\frac{w(t) - w_0}{z(t) - z_0}\right| = |f'(z_0)| \neq 0$$

which is again independent of $z(t)$. Hence f is conformal of the second kind also. Thus f is conformal at z_0.

(ii) If f is conformal of both kinds and $f(z_0) = w_0$ we shall first prove that as $z \to z_0$ through any curve $z = z(t)$ with $z(t_0) = z_0$

$$\lim_{t \to t_0} \frac{f(z(t)) - w_0}{z(t) - z_0}$$

exists and is independent of $z = z(t)$. Indeed

$$\lim_{t \to t_0} \frac{f(z(t)) - w_0}{z(t) - z_0} = \lim_{t \to t_0}\left\{\left|\frac{f(z(t)) - w_0}{z(t) - z_0}\right|\exp\left(i\arg\left(\frac{f(z(t)) - w_0}{z(t) - z_0}\right)\right)\right\}$$

$$= \rho e^{i\alpha}$$

where $\quad \rho = \lim_{t \to t_0}\left|\frac{f(z(t)) - w_0}{z(t) - z_0}\right| \neq 0 \quad$ and $\quad \alpha = \lim_{t \to t_0} \arg\left(\frac{w(t) - w_0}{z(t) - z_0}\right).$

Here ρ and α are independent of $z(t)$ by conformality of f at z_0.
We now show that $f'(z_0)$ exists and is non zero. For this, it suffices to show that given any sequence $z_n \to z_0$, $z_n \in \Omega$

$$\lim_{n \to \infty} \frac{f(z_n) - f(z_0)}{z_n - z_0} = \rho e^{i\alpha}.$$

Without loss of generality we may assume that all z_n's belong to an open disc around z_0, which is completely contained in Ω.

We first construct a curve $\gamma = \gamma(t)$, $0 \leq t \leq 1$ such that $\gamma(t)$ passes through all the points z_n and $\gamma(1) = z_0$. Consider the interval $[0, 1]$. Put $\gamma(0) = z_1$ and $\gamma(1) = z_0$. Now

$$[0, 1) = \bigcup_{m=1}^{\infty}\left[1 - \frac{1}{m}, 1 - \frac{1}{m+1}\right]. \tag{1}$$

Indeed $\left[1 - \dfrac{1}{m}, 1 - \dfrac{1}{m+1}\right] \subseteq [0, 1)$ for every m, one inclusion is obvious.

On the other hand, for $0 < t < 1$, there exists a unique integer $m \geq 1$ such

that $m \leq \dfrac{1}{1-t} < m+1 \Rightarrow \dfrac{1}{m+1} < 1 - t \leq \dfrac{1}{m} \Rightarrow 1 - \dfrac{1}{m} \leq t \leq 1 - \dfrac{1}{m+1}.$

Thus t belongs to the union on the right hand side. This proves (1). (Note that $0 \in [0, 1/2)$, the interval for $m = 1$.)

Define for $t \in \left[1 - \dfrac{1}{m}, 1 - \dfrac{1}{m+1}\right]$

$$\gamma(t) = \left[\left(t - 1 + \frac{1}{m}\right) z_{m+1} + \left(1 - \frac{1}{m+1} - t\right) z_m \right] m(m+1)$$

Observe that $\gamma(t)$ divides the line segment $[z_m, z_{m+1}]$ in the same ratio as t divides the interval $\left[1 - \dfrac{1}{m}, 1 - \dfrac{1}{m+1}\right]$. This observation enables us to conclude that $\gamma(t)$ is continuous at the interior points of the subintervals and the values coincide at the intersection of these intervals. Thus $\gamma(t)$ is continuous in $[0, 1)$. Further $\gamma(t_m) = z_m$ where $t_m = 1 - \dfrac{1}{m}$ $(m = 1, 2, \ldots)$.

(By adding more points to the sequence z_m, if necessary, we can even assume that $\gamma(t) \neq z_0$ for $t \neq 1$.) We now prove that $\gamma(t)$ is continuous at 1. Let $s_k \to 1$ as $k \to \infty$. We have to prove that given $\varepsilon > 0$ there exists a stage M such that for $k \geq M$, $\gamma(s_k) \in B_\varepsilon(z_0) = \{z : |z - z_0| < \varepsilon\}$. First choose a stage N such that $n \geq N$ implies $|z_n - z_0| < \varepsilon$. Now since $s_k \to 1$, there exists a stage M such that $s_k \in [1 - (1/N), 1)$ $(k \geq M)$. But

$$\left[1 - \frac{1}{N}, 1\right) = \bigcup_{m \geq N}^{\infty} \left[1 - \frac{1}{m}, 1 - \frac{1}{m+1}\right]$$

(The proof is similar to the previous analogous result.) Hence, s_k belongs to $\left[1 - \dfrac{1}{m}, 1 - \dfrac{1}{m+1}\right]$ for some $m \geq N$. So $\gamma(s_k)$ lies in the line segment joining z_m and z_{m+1} for this m. Since for these $m \geq N$, $z_m \in B_\varepsilon(z_0)$ and $z_{m+1} \in B_\varepsilon(z_0)$, $\gamma(s_k) \in B_\varepsilon(z_0)$ as $B_\varepsilon(z_0)$ is convex. This is true for every $k \geq M$ and hence γ is continuous at $t = 1$. We know that $\lim\limits_{t \to t_0} \dfrac{f(z(t)) - w_0}{z(t) - z_0} = \rho e^{i\alpha}$

exists $(z(t_0) = z_0)$ and is independent of $z(t)$. Thus $\lim\limits_{t \to 1} \dfrac{f(\gamma(t)) - w_0}{\gamma(t) - z_0} = \rho e^{i\alpha}$.

Take $t_n = 1 - \dfrac{1}{n}$. Then $\lim\limits_{t_n \to 1} \dfrac{f(\gamma(t_n)) - w_0}{\gamma(t_n) - z_0} = \rho e^{i\alpha}$, that is

$\lim\limits_{z_n \to z_0} \dfrac{f(z_n) - w_0}{z_n - z_0} = \rho e^{i\alpha}$ $(\rho \neq 0)$. Thus for every sequence $z_n \to z_0$,

$\lim\limits_{z_n \to z_0} \dfrac{f(z_n) - w_0}{z_n - z_0} = \rho e^{i\alpha}$ $(\rho \neq 0)$. This implies that $f'(z_0) = \rho e^{i\alpha}$ exists and is non-zero.

(iii) Since f is real differentiable at z_0 we can write

$$f(x_0 + h, \ y_0 + k) - f(x_0, \ y_0) = \left[A \binom{h}{k} \right]^T + R(h, \ k)$$

where $\dfrac{R(h, k)}{|h + ik|} \to 0$ as $(h, \ k) \to 0$. Here $A = \begin{pmatrix} a & b \\ c & d \end{pmatrix}$ and $a + ic = (f_x)_{z_0} = \alpha$ and $b + id = (f_y)_{z_0} = \beta$ with at least one of a, b, c, d being non-zero. In the complex notation we rewrite this as

$$f(z_0 + h + ik) - f(z_0) = \alpha h + \beta k + R(h, \ k). \tag{2}$$

Since f is conformal of the first kind

$$\lim_{t \to 0} \arg \left(\frac{f(z_0 + t e^{i\theta}) - f(z_0)}{t e^{i\theta}} \right)$$

is independent of θ. Here we are taking $z = z(t) = z_0 + t e^{i\theta}$ for each θ real as a curve and applying the definition of conformality of the first kind for this $z(t)$. We use (2) to conclude that $\lim\limits_{t \to 0} \arg \left(\dfrac{\alpha t \cos \theta + \beta t \sin \theta + R(t)}{t e^{i\theta}} \right)$ is

independent of θ. i.e., $\arg (e^{-i\theta} [\alpha \cos \theta + \beta \sin \theta])$ is independent of θ. This implies that if we put $\cos \theta = (e^{i\theta} + e^{-i\theta})/2$ and $\sin \theta = (e^{i\theta} - e^{-i\theta})/$

$2i$, then $\arg \left(\dfrac{\alpha - i\beta}{2} + \dfrac{\alpha + i\beta}{2} e^{-2i\theta} \right)$ is independent of θ. But this complex

number $\dfrac{\alpha - i\beta}{2} + \dfrac{\alpha + i\beta}{2} e^{-2i\theta}$ (whose locus as θ varies is evidently a

circle) will have different arguments for several different θ's unless its locus which is a circle has radius 0. Hence $\alpha + i\beta = 0$ and hence $\beta = i\alpha$. But this is precisely the Cauchy-Riemann equation for f at z_0. Together with the real differentiability of f, this means that f is analytic at z_0 by Theorem 1.9.12. Further $f'(z_0) = \alpha \neq 0$ (in fact $\alpha = 0 \Rightarrow \beta = 0$ and so $A = df|_{z_0}$ $= 0$, a contradiction).

We will combine the proof of (iv) and (v). Let f be continuous with its first order partial derivatives. Put $w(t) = f(z(t))$ for any differentiable curve $z(t)$ through z_0 with $z(t_0) = z_0$. Therefore w is a differentiable function of t. Further in a neighbourhood of $t = t_0$

$$w'(t) = \left(\frac{\partial f}{\partial x} \right) x'(t) + \left(\frac{\partial f}{\partial y} \right) y'(t)$$

(by several variable calculus). Put $z'(t_0) = x'(t_0) + iy'(t_0)$.
Therefore

$$w'(t_0) = \left(\frac{\partial f}{\partial x} \right)_{z_0} \frac{z'(t_0) + \overline{z'(t_0)}}{2} + \left(\frac{\partial f}{\partial y} \right)_{z_0} \frac{z'(t_0) - \overline{z'(t_0)}}{2i}$$

Thus $\quad \dfrac{w'(t_0)}{z'(t_0)} = \left(\dfrac{f_x - i f_y}{2} \right)_{z_0} + \left(\dfrac{f_x + i f_y}{2} \right)_{z_0} \dfrac{\overline{z'(t_0)}}{z'(t_0)}.$

As $z(t)$ varies over all differentiable curves through z_0 with $z'(t_0) \neq 0$, the right hand side of the above equality describes a circle with centre

$$\left(\frac{f_x - if_y}{2}\right)_{z_0} \quad \text{and radius} \quad \left|\left(\frac{f_x + if_y}{2}\right)_{z_0}\right|.$$

(iv) In this we assume f is conformal of the first kind, therefore arg $\dfrac{w'(t_0)}{z'(t_0)}$ is independent of $z(t)$. $\left(\text{since } \lim_{t \to t_0} \arg\left(\dfrac{w(t) - w_0}{z(t) - z_0}\right) = \arg \dfrac{w'(t_0)}{z'(t_0)}\right).$
But on a circle argument is constant if and only if the circle is a point circle (i.e., a circle of radius zero). Thus $f_x = -if_y$ at z_0 which is precisely the Cauchy-Riemann equation. From this and the continuity of the partial derivatives of f at z_0, the analyticity of f at z_0 follows. This proves (iv).

(v) On the other hand if f is conformal of the second kind then $\left|\dfrac{w'(t_0)}{z'(t_0)}\right|$

is independent of $z(t)$. But on a circle modulus is constant if and only if either the centre is origin or the radius is zero i.e. either $f_x = -if_y$ or $f_x = if_y$ at z_0. In the first case f is analytic at z_0 as before. In the second case $\bar{f}$ is analytic at z_0. Hence (v) is proved.

Note 3.1.11
As an illustration of the above considerations we shall first observe the following: The function f rotates the direction of each smooth curve $z = z(t)$

at z_0 by the quantity given by $\displaystyle\lim_{t \to t_0} \arg\left(\frac{f(z(t)) - w_0}{z(t) - z_0}\right)$ and the linear change

of scale effected by f at z_0 is given by

$$\lim_{t \to t_0}\left|\frac{f(z(t)) - f(z_0)}{z(t) - z_0}\right|$$

These quantities can vary in general from point to point for a function f which is analytic and whose derivative is not equal to zero in a region Ω. But these values will be approximately arg $f'(z_0)$ and $|f'(z_0)|$ at points z near z_0. (Implicitly, we are assuming here the continuity of $f'(z)$ which is true as we shall see later). Hence the image of a small region in a neighbourhood of z_0, conforms to the original region in the sense that it has almost the same shape (i.e., in a very small neighbourhood of z_0 as a point P moves along a curve C enclosing z_0 in the z plane the image P' moves along a curve C' enclosing $f(z_0)$ in such a way that as and when the arg of the vector $\overline{z_0 P}$ decreases or increases the arg of the vector $\overline{f(z_0)P'}$ also decreases or increases respectively and moreover the modulus of $\overline{z_0 P}$ bears the same constant ratio to $\left|\overline{f(z_0)P'}\right|$. Thus the shape of the curve C'

looks very much the same as the shape of the curve C in a neighbourhood of z_0). This is the reason why we call these mappings as conformal. However it should be cautioned that larger regions may be transformed into regions that need not bear any resemblance to the original ones.

Example 3.1.12
As another illustration let us take $f(z) = z^2$ which is conformal at $z = 1 + i$. Consider C_1 to be the curve $y = x$, $y \geq 0$ and C_2 to be the curve $x = 1$, $y \geq 0$ (intersecting at $1 + i$). The angle measured from C_1 to C_2 is $\pi/4$ at $z = 1 + i$. Under this transformation C_1 is transformed into the curve C_1' with parametric equations $u = 0$, $v = 2y^2$, $0 \leq y < \infty$. Similarly C_2 is transformed into the curve C_2' with parametric equations $u = 1 - y^2$, $v = 2y$, $0 \leq y \leq \infty$. Note that these observations are consequences of $u + iv = w = f(z) = (x^2 - y^2) + i\, 2xy$. Thus C_1' is the upper half of the v axis and C_2' is the upper half of the parabola $v^2 = -4(u - 1)$. Hence $\dfrac{dv}{du} = -\dfrac{2}{v}$ for C_2'. In particular $\dfrac{dv}{du} = -1$ when $v = 2$. Therefore the angle from C_1' to C_2' at $w = f(1 + i) = 2i$ is $\pi/4$ as required by the conformality. The linear change of scale at this point is $|f'(1 + i)| = |2(1 + i)| = 2\sqrt{2}$.

In the above example, the angle of rotation and the linear change of scale at $(1 + i)$ are $\pi/4$ and $2\sqrt{2}$ and those at $z = 1$ are 0 and 2. Again if C_2 is same as above and C_3 is the non-negative x axis, then C_3' (the image of C_3) is the non-negative u axis in the w plane. Again the parabola mentioned above has the line $u = 1$ as its tangent at $z = 1$. Hence, angle between C_3 and C_2 at $z = 1$ which is equal to $\pi/2$ is also equal to the angle between C_3' and C_2' at $w = 1$. The above geometric properties of the mapping $f(z) = z^2$ can be illustrated by Fig. 3.1(ii).

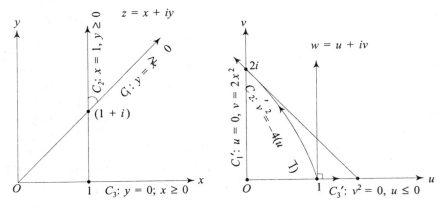

Fig 3.1 (ii)

3.2 Elementary Conformal Mappings

In this section we shall describe the mapping properties of some elementary conformal mappings.

Example 3.2.1

$$f(z) = z^n.$$

The function $f(z) = z^n$ is conformal at all $z \neq 0$, since the derivaties $f'(z)$ = $nz^{n-1} \neq 0$. Now the image of the angular sector $0 < \arg z < \frac{2}{n}\pi$ is the whole plane except for the slit $0 \leq x < \infty$.

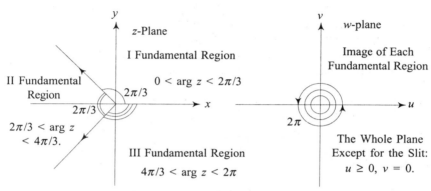

Fig 3.2(i) $w = z^3$

Any region which is mapped one-to-one onto the whole plane except for slits in the w-plane is called a **fundamental region.** Thus for $w = z^n$ there are n fundamental regions $(k - 1)\frac{2\pi}{n} < \arg z < k\frac{2\pi}{n}$, $(k = 1, 2, \ldots, n)$. The function $z = w^{1/n}$ can be uniquely defined as a single-valued function from $\mathbb{C}\backslash[0, \infty)$ to any one of the fundamental regions by choosing the appropriate values for $\arg w$ and $\arg z$, for example $(k - 1)\,2\pi < \arg w < k2\pi$ $(k = 1, 2, \ldots n)$ corresponds to $(k - 1)\frac{2\pi}{n} < \arg z < k\frac{2\pi}{n}$ $(k = 1, 2, \ldots, n)$. The fact that circles with centres at origin and straight lines through origin in the z-plane are getting mapped onto circles centered at origin and straight lines through origin in the w-plane reflects the simplicity of this map and also testifies to the conformality (by mapping orthogonal curves onto orthogonal curves) of this mapping.

If $z = z_1(t)$ and $z = z_2(t)$ are the two smooth curves through origin with $z_1(t_0) = 0 = z_2(t_0)$, $w = z^n$ maps $z_1(t)$ and $z_2(t)$ onto $w_1(t) = z_1^n(t)$, $w_1(t_0) = 0$ and $w_2(t) = z_2^n(t)$, $w_2(t_0) = 0$ in such a way that

$$\lim_{t \to t_0} \arg \frac{w_1(t)}{w_2(t)} = n \lim_{t \to t_0} \arg \frac{z_1(t)}{z_2(t)}.$$

Thus the angles at 0 between curves in the z-plane is magnified n times in the w-plane by $f(z)$ at $f(0)$. In fact this is a characteristic behaviour of analytic functions $f(z)$ at z_0 where $f^{(j)}(z_0) = 0$ for $1 \leq j \leq n - 1$ and $f^{(n)}(z_0) \neq 0$. This we shall see later.

Example 3.2.2

$$w = e^z$$

The function $w = e^z$ is conformal at all points in the finite complex plane as $f'(z) = e^z \neq 0$ everywhere. If $z = x + iy$, $w = Re^{i\phi}$ with $R = e^x$ and $\phi = y$. Evidently the level curves $x = a$ and $y = b$ are mapped onto circles $|w| = e^a$ and rays $\arg w = b$, again preserving the orthogonality of curves. On the other hand any other line (not parallel to the co-ordinate axes) is mapped onto a logarithmic spiral of the form $|w| = \rho_0 e^{k \arg w}$ (ρ_0 positive and k real) (verify). Here the fundamental regions are given by $(k - 1) 2\pi < \operatorname{Im} z < k 2\pi$ ($k = 0, \pm 1, \pm 2, \ldots$). Each of these regions is mapped onto the whole w-plane except for $[0, \infty)$. The inverse function $z = g(w)$ gives a one-to-one correspondence between $\mathbb{C} \setminus [0, \infty)$ and $(k - 1) 2\pi < \operatorname{Im} z < k2\pi$ by the map which sends $w = |w|e^{i \arg w}$ with $(k - 1) 2\pi < \arg w < k2\pi$ to $z = \log |w| + i \arg w$. Also $w = e^z$ maps a horizontal strip $y_1 < y < y_2$ with $y_2 - y_1 \leq 2\pi$ onto an angular sector $y_1 < \arg w < y_2$ and if $y_2 - y_1 = \pi$ this sector reduces to a half plane.

Example 3.2.3

$$w = f(z) = \frac{az + b}{cz + d} \text{ with } ad - bc \neq 0.$$

These are all one-to-one mappings of the extended complex plane onto itself. Since $f(z)$ is analytic except at $z = -d/c$ and $f'(z) \neq 0$ for $z \neq -d/c$, it is conformal at all points $z \neq -d/c$. As we saw already f maps the family of all circles and straight lines onto itself. In general to investigate the image of any domain in the plane usually it is necessary and sufficient to determine the image of its boundary.

In order to study the mapping properties of these functions we shall observe the following facts.

Each linear fractional transformation has either two distinct finite fixed points or two coincident finite fixed points or one finite fixed point and one fixed point at ∞ or has both fixed points coincident at ∞. By a fixed point p for f we mean a point p for which $f(p) = p$.

The last of these happens if and only if $(a - d)^2 + 4bc = 0$ and $c = 0$ $\left(\text{where } w = \frac{az + b}{cz + d}\right)$ (i.e) $a = d$ and $c = 0$ in which case $w = f(z) = z + (b/d)$ which is a mere translation. Since the properties of translations are well known, we omit this case.

Case 1 (w has two distinct finite fixed points). To start with we can discuss more generally the case when a, b (two distinct finite points) are mapped onto two distinct finite points say a', b' respectively and specialize by putting $a' = a$ and $b' = b$ to study our case.

A linear fractional transformation w_1 sends a, b to 0, ∞ if and only if

$$w_1(z) = k_1\left(\frac{z-a}{z-b}\right) \quad (k_1 \neq 0).$$ Similarly w_2 sends a', b' to 0, ∞ if and only

if $w_2(z) = k_2\left(\frac{z-a'}{z-b'}\right)\left(w_2^{-1}(z) = \frac{b'z - k_2 a'}{z - k_2}\right)$ $(k_2 \neq 0)$. Thus our required

transformation $w = w_2^{-1}w_1$ is given by

$$w(z) = \frac{b'\left(k_1\left(\dfrac{z-a}{z-b}\right)\right) - a'k_2}{\left(k_1\left(\dfrac{z-a}{z-b}\right)\right) - k_2}$$

so that $\dfrac{z-a}{z-b} = \dfrac{wk_2 - a'k_2}{k_1 w - k_1 b'}$. Hence $\dfrac{w-a'}{w-b'} = \dfrac{k_1}{k_2}\left(\dfrac{z-a}{z-b}\right) = k\left(\dfrac{z-a}{z-b}\right)$.

Now we denote the system of circles through a and b as C_1 and the Apollonius circles with limit points a and b as C_2. (These circles are locus

of points z such that $\left|\dfrac{z-a}{z-b}\right| = \rho$ as ρ varies. It can be shown that this locus

is indeed a circle whose equation is given by

$$z\bar{z}(1-\rho^2) + |a^2| - \rho^2|b^2| - 2\,\mathrm{Re}\,z\bar{a} + 2\rho^2\,\mathrm{Re}\,z\bar{b} = 0.$$

Note that $\rho = 1$ corresponds to a unique line in C_2. There is also a unique line in C_1.)

Now if circles through a', b' are denoted by C_1', Apollonius circles with

limit points a', b' are denoted by C_2', then w transforms each $c_1 \in C_1$ into

a $c_1' \in C_1'$ and each $c_2 \in C_2$ into a $c_2' \in C_2'$. $\Big($The first is obvious as any linear fractional transformation carries circles to circles. The second holds

because $\left|\dfrac{z-a}{z-b}\right| = \rho$ implies $\left|\dfrac{w-a'}{w-b'}\right| = |k|\rho.\Big)$ With different points a, b and

their images a', b' we can have a better understanding of the mapping properties of w. If $a' = a$ and $b' = b$ (i.e. when w has two distinct finite fixed points) $C_1 = C_1'$ and $C_2 = C_2'$. Thus such a linear fractional transformation carries the family C_1 onto C_1 and C_2 onto C_2. Moreover in the case

when $k > 0$, $\arg\left(\dfrac{z-a}{z-b}\right) = \arg\left(\dfrac{w-a}{w-b}\right)$ which implies that z, w, a, b are concyclic and so each member of C_1 is carried onto itself. Also if $|k| = 1$ each member of C_2 is carried onto itself. Note that linear fractional transformations of the form $\dfrac{w-a}{w-b} = k\dfrac{z-a}{z-b}$ with 'a' and 'b' finite and $|k| = 1$ or of the form $\dfrac{1}{w-a} = \dfrac{k}{z-a}$ with 'a' finite and $|k| = 1$ are referred to as elliptic transformations. Under the same situation if k is real and positive the transformations are called hyperbolic.

Case 2 (w has one finite fixed point). Let the fixed point be 'a'. We shall more generally consider maps that take 'a' to a'. Any linear fractional transformation w_1 that takes 'a' to ∞ is given by

$$w_1(z) = k_1/(z-a) + c_1$$

and the linear fractional transformation w_2 which sends a' to ∞ is given by

$$w_2(z) = \frac{k_2}{z-a'} + c_2 \quad \left[w_2^{-1}(z) = \frac{a'(z-c_2)+k_2}{z-c_2} \right].$$

Hence our required mapping $w = w_2^{-1}w_1$ is given by

$$w(z) = \frac{a'\left(\dfrac{k_1}{z-a} + c_1 - c_2\right) + k_2}{\left(\dfrac{k_1}{z-a} + c_1\right) - c_2}.$$

Thus

$$\left(\frac{k_1}{z-a} + c_1 - c_2\right)(w - a') = k_2 \quad \text{or} \quad \frac{k_2}{w-a'} = \frac{k_1}{z-a} + c.$$

That is, each such map is given by $\dfrac{k'}{w-a'} = \dfrac{k}{z-a} + c$ where k, k' and c are constants. If we now let the system of circles through 'a' which are tangent to each other at 'a' with the common tangent having the direction given by arg k as C_1 and the system of circles orthogonal to C_1 at 'a', as C_2 and similarly if we let the corresponding system of circles through a' as C_1' and its orthogonal system at a' as C_2' then w maps C_1 onto C_1' and C_2 onto C_2'. (If S is a circle through 'a' and $\lim\limits_{\substack{z\to a \\ z\in S}} \arg(z-a) = \arg k$ and S' is its image then

$$0 = \lim_{\substack{z\to a \\ z\in S}} \arg\left(\frac{k}{z-a}\right) = \lim_{\substack{w\to a' \\ w\in S'}} \arg\left(\frac{k'}{w-a'} - c\right)$$

$$= \lim_{\substack{w\to a' \\ w\in S'}} \arg\left(\frac{k'-c(w-a')}{w-a'}\right) = \lim_{\substack{w\to a' \\ w\in S'}} \arg\left(\frac{k'}{w-a'}\right)$$

proving that w maps a member of C_1 onto a member of C_1'. By conformality a member of C_2 gets mapped onto a member of C_2'.)

For different 'a' and their corresponding images a' the mapping properties of w can be visualized. In particular if $a' = a$ (i.e w has a finite fixed point at a) then $C_1 = C_1'$ and $C_2 = C_2'$ and so each circle in any one family is mapped into the same family. The corresponding mapping functions are given by $\dfrac{k'}{w-a} = \dfrac{k}{z-a} + c$.

Subcase (i) (w has coincident fixed points at a finite point "a"). This is obtained from the previous one with the additional information that $k' = k$.

$\left(\text{If } \dfrac{k'}{w-a} = \dfrac{k}{z-a} + c \text{ has coincident finite fixed points at } z = a \text{ and if}\right.$

$k' \neq k$ we have another fixed point at $z = a + \dfrac{k'-k}{c}$ which is a contra-

diction. Note that $c \neq 0$ as otherwise ∞ is a fixed point.$\left.\right)$ The corresponding

mapping functions are given by $\dfrac{k}{w-a} = \dfrac{k}{z-a} + c$. These linear fractional

transformations are also called parabolic.

Subcase (ii) (w has one finite fixed point at "a" and one fixed point at ∞). As before this is also a specialization of case 2 with the additional information that $c = 0$ (∞ is a fixed point implies $c = 0$). The corresponding

functions are given by $\dfrac{k'}{w-a} = \dfrac{k}{z-a}$.

In all these considerations, orthogonal families of "circles" in the z-plane under the map $w = f(z)$ correspond to orthogonal families of "circles" in the w-plane, is the result of conformality of these mappings at all points.

Theorem 3.2.4

A linear fractional transformation $w(z)$ carries $\mathbb{R} \cup \{\infty\}$ onto itself if and only if it can be written with real coefficients. (i.e., w can be written as

$\dfrac{az+b}{cz+d}$ with a, b, c, d real).

Proof

If w maps ∞ to ∞ then w should be of the form $\alpha z + \beta$ where α, β are complex. But then 0 and 1 are mapped onto real numbers implies α, β are real and the proof is over in this case. Let ∞ be mapped to a finite complex number. Since w maps 0 to α (real) and 1 to β (real) it can be written in

the form $\dfrac{w-\alpha}{w-\beta} = k\dfrac{z}{z-1}$ where k is complex for the present. Now as

$z \to \infty$ along the real axis w tends to the image of ∞ (which is not ∞ by our assumption) which should be real say γ. This $\gamma \neq \alpha$ or β as w is one-to-one. Thus

$$\frac{\gamma - \alpha}{\gamma - \beta} = k = \lim_{z \to \infty} \frac{kz}{z-1} \quad (z \in \mathbb{R})$$

Hence k is also real and so $\dfrac{w - \alpha}{w - \beta} = k\dfrac{z}{z-1}$ where α, β, k are real, i.e.

$$w = \frac{(\alpha - \beta k)z - \alpha}{(1-k)z - 1}$$ or that w can be written with real coefficients.

The converse is obvious.

Example 3.2.5

$$w = \frac{az + b}{cz + d}$$ with a, b, c, d real $ad - bc > 0$.

w maps $\mathbb{R} \cup \{\infty\}$ onto itself by the previous theorem. By continuity of w in the extended plane it is clear that each of the two components of the complement of $\mathbb{R}$ in the z plane should be mapped onto each of the two components of the complement of $\mathbb{R}$ in the w plane, i.e., $\text{Im } z > 0$ goes into (infact onto) either $\text{Im } w > 0$ or $\text{Im } w < 0$. But the particular point i in $\text{Im } z > 0$ is mapped to the point $w = \dfrac{ai + b}{ci + d} = \dfrac{(ai + b)(d - ic)}{c^2 + d^2}$ so that $\text{Im } w$

$$= \frac{(ad - bc)}{c^2 + d^2} > 0.$$ Hence $\{z: \text{Im } z > 0\}$ corresponds to $\{w: \text{Im } w > 0\}$ and $\{z: \text{Im } z < 0\}$ corresponds to $\{w: \text{Im } w < 0\}$. Hence any subdomain of $\text{Im } z > 0$ should correspond to some subdomain of $\text{Im } w > 0$ only and vice versa. If

$$w_1 = \frac{a_1 z + b_1}{c_1 z + d_1} \quad \text{and} \quad w_2 = \frac{a_2 z + b_2}{c_2 z + d_2}$$

then $$w_1 \circ w_2 = \frac{(a_1 a_2 + b_1 c_2)z + a_1 b_2 + b_1 d_2}{(c_1 a_2 + d_1 c_2)z + c_1 b_2 + d_1 d_2} = \frac{Az + B}{Cz + D}$$

with $AD - BC = (a_1 d_1 - c_1 b_1)(a_2 d_2 - c_2 b_2) > 0$. Further the inverse $z = \dfrac{dw - b}{-cw + a}$ is again a transformation of the above type. As $w = z = \dfrac{1 \cdot z + 0}{0 \cdot z + 1}$ is also of this type, the set of all linear fractional transformations

$$w = \frac{az + b}{cz + d}$$ with a, b, c, d real and $ad - bc > 0$

forms a subgroup within the group of all linear fractional transformations with the composition as the group operation.

Example 3.2.6

$$w = e^{i\alpha} \frac{(z - z_0)}{(z - \bar{z}_0)} \quad \text{with } z_0 = x_0 + iy_0, \text{ not real.}$$

Evidently the real axis goes onto $|w| = 1$. Thus the components of the complement of $\mathbb{R}$ in the z plane are mapped onto the components of the complement of $|w| = 1$ in the w plane.

$$\text{Since at } z = i, \; |w|^2 = \left| \frac{(i - z_0)}{(i - \bar{z}_0)} \right|^2 = \frac{x_0^2 + (y_0 - 1)^2}{x_0^2 + (y_0 + 1)^2} \quad \text{which is less than 1}$$

if $y_0 > 0$ (and greater than 1 if $y_0 < 0$), w maps $\{z: \text{Im } z > 0\}$ onto $|w| < 1$ if $\text{Im } z_0 > 0$ and onto $|w| > 1$ if $\text{Im } z_0 < 0$. Similarly w maps $\{z: \text{Im } z < 0\}$ onto $|w| > 1$ if $\text{Im } z_0 > 0$ and onto $|w| < 1$ if $\text{Im } z_0 < 0$. When $\text{Im } z_0 > 0$ by conformality parallel lines (which meet at ∞) on the upper half plane are mapped onto circles touching internally each other at $e^{i\alpha}$ (the image of ∞ under w) with the tangent to the unit circle at $e^{i\alpha}$ being the common tangent. All of these circles lie within the open unit disc. Similarly parallel lines in the lower half plane (not through $\bar{z}_0$) are mapped onto circles touching internally each other at $e^{i\alpha}$, with the tangent to the unit circle at $e^{i\alpha}$ being the common tangent. All of these circles lie outside the open unit disc.

Example 3.2.7

$$w = \Phi_\alpha (z) = e^{i\psi} \frac{z - \alpha}{1 - \bar{\alpha}z} \quad |\alpha| < 1, \; \psi \text{ real}$$

These mapping are analytic in the open unit disc U of the complex plane since the only pole $z = 1/\bar{\alpha}$ lies outside U. Further as $z = e^{i\theta}$ varies on the unit circle T

$$|w| = |\Phi_\alpha(z)| = \left| \frac{e^{i\theta} - \alpha}{e^{-i\theta} - \bar{\alpha}} \right| = 1.$$

Hence, w maps T into T and since $w^{-1} = \Phi_\alpha^{-1} = e^{-i\psi} \Phi_{-\alpha e^{i\psi}} (-\alpha \in U)$, w^{-1} maps T into T and so $w(T) = T$. For similar reasons w maps U into U

$$\left(|w|^2 = \left| \frac{z - \alpha}{1 - \bar{\alpha}z} \right|^2 = \frac{|z|^2 + |\alpha|^2 - 2\,\text{Re}\,\bar{\alpha}z}{1 + |z|^2 \, |\alpha|^2 - 2\,\text{Re}\,\bar{\alpha}z} < 1 \text{ for } |z| < 1 \right)$$

and w^{-1} maps U into U and hence $w(U) = U$. Note that $\Phi_\alpha(\alpha) = 0$. We shall now prove that any linear fractional transformation f that takes U onto U and consequently T onto T (why?) should be Φ_α for some α and some ψ.

Let $f(\alpha) = 0$ for some α and $f\left(\dfrac{1 + \alpha}{1 + \bar{\alpha}} \right) = e^{i\psi}$ as $\dfrac{1 + \alpha}{1 + \bar{\alpha}} \in T$. By symmetry principle, as α, $1/\bar{\alpha}$ are symmetric with respect to T, 0 and $f(1/\bar{\alpha})$ should be symmetric with respect to $f(T) = T$. Now as $f(\alpha) = 0$ and ∞ is the only point symmetric to 0 with respect to T, $f(1/\bar{\alpha}) = \infty$. Thus this f takes α to 0, $\dfrac{1+\alpha}{1+\overline{\alpha}}$ to $e^{i\Psi}$ and $1/\overline{\alpha}$ to ∞. For this α and ψ, $\Phi_\alpha(z)$

$$= e^{i\psi}\,\frac{z-\alpha}{1-\overline{\alpha}z}\ \text{also satisfies}\ \Phi_\alpha(\alpha)=0,\ \Phi_\alpha\!\left(\frac{1+\alpha}{1+\overline{\alpha}}\right)=e^{i\psi}\ \text{and}\ \Phi_\alpha(1/\overline{\alpha})=\infty.$$

Now since $f\circ\Phi_\alpha^{-1}$ fixes 0, $e^{i\psi}$ and ∞, $f\circ\Phi_\alpha^{-1}=I$ or $f=\Phi_\alpha$.

Later on (see Theorem 4.4.34) we shall prove a very deep result that any (not necessarily a linear fractional transformation) one-to-one analytic (and hence conformal) map of U onto U should be Φ_α for some α and Ψ.

Example 3.2.8

$$w = f(z) = \frac{1+z}{1-z}$$

f maps $|z| = 1$ into Re $w = 0$, that is, the imaginary axis ($z = e^{i\theta}$ implies Re $w = 0$). In fact it is even onto as $iv = \dfrac{1+z}{1-z} \Rightarrow z = \dfrac{iv-1}{1+iv} \Rightarrow |z| = 1$.

Thus $|z| < 1$ is mapped onto Re $w > 0$ (as $f(0) = 1$) and $|z| > 1$ is mapped onto Re $w < 0$.

It will be interesting to find out the image of $|z| = r$ ($0 < r < 1$) which should be a circle. $|z| = r \Rightarrow \left|\dfrac{1-w}{1+w}\right| = r \Rightarrow w$ varies over the Apollonius circle with limit points 1 and -1 (see Example 3.2.3 case 1 for its equation from which one can find the centre and radius). Its equation is thus given by $\left(\text{the extremities of a diameter should be } \dfrac{1-r}{1+r}, \dfrac{1+r}{1-r}\right)$

$$\left| w - \frac{1+r^2}{1-r^2}\right| = \frac{2r}{1-r^2}.$$

The lines through the origin except the real axis in the z plane are mapped onto circle through 1 and -1. While if $r > 1$, the circles are given by

$$\left| w - \frac{r^2+1}{r^2-1}\right| = \frac{2r}{r^2-1}$$

The common region between Re $z < 1$ and $|z| > 1$ is mapped onto $-1 <$ Re $w < 0$. $\left(\text{Since Re}\dfrac{1+x+iy}{1-x-iy} = \dfrac{1-x^2-y^2}{1+x^2+y^2-2x}\right)$. Developing w in a power series around origin we see that $w = 1 + 2z + 2z^2 + \dots$. We shall see later how this function maximizes the modulus of coefficients around origin among one-to-one conformal mappings that take $|z| < 1$ into Re $w > 0$.

Example 3.2.9

$$w = f(z) = z + \frac{1}{z}$$

Put $z = x + iy$, $w = u + iv$. f maps $|z| = 1$, $0 \leq \theta \leq \pi$ onto $-2 \leq u \leq 2$; $v = 0$ and $|z| = 1$, $\pi \leq \theta \leq 2\pi$ onto $-2 \leq u \leq 2$; $v = 0$. It also maps $-\infty < x \leq -1$; $y = 0$ onto $-\infty < u \leq -2$; $v = 0$ and $1 \leq x < \infty$; $y = 0$ onto $2 \leq u < \infty$; $v = 0$. Further on the x-axis, $-1 \leq x < 0$ is mapped onto $-\infty < u \leq -2$ and $0 < x \leq 1$ onto $2 \leq u < \infty$ (on u-axis). Thus the complement of $|z| \leq 1$, $0 \leq \theta \leq \pi$ in the upper half of the z-plane gets mapped onto $v > 0$ (for example, $f(2i) = 3i/2$) and $|z| < 1$ with $0 \leq \theta \leq \pi$ in the upper half is mapped onto $v < 0$. Similarly, the complement of $|z| \leq 1$, $\pi \leq \theta \leq 2\pi$ in the lower half of the z-plane gets mapped onto $v < 0$ and $|z| < 1$ with $\pi \leq \theta \leq 2\pi$ is mapped onto $v > 0$.

Since $f'(z) = 1 - (1/z^2) \neq 0$ for $z \neq \pm 1$, f is conformal at all points $z \neq 0$ or ± 1. Further, $|z| = r < 1$ is mapped onto an ellipse with semi major axis $a = r + \dfrac{1}{r}$ and semi minor axis $b = r - \dfrac{1}{r}$ $\left(\text{since } z = re^{i\theta} \text{ implies}\right.$

$u = \left(r + \dfrac{1}{r}\right) \cos\theta$, $v = \left(r - \dfrac{1}{r}\right) \sin\theta$ and so $\left. \dfrac{u^2}{a^2} + \dfrac{v^2}{b^2} = 1\right).$

Also $e^{i\alpha}t$, $-1 \leq t \leq 1$ is mapped onto a hyperbola with foci at ± 2 and semi-conjugate and semi-transverse axes as $2\cos\alpha$ and $2\sin\alpha$ $\left(\text{since } z = te^{i\alpha}\right.$

implies $u = \left(t + \dfrac{1}{t}\right)\cos\alpha$, $v = \left(t - \dfrac{1}{t}\right) \sin\alpha$ and so $\left. \dfrac{u^2}{4\cos^2\alpha} - \dfrac{v^2}{4\sin^2\alpha} = 1\right)$

These confocal ellipses and hyperbolas are orthogonal to each other which is the requirement for conformality. The fact that the boundary $\{x: -\infty < x \leq 1\} \cup \{z : |z| = 1\} \cup \{x: 1 \leq x < \infty\}$ is mapped onto the u-axis, repeated twice, reflects the fact that $f\left(\dfrac{1}{z}\right) = f(z)$ and the image of $|z| < 1$, $0 \leq \theta \leq \pi$ and $|z| > 1$, $\pi \leq \theta \leq 2\pi$ under f are the same, namely the lower half plane $v < 0$. $f(z)$ is one-to-one both in $|z| < 1$ and in $|z| > 1$ because $f(z_1) = f(z_2)$, $|z_1| < 1$ and $|z_2| < 1$ or $|z_1| > 1$ and $|z_2| > 1$ implies $(z_1 - z_2) = \dfrac{1}{z_1} - \dfrac{1}{z_2}$ or that $(z_1 z_2 - 1)(z_1 - z_2) = 0$. Thus $z_1 = z_2$ as $z_1 z_2 \neq 1$ in both the cases $|z| < 1$ or $|z| > 1$.

Example 3.2.10

$$w = \sin z.$$

Let $z = x + iy$ and $w = u + iv$. Then $u = \sin x \cos hy$ and $v = \cos x \sin hy$. If $y (\neq 0)$ is fixed (i.e., lines parallel to x-axis) the locus of w becomes

$\dfrac{u^2}{\cosh^2 y} + \dfrac{v^2}{\sinh^2 y} = 1$. Thus lines parallel to x-axis are mapped onto a family of ellipses with foci at ± 1. On the other hand if $x\left(\neq 0, \dfrac{\pi}{2}\right)$ is fixed

(i.e., lines parallel to y-axis) the locus of w becomes $\dfrac{u^2}{\sin^2 x} - \dfrac{v^2}{\cos^2 x} = 1$.

Thus these lines are mapped onto a family of hyperbolas with foci at ± 1.
If z varies in $-\dfrac{\pi}{2} < \mathrm{Re}\, z < \dfrac{\pi}{2}$ with $y > 0$, w varies in the upper half plane
as $v = \cos x \sin hy > 0$. Similarly for $y < 0$, $|\mathrm{Re}\, z| < \pi/2$ we have $v < 0$ and it can be easily verified that $|\mathrm{Re}\, z| < \pi/2$ is mapped onto the whole w plane except for the slits $-\infty < u \le -1$ and $1 \le u < \infty$.

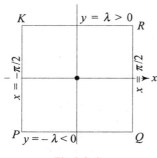

Fig 3.2 (i)

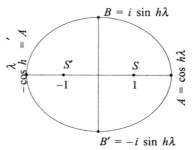

Fig 3.2 (ii)

Consider the rectangle PQRK given by $-\dfrac{\pi}{2} \le x \le \dfrac{\pi}{2}$ and $-\lambda \le y \le \lambda$ in
the z-plane (Fig. 3.2 (i)). The line QR is mapped onto the doubly traversed slit AS (Fig. 3.2 (ii)). The line RK is mapped onto that arc of the ellipse given by ABA'. The line KP is mapped onto the doubly traversed slit $A'S'$. The line PQ is mapped onto that arc of the ellipse given by $A'B'A$. Hence the interior of the rectangle corresponds to the interior of this ellipse except for the slit AS and $A'S'$. (Note that $f(0) = 0$.)

So far we have been dealing with the properties of specific elementary conformal mappings. We now observe that by using composition of many of these mappings we can describe more complicated mapping properties of other conformal mappings.

Example 3.2.11
Map the strip $|\mathrm{Im}\, z| < \pi/2$ onto $|w| < 1$.
We know (see Example 3.2.2) that $\zeta = e^z$ maps $|\mathrm{Im}\, z| < \pi/2$ onto $\mathrm{Re}\, \zeta > 0$ and $w = \dfrac{\zeta - 1}{\zeta + 1}$ maps $\mathrm{Re}\, \zeta > 0$ onto $|w| < 1$ (use the inverse in Example 3.2.8). Combining these two and using composition we get $w = \dfrac{e^z - 1}{e^z + 1} = \dfrac{e^{z/2} - e^{-z/2}}{e^{z/2} + e^{-z/2}} = \tanh \dfrac{z}{2}$ maps $|\mathrm{Im}\, z| < \pi/2$ onto $|w| < 1$.

Example 3.2.12

Find a one-to-one conformal map of the semi disc $|z| < 1$, $\text{Im}\,z > 0$ onto $\text{Im}\,w > 0$.

$\zeta = \dfrac{1+z}{1-z}$ (Example 3.2.8) maps the given semi-disc onto a quadrant $0 < \arg \zeta < \pi/2$ (upper semicircle corresponds to upper half of the imaginary axis and $-1 < x < 1$; $y = 0$ corresponds to the positive real axis) and $w = \zeta^2$ (Example 3.2.1) maps this quadrant onto the half plane $\text{Im}\,w > 0$. Hence the required mapping is $w = \left(\dfrac{1+z}{1-z}\right)^2$. The same objective is obtained by

$z \to \zeta = z + \dfrac{1}{z} \to -\left(z + \dfrac{1}{z}\right)$ which successively map $|z| < 1$, $\text{Im}\,z > 0$ onto

$\text{Im}\,\zeta < 0$ and $w = -\left(z + \dfrac{1}{z}\right) = i^2\left(z + \dfrac{1}{z}\right)$ maps $\text{Im}\,\zeta < 0$ onto $\text{Im}\,w > 0$.

It also follows from the above observations that the conformal mappings are not uniquely determined by the given regions.

Example 3.2.13

Map $-\dfrac{\pi}{2} < x < \dfrac{\pi}{2}$; $y > 0$ in the z-plane conformally onto $0 < v < \pi$ in the w-plane ($z = x + iy$; $w = u + iv$).

We first observe that $\zeta = x_1 + iy_1 = \sin z$ maps $-\dfrac{\pi}{2} < x < \dfrac{\pi}{2}$; $y = 0$ onto $-1 < x_1 < 1$; $y_1 = 0$ and $x = -\pi/2$; $y > 0$ onto $-\infty < x_1 < -1$; $y_1 = 0$ and $x = \pi/2$, $y > 0$ onto $1 < x_1 < \infty$; $y_1 = 0$. Thus the boundary of the required domain in the plane gets mapped onto the entire real axis in the ζ-plane. Further $\sin i = \dfrac{i}{2}\left(e - \dfrac{1}{e}\right)$ is a point on the upper half plane. Thus $\zeta = \sin z$ maps the given region into the upper half plane (Verify onto; see also Example 3.5.3).

Now $w = u + iv = \log\left(\dfrac{\zeta - 1}{\zeta + 1}\right)$ (with $0 < \arg(\zeta - 1) < \pi$, $0 < \arg$ $(\zeta + 1) < \pi$), where the logarithm here has its principal value (see also Theorem 3.4.5), maps the upper half of the $\zeta = x_1 + iy_1$ plane onto the strip $0 < v < \pi$. To see this we observe that $-1 < x_1 < 1$; $y_1 = 0$ is mapped onto $v = \pi$

(on this line $x_1 = -1\lambda + (1 - \lambda)1 = 1 - 2\lambda$, $0 < \lambda < 1$ and $v = \arg\left(\dfrac{1 - 2\lambda - 1}{1 - 2\lambda + 1}\right)$

$= \pi$ and so as ζ tends to a point on $(-1, 1)$, $\text{Arg}\left(\dfrac{\zeta - 1}{\zeta + 1}\right)$ tends to π) and

$\{x_1: -\infty < x_1 < -1\} \cup \{x_1: 1 < x_1 < \infty\}$ is mapped onto $v = 0$. (On these

rays $\left(\dfrac{\zeta - 1}{\zeta + 1}\right)$ is real and positive and so $\mathrm{Arg}\left(\dfrac{\zeta - 1}{\zeta + 1}\right)$ equals 0.) Further

$y_1 > 0$ implies $v > 0$ because

$$\mathrm{Im}\left(\frac{z_1 - 1}{z_1 + 1}\right) = \frac{2y_1}{(x_1 + 1)^2 + y_1^2} > 0$$

and so its principal argument $\mathrm{Arg}\left(\dfrac{z_1 - 1}{z_1 + 1}\right)$ lies between 0 and π i.e. $0 <$

$v < \pi$. Thus combining these two transformations using composition we get

$$w = \log\left(\frac{\zeta - 1}{\zeta + 1}\right) = \log\left(\frac{\sin z - 1}{\sin z + 1}\right) \text{ maps } -\frac{\pi}{2} < x < \frac{\pi}{2}; y > 0 \text{ in the } z\text{-plane}$$

onto $0 < v < \pi$ in the w-plane. Since at each stage the mapping is analytic with derivative non-zero, the composition is conformal.

3.3 Physical Applications of Conformal Mappings

Introduction 3.3.1

In the theory of heat conduction, the flux across a surface in a solid homogeneous body at a point on the surface is defined to be the quantity of heat flowing in a direction normal to the surface per unit time and per unit area at that point. The unit of measurement can be taken as calories per second per square centimeter. If we denote this flux by Φ, from the theory of heat conduction we can write $\overline{\Phi} = -k\dfrac{dT}{dN}$ where the constant k is known as the thermal conductivity of the material of the solid and $\dfrac{dT}{dN}$ is the normal derivative of temperature T at the given point on the surface. It can be proved without much difficulty that when we restrict our attention to the case were T does not vary along the axis perpendicular to the xy-plane (i.e., the flow of heat is two dimensional and parallel to the xy-plane) and the flow is in a steady rate (i.e., T does not vary with time) and no heat sources or sinks are present to create or destroy the thermal energy and further that the temperature function $T(x, y)$ is continuous together with its first and second order partial derivatives inside the solid, then $T(x, y)$ will be a harmonic function of x and y in the interior of the solid body. (For the present we shall take this for granted.)

In physical problems, it is necessary to find an expression for the steady temperature $T(x, y)$ with given boundary conditions. We shall illustrate this with the following example.

Example 3.3.2

Consider a thin semi-infinite plate in $\mathbb{R}^2$ bounded by $x = \pm \pi/2$ and $y = 0$ in the upper half plane. We assume that the temperature on $x = \pm \pi/2$ are kept at 0 and the temperature on $y = 0$ is kept at unity. The problem is to find an expression for $T(x, y)$ at any interior point of this slab (the other assumptions are as listed in the Introduction 3.3.1).

Mathematically, this can be formulated as a problem of finding a function

$T(x, y)$ harmonic in $-\dfrac{\pi}{2} \le x \le \dfrac{\pi}{2}$; $y > 0$ with the boundary conditions

$T\left(\pm\dfrac{\pi}{2}, y\right) = 0 \ \forall \ y > 0$. $T(x, 0) = 1$, $|x| < \dfrac{\pi}{2}$ and $T(x, y)$ is bounded.

First we observe that we can easily find a function T_1 in the $w = u + iv$ plane, harmonic in $0 < v < \pi$ and with the boundary conditions $T_1 = 0$ on

$v = 0$ and $T_1 = 1$ on $v = \pi$. In fact $T_1 = \dfrac{1}{\pi} \operatorname{Im} w$ solves this problem because

imaginary part of the identity function w is harmonic and it evidently satisfies the boundary conditions. Using the properties of the conformal map-

ping $w = f(z) = \log\left(\dfrac{\sin z - 1}{\sin z + 1}\right)$ (see Example 3.2.13) we see that

$T(z) = T_1(f(z))$ satisfies our requirements.

$$T(z) = \dfrac{1}{\pi} \operatorname{Im} f(z) = \dfrac{1}{\pi} \tan^{-1}\left(\dfrac{2\cos x \sinh y}{\sin^2 x \cosh^2 y + \cos^2 x \sinh^2 y - 1}\right)$$

$$= \dfrac{2}{\pi} \tan^{-1}\left(\dfrac{\cos x}{\sinh y}\right)$$

$\left(\text{Since } \sin^2 x \cosh^2 y + \cos^2 x \sinh^2 y - 1 = \sinh^2 y - \cos^2 x \text{ and so}\right.$

$\tan^{-1}\left(\dfrac{2\cos x \sinh y}{\sinh^2 y - \cos^2 x}\right) = \tan^{-1}\left(\dfrac{2\cos x/\sinh y}{1 - (\cos x/\sinh y)^2}\right) = 2 \ \tan^{-1}\left(\dfrac{\cos x}{\sinh y}\right)\left.\right)$

In the above we have taken up the study of temperature flow in a thin semi-infinite plate in the two dimensional space to illustrate how conformal mapping is essentially used in physical applications. On the other hand several other problems in physical applications, like electrostatic potential, two dimensional fluid flow, hydrodynamics and aerodynamics, require conformal mappings for their solutions (See 'Notes' at the end of this chapter).

3.4 Single-Valued Branches for Multi-Valued Functions

Introduction 3.4.1

We first observe that if $f(z)$ is a one-many map from $\mathbb{C}$ into $\mathbb{C}$ we can not even call these maps as functions. But more often we come across such one-many maps in analysis, e.g., $f(z) = z^{1/n}$ $(n = 1, 2, \ldots)$ or $f(z) = \log z$. These mappings have their inverses which are well-behaved as functions. In this section and in the next, we shall devise ways and means by which these one-many maps can be made one-to-one and show that these functions can be studied from the stand-point of analysis (i.e., these can be made continuous and analytic in specific regions or surfaces).

Let $f(z)$ be a one-many map from $\mathbb{C}$ into $\mathbb{C}$. By a single-valued branch of f we mean a triplet (Ω, R, f) where Ω is a region in $\mathbb{C}$, R is a rule which specifies that only one of the many values of $f(z)$ at z be taken and f denotes the resulting one-to-one correspondence from Ω into $\mathbb{C}$. (Recall that a group is an ordered pair $(G,.)$ and a topological space is an ordered pair (X, τ) and similarly a single-valued branch of a multi-valued function f can be defined as a triplet (Ω, R, f).)

In the following we shall describe single-valued branches for the following two multi-valued functions

(i) $f(z) = z^{1/n}$ $(n = 1, 2, \ldots)$ $(z \in \mathbb{C})$

(ii) $f(z) = \log z$ $(z \in \mathbb{C})$

These two examples are taken as illustrations of constructing the branches of (i) an algebraic function and (ii) transcendental function. We shall later see how the construction of a single-valued branch for $\log z$ alone will suffice for other multi-valued functions that we have opportunities to deal with (for example a branch of $\log z$ gives rise to a branch of $z^{1/n}$ also).

Theorem 3.4.2

A single-valued analytic branch for $w = z^{1/n}$ $(n = 2, 3, \ldots)$ is given by (Ω, R, f) where Ω is the complement of non-positive real axis in the finite complex plane, R the rule which stipulates that for each $z \in \Omega$ (represented uniquely by $z = re^{i\theta}$, $r > 0$, $-\pi < \theta < \pi$) the value $w = z^{1/n}$ is the unique number w for which $|\arg w| < \dfrac{\pi}{n}$ and $f: \Omega \to \mathbb{C}$ is given by $f(z) = w = z^{1/n}$ with $|\arg w| < \dfrac{\pi}{n}$.

This branch is usually referred to as the principal branch for $w = z^{1/n}$. To prove the above theorem we need the following lemmas.

Lemma 3.4.3

Given $x > 0$, there exists a unique $y > 0$ such that $y^n = x$ (we denote this y as $x^{1/n}$) and the map $x \to x^{1/n}$ is continuous in $(0, \infty)$.

Proof of Lemma 3.4.3

The existence and uniqueness of $x^{1/n}$ has already been established in real analysis. We shall prove that $y = g(x) = x^{1/n}$ is continuous at each $x_0 \in (0, \infty)$. Let $y_0 = x_0^{1/n}$. Let $\varepsilon > 0$ be given. We have

$$x - x_0 = (y - y_0)(y^{n-1} + y^{n-2} y_0 + \ldots + y_0^{n-1})$$

Hence
$$|g(x) - g(x_0)| = |x - x_0|/|y^{n-1} + y^{n-2} y_0 + \ldots + y_0^{n-1}|$$

$$\leq |x - x_0|/y_0^{n-1}.$$

If $|x - x_0| < \varepsilon y_0^{n-1} = \delta$, we have $|g(x) - g(x_0)| < \varepsilon$, proving continuity of g at x_0.

Lemma 3.4.4

The map $z \to w = z^{1/n}$ with $|\text{Arg } w| < \dfrac{\pi}{n}$ is one-to-one and maps Ω onto the domain Ω_1 in the w plane bounded by the lines $\text{Arg } w = \pm \dfrac{\pi}{n}$. Further its inverse $w \to w^n$ is also one-to-one from Ω_1 onto Ω.

Proof of Lemma 3.4.4

Given any $z \in \Omega$ we have $|\text{Arg } w| \neq \pi$. Hence if we choose $z = re^{i\theta}$, $-\pi < \theta < \pi$, all the n different values of $w = z^{1/n}$ are given by $w_k = r^{1/n} e^{i(\theta + 2k\pi)/n}$ $k = 0. 1, \ldots n - 1$. We now see that there is a unique $w_0 = z^{1/n}$ with $|\text{Arg } w_0| < \dfrac{\pi}{n}$ $\left(\text{This corresponds to } k = 0 \text{ and other roots } w_k \ (k \neq 0) \text{ are such} \right.$

that $|\text{arg } w_k| = |(\theta + 2k\pi)/n| \geq \dfrac{2\pi}{n} - \dfrac{\pi}{n} = \dfrac{\pi}{n} \Big)$. Hence if we restrict to this unique value of $z^{1/n}$ for each z, $w = z^{1/n}$ becomes one-to-one, and maps Ω into $\left\{ w \in \mathbb{C} : |\text{Arg } w| < \dfrac{\pi}{n} \right\}$. To show that this map is onto we first observe that if $|\text{Arg } w_0| < \dfrac{\pi}{n}$, $z_0 = w_0^n$ is such that $|\text{Arg } z_0| < \pi$ and so $z_0 \in \Omega$.

Further w_0 is one of the many values of $z_0^{1/n}$. But $|\text{Arg } w_0| < \dfrac{\pi}{n}$ and so w_0 is the unique n-th root of z_0 under our correspondence and so $z_0^{1/n} = w_0$. It now follows that its inverse is also a bijection from Ω_1 onto Ω.

Proof of the Theorem 3.4.2

From Lemma 3.4.4. $w = f(z) = z^{1/n}$ with $|\text{Arg } w| < \dfrac{\pi}{n}$ $(z \in \Omega)$, is a one-to-one map of Ω onto Ω_1. Write $f(z) = |f(z)| e^{i \, \text{Arg } f(z)}$ where $\text{Arg } f(z) = \dfrac{1}{n}$ $\text{Arg } z$ and $\text{Arg } z$ represents the principal branch of $\arg z$. By Lemma 3.4.3 $z \to |f(z)|$ is continuous in Ω (indeed it is a composition of $z \to |z|$ and $|z|$ $\to |z|^{1/n}$). Further by Theorem 3.1.4, $z \to \dfrac{1}{n} \text{Arg } z = \text{Arg } f(z)$ is continuous

in Ω and so $z \to e^{i \, \arg f(z)}$ is also continuous. Thus $f(z)$ is continuous. Note that the continuity of $f(z)$ is equivalent to
$z \to z_0 \Rightarrow w \to w_0$. Thus we have

$$\lim_{z \to z_0} \left(\frac{w - w_0}{z - z_0} \right) = \lim_{z \to z_0} \left(\frac{z - z_0}{w - w_0} \right)^{-1} = \lim_{z \to z_0} \left(\frac{w^n - w_0^n}{w - w_0} \right)^{-1}$$

$$= \lim_{w \to w_0} (w^{n-1} + w^{n-2} w_0 + \ldots + w_0^{n-1})^{-1}$$

$$= \frac{1}{n} w_0^{-(n-1)}$$

$$= \frac{1}{n} z_0^{-(n-1)/n} .$$

Hence the derivative of w at z_0 exists and is equal to $\dfrac{1}{n} z_0^{-(n-1)/n}$. This proves the theorem.

Theorem 3.4.5
An analytic branch of $w = \log z$ is given by (Ω, R, f) where Ω is the complement of the non-positive real axis in the finite complex plane, R is the rule which specifies that for each $z \in \Omega$ with $-\pi < \text{Arg } z < \pi$ there exists a unique $\log z$ with $|\text{Im} \log z| < \pi$ and $f(z) = \log z$ with $|\text{Im} \log z| < \pi$, or that $f(z) = \log |z| + i \text{ Arg } z$.

This branch of $\log z$ is usually referred to as the principal branch of $\log z$.

Proof
We first observe that for each $z \in \Omega$, there exists a unique argument θ of z with $|\theta| < \pi$. With this value for the argument of z we write $z = re^{i\theta}$ and get all the infinite values of $\log z = \log r + i(\theta + 2\pi k)$, $k = 0, \pm 1, \ldots$ For $k = 0$, we get one $\log z$ with $|\text{Im} \log z| < \pi$ and for all other values of $\log z$, we have $|\text{Im} \log z| = |\theta + 2\pi k|$, $k = \pm 1 \ldots$ Thus for these values $|\text{Im} \log z| \geq 2|k|\pi - |\theta| > 2\pi - \pi = \pi$. Thus there exists one and only one value for $\log z$ with $|\text{Im} \log z| < \pi$. Thus $w = f(z) = \log z$ with $|\text{Im} \log z| < \pi$ becomes a one-to-one map of Ω into $\{w : |\text{Im } z| < \pi\}$. This map is also onto $\{w : |\text{Im } w| < \pi\}$ as can be seen by taking w_0 with $|\text{Im } w_0| < \pi$ and considering $z_0 = e^{w_0}$. Now this w_0 is one of the infinitely many values of $\log z_0$ and in as much as $|\text{Im } w_0| < \pi$, we see that $f(z_0) = w_0$. Thus $w = f(z)$ gives a one-to-one onto correspondence between Ω and $\{w : |\text{Im } w| < \pi\}$.

Our next step is to show that this branch $w = f(z) = \log z$ with $|\text{Im} \log z| < \pi$ is continuous in Ω. From the definition, the principal branch described above can be written as $f(z) = \log z = \log |z| + i \text{ Arg } z$ with $|\text{Arg } z| < \pi$. The above definition is valid in Ω. From real analysis we know that $x \to \log x$ is continuous from $(0, \infty)$ onto $\mathbb{R}$. Further $z \to |z|$ is always continuous. Thus $\text{Re } f(z) = \log |z|$ is continuous in Ω. On the other hand $\text{Im } f(z)$ represents the principal branch of $\arg z$ and as such it is continuous in Ω by Theorem 3.1.4. Hence $f(z)$ is continuous in Ω.

The continuity of the principal branch of $w = f(z) = \log z$ in Ω tells us that as $z \to z_0$ ($z, z_0 \in \Omega$) $w \to w_0$. Thus

$$\lim_{z \to z_0} \left(\frac{w - w_0}{z - z_0} \right) = \lim_{z \to z_0} \left(\frac{z - z_0}{w - w_0} \right)^{-1} = \lim_{w \to w_0} \left(\frac{e^w - e^{w_0}}{w - w_0} \right)^{-1} = e^{-w_0} = \frac{1}{z_0}.$$

Therefore $f'(z_0)$ exists and is equal to z_0^{-1}, $\forall\ z_0 \in \Omega$. Hence f defines a single-valued analytic branch of $\log z$.

This completes the proof of the theorem.

Remark 3.4.6

Having defined a single-valued analytic branch for $w = \log z$ we can easily deduce the existence of a single-valued analytic branch for $w = z^\alpha$ where α is any complex number. (Compare Remark 2.4.6). Indeed if $w_1(z) = \log z$ denotes the principal branch of $\log z$ in $\Omega = \mathbb{C} \backslash (-\infty, 0)]$ then the function $g(z) = e^{\alpha w_1(z)}$ is single-valued in Ω and is analytic also. We take this as the definition of a single-valued analytic branch for $w = z^\alpha$. In fact $g(z) = e^{\alpha \log z}$ is one of the values of z^α (consistent with our Definition 2.4.4 and with the usual laws of logarithm and is also analytic). This method also gives the principal branch for $w = z^{1/n}$ when $\alpha = 1/n$.

Another remark that we would like to mention here is the following: There is no uniqueness in defining a branch for $\log z$. Actually we can cut off any line segment (or even a smooth curve) from 0 to ∞ and in the complement we can define a single-valued analytic branch for $\log z$. We can see how this can be done later.

3.5 Elementary Riemann Surfaces

Introduction 3.5.1

In the previous section we have defined single-valued branches for multi-valued functions such as $z^{1/n}$, $\log z$ and proved that these branches are analytic functions of a single complex variable in their respective domains of definitions. Another type of a map that is not a function is $z \to \sqrt{z^2 - 1}$. This map can be viewed both as a one-many map as well as a many-one map, since the set $\{z, -z\}$ goes to the same set of values $\left\{ +\sqrt{z^2 - 1}, -\sqrt{z^2 - 1} \right\}$. See the following diagram:

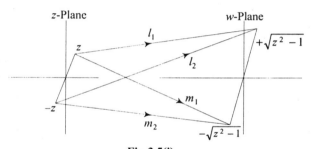

Fig 3.5(i)

If we concentrate on the arrows (l_1, l_2) or (m_1, m_2) the map is two-one and if we look at (l_1, m_1) or (l_2, m_2) the map is one-two. In fact, its inverse is also a similar map $w \rightarrow \sqrt{w^2 + 1}$. As before we can define single-valued analytic branches for both $w = \sqrt{z^2 - 1}$ and its inverse $z = \sqrt{w^2 + 1}$.

Recall that a branch for f was defined as a triplet (Ω, R, f) (where Ω is a chosen region in the plane, R is a rule which helps us in choosing one out of the many values and f is the resulting one-to-one correspondence.)

However Riemann's idea of making these maps one-to-one is totally different. He thought that instead of omitting all but one value in the correspondence we could as well add points in the domain, corresponding to the many values of the map. More precisely we shall ask the question whether we can add points to the domain or the range or both so that these multi-valued functions become one-to-one and onto in their respective domains and ranges. We call the resulting surfaces as Riemann surfaces either for f or for f^{-1}. (If we say that it is a Riemann surface of a one-many map, its domain has to be that surface and if we call it as a Riemann surface of a many-one map its range has to be that surface and for a map which is both one-many and many-one, both the domain and the range have to be Riemann surfaces).

We shall illustrate these ideas by constructing the Riemann surfaces associated with the following maps

(i) $w = z^3$ (ii) $w = \cos z$ (iii) $w = \sqrt{z^2 - 1}$.

In each case we shall analyse the nature of the surface in detail and see whether we can give a topological structure for these surfaces so that we can study the continuity properties of the resulting mappings which are now one-to-one functions between these surfaces.

Example 3.5.2

Elementary Riemann surface associated with the function $w = z^3$.

Under this mapping the positive real axis in the z plane goes into the positive real axis in the w plane. Moreover, this is onto because to every positive real number there exists one and only one positive cube root. To understand the mapping further let us put $z = re^{i\theta}$ and $w = Re^{i\phi}$. Thus the equation $w = z^3$ can be split into $R = r^3$ and $\phi = 3\theta$. From this relationship, it is clear that as θ varies from $\theta = 0$ to $\theta = 2\pi/3$ (r arbitrary), ϕ sweeps the entire w plane, i.e., from $\phi = 0$ to $\phi = 2\pi$ (R arbitrary). Also if θ varies from $\theta = 2\pi/3$ to $\theta = 4\pi/3$ again ϕ sweeps the entire plane from $\phi = 2\pi$ to $\phi = 4\pi$. This can also be understood by the many-one nature of $w = z^3$ as follows: $R e^{i\phi} = r^3 e^{3i\theta}$; $\phi + 2k\pi \cong 3\theta$ ($k \in \mathbb{Z}$); $\theta = (\phi + 2k\pi)/3$. For $k = 0, 1, 2$, these are distinct in general and from $k = 3$ onwards the same triad of values recur. Thus the region r arbitrary and $0 < \theta < 2\pi/3$ is getting mapped onto the whole w-plane except for a slit from 0 to ∞ on the positive side of the real axis. This slit corresponds to both $\theta = 0$ and $\theta = 2\pi/3$. The same is true with reference to the regions $2\pi/3 < \theta < 4\pi/3$ and

$4\pi/3 < \theta < 2\pi$. These three disjoint regions (whose union is the whole z-plane except for $\theta = 0$, $2\pi/3$, $4\pi/3$) which correspond to the whole w-plane except for the slit from 0 to ∞ on the real axis will be the fundamental regions.

The construction of the Riemann surface for $w = z^3$ is done as follows: Take three copies of the complex plane and in each make a cut along the positive side of the real axis. Let us distinguish in each cut an upper edge and a lower edge. In the first plane let them be marked respectively by $\phi = 0$ and $\phi = 2\pi$. In the second by $\phi = 2\pi$ and $\phi = 4\pi$ and in the third plane by $\phi = 4\pi$ and $\phi = 6\pi$. Since points belonging to $\theta = 0$ corresponds to $\phi = 0$ and points belonging to $\theta = 2\pi/3$ corresponds to $\phi = 2\pi$ the images of $\theta = 0$ and $\theta = 2\pi/3$ (which are same originally) can now be distinguished as points belonging to the upper edge and the lower edge of the first plane. As the mapping $w = z^3$ is one-to-one and onto from $0 < \theta < 2\pi/3$ to $0 < \phi < 2\pi$, the entire first fundamental region could be mapped in a one-to-one fashion onto the whole w-plane except for these two edges.

Similarly the second fundamental region could be mapped in a one-to-one fashion onto the whole second w-plane except for the two edges that correspond to $\phi = 2\pi$ and $\phi = 4\pi$ and the third fundamental region could be mapped in a one-to-one fashion onto the third w-plane except for the two edges that correspond to $\phi = 4\pi$ and $\phi = 6\pi$.

However since the common boundary $\theta = 2\pi/3$ is the same for the first and second fundamental regions it is but natural that we identify the edges $\phi = 2\pi$ in the first plane with $\phi = 2\pi$ in the second plane. Similarly we identify the edges $\phi = 4\pi$ in the second plane with $\phi = 4\pi$ in the third plane and $\phi = 6\pi$ in the third plane with $\phi = 0$ in the first. The reason for the last identification being however that $\theta = 0$ and $\theta = 2\pi$ are one and the same as far as the z-plane is concerned. This also enables us to move from the third plane to the first plane as z moves from the third fundamental region to the first fundamental region. We also note that the mapping $w = z^3$ is three-one at all points except at 0 where it is one-to-one. This aspect could be reflected in our construction by gluing all the three sheets at origin so that origin is a single point on the resulting surface. The special position occupied by the origin in the Riemann surface will be described by saying that origin is a branch point of order two as it connects three sheets.

Geometrically the one-to-one correspondence between the z-plane and the Riemann surface so constructed could be exhibited by following a sequence of positions of the variable z in the z-plane and the variable w in the surface as the following diagram shows: (Here 1, 2, 3 etc denote the various positions of z and 1', 2', 3' etc are those of w).

Let us closely look at the domain and the range namely the z-plane and the constructed Riemann surface for $w = z^3$. This mapping provides a one-to-one correspondence between the z-plane and the Riemann surface. The first fundamental region corresponds to the first sheet except for the rays

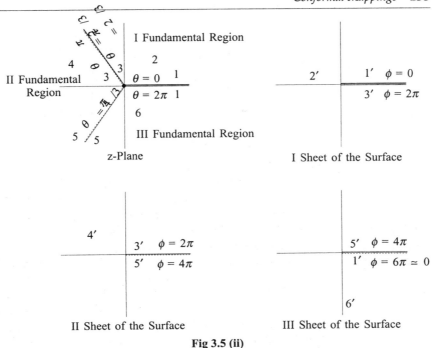

I Fundamental Region

II Fundamental Region

III Fundamental Region

z-Plane

I Sheet of the Surface

II Sheet of the Surface

III Sheet of the Surface

Fig 3.5 (ii)

$\phi = 0$ and $\phi = 2\pi$. The ray $\theta = 2\pi/3$ corresponds to the identified ray $\phi = 2\pi$ in the first and the second sheets. Similarly, the image of other fundamental regions and their boundaries could be described. The inverse map under this correspondence deserves a special scrutiny. The first sheet is mapped by the first branch of $z = w^{1/3}$ onto the first fundamental region. The ray $\phi = 2\pi$ (common to the first and the second) gets mapped onto $\theta = 2\pi/3$ and so on. In fact for points w belonging to any one of the sheets but not common to two or three sheets we can choose a small neighbourhood of w which can be mapped onto an open neighbourhood of its image $z = w^{1/3}$ in the corresponding fundamental region by using a suitable branch for $z = w^{1/3}$. If however point w is common to two sheets, a small neighbourhood of w can always be found, part of which gets mapped onto one fundamental region and the complementary part gets mapped onto another fundamental region by using suitable branches of $z = w^{1/3}$ in such a way that the resulting neighbourhood of its image $z = w^{1/3}$ is an ordinary neighbourhood in the z plane. The only exceptional point namely the origin in the surface has the following interesting property: The region enclosed by a simple closed curve around it on the surface gets mapped onto an ordinary neighbourhood of the origin in the z-plane using all the three branches of $z = w^{1/3}$. Thus we can set up a canonical neighbourhood around each point in the surface in such a way that these neighbourhoods are mapped onto their images in a bijective bi-analytic way (analytic both ways). The analyticity of these maps are easily obtained because we use one and only one analytic branch for $z = w^{1/3}$ at generic points and these maps also coincide when two or more branches are used at points common to two or three sheets. It is also clear from the construction of our canonical neighbourhoods and the local homeomorphisms that if two canonical neighbourhoods U and V (with local homeomorphism f_u and f_v) intersect then the transition functions $f_v \circ f_u^{-1}$ defined from $f_u(U \cap V)$ to $f_v (U \cap V)$ are nothing but identity transformations.

Example 3.5.3
Elementary Riemann surface associated with $w = \cos z$.

Let $z = x + iy$ and $w = u + iv$. Then $w = \cos z$ is equivalent to

$$u = \cos x \cosh y; \quad v = -\sin x \sinh y \qquad (1)$$

Observe that the upper half of the line $x = 0$ (i.e., $x = 0$, $y \geq 0$) is mapped onto the u axis from 1 to ∞ with the origin corresponding to 1. This is because $\cosh y = (e^y + e^{-y})/2 \geq 1$ and assumes all positive real values ≥ 1. Similarly the lower half ($x = 0$, $y \leq 0$) is also mapped onto the u-axis from 1 to ∞. We also note that the part of the line $x = \pi$ described by $x = \pi$, $y \geq 0$ is mapped onto the negative u-axis from -1 to $-\infty$ and $x = \pi$, $y \leq 0$ is also getting mapped onto, the same part of the u-axis. We now show that the entire vertical strip described by $0 < x < \pi$ in the z-plane is getting mapped onto the whole w-plane with two slits on the u-axis one from 1 to ∞ on the positive side and another from -1 to $-\infty$ on the negative side (call this region as G), in a one-to-one onto fashion, i.e., we claim that $0 < x < \pi$ is a fundamental region for $w = \cos z$. To do this, we observe that the mapping equations (1) clearly show that $0 < x < \pi$, $y > 0$ is getting mapped into $v < 0$ and $0 < x < \pi$, $y < 0$ is getting mapped into $v > 0$. Further $0 < x < \pi$, $y = 0$ is mapped onto $v = 0$ and $|u| < 1$. However $w = \cos z$ is one-to-one in this region because $\cos z_1 = \cos z_2$ implies that $z_1 = z_2 \pm 2k\pi$ or $z_1 = -z_2 \pm 2k\pi$ and these conditions force us to conclude that $z_1 = z_2$. Thus we need only to prove that the map is onto. In fact if $w \in G$, the many values of $\cos^{-1} w$ are given by $-i \log \left(w \pm \sqrt{w^2 - 1} \right)$ and their various real parts are nothing but various values of the argument of $w \pm \sqrt{w^2 - 1}$. If for a certain w one value of $\arg \left(w \pm \sqrt{w^2 - 1} \right) = \theta$ is such that $0 < \theta < \pi$, we can take $z = \theta - i \log |w \pm \sqrt{w^2 - 1}|$ so that $\cos z = w$ and also z belongs to the fundamental region. Note that if $w \in G$, θ cannot be 0 or π. However if θ satisfies $\pi < \theta < 2\pi$, then we can take $z = 2\pi - \theta - i \log |w \pm \sqrt{w^2 - 1}|$ so that z belongs to the fundamental region and again $\cos z = w$. Exactly the same way, we can conclude that $(k - 1)\pi < x < k\pi$ where k is an integer, is (called the k-th) a fundamental region for $\cos z$. For the sake of description we call $(k - 1)\pi < x < k\pi$, k odd as a fundamental region of the first type and $(k - 1)\pi < x < k\pi$, k even as a fundamental region of the second type. The vital difference between these two types of fundamental regions is that in the case of fundamental regions of the first type the upper and lower half planes correspond to lower and upper half planes respectively in the image whereas in the case of fundamental regions of the second type the upper and lower half planes correspond to themselves. Let us now come to the construction of the Riemann surface for $w = \cos z$. As before we make cuts from 1 to ∞ and from -1 to $-\infty$ in each of the infinite number of complex planes that correspond to infinitely many fundamental regions. We identify two pairs of edges in both these cuts to be mentioned precisely as right upper edge, right lower side, left upper edge and left lower edge. Let us call the fundamental region $0 < x < \pi$ as the first fundamental region and its image as the first sheet. Similarly the fundamental region $\pi < x < 2\pi$ will be called the second fundamental region and its image the second sheet. Since both the upper half and the lower half of the ray $x = 0$ correspond to the cut from 1 to ∞ in the first sheet and the upper ray of $x = 0$ can be considered as the boundary of the upper half plane in $0 < x < \pi$ (which in turn goes to the lower half of G in the first sheet) it is reasonable to map the upper part of the ray $x = 0$ with the right lower edged in the first sheet. For similar reasons the lower part of the ray $x = 0$ will be mapped to the right upper edge in the first sheet. In the same way the upper part of the ray $x = \pi$ will correspond to the left lower edge and the lower part of the ray $x = \pi$ will correspond to the left upper edge in the first sheet. However the ray $x = \pi$ can also be considered as boundary of the second fundamental region also. But then the correspondence should be described as follows: the upper half of $x = \pi$ corresponds to left upper edge and the lower half of $x = \pi$ to left lower edge of the second sheet. This is because as far as the second fundamental region and its image, the second sheet, is concerned upper half plane goes to upper half plane and the lower half plane goes to the lower half plane. These considerations show that in the construction of the Riemann surface, the first and second sheets (corresponding to first and second fundamental regions) should be connected in such a way that the left upper edge of the first sheet must be identified with left lower edge of the second sheet and the left lower edge of the first sheet must be identified with the left upper edge of the second sheet. Similarly the next two sheets (Corresponding to the next two fundamental regions) should be joined along 1 to ∞ cross-wise. The construction with regard to other fundamental regions are similar and the resulting surface thus obtained generates simple branch points at -1 and $+1$ alternatively. As before the one-to-one correspondence between the z-plane and the Riemenn surface so constructed could be exhibited geometrically by following a sequence of positions of the variable z in the z-plane and the variable w in the surface as the following diagram shows (Here z and the corresponding w are marked by the same set of integers).

z-Plane:

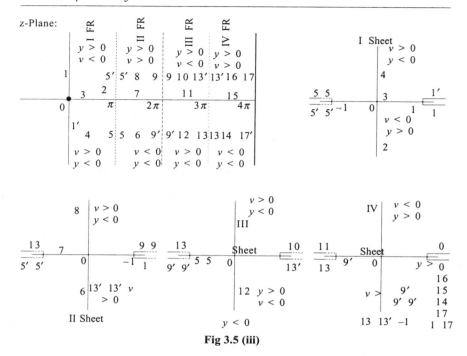

Fig 3.5 (iii)

The above construction of the Riemann surface of $w = \cos z$ consists of infinitely many sheets connected at ± 1 alternately along a slit consisting of upper and lower edges from $-\infty$ to -1 and 1 to ∞. As before we can prescribe neighbourhood systems as follows. For a point w belonging to a single sheet an ordinary open neighbourhood of w inside that sheet can be mapped by a branch of $z = \cos^{-1} w$ onto an open neighbourhood of its image $\cos^{-1}w$. For points on the edges other than ± 1, we can use a neighbourhood part of which belongs to one sheet and the complementary part in the next sheet and use two consecutive branches of $z = \cos^{-1}w$ to map this neighbourhood onto an usual neighbourhood of their images. For ± 1 in all these sheets we can use closed curves around ± 1 which starts in that sheet, moves to the connecting next sheet encircles the point in that sheet and comes back to the original point via the connecting edges. Again we can use corresponding branches of $z = \cos^{-1}w$ to map these neighbourhoods onto ordinary neighbourhoods of $\cos^{-1}(\pm 1)$ in the z-plane. With this correspondence we can see that the transition mappings are analytic.

Example 3.5.4

The elementary Riemann surfaces associated with $w = \sqrt{z^2 - 1}$.

As we saw already this map can be considered both as a one-two map and a two-one map and its "inverse" $z = \sqrt{w^2 + 1}$ is again such a map. In order to make this correspondence one-to-one and onto we need to construct two Riemann surfaces one for w and another for z. We first begin by proving the following result.

Theorem 3.5.5

There exists a single valued one-to-one analytic branch for $w = f(z) = u + iv = \sqrt{z^2 - 1}$ which is defined on $\Omega = \mathbb{C}\backslash[-1, 1]$ (the finite complex plane except for the slit $-1 \leq x \leq 1$ on the real axis) and maps Ω onto the finite complex plane except for the slit $-1 \leq v \leq 1$ on the imaginary axis.

Proof

We start defining the principal branches F_1, F_2 for $\sqrt{z-1}$ and $\sqrt{z+1}$ respectively. The principal branch F_1 of $\sqrt{z-1}$ is defined in the region Ω_1 which is the complement of the set of points z for which $z - 1$ is real and less than or equal to zero. We have to choose the unique $\sqrt{z-1}$ having positive real part in Ω_1.

Then $F_1(z) = \sqrt{r_1}\, e^{i\theta_1/2}$ where $|z - 1| = r_1$ and $\theta_1 = \text{Arg}\,(z - 1) = \text{Arg}$ (vector from 1 to z) so that $-\pi < \theta_1 < \pi$ for $z \in \Omega_1 = \mathbb{C}\backslash(-\infty, 1]$. Similarly $F_2(z) = \sqrt{r_2}\, e^{i\theta_2/2}$ where $|z + 1| = r_2$ and $\theta_2 = \text{Arg}\,(z + 1) = \text{Arg}$ (vector from -1 to z) so that $-\pi < \theta_2 < \pi$ for $z \in \Omega_2 = \mathbb{C}\backslash(-\infty, -1]$. Hence for $z \in \Omega_1 \cap \Omega_2 = \mathbb{C}\backslash(-\infty, 1]$ we can define a single valued analytic branch for $\sqrt{z^2 - 1} = \sqrt{z-1}\sqrt{z+1}$ as $F(z) = F_1(z)\, F_2(z)$. To extend this to $(-\infty, -1)$ we consider another branch for $\sqrt{z^2 - 1}$ given by $G(z) = G_1(z)\, G_2(z)$ for $z \in \mathbb{C}\backslash[-1, \infty)$ where $G_1(z)$ and $G_2(z)$ are the principal branches of $\sqrt{1-z}$ and $\sqrt{-1-z}$ respectively) given by $G_1(z) = \sqrt{s_1}\, e^{i\theta_1/2}$ where $|1 - z| = s_1$ and $\phi_1 = \text{Arg}\,(1 - z) = \text{Arg}$ (vector from z to 1) so that $-\pi < \phi_1 < \pi$ for $z \in \Omega_3 = \mathbb{C}\backslash[1, \infty)$. $G_2(z) = \sqrt{s_2}\, e^{i\theta_2/2}$ where $|z + 1| = s_2$ and $\phi_2 = \text{Arg}\,(-z -1) = \text{Arg}$ (vector from z to -1) so that $-\pi < \phi_2 < \pi$ for $z \in \Omega_4 = \mathbb{C}\backslash[-1, \infty)$. This latter function is well defined, one-to-one and analytic in $\Omega_3 \cap \Omega_4 = \mathbb{C}\backslash[-1, \infty)$ and this domain of definition $\Omega_3 \cap \Omega_4$ includes a part $(-\infty, -1)$ of the omitted portion $(-\infty, 1]$ in $\Omega_1 \cap \Omega_2$. Our endeavour will be to show that

$$f(z) = \begin{cases} F(z) & z \in \Omega_1 \cap \Omega_2 \\ -G(z) = -\sqrt{r_1 r_2} & (z \in (-\infty, -1),\, r_1 = |z - 1|,\, r_2 = |z + 1|) \end{cases}$$

will define an analytic branch of $\sqrt{z^2 - 1}$ in $\Omega = \mathbb{C}\backslash[-1, 1]$. For this we must first prove that $F(z)$ and $-G(z)$ agree in the intersection of their respective domains. This common domain consists of the union of the upper half plane Π^+ and the lower half plane Π^-. Let us take $z \in \Pi^+$.

$$F_1(z) = \sqrt{r_1}\, e^{i\theta_1/2}, \quad |z - 1| = r_1 > 0,\ 0 < \theta_1 < \pi$$

$$F_2(z) = \sqrt{r_2}\, e^{i\theta_2/2}, \quad |z + 1| = r_2 > 0,\ 0 < \theta_2 < \pi.$$

Note that $\theta_1 = \text{Arg } (z - 1)$ and $\theta_2 = \text{Arg } (z + 1)$ will vary only between 0 and π for $z \in \Pi^+$. See Figure 3.5 (iv).

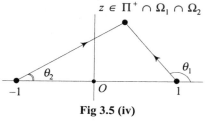

$z \in \Pi^+ \cap \Omega_1 \cap \Omega_2$

Fig 3.5 (iv)

Thus $F(z) = \sqrt{r_1 r_2}\, e^{i(\theta_1 + \theta_2)/2}$.

Similarly

$$G_1(z) = \sqrt{s_1}\, e^{i\phi_1/2},$$
$$|z - 1| = s_1 > 0,\ -\pi < \phi_1 < 0$$

$$G_2(z) = \sqrt{s_2}\, e^{i\phi_2/2},\ |z + 1| = s_2 > 0,\ -\pi < \phi_2 < 0$$

Note that $\phi_1 = \text{Arg } (1 - z)$ and $\phi_2 = \text{Arg } (-1 - z)$ will vary only between $-\pi$ and 0 for $z \in \Pi^+$. See Figure 3.5 (v).

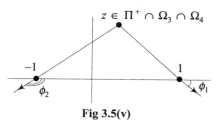

$z \in \Pi^+ \cap \Omega_3 \cap \Omega_4$

Fig 3.5(v)

Thus $G(z) = \sqrt{s_1 s_2}\, e^{i(\phi_1 + \phi_2)/2}$. Now for a single $z \in \Pi^+$, $s_1 = r_1$ and $s_2 = r_2$, $\theta_1 + |\phi_1| = \theta_1 - \phi_1 = \pi$; $\theta_2 + |\phi_2| = \theta_2 - \phi_2 = \pi$ and hence

$$G(z) = \sqrt{r_1 r_2}\, e^{i([(\theta_1 + \theta_2)/2] - \pi)} = -\sqrt{r_1 r_2}\, e^{i(\theta_1 + \theta_2)/2} = -F(z).$$

Similarly we can prove that $F(z) = -G(z)$ for $z \in \Pi^-$. During the course of the above proof, we have also proved that if $z \in \Pi^+$ then $0 < \theta_1 < \pi$ and $0 < \theta_2 < \pi$ and so $\text{Arg } F(z) = (\theta_1 + \theta_2)/2$ also satisifes $0 < \arg F(z) < \pi$ which means F maps Π^+ into Π^+. As $f(z) = F(z)$ we see that f maps Π^+ into Π^+. A similar proof can be given to show that $z \in \Pi^- \Rightarrow f(z) = F(z) \in \Pi^-$. Since $f(z) = F(z)$ in $\Omega_1 \cap \Omega_2$ and $F(z)$ is a single-valued analytic branch of $\sqrt{z^2 - 1}$ we see that $f(z)$ is analytic in $\Omega_1 \cap \Omega_2$. Hence in order to prove the analyticity of f in Ω we need only prove that $f(z)$ is analytic in $(-\infty, -1)$. Let $x \in (-\infty, -1)$. Let $z \to x$ along the negative real axis.

$$\lim_{z \to x} \frac{f(z) - f(x)}{z - x} = \lim_{z \to x} \frac{-G(z) + G(x)}{z - x} = -G'(x)$$

(G is analytic in $(-\infty, -1)$). Let $z \to x$ with $z \in \Pi^+$ or Π^-

$$\frac{f(z) - f(x)}{z - x} = \lim_{z \to x} \frac{F(z) + G(x)}{z - x} = \lim_{z \to x} \frac{-G(z) + G(x)}{z - x} = -G'(x)$$

($F = -G$ in $\Pi^+ \cup \Pi^-$).

Thus, whichever way $z \to x$, the required limit exists and is the same. So $f(z)$ is analytic in $(-\infty, -1)$. Next we show that f is one-to-one on Ω. $F(z)$ and $G(z)$, being principal branches, are one-to-one in their respective domains. Since $f(z) = F(z)$ for $z \in \Pi^+$ or Π^- and $f(z) = -G(z)$ for

$z \in (-\infty, -1)$ we can easily get $f(z_1) = f(z_2)$ implies $z_1 = z_2$ whenever both z_1 and z_2 belong to Π^+ or Π^- or $(-\infty, -1)$. Let now $z_1 \in \Pi^+$ or Π^- and $z_2 \in (-\infty, -1)$. This implies

$$F(z_1) = -G(z_2) \Rightarrow -G(z_1) = -G(z_2) \Rightarrow z_1 = z_2$$

(as G is one-to-one in its domain). Thus f is analytic, one-to-one in Ω and maps Ω into the w-plane.

To complete the proof of our theorem it only remains to prove that f maps Ω onto Ω_1 = the whole w-plane with a slit $-1 \le v \le 1$ on the imaginary axis. This will be proved by constructing its inverse map. Put $g(w) = if(-iw)$. As w varies over Ω_1, $-iw \in \Omega$ and so $g(w)$ is well defined. Further g is injective as f is injective and we claim that $f(g(w)) = w, \forall w \in \Omega_1$. This will prove that f is onto. First of all it is clear that $f(g(w))$ is one of the values of $\sqrt{(g(w))^2 - 1}$ i.e.,

$$f(g(w)) = \pm\sqrt{-(f(-iw))^2 - 1} = \pm\sqrt{-(-w^2 - 1) - 1} = \pm\sqrt{w^2} = \pm w.$$

It suffices to prove that Im $w > 0 \Rightarrow$ Im $g(w) > 0$ and Im $w < 0 \Rightarrow$ Im $g(w) < 0$. Indeed if this is proved, by the mapping properties of f, Im $g(w) > 0 \Rightarrow$ Im $f(g(w)) > 0$ and so for $w \in \Pi^+$, $f(g(w) \ne -w$ and so $f(g(w)) = w$. A similar argument proves that for $w \in \Pi^-$, $f(g(w)) = w$. By continuity of f and g, $f(g(w)) = w$ holds for all $w \in \Omega_1$. We shall consider the following three cases for $w \in \Pi^+ \cap \Omega_1$.

Case 1

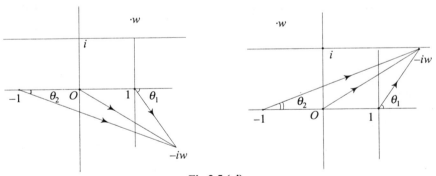

Fig 3.5 (vi)

$w = u + iv$; u arbitrary $v > 1$; $-iw = -iu + v$. Thus Re $(-iw) = v > 1$ and Im $(-iw) = -u$ arbitrary. To find $f(-iw)$ we first find out

$$\theta_1 = \arg (-iw - 1) \quad \text{and} \quad \theta_2 = \arg (-iw + 1).$$

From the above figures it is clear that

$$-\frac{\pi}{2} < \theta_1 < \frac{\pi}{2} \quad \text{and} \quad -\frac{\pi}{2} < \theta_2 < \frac{\pi}{2}.$$

Thus $-\dfrac{\pi}{2} < (\theta_1 + \theta_2)/2 < \dfrac{\pi}{2}$ and we see that $\text{Arg } f(-iw) = (\theta_1 + \theta_2)/2$.

Hence $f(-iw)$ belongs to either IV or I quadrant. Thus $g(w) = i\, f(-iw)$ belongs to either I or II quadrants and so Im $g(w) > 0$. (In addition using the signs of $\tan \theta_1 = -u/(v-1)$ and $\tan \theta_2 = -u/(v+1)$ we can conclude that for $u > 0$, $(\theta_1 + \theta_2)/2$ belongs to the fourth quadrant and for $u < 0$, $(\theta_1 + \theta_2)/2$ belongs to the first quadrant. Thus g maps $v > 1$, $u > 0$ into the first quadrant and maps $v > 1$, $u < 0$ into the second quadrant.

Case 2

$w = u + iv$; $u > 0$, $0 < v < 1$, $-iw = -iu + v$, Im $(-iw) < 0$ and $0 < $ Re $(-iw) = v < 1$. From the adjacent figure it is clear that
$\theta_1 = \text{Arg } (-iw - 1)$ and $\theta_2 = \text{Arg } (-iw + 1)$ satisfy

$-\pi < \theta_1 < -\dfrac{\pi}{2}$ and $-\dfrac{\pi}{2} < \theta_2 < 0$.

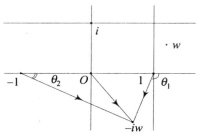

Fig 3.5 (vii)

Hence $-\dfrac{3\pi}{2} < \theta_1 + \theta_2 < -\dfrac{\pi}{2}$. But $\tan \theta_1 = -u/(v-1)$ and $\tan \theta_2 = -u/(v+1)$ imply that $\tan (\theta_1 + \theta_2) = 2uv/(1 + u^2 - v^2) > 0$. Thus $-\pi < \theta_1 + \theta_2 < -\dfrac{\pi}{2}$ or $-\dfrac{\pi}{2} < (\theta_1 + \theta_2)/2 < -\dfrac{\pi}{4}$. Hence $f(-iw)$ belongs to the fourth quadrant and so $g(w) = if(-iw)$ belongs to the first quadrant and so Im $g(w) > 0$.

Case 3

$w = u + iv$; $u < 0$, $0 < v < 1$, $-iw = -iu + v$. Im $(-iw) = -u > 0$ and $0 < $ Re $(-iw) = v < 1$. From the adjacent figure
$\theta_1 = \text{Arg } (-iw - 1)$ and $\theta_2 = \text{Arg } (-iw + 1)$ satisfy
$\dfrac{\pi}{2} < \theta_1 < \pi$ and $0 < \theta_2 < \dfrac{\pi}{2}$.

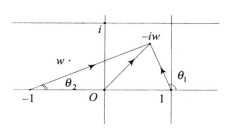

Fig 3.5 (viii)

Hence $\dfrac{\pi}{2} < \theta_1 + \theta_2 < \dfrac{3\pi}{2}$. But $\tan \theta_1 = \dfrac{-u}{v-1}$ and $\tan \theta_2 = \dfrac{-u}{v+1}$ imply that $\tan (\theta_1 + \theta_2) = \dfrac{2uv}{1 + u^2 - v^2} < 0$. Thus $\dfrac{\pi}{2} < \theta_1 + \theta_2 < \pi$. i.e.,

$\dfrac{\pi}{4} < \dfrac{\theta_1 + \theta_2}{2} < \dfrac{\pi}{2}$. Hence $f(-iw)$ belongs to the first quadrant and so $g(w)$ $= if(-iw)$ belongs to the second quadrant and again Im $g(w) > 0$.

A similar argument ensures us that Im $w < 0 \Rightarrow \operatorname{Im} g(w) < 0$. This completes the proof of the fact that f is onto. Incidentally we see that $g(w)$ is an analytic branch of $\sqrt{w^2 + 1}$ and this maps G onto Ω. The following figure illustrates the mapping properties of g in the upper half plane.

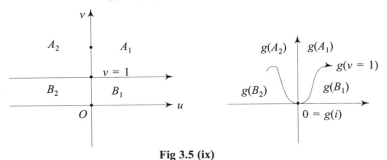

Fig 3.5 (ix)

Here A_i, B_i, are regions excluding their linear boundaries and A_i', B_i', are the image regions excluding their boundaries.

Now we shall construct the Riemann surface for the domain of $w = \sqrt{z^2 - 1}$. We shall take two different copies of the z-plane and cut them along $[-1, 1]$. Distinguish the upper and the lower edge in both the sheets. Join the lower edge of the first sheet to the upper edge of the second sheet and lower edge of the second sheet to the upper edge of the first. Keep both the sheets glued at $+1$ and -1. This is the required Riemann Surface. We also observe that using $f(z)$ we can map the first sheet onto G. Similarly we can use $-f(z)$ to map the second sheet onto G. We shall now describe neighbourhood system for points on this surface. Call points on the surface not equal to ±1 and not on the edges of the cuts as points of the first category. Call points belonging to the edges but not equal to ±1 as points of the second category. To each z in the first category we describe small circles around them contained in the same sheet. To a point z in the second category a neighbourhood will be the inside of a circle around that point half of which belongs to one sheet and the other half to the other sheet. To -1, we describe a circle starting on the first sheet to the left of -1 moving in the lower half of the first sheet, going to the second sheet via the lower edge, encircling -1 in the second sheet and coming back to the first sheet via the lower edge of the second sheet and returning to the original point. A similar description applies at $+1$ by starting to its right. For points in the first or second category their basic neighbourhoods can be mapped by either f or, f and $-f$ to usual neighbourhoods of points in the complex plane. At -1 the basic neighbourhood described above can be mapped onto an usual neighbourhood of 0 in the complex plane as follows: We shall keep in mind the mapping properties of the branch $f(z)$ described in the above theorem. Using $f(z)$ for points in the first sheet and $-f(z)$ for points in the second sheet, the basic neighbourhood of -1 on the surface corresponds to an usual neighbourhood of 0 in the complex plane. The following will illustrate the basic neighbourhood of -1 on the surface and an usual neighbourhood of origin in the complex plane.

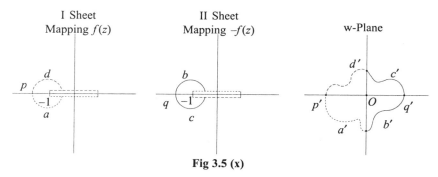

Fig 3.5 (x)

The points on the surface are denoted by a, b, c etc., and the images are denoted by a', b', c' etc. As usual we can also prove that the transition functions are analytic.

A similar surface for $z = \sqrt{w^2 + 1}$ can be constructed by gluing two complex planes at $\pm i$ and making a cut along the imaginary axis on the both the sheets from i to $-i$ and by joining the right edge of the first sheet to the left edge of the second one and the right edge of the second sheet to the left edge of the first one.

We can also use $f(z)$ to map the first sheet of the Riemann surface associated to $w = \sqrt{z^2 - 1}$ to the first sheet of the Riemann surface associated to $z = \sqrt{w^2 + 1}$ and use $-f$ to map the second sheet of the Riemann surface of $w = \sqrt{z^2 - 1}$ to the second sheet of the Riemann surface of $z = \sqrt{w^2 + 1}$. The edges of the first surface can be made to correspond to the edges of the second surface as illustrated in the Figure 3.5(xi).

Here edges of the first surface are denoted by PAQ, PBQ etc., and the corresponding edges of the second surface are denoted by $P'A'Q'$, $P'B'Q'$ etc. One can verify that this association exhibits a one-to-one correspondence between these surfaces.

Looking at the properties of the various Riemann surfaces described above we can unify the common features and formulate the following definitions.

A topological space X in which around each point x there is a neighbourhood U and a local homeomorphism f_U mapping U onto an open

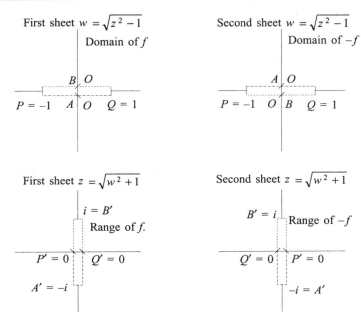

Fig 3.5 (xi)

subset of $\mathbb{C}$ is said to be a complex manifold of dimension 1 or a Riemann surface if the transition maps $f_V \circ f_U^{-1}$ from $f_U(U \cap V)$ into $f_V(U \cap V)$ (whenever they make sense) are analytic as mappings from $\mathbb{C}$ to $\mathbb{C}$. We can also talk about analytic maps between two Riemann surface as follows: If $h: R \to R'$ is a continuous map between two Riemann surfaces R and R' we say h is analytic if given $x \in R$, a canonical neighbourhood U of x, a homeomorphism $f_U: U \to \mathbb{C}$, a canonical neighbourhood U' of $f(x) = x' \in R'$, a homeomorphism $f_{U'}: U' \to \mathbb{C}$, we have $f_{U'} \circ f_V \circ f_U^{-1} : f_U(U) \to f_{U'}(U')$ analytic as a map between two open subsets of $\mathbb{C}$. It is interesting to note that this notion is independent of the chart (U, f_U) at x and the chart $(U', f_{U'})$ at x'. Several properties of complex valued functions of a complex variable could be generalized to this set up.

SOLVED EXERCISES

Problem 1   Describe the mapping properties of $w = \dfrac{z}{1-z}$.

Solution w maps $|z| = 1$ onto Re $w = -\dfrac{1}{2}$ $\left(\text{as Re}\dfrac{e^{i\theta}}{1-e^{i\theta}} = -\dfrac{1}{2}\right)$ and so $|z| < 1$ is mapped onto Re $w > -1/2$ and $|z| > 1$ is mapped onto Re $w < -1/2$. (Note that $w(0) = 0$).

Further the crescent region described by $|z| < 1$ and $\left|z - \dfrac{1}{2}\right| > \dfrac{1}{2}$ is

mapped onto $-\dfrac{1}{2} < \operatorname{Re} w < 0$ (Because $|z| = 1$ is mapped onto $\operatorname{Re} w = -\dfrac{1}{2}$

and $\left|z - \dfrac{1}{2}\right| = \dfrac{1}{2}$ is mapped onto $\operatorname{Re} w = 0$ and one point in the crescent

region namely $z = -\dfrac{1}{2}$ corresponds to $w = -1/3$ which lies in $-\dfrac{1}{2} < \operatorname{Re} w < 0$).

Developing w in a power series around origin we see that $w = z + z^2 + z^3 + \ldots$ and we shall see later how this function maximizes the modulii of coefficients around origin among one-to-one conformal mappings that take $|z| < 1$ onto regions that are convex.

Problem 2 Describe the mapping properties of $w = \dfrac{z}{(1-z)^2}$.

Solution $w = f(z)$ maps $|z| = 1$ onto the part of the real axis from $-\infty$ to $-1/4$ since

$$\frac{e^{i\theta}}{(1 - e^{i\theta})^2} = \frac{e^{i\theta} + e^{-i\theta} - 2}{4(1 - \cos \theta)^2} \leq -\frac{1}{4}.$$

Further $f(z)$ is one-to-one in $|z| < 1$ since $f(z_1) = f(z_2)$, $|z_1| < 1$, $|z_2| < 1$ implies $z_1(1 - z_2)^2 = z_2(1 - z_1)^2$ or $(z_1 - z_2)(1 - z_1 z_2) = 0$ which holds if and only if $z_1 = z_2$. f is conformal at all points except at $z = \pm 1$ since $f'(z) = \dfrac{1+z}{(1-z)^3}$. Similarly $f(z)$ is one-to-one in $|z| > 1$ and f maps $|z| > 1$ onto

the whole plane except for $(-\infty, -1/4]$. Note that $f(z) = f\left(\dfrac{1}{z}\right)$. f maps $|z|$

< 1 or $|z| > 1$ onto the region $\Omega = \mathbb{C}\backslash(-\infty, -1/4]$ which is such that each point in Ω can be seen from the origin within Ω. i.e., a region star-shaped with respect to origin (Again $w = z + 2z^2 + 3z^3 + \ldots$ maximizes the modulii of the coefficients of mappings that maps $|z| < 1$ injectively into the plane with $f(0) = 0 = 1 - f'(0)$, as we shall see later).

Problem 3 Map inside of a circle centre a and radius R onto the inside of the circle centre a' and radius R'.

Solution To find such a map we first take a linear fractional transformation w_1 that maps the circle with centre a and radius R onto the unit circle. Similarly let w_2 map the circle with centre a' and radius R' onto the unit circle. The required transform w is then given by $w_2^{-1} w_1$. In fact w_1

$= \dfrac{z - a}{R}$ and $w_2 = \dfrac{z - a'}{R'}$ and so $w_2^{-1} = R'z + a'$ and $w = R'\dfrac{z - a}{R} + a'$.

Problem 4   Prove the continuity of the principal branch of $w = z^{1/n}$ without using Theorem 3.1.4 (See Theorem 3.4.2).

Solution   We shall first prove the following Lemma.

Lemma   Let $z_0 \in \Omega = \mathbb{C} \setminus (-\infty, 0]$. Let $w_0 = z_0^{1/n}$ with $\theta_0 = \arg z_0$ satisfying $|\theta_0| < \pi$. As z varies in a sufficiently small neighbourhood of z_0 in Ω and $w = z^{1/n}$ with $|\operatorname{Arg} w| < \dfrac{\pi}{n}$ the expressions $|w^{n-1} + w^{n-2} w_0 + \ldots + w_0^{n-1}|$ have a positive minimum ρ.

Proof   Choose β such that $-\pi < \theta_0 - \beta < \theta_0 + \beta < \pi$ and consider the sector $S = \{z = re^{i\theta}: \theta_0 - \beta \le \theta \le \theta_0 + \beta\} \subset \Omega = \mathbb{C} \setminus (-\infty, 0]$.

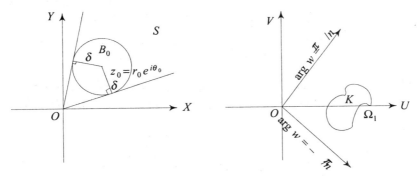

Fig 3.5.E(i)

Let B be the closed ball centre z_0 and radius $\delta = |z_0| \sin \beta$ and denote the corresponding open ball as B_0. From the above figure it is clear that $B_0 \subset B \subset S \subset \Omega$. As z varies in B the values $w = z^{1/n}$, with $|\operatorname{Arg} w| < \dfrac{\pi}{n}$, are such that $|\operatorname{Arg} w| = \dfrac{1}{n} |\operatorname{Arg} z| \le \dfrac{1}{n} |\theta_0 + \beta| < \dfrac{\pi}{n}$. Further by the continuity of the mappings $z \to |z| \to |z|^{1/n}$ (Lemma 3.4.3) as z varies over B, a compact set in Ω, $|w| = |z|^{1/n}$ are bounded say by M. Hence these values w lie in a compact subset K of $\Omega_1 = \{w \in \mathbb{C}: |\operatorname{Arg} w| < \pi/n\}$. We now claim that the function $g(w) = w^{n-1} + w^{n-2} w_0 + \ldots + w_0^{n-1}$ is continuous and never vanishes in K. Since $g(w)$ is a polynomial its continuity is clear. First $g(w_0) = n w_0^{n-1} \ne 0$. By Lemma 3.4.4 the mapping $w \to w^n$ with $|\operatorname{Arg} w| < \dfrac{\pi}{n}$ is one-to-one on Ω_1. Thus if $g(w) = 0$ for some $w \ne w_0$, $w \in K$ then $w^n - w_0^n = (w - w_0) g(w) = 0$, a contradiction. Thus on this compact set K, $|g(w)| \ge \rho > 0$ proving our lemma.

We shall now prove that the principal branch of $w = z^{1/n}$ is continuous in Ω. Let $w_0 = z_0^{1/n}$ with $|\text{Arg } w_0| < \dfrac{\pi}{n}$. Let $\varepsilon > 0$ be given. Choose β, δ and ρ as in the above lemma. Now

$$z - z_0 = w^n - w_0^n = (w - w_0)(w^{n-1} + w^{n-2}w_0 + \ldots + w_0^{n-1})$$

or that

$$|w - w_0| = \frac{|z - z_0|}{|w^{n-1} + w^{n-2}w_0 + \ldots + w_0^{n-1}|} \leq \frac{|z - z_0|}{\rho}.$$

Thus choosing $\eta < \min\{\delta, \rho\varepsilon\}$ we see that $|z - z_0| < \eta \Rightarrow |w - w_0| < \varepsilon$. This proves the continuity of w at z_0. Since z_0 is arbitrary in Ω we see that $w = z^{1/n}$ is continuous in Ω.

Problem 5   Prove the continuity of the principal branch of $\log z$ without using Theorem 3.1.4 (See Theorem 3.4.7).

Solution   Let $w = f(z) = \log z$ with $|\text{Im} \log z| < \pi$ be the principal branch defined in $\Omega = \mathbb{C} \backslash (-\infty, 0]$. Let $z_0 \in \Omega$ and $\varepsilon > 0$ be given and $w_0 = f(z_0) = u_0 + iv_0$ with $|v_0| < \pi$. We have to find a $\delta > 0$ such that $|z - z_0| < \delta$, $z \in \Omega \Rightarrow |f(z) - f(z_0)| < \varepsilon$. Consider the figure below.

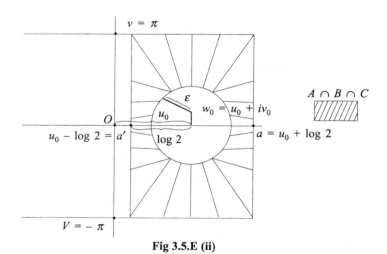

Fig 3.5.E (ii)

For sufficiently small $\varepsilon > 0$ ($\varepsilon < \min\{\log 2, \pi + v_0, \pi - v_0\}$) the set $A \cap B \cap C \neq \phi$ where $A = \{w : |w - w_0| \geq \varepsilon\}$, $B = \{w = u + iv : |v| \leq \pi\}$ and $C = \{w = u + iv : |u - u_0| \leq \log 2\}$. Since each of A, B, C is closed $A \cap B \cap C$ is closed. Obviously $A \cap B \cap C$ is bounded and so is compact. Since $e^w - e^{w_0}$ is continuous on this compact set $A \cap B \cap C$ and $e^w \neq$

e^{w_0} in $A \cap B \cap C$ (if $e^w = e^{w_0}$ then $w = w_0$ or $w = w_0 + 2n\pi i$ with $|n| \geq 1$. i.e., if $w \neq w_0$, $|\text{Im } w| = |2n\pi + \text{Im } w_0| \geq |2n\pi| - |\text{Im } w_0| > 2\pi - \pi = \pi$ which is a contradiction to the fact that $w \in B$) we have

$$0 < \rho = \text{Min } \{|e^w - e^{w_0}| : w \in A \cap B \cap C\}.$$

Now choose $\delta < \min \left\{\rho, \frac{1}{2} e^{u_0}\right\}$ and assume $|z - z_0| < \delta$. We claim first that the corresponding $w = f(z) = \log z$ does not belong to $A \cap B \cap C$. In fact, if it did, $|z - z_0| = |e^w - e^{w_0}| \geq \rho > \delta$, a contradiction. But w does belong to B because $|\text{Im } w| < \pi$. Also $w \in C$ because, if not $w = u + iv$ with $|u - u_0| > \log 2$. There are two cases to consider (i) $u > u_0 + \log 2$ (ii) $u < u_0 - \log 2$.

In case (i) $u > u_0 + \log 2$, $e^u > 2e^{u_0}$,

$$|z - z_0| = |e^w - e^{w_0}| \geq |e^w| - |e^{w_0}| = e^u - e^{u_0} > e^{u_0} > \frac{1}{2} e^{u_0} > \delta$$

which is a contradiction. In case (ii)

$$|z - z_0| = |e^w - e^{w_0}| \geq |e^{w_0}| - |e^w| = e^{u_0} - e^u > \frac{1}{2} e^{u_0} > \delta$$

which is again a contradiction. Now $w \notin A \cap B \cap C$ but $w \in B \cap C \Rightarrow w \notin A$. Therefore $|w-w_0| < \varepsilon$, proving continuity of $w = f(z)$.

Problem 6 Give a precise definition of a single-valued branch of log (log z) in a suitable region, and prove that it is analytic.

Solution We note that the principal branch of $\log z$ is defined in Ω which is the complement of non-positive real axis in the z-plane and maps Ω onto the strip $|v| < \pi$ in the $w = u + iv$ plane. Further, the principal branch of $\log w$ in the w plane on this image can also be defined, provided we exclude the non-positive u-axis. Thus in order to get a single valued analytic branch for log (log z) we have to remove from Ω, the closed segment $[0, 1]$ in the z-plane which corresponds under principal branch of $\log z$ to the non-positive u-axis. Thus a single-valued analytic branch of log (log z) exists in $\Omega \setminus [0, 1] = \mathbb{C} \setminus (-\infty, 1]$ as composition of both the principal branches of $w = \log z$ and $\zeta = \log w$.

Problem 7 Show that a linear fractional transformation which satisfies $S^n z = z$ for some positive integer n is necessarily elliptic.

Solution If S is the identity map then it is necessarily elliptic. Let S be not the identity map but satisfy $S^n z = z$, $\forall z$, $(n \neq 0)$. Claim: S can not have coincident fixed points. Indeed, if not, $w = Sz$ has the form $\dfrac{k}{w - a} = \dfrac{k}{z - a}$

$+ c$ $(c \neq 0)$. Thus putting $w^m = S^m z$ we have $\dfrac{k}{w^2 - a} = \dfrac{k}{w - a} + c$

$$= \frac{k}{z-a} + 2c.$$ Similarly $\dfrac{k}{w^n - a} = \dfrac{k}{z-a} + nc.$ If w^n is identity then $nc = 0$

$\Rightarrow n = 0$, a contradiction. Hence the claim follows. If S has one fixed point at ∞ and another fixed point at a finite point "a", $w = Sz$ has the form

$\dfrac{k'}{w-a} = \dfrac{k}{z-a}$. Thus with same notations as before, $\dfrac{k'^2}{w^2 - a} = k' \dfrac{k}{w-a} =$

$\dfrac{k^2}{z-a}$. Similarly $\dfrac{k'^n}{w^n - a} = \dfrac{k^n}{z-a}$. If $w^n = z$ we have $k'^n = k^n$. Thus

$\dfrac{1}{w-a} = \dfrac{k}{k'} \dfrac{1}{z-a}$ with $\left| \dfrac{k}{k'} \right| = 1$. Hence S is an elliptic transformation.

Now let a and b be two finite fixed points of $w = Sz$.

Then $$\frac{w-a}{w-b} = k \frac{z-a}{z-b}.$$

We have to prove that $|k| = 1$. (i.e., S is elliptic) $\dfrac{w^2 - a}{w^2 - b} = k \dfrac{w-a}{w-b}$

$= k^2 \dfrac{z-a}{z-b}$. Similarly $\dfrac{w^n - a}{w^n - b} = k^n \dfrac{z-a}{z-b}$. w^n is identity $\Rightarrow k^n = 1 \Rightarrow |k|$

$= 1$. Hence the result follows.

EXERCISES

(1) Give a precise definition of a single valued branch of $\sqrt{1+z} + \sqrt{1-z}$ in a suitable region and prove that it is analytic.

(2) Suppose that $f(z)$ is analytic and satisfies the condition $|f(z)^2 - 1| < 1$ in a region Ω. Show that either Re $f(z) > 0$ or Re $f(z) < 0$ throughout Ω. (**Hint:** Use connectedness of Ω after proving Re $f(z) \neq 0$ in Ω)

(3) Suppose that the coefficients of the transformation $Sz = \dfrac{az+b}{cz+d}$ are normalized by $ad - bc = 1$. Show that Sz is
   (i) elliptic iff $-2 < a + d < 2$.   (ii) parabolic iff $a + d = \pm 2$.
   (iii) hyperbolic iff $a + d < -2$ or $a + d > 2$.

(4) Map the common part of the discs $|z| < 1$ and $|z - 1| < 1$ conformally onto the inside of the unit circle. Choose the mapping so that the symmetries are preserved.

(5) Map the region between $|z| = 1$ and $|z - (1/2)| = 1/2$ on a half plane.

(6) Map the outside of the ellipse $\left(\dfrac{x}{a}\right)^2 + \left(\dfrac{y}{b}\right)^2 = 1$ onto $|w| < 1$ with preservations of symmetries.

(**Hint:** $-w = \dfrac{1}{2}(ze^{-\alpha} + z^{-1}e^{\alpha})$, $\sinh \alpha = b$, $\cosh \alpha = a$).

(7) Describe the Riemann surface associated with $w = e^z$.

(8) Describe the Riemann surface associated with the function

$$w = \frac{1}{2}\left(z + \frac{1}{z}\right).$$

(9) Describe the Riemann surface associated with the function $w = (z^2 - 1)^2$.

(10) Show that the linear fractional transformation $z = X + iY = \dfrac{z - 1}{z + 1}$

($z = x + iy$) maps the x-axis onto the X-axis and the half planes $y > 0$ and $y < 0$ onto the half planes $Y > 0$ and $Y < 0$ respectively. Show also that in particular it maps the segment $-1 \le x \le 1$ of the x-axis onto the segment $X \le 0$ of the X-axis. Then show that when the principal branch of the square root is used, the composite function w

$$= Z^{1/2} = \left(\frac{z - 1}{z + 1}\right)^{1/2} \quad \text{maps the } z\text{-plane except for the segment } -1 \le x$$

≤ 1 of the x-axis onto the half plane $u > 0$.

NOTES

Gauss [4] was the first to deal with angle preserving mappings. In fact it was he who discovered the relationship between angle preserving mappings and analytic or anti-analytic functions (f is anti-analytic if $\bar{f}$ the complex conjugate of f is analytic). Riemann [10] on the other hand recognizes that analytic functions preserve the shapes of regions between small parts of the z-plane and their images in the w-plane. He also recognizes the dependence of $|w|$ on $|z|$ locally if w is an analytic function of z. Curiously angle preserving maps play no role at all in the works of Cauchy and Weierstrass.

Regarding other physical applications of the Theory of Conformal mappings like Temperature flow in a quadrant, electrostatic potential, two dimensional fluid flow, stream functions etc., we refer to [2]. Besides physical applications, conformal mappings are essentially used in getting solutions of partial differential equations with given boundary conditions [3].

Several conformal mappings with given domains and ranges have been complied by H. Kober [7]. Another book containing a list of several conformal mappings with their domains and ranges is [8]. A few more results concerning conformal mapping properties of analytic functions will be given at the end of chapter **4**.

Regarding single-valued analytic branches for log z one should mention that there is no branch of log z which is continuous through out $\mathbb{C}\backslash\{0\}$. Any such branch must be of the form log $|z| + i$ arg z where $n\pi \le$ arg $z < (n + 2)\,\pi\ (n \in \mathbb{Z})$. However this branch is discontinuous on the line arg $z = n\pi$ as can be easily verified.

While defining abstract Riemann surfaces we mentioned that several properties of complex functions of a complex variable could be extended to this set up. A few interesting questions are the following: (i) How many simply connected Riemann surfaces are there up to conformal equivalence (Bijective, Biholomorphic maps between Riemann surfaces)? (ii) How to characterize compact Riemann surfaces? The answer for (i) above is that there are only three surfaces namely the finite complex plane (parabolic case), the extended complex plane or the Riemann sphere (elliptic case) and the open unit disc (hyperbolic case). The proof is complicated and the reader can refer to [1]. The second question is answered by the characterization that any compact Riemann surface is equivalent to either a sphere or a sphere with a finite number of handles (The number of handles is called the genus of the surface). One example of a compact Riemann surface of genus one is the Torus (Geometrically it is easy to see how a Torus is equivalent to a sphere with one handle by blowing air into a ring ball keeping the pressure concentrated at some portion of the ball). For more details about Riemann surfaces we refer to [5, 6, 9].

It is interesting to note that one can construct crude models for the cross sections of the Riemann surfaces associated with the functions $w = z^3$ and $w = \cos z$. In these constructions we come across the following difficulties. (i) Gluing branch points belonging to different sheets. (ii) Making a fine cut on a ray without loss of points. (iii) Identifying the edges from one sheet to the other. However, symbolically we can get rid of some of these difficulties in the following ways: (i) Making holes in different sheets and representing the gluing of branch points by a rod. (ii) Loss of points on edges can not however be rectified. (iii) The edges can be joined cross wise by drilling holes on the edges and allowing a thread to pass through the various holes.

For $w = z^3$, one sheet of cardboard for the domain and three sheets of card boards for the surface could be used. For $w = \cos z$, one sheet of cardboard for the domain and four sheets of card boards for the images of four successive fundamental regions in the surface could be used. Colour codes could be used to distinguish various types of fundamental regions and their images. A separate chart indicating the correspondence, the colour codes used and all the details pertaining to the understanding of the two surfaces can also be kept along with the models. These models can be mounted on wooden planks for easy demonstration. These models could be used as teaching aids for graduate students.

References

1. L. Ahlfors, Conformal Invariants, McGraw-Hill Book Company, 1973.
2. R. V. Churchill and J.W. Brown, Complex variables and applications, McGraw-Hill Book Company, V Edition, 256-278, 1990.
3. S.J. Farlow, Partial differential equations for Scientists and Engineers, John Wiley and Sons, Lesson 47, 379-388, 1982.

4. Gauss, Math, Werke, 4, 189-216, 1825.

5. Gilbert Ames Bliss, Algebraic functions, American Mathematical Society, 1933.

6. Hermann Weyl, The concept of a Riemann surface, Addison-Wesley Publishing Company, Inc.

7. H. Kober, Dictionary of Conformal Mappings, Dover, 1960.

8. M. Krasnov et al, Mathematical analysis for engineers, II, Mir Publishers, Moscow, 1990.

9. Mathematical pamphlets-1, Riemann surfaces, TIFR, 1963.

10. B. Riemann, "Grundlagenfür eine allgemeine Theories der Functionen einer veränderlichen complexen Grösse", Inaugural Dissertation (1851), Göttingen; Werke, 5-43.

Complex Integral Calculus

4.1 Basic Definition and Properties of Complex Integration

Introduction 4.1.1

The integral calculus for complex-valued function of a complex variable forms an integral part of complex analysis and it is central to this theory. Many significant results in complex function theory can be proved with the aid of integral calculus.

The study of line integrals for complex-valued functions of a complex variable can be easily developed based on Riemann's theory of integration of bounded real-valued function of a real variable defined on a finite interval on the real axis.

We shall assume that the reader is familiar with this part of real analysis. However for the sake of completeness we shall quickly recall the basic definitions and properties.

Let $f(x)$ be a bounded real-function defined on $[a, b]$, $-\infty < a < b < \infty$. We say f is Riemann integrable or integrable on $[a, b]$ if

$$\int_{\underline{a}}^{b} f(x)\,dx = \sup_{P} L(P, f) = \inf_{P} U(P, f) = \int_{a}^{\overline{b}} f(x)\,dx$$

where for any partition $P = \{a = x_0 < \ldots < x_n = b\}$ of $[a, b]$ and

$$M_i = \sup_{x_{i-1} \le x \le x_i} f(x); \quad m_i = \inf_{x_{i-1} \le x \le x_i} f(x)$$

$$U(P, f) = \sum_{i=1}^{n} M_i (x_i - x_{i-1}) \quad \text{and} \quad L(P, f) = \sum_{i=1}^{n} m_i (x_i - x_{i-1}).$$

The common value of $\int_{\underline{a}}^{b} f(x)dx$ and $\int_{a}^{\overline{b}} f(x)dx$ is usually denoted by $\int_{a}^{b} f(x)dx$ and is called the Riemann integral of f over $[a, b]$ or simply integral of f over $[a, b]$ and it has the following properties. We leave the proofs.

Theorem 4.1.2

(i) If f and g are integrable on $[a, b]$ then so is $\alpha f + \beta g$ ($\alpha, \beta \in \mathbb{R}$) and

$$\int_a^b (\alpha f + \beta g)(x)\,dx = \alpha \int_a^b f(x)\,dx + \beta \int_a^b g(x)\,dx.$$

(ii) If f and g are integrable on $[a, b]$ then $f \le g$ on $[a, b]$ implies

$$\int_a^b f(x)\,dx \le \int_a^b g(x)\,dx.$$

(iii) If f is integrable on $[a, b]$, $m \le f \le M$ and if ϕ is continuous from $[m, M]$ into $\mathbb{R}$, then $\phi \cdot f$ (the composition) is integrable on $[a, b]$.

(iv) f is integrable on $[a, b]$ if and only if for each $a < c < b$, f is integrable on $[a, c]$ and $[c, b]$ and

$$\int_a^b f(x)\,dx = \int_a^c f(x)\,dx + \int_c^b f(x)\,dx .$$

(v) Any continuous function on $[a, b]$ is integrable on $[a, b]$.

(vi) If f is integrable on $[a, b]$ then $|f|$ is also integrable on $[a, b]$ and

$$\left| \int_a^b f(x)\,dx \right| \le \int_a^b |f(x)|\,dx.$$

(vii) If ϕ is a strictly increasing differentiable function from $[\alpha, \beta]$ onto $[a, b]$ and f is integrable on $[a, b]$ and ϕ' is integrable on $[\alpha, \beta]$ then

$$\int_a^b f(x)\,dx = \int_\alpha^\beta f(\phi(y))\,\phi'(y)\,dy.$$

(viii) If γ is a differentiable curve given by $z = z(t)$, ($a \le t \le b$) then γ is rectifiable and its length $l(\gamma) = \int_a^b |z'(t)|\,dt$.

Definition 4.1.3

Let $f(t) = u(t) + iv(t)$ be a complex-valued function of a real variable t in $[a, b]$. We say f is integrable if both u, v are integrable on $[a, b]$ and write

$$\int_a^b f(t)\,dt = \int_a^b u(t)\,dt + i \int_a^b v(t)\,dt.$$

Theorem 4.1.4

Let f, g be complex-functions of a real variable x in $[a, b]$. Then

(i) If f and g are integrable on $[a, b]$ then for complex α, β, $\alpha f + \beta g$ is integrable on $[a, b]$ and

$$\int_a^b (\alpha f + \beta g)(x)\,dx = \alpha \int_a^b f(x)\,dx + \beta \int_a^b g(x)\,dx.$$

(ii) f is integrable on $[a, b]$ if and only if for each $a < c < b$, f is integrable on $[a, c]$ and on $[c, b]$, and

$$\int_a^b f(x)\,dx = \int_a^c f(x)\,dx + \int_c^b f(x)\,dx.$$

(iii) If f is a continuous function on $[a, b]$ then f is integrable on $[a, b]$.

(iv) If f is integrable on $[a, b]$ then $|f|$ is also integrable on $[a, b]$ and

$$\left|\int_a^b f(x)\,dx\right| \le \int_a^b |f(x)|\,dx.$$

(v) If $f'(x)$ exists and is continuous on $[a, b]$ then

$$\int_a^b f'(x)\,dx = f(b) - f(a).$$

Proof

The proofs of (i), (ii) and (iii) follow easily from the corresponding results of Theorem 4.1.2. (v) is easily verified using the fundamental theorem of integral calculus for real and imaginary parts of $f'(x)$. We shall prove (iv). Indeed if f is integrable then so are u, v, u^2 and v^2 and hence $|f| = \sqrt{u^2 + v^2}$ by Theorem 4.1.2 (iii).

Let $\int_a^b f(t)\,dt \ne 0$. Otherwise, there is nothing to prove.

Put $\theta = \arg\left(\int_a^b f(t)dt\right)$ and consider

$$\int_a^b \mathrm{Re}\,[e^{-i\theta} f(t)]\,dt = \mathrm{Re}\int_a^b e^{-i\theta} f(t)\,dt$$

$$= \mathrm{Re}\,e^{-i\theta}\int_a^b f(t)\,dt = \left|\int_a^b f(t)\,dt\right|.$$

We know that both $\mathrm{Re}\,[e^{-i\theta} f(t)]$ and $|f(t)|$ are real and Riemann integrable and also

$$\mathrm{Re}\,[e^{-i\theta} f(t)] \le |e^{-i\theta} f(t)| = |f(t)|.$$

Thus using Theorem 4.1.2 (ii) we get that $\left|\int_a^b f(t)dt\right| = \int_a^b \mathrm{Re}\,[e^{-i\theta} f(t)]\,dt$

$$\le \int_a^b |f(t)|\,dt.$$

With the above definitions and properties we are now in a position to define the following:

Definitions 4.1.5

(a) By a path we mean a sequence of curves $\{\gamma_1, \gamma_2, \ldots, \gamma_n\}$ with $\gamma_j(a_{j+1})$ $= \gamma_{j+1}(a_{j+1})$ for $j = 1, 2, \ldots, n - 1$, where each γ_j is defined in the parametric interval $[a_j, a_{j+1}]$. If in addition each γ_j is a horizontal or a vertical line segment (line segments parallel to co-ordinate axes) then we say γ is a rectangular path.

(b) A curve $w = w(s)$ $(\alpha \leq s \leq \beta)$ is called a reparametrization of $z = z(t)$ $(a \leq t \leq b)$, if there exists a strictly increasing function $t = t(s)$ from $[\alpha, \beta]$ onto $[a, b]$ such that $w(s) = z(t(s))$ $(s \in [\alpha, \beta])$.

(c) A curve γ, given by $z = z(t)$ defines an opposite curve $-\gamma$, given by $w = w(s) = z(-s)$ $-b \leq s \leq -a$.

(d) If γ is a piece-wise differentiable curve given by the equation $z = z(t)$ $a \leq t \leq b$ and $f(z)$ is defined and continuous on γ then

$$\int_\gamma f(z)\,dz = \int_a^b f(z(t))\,z'(t)\,dt.$$ In the right hand side we assume that the

integral is split into sum of integrals over subintervals of $[a, b]$ in each of which $z'(t)$ is continuous. In the same context we also define

$$\int_\gamma f(z)\,|dz| = \int_a^b f(z(t))\,|z'(t)|\,dt., \text{ and some times also write it as } \int_\gamma f\,ds$$

where s is the arc-length parameter on γ.

(e) If $P(x, y)\,dx + Q(x, y)\,dy$ is a general differential 1– form, with P and Q being continuous complex-functions defined on a piece-wise differentiable curve γ given by $x = x(t)$ and $y = y(t)$, $a \leq t \leq b$, then

$$\int_\gamma P dx + Q dy = \int_a^b [P(x(t), y(t))] x'(t)\,dt + \int_a^b [Q(x(t), y(t))]\,y'(t)\,dt.$$

Theorem 4.1.6

Let γ be a differentiable curve given by $z = z(t)$ $(a \leq t \leq b)$. Let $f(z)$ be a continuous function on γ. Then

(i) If σ is a reparametrization of γ then

$$\int_\gamma f(z)\,dz = \int_\sigma f(z)\,dz$$

(ii) $$\int_{-\gamma} f(z)\,dz = -\int_\gamma f(z)\,dz$$

(iii) $$\int_{-\gamma} f(z)\,|dz| = \int_\gamma f(z)\,|dz|$$

(iv) $$\left| \int_\gamma f(z)\,dz \right| \leq \int_\gamma |f(z)|\,|dz|$$

(v) For any complex constants α and β and continuous functions f and g on γ we have

$$\int_\gamma (\alpha f + \beta g)(z)\,dz = \alpha \int_\gamma f(z)\,dz + \beta \int_\gamma g(z)\,dz$$

(vi) If $a < c < b$ then with $\gamma_1 = \gamma|_{[a,\,c]}$, $\gamma_2 = \gamma|_{[c,\,b]}$ (restrictions of γ (of $z(f)$) to $[a,\,c]$ and $[c,\,b]$]

$$\int_\gamma f(z)\,dz = \int_{\gamma_1} f(z)\,dz + \int_{\gamma_2} f(z)\,dz.$$

(vii) If the end points of γ are α and β then $\int_\gamma dz = \beta - \alpha$.

(viii) Let $g(z)$ be a continuous function defined in a region Ω. If $g(z) = f'(z)$ where $f(z)$ is analytic in Ω then

$$\int_\gamma g(z)\,dz = f(\beta) - f(\alpha)$$

where α and β are the end points of γ. In particular if γ is a closed curve $\int_\gamma g(z)\,dz = 0$.

(ix) $\int_\gamma |dz| = \int_a^b |z'(t)|\,dt = l(\gamma)$ (= length of the curve γ).

Proof

Since (i)-(vii) and (ix) can be verified using the definitions and Theorem 4.1.2 and Theorem 4.1.4 we merely prove (viii). We are assuming that γ is a differentiable curve given by $z = z(t)$ ($a \le t \le b$) with $z(a) = \alpha$ and $z(b) = \beta$. From the definition we have $\int_\gamma g(z)\,dz = \int_a^b g(z(t))z'(t)\,dt$. But $f'(z) = g(z)$

$\Rightarrow \dfrac{d}{dt}(f(z(t)) = f'(z(t))z'(t) = g(z(t))z'(t)$. Thus $\int_\gamma g(z)\,dz = \int_a^b \dfrac{d}{dt}(f(z(t))$

$dt = f(z(b)) - f(z(a)) = f(\beta) - f(\alpha)$. In particular if γ is closed we have $\alpha = \beta$ and so $\int_\gamma g(z)\,dz = 0$.

Remarks 4.1.7

(i) If $f(z) = u(x,\,y) + iv(x,\,y)$ is continuous on a piece-wise differentiable curve γ given by $z = z(t) = x(t) + iy(t)$ ($a \le t \le b$) then $\int_\gamma f(z)\,dz$ can be viewed in two different ways, one using the definition and another viewing $f(z)\,dz$ as $P\,dx + Q\,dy$ where $P = f = u + iv$ and $Q = if = -v + iu$. For this P and Q

$$\int_\gamma P\,dx + Q\,dy = \int_a^b [u(x(t),\,y(t)) + iv\,(x(t),\,y(t))]\,x'(t)\,dt$$

$$+ \int_a^b [-v(x(t),\,y(t)) + iu(x(t),\,y(t))]y'(t)\,dt$$

$$= \int_a^b (u + iv) \, (x' + iy') \, dt = \int_a^b f(z(t))z'(t) \, dt = \int_\gamma f(z) \, dz.$$

(ii)   If γ_1 is given by $z = z(t)$, $a \le t \le b$ and γ_2 is given by $w = w(s)$, $b \le s \le c$ and if $z(b) = w(b)$ then we can define a new curve which we call as $\gamma_1 + \gamma_2$ whose equation $\zeta = \zeta(\lambda)$ ($a \le \lambda \le c$) is given by

$$\zeta(\lambda) = \begin{cases} z(\lambda) \text{ for } a \le \lambda \le b \\ w(\lambda) \text{ for } b \le \lambda \le c. \end{cases}$$

It follows easily that if γ_1 and γ_2 are piece-wise differentiable then so is $\gamma_1 + \gamma_2$ and

$$\int_\gamma f(z) \, dz = \int_{\gamma_1} f(z) \, dz + \int_{\gamma_2} f(z) \, dz.$$

This can be extended by induction to a finite number of curves. Thus we can regard every path as defined in 4.1.5 as a curve. In fact this observation enables us to use the formal notation $\gamma = \sum_{i=1}^{n} \gamma_i$ for disjoint piece-wise differentiable curves γ_i and to define $\int_\gamma f(z) \, dz = \sum_{i=1}^{n} \int_{\gamma_i} f(z) \, dz$.

Theorem 4.1.8

If $f(z)$ is continuous on a differentiable curve γ then we call $\sum_{k=1}^{n} f(\zeta_k)\Delta z_k$ [where γ is divided into subarcs given the division $a = z_0, z_1, \ldots, z_n = b$ where a and b are end points of γ and $\Delta z_k = (z_k - z_{k-1})$ and ζ_k's are points in the subarcs $[z_{k-1}, z_k]$] as a complex integral sum of f along γ. For various partitions of γ such that $\max_k |\Delta z_k| \to 0$ and for arbitrary choice of ζ_k in the sub arc $[z_{k-1}, z_k]$ the limit of $\sum_{k=1}^{n} f(\zeta_k) \, \Delta z_k$ exists and is equal to $\int_\gamma f(z) \, dz$.

Proof

We put $f(z) = u(x, y) + iv(x, y)$, $z_k = x_k + iy_k$, $\Delta z_k = \Delta x_k + i\Delta y_k = z_k - z_{k-1}$, $\Delta x_k = x_k - x_{k-1}$, $\Delta y_k = y_k - y_{k-1}$, $\zeta_k = \xi_k + i\eta_k$, $u_k = u_k(\xi_k, \eta_k)$, $v_k = v_k(\xi_k, \eta_k)$,

$$\sum_{k=1}^{n} f(\zeta_k) \, \Delta z_k = \sum_{k=1}^{n} (u_k \Delta x_k - v_k \Delta y_k) + i \sum_{k=1}^{n} (v_k \Delta x_k + u_k \Delta y_k)$$

Using the Riemann-Stieltjes integration theory we can easily see that

$$\left(\max_k |\Delta z_k| \to 0 \Rightarrow \max_k |\Delta x_k| \to 0 \text{ and } \max_k |\Delta y_k| \to 0 \right) \text{ as } \max_k |\Delta z_k| \to 0$$

the right hand side tends to

$$\int_\gamma (u\,dx - v\,dy) + i\int_\gamma (v\,dx + u\,dy).$$

On the other hand $\int_\gamma f(z)\,dz$ (using the definition and Remark 4.1.7) equals

$$\int_\gamma (u + iv)\,(dx + i\,dy) = \int_\gamma (u\,dx - v\,dy) + i\int_\gamma (v\,dx + u\,dy)$$

This completes the proof.

Theorem 4.1.9

Let $\{f_n\}$ be a sequence of complex-valued continuous functions defined on a differentiable curve γ. Let $f_n \to f$ uniformly on γ. Then $\int_\gamma f(z)\,dz =$

$$\lim_{n\to\infty} \int_\gamma f_n(z)\,dz$$

Proof

$\left| \int_\gamma (f_n - f)\,(z)\,dz \right| \le \int_\gamma |(f_n - f)\,(z)|\,|dz|$. Given $\varepsilon > 0$ there exists N such that $\forall\, n \ge N$, $|(f_n - f)\,(z)| < \varepsilon\ \forall\, z \in \gamma$. Hence the right side of the above inequality becomes less than $\varepsilon l(\gamma)$ where $l(\gamma)$ is the length of the curve γ. Since ε is arbitrary the proof follows.

Corollary 4.1.10

Let $\sum_{n=1}^{\infty} f_n$ be a series of complex continuous functions defined on a differentiable curve γ converging uniformly on γ. Then

$$\int_\gamma \left(\sum_{n=1}^{\infty} f_n(z) \right) dz = \sum_{n=1}^{\infty} \int_\gamma f_n(z)\,dz.$$

Proof

Apply the previous theorem to the sequence of partial sums of $\sum_{n=1}^{\infty} f_n$.

4.2 Cauchy's Theorems

Introduction 4.2.1

Let Ω be a region in the plane and γ a piece-wise differentiable closed curve in Ω. Theorems which describe conditions on Ω or γ so that $\int_\gamma f(z)\,dz = 0$ for all analytic functions on Ω are usually referred to as Cauchy's theorems.

Cauchy's theorems are powerful tools in the study of local properties of analytic functions. We now proceed to obtain preliminary versions of this theorem and obtain the most general form in the next section. We begin with a few preliminary observations. Throughout the rest of this chapter by a 'curve' we always mean a 'piece-wise differentiable curve'.

Definition 4.2.2

A differential or a differential form $p\,dx + q\,dy$ where p and q are continuous complex-functions defined in a region in Ω is said to be exact in Ω if there exists a continuous function U defined on Ω such that $\dfrac{\partial U}{\partial x} = p,\ \dfrac{\partial U}{\partial y} = q$.

Theorem 4.2.3

Let $f(z)$ be a continuous function defined in a region Ω. The differential $f(z)\,dz$ is exact if and only if there exists an analytic function $U(z)$ defined on Ω such that $U'(z) = f(z)$ (i.e., $f(z)$ has a primitive in Ω).

Proof

Let $U(z)$ be analytic with $U'(z) = f(z)$ in Ω. Now $f(z)\ dz = f(z)\ dx +$ $if(z)\ dy$. Since $U(z)$ is analytic in Ω we have $\partial U = U'(z) = \dfrac{\partial U}{\partial x} = -i\dfrac{\partial U}{\partial y}$

(See Theorem 2.1.3). Thus $\dfrac{\partial U}{\partial x} = U'(z) = f(z)$, $\dfrac{\partial U}{\partial y} = iU'(z) = if(z)$. Thus $f(z)\ dz$ is exact. On the other hand if $f(z)\ dz$ is exact we have a continuous function $U(z)$ defined on Ω such that $\dfrac{\partial U}{\partial x} = f(z)$, $\dfrac{\partial U}{\partial y} = if(z)$. But then

$U(z)$, $\dfrac{\partial U}{\partial x}, \dfrac{\partial U}{\partial y}$ are continuous and $\dfrac{\partial U}{\partial x} = -i\dfrac{\partial U}{\partial y}$. Using Theorem 2.1.3 it follows that $U(z)$ is analytic with derivative $\partial U = U'(z) = \dfrac{\partial U}{\partial x} = f(z)$. This completes the proof.

Theorem 4.2.4

Let $f(z)$ be continuous in a region Ω. The differential $f(z)\ dz$ is exact if and only if $\int_\gamma f(z)\ dz = 0$ for all closed curves γ in Ω.

Proof

The necessary part follows from Theorem 4.2.3 and Theorem 4.1.6 (viii). To prove the other part let us assume that $\int_\gamma f(z)\ dz = 0$ for all closed curves γ in Ω. Write $f(z)\ dz = f(z)\ dx + if(z)\ dy = p\,dx + q\,dy$ and choose some base point $(x_0, y_0) \in \Omega$. Define $g(x, y) = \int_\gamma p\,dx + q\,dy$ where γ is any path that joins (x_0, y_0) and (x, y) in Ω. This function g is well defined because if γ_1 and γ_2 are any two paths joining (x_0, y_0) and (x, y) in Ω then $\gamma_1 - \gamma_2$ is a closed curve at (x_0, y_0) and by our hypothesis $\int_{\gamma_1 - \gamma_2} f(z)\ dz = 0$ or that

$$\int_{\gamma_1} p \; dx + q \; dy = \int_{\gamma_2} p \; dx + q \; dy.$$

On the other hand fixing (x_1, y_1) if we choose polygonal paths (i.e., a polygon) from (x_0, y_0) to (x_1, y_1) and (x_0, y_0) to $(x_1 + h, y_1)$ wherein the last line segments are horizontal (such a path is always possible. See Theorem 1.9.3) then

$$g(x_1 + h, y_1) - g(x_1, y_1) = \int_{\gamma} p\,dx + q\,dy$$

where γ is a horizontal line segment joining (x_1, y_1) and $(x_1 + h, y_1)$. On this line segment $x = x_1 + t, y = y_1$ where t varies from h to 0 or 0 to h (depending on whether $h < 0$ or $h > 0$). Thus $dx = dt$, $dy = 0$ on γ and we get

$$\left| \frac{g(x_1 + h, y_1) - g(x_1, y_1)}{h} - p(x_1, y_1) \right| \leq \frac{1}{|h|} \int_0^{|h|} |p(x_1 + t, y_1) - p(x_1, y_1)| \; dt.$$

Using continuity of p at (x_1, y_1) it is easy to see that the right side of the above inequality tends to zero as $|h| \to 0$. Thus $\dfrac{\partial g}{\partial x}$ at (x_1, y_1) equals $p(x_1, y_1)$. Since (x_1, y_1) is arbitrary we have $\dfrac{\partial g}{\partial x} = p$. Similarly by taking the last line segment as vertical (g is not altered by this choice) we can prove that $\dfrac{\partial g}{\partial y} = q = ip = i\dfrac{\partial g}{\partial x}$. By Theorem 2.1.3, g is analytic and $\partial g = g' = $

$\dfrac{\partial g}{\partial x} = p = f$ and an application of Theorem 4.2.3 tells us that $f(z)\,dz$ is exact.

Theorem 4.2.5 (Cauchy's theorem for a rectangle)
Let $f(z)$ be analytic in a region Ω. Let R be a rectangle such that its interior and the boundary ∂R are contained in Ω then $\int_{\partial R} f(z) \; dz = 0$.

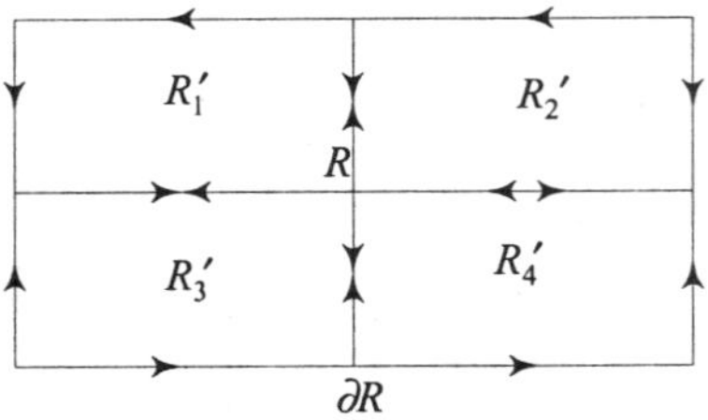

Fig 4.2 (i)

Proof

We shall divide the rectangle R into four equal parts and call the individual rectangles as R_i' and their boundaries as $\partial R_i'$. See the adjacent figure.

Let $\eta(R) = \int_{\partial R} f(z) \; dz$ and $\eta(R_i') = \int_{\partial R_i'} f(z) \; dz$. It is clear that (with orientations anti-clockwise) $\eta(R) = \sum_{i=1}^{4} \eta(R_i')$. If for each i,

$|\eta(R_i')| < \dfrac{1}{4}\,|\eta(R)|$ then by triangle inequality we will have $|\eta(R)| \le \displaystyle\sum_{i=1}^{4}|\eta(R_i')| < |\eta(R)|$ which is a contradiction. Hence for at least one i we must have $|\eta(R_i')| \ge \dfrac{1}{4}\,|\eta(R)|$. We shall fix one such rectangle and call it as R_1 so that $|\eta(R_1)| \ge \dfrac{1}{4}\,|\eta(R)|$. Divide again R_1 in the same way and

get one particular rectangle R_2 so that $|\eta(R_2)| \ge \dfrac{1}{4}\,|\eta(R_1)| \ge \dfrac{1}{4^2}\,|\eta(R)|$. Continuing this process we get a sequence of rectangles R_1 such that $R_1 \supset \ldots \supset R_k \supset \ldots$ and that $|\eta(R_k)| \ge \dfrac{1}{4}\,|\eta(R_{k-1})| \ge \ldots \ge \dfrac{1}{4^k}|\eta(R)|$. Note that the diameter of R_k denoted by d_k and the perimeter of R_k denoted by l_k satisfy $d_k = \dfrac{d}{2^k}$ and $l_k = \dfrac{l}{2^k}$ where d and l are the diameter and the perimeter of R, the original rectangle.

By Cantor intersection theorem $\bigcap_{1}^{\infty} R_k$ is a singleton set say $\{z*\}$ and so $z*$ belongs to every R_k. Now given any neighbourhood B of z^* (a circle with centre z^* and radius say ρ) we can find a large N such that $k \ge N$ implies $R_k \subset B$ [In fact if we choose $d_k = \dfrac{d}{2^k} < \rho$ for $k \ge N$ then for any $z \in R_k$, $|z - z^*| \le d_k < \rho$ and so $R_k \subset B$]. Now z^* belongs to Ω and so $f(z)$ is analytic at z^*. Hence given $\varepsilon > 0$ we can find a $\delta > 0$ so that

$0 < |z - z^*| < \delta$ implies $\left| \dfrac{f(z) - f(z^*)}{z - z*} - f'(z*) \right| < \dfrac{\varepsilon}{2dl}$. Since $f(z^*)$ and $f'(z^*)$

are constants and $1dz$ and zdz are exact differentials (Note that $U(z) = z$ implies $U'(z) = 1$ and $U(z) = z^2/2$ implies $U'(z) = z$) we have by Theorem 4.2.4 $\displaystyle\int_{\partial R_k} 1dz = 0$ and $\displaystyle\int_{\partial R_k} zdz = 0$. Therefore

$$\eta(R_k) = \int_{\partial R_k} f(z)\ dz = \int_{\partial R_k} [f(z) - f(z^*) - f'(z^*)\ (z - z^*)]\ dz$$

and hence

$$|\eta(R_k)| \le \int_{\partial R_k} |[f(z) - f(z^*) - f'(z^*)\ (z - z^*)]|\ |dz|$$

$$\le \dfrac{\varepsilon}{2dl} \int_{\partial R_k} |(z - z^*)|\ |dz|$$

$$\le \dfrac{\varepsilon}{2dl}\, d_k\, l_k = \dfrac{\varepsilon}{2dl}\, \dfrac{d}{2^k}\, \dfrac{l}{2^k} = \dfrac{\varepsilon}{2}\left(\dfrac{1}{4^k}\right).$$

Thus $|\eta(R)| \le 4^k|\eta(R_k)| \le \varepsilon/2 < \varepsilon$. Since $\varepsilon > 0$ is arbitrary it follows that $\eta(R) = 0$ proving our theorem.

Definition 4.2.6
Let $f(z)$ be analytic in a region Ω except at $a \in \Omega$. We say "a" is exceptional to $f(z)$ if $\lim_{z \to a} (z - a) f(z) = 0$.

Theorem 4.2.7 (Cauchy's theorem for a rectangle in the presence of exceptional points)
Let $f(z)$ be analytic in a region Ω except for a finite number of exceptional points. Let R be a rectangle such that its interior and the boundary ∂R are contained in Ω and ∂R does not pass through any of the exceptional points then $\int_{\partial R} f(z) \, dz = 0$.

Proof
Since the rectangle can be subdivided into smaller rectangles and squares in such a way that each exceptional point occurs as the centre of one and only one square and that the integral over the boundary of the original rectangle equals the sum of integrals over the boundaries of smaller rectangles and the boundaries of the squares, there is no loss of generality in assuming that R is a square, that it contains one and only one exceptional point "a" which lies at its centre (In fact one can first divide the rectangle into subrectangles in such a way that each subrectangle contains atmost one exceptional point. Again each subrectangle having one exceptional point can be further subdivided in such a way that it lies at the centre of a square). See the following figure for guidance.

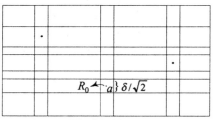

Fig 4.2 (ii)

Let $\varepsilon > 0$ be given. Since "a" is exceptional there exists a $\delta > 0$ such that $0 < |z - a| \leq \delta$ implies $|f(z)| \, |z - a| < \dfrac{\varepsilon}{8}$. Also this square can further be subdivided so that for this $\delta > 0$ "a" lies at the centre of a square R_0 ($\subset R$) of diameter $\delta \left(\text{its side being } \dfrac{\delta}{\sqrt{2}}\right)$ and that $\int_{\partial R} f(z) \, dz = \int_{\partial R_0} f(z) \, dz$.

As z varies over ∂R_0, $|f(z)| < \dfrac{\varepsilon}{8|z - a|}$ and so

$$\left| \int_{\partial R} f(z) \, dz \right| = \left| \int_{\partial R_0} f(z) \, dz \right|$$

$$\le \frac{\varepsilon}{8} \int_{\partial R_0} \frac{1}{|z-a|} |dz|$$

$$\le \frac{\varepsilon}{8} \frac{2\sqrt{2}}{\delta} \int_{\partial R_0} |dz| = \frac{\varepsilon}{8} \frac{2\sqrt{2}}{\delta} \left(\frac{4\delta}{\sqrt{2}} \right) = \varepsilon.$$

$\left(\text{Note that for points } z \text{ in } \partial R_0, |z-a| \ge \dfrac{\delta}{2\sqrt{2}} \right)$. Since $\varepsilon > 0$ is arbitrary we have $\int_{\partial R} f(z) \, dz = 0$. This completes the proof.

Corollary 4.2.8
Let $f(z)$ be analytic in a region Ω except for a finite number of exceptional points. Let γ be a closed rectangular path (consisting of a finite number of line segments which are parallel to the co-ordinate axes). Assume that γ and its interior lie in Ω and that γ does not pass through any of the exceptional points. Then $\int_{\gamma} f(z) \, dz = 0$.

Proof
Let γ be given the anti-clockwise orientation. Divide the interior of γ into a finite mesh of rectangles R_i with boundaries ∂R_i satisfying the following conditions. (i) None of the exceptional points lie on any ∂R_i (ii) Each line segment in·γ is either a side or part of a side of one and only one rectangle R_i (iii) Any side of any R_i which is not part of γ appears as common part of two adjacent rectangles in the mesh (iv) ∂R_i's are given the anti-clockwise orientation. See Fig. 4.2 (iii) for guidance.

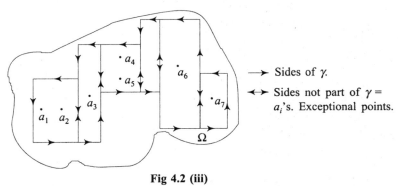

$\longrightarrow$ Sides of γ.

$\longleftrightarrow$ Sides not part of $\gamma =$ a_i's. Exceptional points.

Fig 4.2 (iii)

Using the above properties it is clear that $\gamma = \sum_i \partial R_i$ and each R_i contains atmost a finite number of exceptional points of f. From Theorem 4.2.7 it follows that

$$\int_{\gamma} f(z) \, dz = \sum_i \int_{\partial R_i} f(z) \, dz = 0.$$

Theorem 4.2.9
Let f be analytic in an open disc U. Then f has a primitive in U. i.e., there exists an analytic function $g(z)$ on U such $g'(z) = f(z)$. Furthermore any two primitives of f in U differ by a constant.

Proof

Let z_0 be the centre of the disc U. For every point z_1 in U define $g(z_1) = \int_\gamma f(z)\,dz$ where the path γ from z_0 to z_1 consists of the horizontal line segment at z_0 and the vertical line segment at z_1. Note that these segments are completely contained in U. Now $g(z_1 + h) = \int_\sigma f(z)\,dz$ where the path σ joining z_0 and $z_1 + h$ consists of the horizontal line segment at z_0 and the vertical line segment at $z_1 + h$. Using Cauchy's theorem for a rectangle we have

$$g(z_1 + h) - g(z_1) = \int_\sigma f(z)\,dz - \int_\gamma f(z)\,dz = \int_\eta f(z)\,dz$$

where the path η joining z_1 and $z_1 + h$ consists of one horizontal line segment and one vertical line segment say σ_1 and σ_2 (σ_1 is at z_1 or at $z_1 + h$ depending on the relative positions of z_1 and $z_1 + h$). See Figure 4.2 (iv) for guidance.

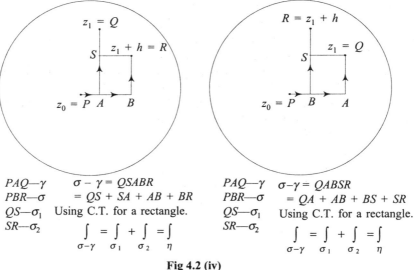

PAQ—γ $\quad$ $\sigma - \gamma = QSABR$
PBR—σ $\quad$ $= QS + SA + AB + BR$
QS—σ_1 $\quad$ Using C.T. for a rectangle.
SR—σ_2

$$\int_{\sigma-\gamma} = \int_{\sigma_1} + \int_{\sigma_2} = \int_\eta$$

PAQ—γ $\quad$ $\sigma-\gamma = QABSR$
PBR—σ $\quad$ $= QA + AB + BS + SR$
QS—σ_1 $\quad$ Using C.T. for a rectangle.
SR—σ_2

$$\int_{\sigma-\gamma} = \int_{\sigma_1} + \int_{\sigma_2} = \int_\eta$$

Fig 4.2 (iv)

By continuity of $f(z)$ at z_1 we can write $\psi(z) = f(z) - f(z_1)$ so that $\lim_{z \to z_1} \psi(z) = 0$. Hence by Theorem 4.1.6 (vii), $g(z_1 + h) - g(z_1) = \int_{\sigma_1 + \sigma_2} [f(z_1) + \psi(z)]\,dz = hf(z_1) + \int_{\sigma_1 + \sigma_2} \psi(z)\,dz$.

Hence

$$\left| \frac{g(z_1 + h) - g(z_1)}{h} - f(z_1) \right| = \frac{1}{|h|} \left| \int_{\sigma_1 + \sigma_2} \psi(z)\,dz \right| \le \frac{1}{|h|} \int_{\sigma_1 + \sigma_2} |\psi(z)|\,|dz|.$$

As z varies over $\sigma_1 + \sigma_2$, $|z - z_1| \le |h|$ and the length of σ_1 and σ_2 are at most $|h|$. Now given $\varepsilon > 0$ there exists $\delta > 0$ such that $|z - z_1| < \delta$

implies that $|\psi(z)| < \dfrac{\varepsilon}{2}$. Choose h so small that $|h| < \delta$. Then

$$\left| \frac{g(z_1 + h) - g(z_1)}{h} - f(z_1) \right| \leq \frac{1}{|h|} \frac{\varepsilon 2|h|}{2} = \varepsilon.$$

Since $\varepsilon > 0$ is arbitrary we have $g'(z_1) = f(z_1)$. Since z_1 is arbitrary g is the required primitive of f.

If h is another primitive of f in U (an open connected set) then $h' = f = g'$. Thus $(h - g)' = 0$ everywhere in U which implies that $h(z) - g(z)$ is a constant.

Theorem 4.2.10 (Cauchy's theorem for a circular disc)

Let U be an open disc and suppose that f is analytic on U. Then for any closed curve γ in U, $\int_\gamma f(z)\, dz = 0$.

Proof

By Theorem 4.2.3 and Theorem 4.2.9, $f(z)\, dz$ is exact. An application of Theorem 4.2.4 completes the proof of this theorem.

Theorem 4.2.11 (Cauchy's theorem for a disc in the presence of exceptional points)

Let f be analytic in an open disc D except for a finite number of exceptional points $a_i \in D$ ($i = 1, 2, \ldots, n$). Let γ be a closed curve in D not passing through any of the points a_i. Then $\int_\gamma f(z)\, dz = 0$.

Proof

Let Ω be the region consisting of all points of the open disc D except the finite number of exceptional points (Is Ω a region ?). We shall show that f has a primitive in Ω. Once this is done, by Theorem 4.2.3 and Theorem 4.2.4 we will have $\int_\gamma f(z)\, dz = 0$ for every closed curve γ in Ω. Choose a base point z_0 in Ω. For $z_1 \in \Omega$, we can always choose a rectangular path γ in

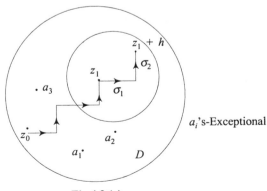

Fig 4.2 (v)

Ω joining z_0 and z_1. Define $g(z_1) = \int_\gamma f(z)\,dz$. If σ is any other rectangular path in Ω joining z_0 and z_1, $\gamma - \sigma$ is a closed rectangular path in Ω and by Corollary 4.2.8, $\int_{\gamma-\sigma} f(z)\,dz = 0$ or that $\int_\gamma f(z)\,dz = \int_\sigma f(z)\,dz$. Thus $g(z_1)$ is well defined for all $z_1 \in \Omega$. Since $f(z)$ is continuous at z_1, given $\varepsilon > 0$ there exists $\delta > 0$ such that $|z - z_1| < \delta$ implies that $|f(z) - f(z_1)| < \dfrac{\varepsilon}{2}$ and we are free to choose δ so that the closed δ neighbourhood of z_1 does not contain exceptional points. Put $\psi(z) = f(z) - f(z_1)$ so that $|\psi(z)| < \dfrac{\varepsilon}{2}$ for $|z - z_1| < \delta$. Let $|h| < \delta$. Having chosen γ, joining z_0 and z_1, choose the rectangular path σ, joining z_1 and $z_1 + h$, to be $\gamma \cup \sigma_1 \cup \sigma_2$ where σ_1 is the horizontal line segment at z_1 and σ_2 is the vertical line segment at $z_1 + h$. [Note that we can choose any rectangular path in Ω joining z_0 and $z_1 + h$ for defining $g(z_1 + h)$].

It is now clear that (the reader can assume various possibilities for the position of $z_1 + h$ in relation to that of z_1 and arrive at the same conclusion) $g(z_1 + h) - g(z_1) = \int_{\sigma_1+\sigma_2} f(z)\,dz$ and that σ_1 and σ_2 are fully contained in a disc of radius δ at z_1 and that for every z in this path $|z - z_1| \le |h| < \delta$. Now

$$\left| \frac{g(z_1 + h) - g(z_1)}{h} - f(z_1) \right| \le \frac{1}{|h|} \int_{\sigma_1+\sigma_2} |\psi(z)|\,|dz| < \varepsilon$$

and therefore $g'(z_1) = f(z_1)$. Since z_1 is arbitrary in Ω we have completed the proof.

Definition 4.2.12
Let U be a non-empty open set in $\mathbb{C}$. Let γ and η be two closed curves in U defined on $[a, b]$. We say that γ and η are close together if there exists a partition $a = a_0 \le a_1 \le \ldots \le a_n = b$ of $[a, b]$ and discs $D_i (i = 0, 1, \ldots, n - 1)$ such that (a) $D_i \subset U$ and (b) $\gamma([a_i, a_{i+1}]) \subset D_i$ and $\eta([a_i, a_{i+1}]) \subset D_i$.

Lemma 4.2.13
Let f be analytic in a non-empty open set U. Let γ and η be two closed curves in U defined on $[a, b] \subset \mathbb{R}$ such that γ and η are close together. Then

$$\int_\gamma f(z)\,dz = \int_\eta f(z)\,dz.$$

Proof
Choose a partition $a = a_0 \le a_1 \le \ldots \le a_n = b$ and disc $D_i (i = 0, \ldots, n - 1)$ in U such that $\gamma([a_i, a_{i+1}]) \subset D_i$ and $\eta([a_i, a_{i+1}]) \subset D_i$. Put $z_i = \gamma(a_i)$

and $w_i = \eta(a_i)$, $i = 0, \ldots, n-1$. Let g_i be a primitive of f on D_i (which exists by Theorem 4.2.9). Now g_i and g_{i+1} are both primitives of f in $D_i \cap D_{i+1}$ which is a connected open set. Hence they differ by a constant in $D_i \cap D_{i+1}$ ($i = 0, \ldots, n-2$). Since z_{i+1}, $w_{i+1} \in D_i \cap D_{i+1}$ we have

$$g_{i+1}(z_{i+1}) - g_i(z_{i+1}) = g_{i+1}(w_{i+1}) - g_i(w_{i+1})$$

or

$$g_{i+1}(z_{i+1}) - g_{i+1}(w_{i+1}) = g_i(z_{i+1}) - g_i(w_{i+1}). \tag{1}$$

We also note that $z_n = \gamma(b) = \gamma(a) = z_0 \in D_0 \cap D_{n-1}$ and similarly $w_n = \eta(b) = \eta(a) = w_0 \in D_0 \cap D_{n-1}$. Since g_{n-1} and g_0 are both primitives of the same f in $D_0 \cap D_{n-1}$ we have

$$g_{n-1}(z_0) - g_0(z_0) = g_{n-1}(w_0) - g_0(w_0). \tag{2}$$

Put $\gamma_i = \gamma/[a_i, a_{i+1}]$, $\eta_i = \eta/[a_i, a_{i+1}]$, $i = 1, \ldots, n-1$. It is now clear from the definition that $\gamma = \sum_{i=0}^{n-1} \gamma_i, \eta = \sum_{i=0}^{n-1} \eta_i$ and that

$$\int_\gamma f(z)\,dz = \sum_{i=0}^{n-1} \int_{\gamma_i} f(z)\,dz = \sum_{i=0}^{n-1} [g_i(z_{i+1}) - g_i(z_i)]$$

$$\int_\eta f(z)\,dz = \sum_{i=0}^{n-1} \int_{\eta_i} f(z)\,dz = \sum_{i=0}^{n-1} [g_i(w_{i+1}) - g_i(w_i)]$$

[see Theorem 4.1.6(viii)]. Thus

$$\int_\gamma f(z)\,dz - \int_\eta f(z)\,dz = \sum_{i=0}^{n-1} ([g_i(z_{i+1}) - g_i(z_i)] - [g_i(w_{i+1}) - g_i(w_i)])$$

$$= \sum_{i=0}^{n-1} ([g_i(z_{i+1}) - g_i(w_{i+1})] - [g_i(z_i) - g_i(w_i)])$$

$$= \sum_{i=0}^{n-2} ([g_{i+1}(z_{i+1}) - g_{i+1}(w_{i+1})])$$

$$- \sum_{i=0}^{n-1} ([g_i(z_i) - g_i(w_i)] + [g_{n-1}(z_n) - g_{n-1}(w_n)])$$

$$= [g_{n-1}(z_0) - g_0(z_0)] - [g_{n-1}(w_0) - g_0(w_0)]$$

$$= 0 \,[\text{using (1) and (2)}]$$

Definition 4.2.14

Let γ and η be two closed curves in a non-empty open set U defined on a common interval $[a, b] \subset \mathbb{R}$. We say that γ is homotopic to η if there exists a continuous function $\psi : [a, b] \times [c, d] \to U$ where $[c, d] \subset \mathbb{R}$ is

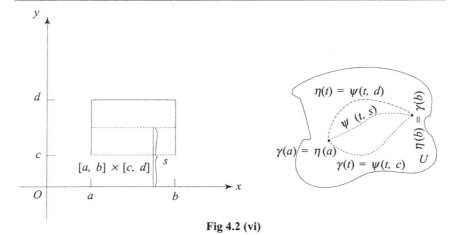

Fig 4.2 (vi)

another closed interval, such that $\psi(t, c) = \gamma(t)$ and $\psi(t, d) = \eta(t)$, $(a \le t \le b)$, $\psi(a, s) = \psi(b, s)$ for $c \le s \le d$. For each s in $[c, d]$ we may view $\psi_s(t) = \psi(t, s)$ as a deformation of γ to η. In particular if η is a constant curve then we say γ is homotopic to a point or to a constant. If in addition γ and η are two closed curves with same end points and if $\psi(a, s) = \psi(b, s) = \gamma(a) = \eta(a)$ for each $s \in [c, d]$ then we say that γ and η are homotopic with the end points fixed.

Theorem 4.2.15 (Homotopic form of Cauchy's theorem)
Let f be analytic on a non-empty open set U. Let γ and η be two closed curves in U defined on $[a, b] \subset \mathbb{R}$ such that γ and η are homotopic. Then

$$\int_\gamma f(z)\, dz = \int_\eta f(z)\, dz.$$ In particular if γ is homotopic to a point then

$$\int_\gamma f(z)\, dz = 0.$$

Proof
Let $\psi : [a, b] \times [c, d] \to U$ be a homotopy between γ and η. Since the image K of $[a, b] \times [c, d]$ under ψ is compact in U, $d(K, U^c) > 0$. Given this $\delta = d(K, U^c) > 0$, by uniform continuity of ψ there exists an $\alpha > 0$ such that $|z - w| < \alpha$ $(z, w \in [a, b] \times [c, d])$ implies that $|\psi(z) - \psi(w)| < \delta$. Now take partitions of $[a, b]$ and $[c, d]$ say $a = a_0 \le a_1 \le \ldots \le a_n = b$ and $c = c_0 \le c_1 \le \ldots \le c_n = d$ so that the diameter of these partitions are less than

$\dfrac{\alpha}{\sqrt{2}}$. Choose discs D_{ij}, $(i = 0, 1, 2, \ldots n - 1, j = 0, 1, 2, \ldots, m - 1)$ so that

its centre is $\psi(a_i, c_j)$ and radius δ. Now $D_{ij} \subset U$ and $\psi([a_i, a_{i+1}] \times [c_j, c_{j+1}])$ $\subseteq D_{ij}$ $((s, t) \in [a_i, a_{i+1}] \times [c_j, c_{j+1}] \Rightarrow |a_i - s|^2 + |c_j - t|^2 < \alpha^2 \Rightarrow |\psi(a_i, c_j) - \psi(s, t)| < \delta \Rightarrow \psi(s, t) \in D_{ij})$. Define $\psi_j(t)$, for $a_i \le t \le a_{i+1}$ as the line segment joining $\psi(a_i, c_j)$ and $\psi(a_{i+1}, c_j)$ (for $0 \leq i \leq n - 1$ and $0 \leq j \leq m - 1$). Then ψ_j and ψ_{j+1} define piece-wise differentiable closed curves such that ψ_j and ψ_{j+1} are close together because $[\psi_j[a_i, a_{i+1}] \cup \psi_{j+1}[a_i, a_{i+1}]$ is nothing but the union of the line segments joining $\psi[a_i, c_j)$, $\psi(a_{i+1}, c_j)$ and $\psi(a_i, c_{j+1})$, $\psi(a_{i+1}, c_{j+1})$ which is $\subseteq D_{ij}$. Hence by Lemma

4.2.13, we have $\int_{\psi_j} f(z)dz = \int_{\psi_{j+1}} f(z)dz$ $(0 \leq j \leq m-1)$. Now ψ_0 and γ

are close together and ψ_m and η are close together. Hence $\int_\gamma f(z)dz =$

$\int_\eta f(z)dz$. It is also clear that if $\eta(t)$ is a constant curve then $\int_\eta f(z)dz =$

0 by definition and our proof is complete.

Note 4.2.16

In the above proof it is tempting to define $\psi_j(t) = \psi(t, c_j)$ for all $t \in [a, b]$ and $0 \leq j \leq m-1$. However we should avoid this temptation at this stage because $\psi(t, c_j)$ considered as a function of t is mearly continuous and may not be piece-wise differentiable and unless ψ_j is piece-wise differentiable it

makes no sense to talk about $\int_{\psi_j} f(z)dz$. It is for this reason that we have

defined $\psi_j(t)$ as union of line segments as in the above proof. However if

$\int_\gamma f(z)dz$ can be defined for any continuous curve (as is done in S.Lang,

Complex Analysis, Springer-Verlag, Second Edition, p.111-113) then of course we can define $\psi_j(t) = \psi(t, c_j)$ as well.

Definition 4.2.12 can be extended even if γ and η are two curves having same initial and end points instead of being closed. Lemma 4.2.13 in still valid for these γ and η. Definition 4.2.14 can also be made applicable if γ and η are two curves having same initial and end points. In this case we need to assume $\psi(a, s) =$ the initial point of γ or η and $\psi(b, s) =$ end point of γ or η for all $s \in [c, d]$. With this new definition Theorem 4.2.15 is still valid for these γ and η [See fig. 4.2 (vi)]. The reader is encouraged to prove these assertions.

Definition 4.2.17
A region Ω is said to be simply connected if every closed curve in Ω is homotopic to a point.

Examples 4.2.18
(a) The finite complex plane and any open disc are simply connected. To prove this let Ω be either the plane or an open disc. Let $\gamma = \gamma(t)$ $(0 \leq t$

≤ 1) be any closed curve in Ω. Consider $H: I \times I \to \Omega$ defined by $H(s, t) = (1 - s) \gamma(t) + s\gamma(0)$. Now $H(0, t) = \gamma(t)$; $H(1, t) \equiv \gamma(0)$; $H(s, 0) = H(s, 1) = \gamma(0)$ for all $s \in I$. Further $H(s, t)$ is obviously continuous on $I \times I$. Thus $H(s, t)$ is a homotopy between γ and the constant curve at $\gamma(0)$.
(b) The punctured plane $\mathbb{C}\backslash\{0\}$ is not simply connected. If it were, then every closed curve would be homotopic to a constant curve. Take for example the unit circle described positively which lies in $\mathbb{C}\backslash\{0\}$. Its equation is $z = z(t) = e^{2\pi i t}$ $(0 \leq t \leq 1)$. If this is homotopic to the constant curve at 1 then by Theorem 4.2.15 $\int\limits_{\gamma} \dfrac{dz}{z} = 0$ as $f(z) = \dfrac{1}{z}$ is analytic in $\dot{\mathbb{C}}\backslash\{0\}$.

A simple computation shows that $\int\limits_{\gamma} \dfrac{dz}{z} = 2\pi i \neq 0$. This contradiction shows that $\mathbb{C}\backslash\{0\}$ is not simply connected. We just remark that intuitively a region is simply connected if and only if it does not contain any "hole".

Theorem 4.2.19 (Cauchy's theorem for a simply connected region)
Let f be analytic in a simply connected region Ω. Let γ be a closed curve in Ω. Then $\int\limits_{\gamma} f(z)\, dz = 0$. Thus every analytic function in a simply connected region has a primitive.

Proof
If γ is a closed curve in Ω then by hypothesis γ is homotopic to a constant curve and by Theorem 4.2.15 $\int\limits_{\gamma} f(z)\, dz = 0$. Since this is true for any closed curve, $f(z)\, dz$ is exact by Theorem 4.2.4. Hence by Theorem 4.2.3 $f(z)$ has a primitive. This completes the proof.

Theorem 4.2.20

Let γ be a closed curve in $\mathbb{C}$ and $a \notin \gamma$. Then $n(\gamma, a) = \dfrac{1}{2\pi i} \int\limits_{\gamma} \dfrac{dz}{z - a}$ is an integer called the winding number of γ with respect to 'a' or the index of γ with respect to 'a' and it has the following properties.
   (i) $n(-\gamma, a) = -n(\gamma, a)$
   (ii) $n(\gamma, a)$ as a function of 'a' is constant in each of the regions determined by γ i.e., in each of the components of the complement of γ.
   (iii) $n(\gamma, a) = 0$ for all 'a' belonging to the unbounded region determined by γ i.e., in the unbounded component of the complement of γ.

Proof
Let γ be parameterized by $z = z(t)$, $\alpha \leq t \leq \beta$, with $z(\alpha) = z(\beta)$. Then

$$n(\gamma, a) = \frac{1}{2\pi i} \int\limits_{\alpha}^{\beta} \frac{z'(t)}{z(t) - a}\, dt.$$ Put $h(s) = \int\limits_{\alpha}^{s} \frac{z'(t)}{z(t) - a}\, dt.$ Then as $\dfrac{z'(t)}{z(t) - a}$ is a continuous function for $\alpha \leq t \leq \beta$, $h(s)$ is derivable as a function of a single variable s and by the fundamental theorem of integral calculus $h'(s) =$

$\dfrac{z'(s)}{z(s)-a}$. Now $F(s) = e^{-h(s)} (z(s) - a)$ is also derivable as a function of s
and $F'(s) = e^{-h(s)} [(z'(s) - h'(s) (z(s) - a)] = 0$ and since F is defined on
$[\alpha, \beta]$, $F(s)$ is a constant. We evaluate this constant at $s = \alpha$ and get that
$F(s) = F(\alpha) = e^{-h(\alpha)} (z(\alpha) - a) = z(\alpha) - a$ as $h(\alpha) = 0$. But $F(s) =$
$e^{-h(s)} (z(s)-a)$ and hence $e^{h(s)} = \dfrac{z(s) - a}{z(\alpha) - a}$. In particular $e^{h(\beta)} = 1$ as $z(\alpha)$
$= z(\beta)$. We can now conclude that $h(\beta)$ is an integral multiple of $2\pi i$ and
so $n(\gamma, a) = \dfrac{1}{2\pi i} h(\beta)$ is an integer. We now observe that as a point set
γ is closed and bounded in $\mathbb{C}$. Hence in the extended plane its complement
is open and so its various components are open and connected (*i.e.,* they
are regions). Evidently they are disjoint. Each such region is called a region
determined by γ and in as much as ∞ belongs to one and only one of these
regions, this region must be unbounded and is called the unbounded region
determined by γ.

Property (i) is clear from Theorem 4.1.6 (ii). To prove (ii), we need to
show that if a and b belong to the same region determined by γ then
$n(\gamma, a) = n(\gamma, b)$. On the other hand these regions are connected and any
two points in the same region can be joined by a rectangular path which
is completely contained in the region. If these line segments are assumed
to be $\overline{a_0 a_1}, \overline{a_1 a_2} ... \overline{a_{n-1} a_n}$ (put $a_0 = a$, $a_n = b$) for some n then $n(\gamma, a)$
$= n(\gamma, b)$ follows if we can prove $n(\gamma, a_i) = n(\gamma, a_{i+1})$ for $i = 0, 1, ..., n$.
Thus it suffices to show that if the line segment joining a and b is contained
in this region then $n(\gamma, a) = n(\gamma, b)$. Now in the complement of the closed
line segment joining a and b, $\dfrac{z - a}{z - b}$ is never real and non-positive. In fact
$\dfrac{z - a}{z - b}$ is real and non-positive if and only if either $z = a$ or z lies on the open
line segment joining a and b (A point z lies on the open line segment joining
a and b if and only if $z = ta + (1 - t) b$ for some $0 < t < 1$. Thus for points
on this segment we have $\dfrac{z - a}{z - b} = \dfrac{ta + (1-t) b - a}{ta + (1-t) b - b} = -\dfrac{(1-t)}{t} < 0$. Also if
$\dfrac{z - a}{z - b} = -k$ $(k > 0)$ then $z = \dfrac{a}{1+k} + \dfrac{kb}{1+k} = ta + (1 - t) b$ for $t = \dfrac{1}{1+k}$
with $0 < t < 1$. Thus the principal branch for $\log \left(\dfrac{z - a}{z - b} \right)$ is analytic in the
complement Ω of the closed line segment joining a and b and its derivative

is $\left(\dfrac{1}{z-a} - \dfrac{1}{z-b}\right)$. Thus $\left(\dfrac{1}{z-a} - \dfrac{1}{z-b}\right)$ has primitive in Ω and our closed

curve $\gamma \subset \Omega$. Hence by Theorem 4.2.4, $\displaystyle\int_{\gamma}\left(\dfrac{1}{z-a} - \dfrac{1}{z-b}\right)dz = 0$ or that

$n(\gamma, a) = n(\gamma, b)$.

(iii) If Ω is the unbounded component determined by γ, choose a large circle containing γ in its interior (Note that as a point set γ is bounded). We can always choose a point α of Ω outside this circle (as Ω is unbounded). Using Cauchy's theorem for a circular disc we can now get that

$\dfrac{1}{2\pi i}\displaystyle\int_{\gamma}\dfrac{dz}{z-\alpha} = 0$ as $\dfrac{1}{z-\alpha}$ is analytic inside C and $\gamma \subset C$. Thus $n(\gamma, \alpha) =$

0. But $n(\gamma, a)$ is constant throughout Ω and so this constant must be equal to 0 only.

4.3 General form of Cauchy's Theorem

We have already seen several forms of Cauchy's Theorems (for example, for a rectangle, for a disc, for simply connected regions etc.). In general, even if the region is not simply connected there are certain closed curves γ for which $\int_{\gamma} f(z)dz = 0$ is valid for arbitrary analytic functions in these regions. In fact we can assert that if Ω is a region and $\gamma \subset \Omega$, is a closed curve such that $\int_{\gamma} f(z)dz = 0$ holds for all $f(z) = \dfrac{1}{z-a}(a \notin \Omega)$ then $\int_{\gamma} f(z)dz = 0$ holds for all analytic functions f in Ω. This is the most generalized version of Cauchy's theorem and we proceed to obtain this theorem.

Definition 4.3.1

By a chain we mean a formal sum of integral multiples of curves. That is, if m_i's are integers ($i = 1, 2, \ldots, n$) positive, negative or zero, and g_i's are curves (defined on $[a_i, b_i]$ ($i = 1, 2, \ldots, n$)) then $\gamma = \displaystyle\sum_{i=1}^{n} m_i \gamma_i$ is a chain

and we define $\displaystyle\int_{\gamma} f(z)dz = \sum_{i=1}^{n} m_i \int_{\gamma_i} f(z)dz$ for every f which is continuous on the compact set $\cup \gamma_i^*$ where $\gamma_i^* = \gamma_i[a_i, b_i]$. Two such chains $\gamma = \displaystyle\sum_{i=1}^{n} m_i \gamma_i$ and $\eta = \displaystyle\sum_{i=1}^{m} n_i \eta_i$ are considered as equal if $\displaystyle\int_{\gamma} f(z)dz = \int_{\eta} f(z)dz$ for every f continuous on $(\cup \gamma_i^*) \cup (\cup \eta_i^*)$. We also define a chain γ to be a cycle if at least in one representation, $\gamma = \sum_{i=1}^{n} m_i \gamma_i$ where each γ_i is

closed. Further if $\gamma = \sum_{i=1}^{n} m_i \gamma_i$, is a chain and if all the curves γ_i are in U

(i.e., $\gamma_i^* \subset U$) for a non-empty open set U, then we call γ as a chain in U. Similarly if each γ_i is a rectangular path or a closed rectangular path then γ is called a rectangular chain or closed rectangular chain.

Note 4.3.2

From our definition it is clear that if $\gamma = \sum_{i=1}^{n} m_i \gamma_i$ is a cycle and $a \notin \bigcup_{1}^{n}$

γ_i^* then $n(\gamma, a) = \sum_{i=1}^{n} m_i n(\gamma_i, a)$ is a well defined integer. We also note that

if $\gamma = \sum_{i=1}^{n} m_i \gamma_i$ is a chain in which $m_i = 0$ for all i then γ can be called a

zero chain and that if $\gamma - \eta$ is a zero chain then $\gamma = \eta$ or that $\int_{\gamma} f(z)dz$

$= \int_{\eta} f(z)dz$ for all continuous function on $(\gamma^* \cup \eta^*)$.

Definition 4.3.3

If γ and η are two cycles in an open set U we say γ is homologous to η(written as $\gamma \sim \eta$) in U if $n(\gamma, a) = n(\eta, a)$ for all $a \notin U$. i.e., $\int_{\gamma} \frac{1}{z-a} dz$

$= \int_{\eta} \frac{1}{z-a} dz$ for $a \notin U$. In particular we say γ is homologous to zero

(written as $\gamma \sim 0$) in U if $n(\gamma, a) = 0$ for all $a \notin U$. Note that $\gamma \sim \eta$ is equivalent to $\gamma - \eta \sim 0$.

Theorem 4.3.4

Let γ be a curve in an open set U. Then there exists a rectangular path η in U with the same end points as γ and such that for every analytic function f in U, $\int_{\gamma} f(z)dz = \int_{\eta} f(z)dz$.

Proof

As a point set γ^* is closed and bounded i.e., compact. Consider $d = \inf\{|z - w| : z \in \gamma, w \in U^c\}$. Since U^c is closed and $\gamma \subseteq U$ this infimum must be greater than zero. Let $\gamma = \gamma(t)$ be defined on $[a, b]$. Now γ being continuous, we can choose a $\delta > 0$ such that $|s - t| < \delta$ implies that $|\gamma(s) - \gamma(t)| < d$. Choose a partition $a = x_0 < x_1 < ... < x_n = b$ of $[a, b]$ in such a way that $|x_i - x_{i-1}| < \delta$ so that $|\gamma(x_i) - \gamma(x_{i-1})| < d$. We shall take open discs D_i centered at each $\gamma(x_{i-1})$ ($i = 1, \ldots, n$) with fixed radius d so that $\gamma[x_{i-1}, x_i] \subset D_i$ for $i = 1, \ldots, n$. Now we construct a rectangular path η as follows: Join $\gamma(x_{i-1})$ and $\gamma(x_i)$ by a horizontal line segment at $\gamma(x_{i-1})$ followed by the vertical line segment at $\gamma(x_i)$. Call this as $\eta_i \subset D_i$ (in some cases one of these segments can be singleton). Let $\eta = \sum\limits_{i=1}^{n} \eta_i$

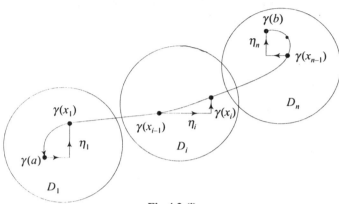

Fig 4.3 (i)

Set $\gamma_i = \gamma|[x_{i-1}, x_i]$ so that $\gamma = \sum\limits_{i=1}^{n}\gamma_i$. Using Cauchy's theorem for circular discs D_i we see that for any analytic function f in U, $\int\limits_{\gamma_i} f(z)dz =$ $\int\limits_{\eta_i} f(z)dz$ (as η_i and γ_i both join $\gamma(x_{i-1})$ and $\gamma(x_i)$ in D_i, $\gamma_i - \eta_i$ is a closed curve in D_i). Hence $\int\limits_{\gamma} f(z)dz = \int\limits_{\eta} f(z)dz$. Note that the end points of both γ and η are $\gamma(a)$ and $\gamma(b)$ only.

Corollary 4.3.5

If γ is a chain in an open set U then there exists a rectangular chain η in U (a formal sum of integral multiples of rectangular paths) such that for every analytic function f in U, $\int\limits_{\eta} f(z)dz = \int\limits_{\gamma} f(z)dz$.

Corollary 4.3.6

If γ is a cycle homologous to zero in an open set U and f is an analytic function in U then there exists a closed rectangular chain η in U which is also homologous to zero in U such that $\int\limits_{\eta} f(z)dz = \int\limits_{\gamma} f(z)dz$.

The proof of Corollary 4.3.5 is obvious by writing γ as a formal sum of integral multiples of curves. On the other hand for the proof of Corollary

4.3.6 we apply theorem 4.3.4 to get a rectangular chain η for which both the following equalities hold:

$$\int_\gamma \frac{1}{z-a}\,dz = \int_\eta \frac{1}{z-a}\,dz \text{ for } a \notin U \tag{1}$$

$$\int_\eta f(z)\,dz = \int_\gamma f(z)\,dz. \tag{2}$$

(1) implies that η is homologous to zero and if we use (2) the proof of Corollary 4.3.6 is complete.

Theorem 4.3.7
Let γ be a closed rectangular chain in an open set U such that $\gamma \sim 0$ in U. Then we can find a finite number of open rectangles $R_i(i = 1, 2, ..., n)$ with $R_i \cup \partial R_i \subset U$ (where ∂R_i is the boundary of R_i) and integers $m_i \neq 0$ $(i = 1, 2, ..., n)$ such that $\gamma = \sum_{i=1}^{n} m_i\, \partial R_i$

Proof
We first observe that the given rectangular chain γ in U can be written as $\gamma = \sum_{i=1}^{n} k_i \gamma_i$ where k_i's are integers and γ_i closed rectangular paths. We are free to assume that in each γ_i the vertical and horizontal line segments do not overlap except for the end points and that possible cancellations are already carried out.

Further we give the positive orientation for each of this γ_i's so that as z varies over each vertical or horizontal line segment in γ_i, every neighbourhood of z contains points a with $n(\gamma_i, a) \neq 0$.

Given the rectangular chain γ draw all possible horizontal and vertical lines through every segment of γ so that the entire plane is divided into many rectangular regions which are either finite or infinite (bounded or unbounded). See the following figure for guidance.

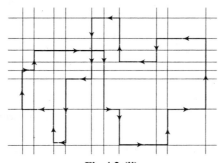

Fig 4.3 (ii)

From each finite open rectangle R in this mesh choose one point $\alpha \in R$ and let $m = n(\gamma, \alpha)$. The other rectangles in this mesh will then be unbounded rectangular regions. Choose all possible finite rectangles R_1, $R_2, \ldots R_n$ and points $\alpha_1, \alpha_2, \ldots \alpha_n$ in these rectangles so that the corresponding $m_i = n(\gamma, \alpha_i)$ $(1 \leq i \leq n)$ are not equal to zero. Put $\eta = \sum_{i=1}^{n} m_i \, \partial R_i$

with positive orientation for each ∂R_i (i.e., the counter clockwise direction).

Consider $\sigma = \gamma - \eta$. We first claim that for every point α not on σ, $n(\sigma, \alpha) = 0$. Now there are four possibilities. (i) α belongs to one of the infinite rectangles. (ii) α belongs to one of the finite rectangles $R_i (i = 1, 2, \ldots n)$. (iii) α belongs to one of the finite rectangles $R \neq R_i (i = 1, 2, \ldots n)$. (iv) α lies on the boundary of an infinite rectangle R but not on σ or on the boundary of a finite rectangle of the type $R \neq R_i$ for any i. If α belongs to the interior of an infinite rectangular region then α also belongs to the unbounded region determined by γ as well as each $\partial R_i (i = 1, 2, \ldots n)$, because this rectangular region being connected should lie in one of the regions determined by γ and in as much as this is unbounded this region determined by γ must also be unbounded. A similar argument applies for ∂R_i. Thus $n(\gamma, \alpha) = 0$ and $n(\partial R_i, \alpha) = 0$ for $(i = 1, 2, \ldots n)$. Thus $n(\sigma, \alpha) = 0$. If α belongs to one of the rectangles of the form R_i then by the properties of the winding number, $n(\partial R_i, \alpha) = n(\partial R_i, \alpha_i) = 1$ and $n(\partial R_j, \alpha) = 0$ for all $j \neq i$ and $n(\gamma, \alpha) = n(\gamma, \alpha_i) = m_i$ (Note that α and α_i belong to the same region determined by γ namely the inside of R_i). Then again

$$n(\sigma, \alpha) = m_i - (m_i \cdot 1 + \sum_{j \neq i} m_j \cdot 0) = 0.$$ If α belongs to one of the finite

rectangles R not of the form R_i then $n(\partial R_i, \alpha) = 0$ for $(i = 1, 2, \ldots n)$ since α does not belong to the interior of any of these R_i's. But $n(\gamma, \alpha) = 0$ (if not $R = R_i$ for some $i = 1, 2, \ldots n$) and hence $n(\sigma, \alpha) = 0$. There is still a last possibility that α lies on the boundary of an infinite rectangle R but not on σ or on the boundary of a finite rectangle of the type $R \neq R_i$ for any i. However in this case $\alpha \notin \sigma$ and there exists a small neighbourhood of α which is completely contained in the same component determined by σ to which α belongs. Yet again in this neighbourhood $n(\sigma, \alpha)$ is constant for all 'a' and some points in its neighbourhood also belong to the interior of R. (Note that α is a boundary point of R). For these a's $n(\sigma, a) = 0$ by the previous case. Thus $n(\sigma, \alpha) = n(\sigma, a) = 0$.

We now show that σ does not contain any horizontal or vertical segment with non-zero coefficient (forcing the chain σ to be identically zero). If possible let $\sigma = m\sigma_1 + \sigma^*$ where σ_1 is a horizontal or vertical line segment in σ, σ^* does not contain σ_1 and $m \neq 0$. Again there are three possibilities. (i) σ_1 can be a common side of two infinite rectangles (ii) σ_1 can be a common side of one infinite and one finite rectangle. (iii) σ_1 can be a common side of two finite rectangles in the mesh.

First we prove that case (i) cannot occur. Indeed in this case it is clear that σ_1 cannot be a part of η. On the other hand if it were to be a part of some γ_i then for $z \in \sigma_1$ and for each sufficiently small neighbourhood of z, $n(\gamma_i, a)$ will be equal to zero for all $a \notin \sigma_1$. This is contrary to our assumption on γ_i. Thus there are only two cases namely (ii) and (iii) and our proof does not distiguish between these cases.

Thus σ_1 is a common boundary of a finite rectangle R and another rectangle R_1 (which may be finite or infinite). We now choose $\alpha \in R$ and $\alpha' \in R_1$ such that the line segment join α and α' does not intersect the closed rectangular chain $\sigma - m\partial R$. Now

$$n(\sigma - m\partial R, \alpha) = n(\sigma - m\partial R, \alpha')$$

But $n(m\partial R, \alpha) = m$ and $n(m\partial R, \alpha') = 0$. Thus $m = 0$ a contradiction. Hence σ is an identically zero chain and so $\gamma = \eta$.

Finally we claim that each rectangle R_i with its boundary ∂R_i is contained

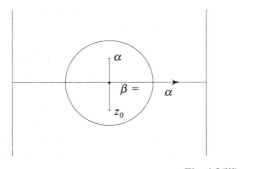
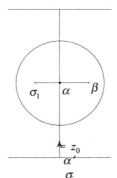

Fig 4.3(iii)

in U. For each point α inside R_i, $n(\gamma, \alpha) = n(\gamma, \alpha_i) = m_i \neq 0$. Hence $\alpha \in U$. If a boundary point α of R_i is on γ then $\alpha \in U$ (since $\gamma \subset U$). If a boundary point α of R_i is not on γ, there exists a neighbourhood N of α in which $n(\gamma, z)$ is constant and is equal to $n(\gamma, \alpha_i) = m_i \neq 0$. Again $\alpha \in U$. Hence $R_i \cup \partial R_i \subset U$ for every $i = 1, 2, \ldots, n$. Hence $\eta \subset U$.

Theorem 4.3.8 (General form of Cauchy's Theorem)
If U is a non-empty open set in the finite complex plane and γ is a cycle homologous to zero in U and $f(z)$ is analytic in U then $\displaystyle\int_\gamma f(z)\,dz = 0$.

Proof

By Theorem 4.3.7, we can find a rectangular chain $\eta = \displaystyle\sum_{i=1}^{n} m_i \partial R_i$ in

U such that $\displaystyle\int_\gamma f(z)\,dz = \int_\eta f(z)\,dz = \sum_{i=1}^{n} m_i \int_{\partial R_i} f(z)\,dz$.

Since for each i $(1 \leq i \leq n)$, $R_i \cup \partial R_i \subset U$, by Cauchy's theorem for a rectangle (Theorem 4.2.5), $\displaystyle\int_{\partial R_i} f(z)dz = 0, 1 \leq i \leq n$. Hence $\displaystyle\int_\gamma f(z)dz$

$= 0$.

Remarks 4.3.9

So far in this section we have stated and proved several versions of Cauchy's theorem. Cauchy's theorem for a rectangle is fundamental to all the other Cauchy's theorem and so can not be dispensed with. It looks as though Cauchy's theorem for simply connected regions includes Cauchy's theorem for a circular disc also, but its proof essentially makes use of the latter. Cauchy's theorem for a rectangle in the presence of exceptional points has been used in proving the analogous theorem for circular disc. For most of the local properties of analytic functions these theorems are sufficient. However Cauchy's theorem for simply connected regions is the most elegant and simple one. Further it has also interesting applications such as providing existence for single-valued analytic branches for $\log f(z)$ for suitable functions $f(z)$. This we will see later on. On the other hand when a given function is analytic in regions which are not simply connected (For example a simply connected region with a finite number of points removed and wherein the behavior of the function at these points is not as simple as in the case of exceptional points) Cauchy's theorem for arbitrary regions plays a crucial role.

In what follows we shall look at simply connected regions more closely and prove that simply connectedness is a topological property of the region. There are several equivalent formulations of simply connectedness a few of which we shall look into in this section and postpone a few more observations to the next chapter.

Difinition 4.3.10 (Fundamental group of a region Ω)

Fix $a \in \Omega$. Consider the set of all closed curves in Ω defined in $[0, 1]$ with the initial and final points at $z = a$. We call them as loops at "a". Let f and g be loops at "a". Define

$$h(s) = \begin{cases} f(2s) & 0 \leq s \leq \dfrac{1}{2} \\ g(2s-1) & \dfrac{1}{2} \leq s \leq 1. \end{cases}$$

Since $f(0) = f(1) = a = g(0) = g(1)$ and $h(0) = f(0) = g(1) = h(1) = a$, h is also a loop at "a" called $f * g$. We write $f \approx g$ if f is homotopic to g with end points fixed. It is easily seen to be an equivalence relation. We shall merely check the transitivity. Let $f \approx g$ and $g \approx h$. There exists $F: I \times I \rightarrow \Omega$ and $G: I \times I \rightarrow \Omega$ such that

$$\begin{array}{ll} F(s, 0) = f(s) & G(s, 0) = g(s) \\ F(s, 1) = g(s) & G(s, 1) = h(s) \\ F(0, t) = F(1, t) = a & G(0, t) = G(1, t) = a \end{array}$$

Define $H: I \times I \to \Omega$ by

$$H(s,t) = \begin{cases} F(s,2t) & 0 \le t \le \dfrac{1}{2} \\ G(s, 2t - 1) & \dfrac{1}{2} \le t \le 1. \end{cases}$$

One can easily verify that H is a homotopy between f and h.

$$H(s, 0) = F(s, 0) = f(s)$$
$$H(s, 1) = G(s, 1) = h(s)$$
$$H(0, t) = F(0, 2t) = G(0, 2t - 1) = a$$
$$H(1, t) = F(1, 2t) = G(1, 2t - 1) = a.$$

The set of all equivalence classes of loops at "a" will be denoted by $\pi(\Omega, a)$. We shall introduce a multiplication $*$ in this set as follows: $[f] * [g] = [f * g]$. We claim that this $*$ is well defined in the sense that if $f \approx f_1$, $g \approx g_1$ then $f * g \approx f_1 * g_1$. Let F be a homotopy between f and f_1 and G between g and g_1. Then one can easily verify that

$$H(s,t) = \begin{cases} F(2s, t) & 0 \le s \le \dfrac{1}{2} \\ G(2s - 1, t) & \dfrac{1}{2} \le s \le 1 \end{cases}$$

is a homotopy between $f * g$ and $f_1 * g_1$. Further $*$ has the following properties:

(1) $([f] * [g]) * [h] = [f] * ([g] * [h])$

(2) If ρ_a denotes the constant path at "a" then

$$[f] * [\rho_a] = [f] = [\rho_a] * [f] \; \forall \; [f] \in \pi(\Omega, a)$$

(3) For $[f] \in \pi(\Omega, a)$, $[f] * [\tilde{f}] = [\rho_a]$ where $\tilde{f}(s) = f(1 - s)$, $0 \le s \le 1$

In verifying the above properties one need to know the exact homotopy involved. Once this is given it is fairly a routine matter to check the details. In the following we merely give the relevant homotopies:

(1) To prove $(f * g) * h$ is homotopic to $f * (g * h)$ define

$$F(s,t) = \begin{cases} f\left(\dfrac{4 + s}{t + 1}\right) & s \in \left[0, \dfrac{1 + t}{4}\right] \\ g(4s - t - 1) & s \in \left[\dfrac{1 + t}{4}, \dfrac{2 + t}{4}\right] \\ h\left(\dfrac{4s - t - 2}{2 - t}\right) & s \in \left[\dfrac{2 + t}{4}, 1\right]. \end{cases}$$

(2)   To show $f \approx f * \rho_a$ take

$$G(s,t) = \begin{cases} f\left(\dfrac{2s}{2-t}\right) & s \in \left[0, \dfrac{2-t}{2}\right] \\[2mm] a & s \in \left[\dfrac{2-t}{2}, 1\right]. \end{cases}$$

(3)   To prove $f * \tilde{f} \approx \rho_a$ define

$$H(s,t) = \begin{cases} f(2ts) & s \in \left[0, \dfrac{1}{2}\right] \\[2mm] f(2t(1-s)) & s \in \left[\dfrac{1}{2}, 1\right]. \end{cases}$$

Note also that by (3) above $\tilde{f} * \tilde{\tilde{f}} \approx \rho_a$ i.e., $\tilde{f} * f \approx \rho_a$. Thus $\pi(\Omega, a)$ is a group called the fundamental group of Ω at "a". Now it looks as though $\pi(\Omega, a)$ is dependent on "a". In fact for any two points a and b belonging to Ω (Ω is a path connected open set) we can easily prove that $\pi(\Omega, a)$ is group isomorphic to $\pi(\Omega, b)$. Once this is done the group is invariant and depends only on Ω and not on the point "a". We now exhibit this isomorphism. We first observe that the notion of $f * g$ or $[f] * [g]$ can also be made applicable for paths f and g such that the initial point of g is the same as final point of f in the plane. With this extended notion we can describe the required isomorphism as follows. If α is a path from a to b in Ω then $\overline{\alpha} : \pi(\Omega, a) \to \pi(\Omega, b)$ given by $\overline{\alpha}([f]) = [\tilde{\alpha} * f * \alpha]$ is an isomorphism between the groups. Indeed

$$\begin{aligned} \overline{\alpha}([f]) * \overline{\alpha}([g]) &= [\tilde{\alpha} * f * \alpha] * [\tilde{\alpha} * g * \alpha] \\ &= [\tilde{\alpha} * f * (\alpha * \tilde{\alpha}) * g * \alpha] \\ &= [\tilde{\alpha} * (f * g) * \alpha] \\ &= \overline{\alpha}([f * g]). \end{aligned}$$

Thus $\overline{\alpha}$ is a homomorphism. Also if $\beta = \tilde{\alpha}$. then $\overline{\beta}$ defined by $\overline{\beta}([h]) = [\overline{\beta} * h * \beta]$ satisfies (can be easily verified)

$$\tilde{\alpha}(\overline{\beta}([h])) = [h], \ \forall \ [h] \in \pi(\Omega, b)$$

and

$$\overline{\beta}(\overline{\alpha}([f])) = [f], \ \forall \ [f] \in \pi(\Omega, a).$$

Thus $\overline{\alpha}$ is both one-to-one and onto and so is an isomorphism. We now come to the main result of this section.

Theorem 4.3.11
A region Ω is simply connected if and only if $\pi(\Omega, a)$ is the identity group.

Proof

Ω is simply connected if and only if all closed curves at "a" are homotopic to ρ_a i.e., if and only if $\pi(\Omega, a) = \{[\rho_a]\}$.

If Ω_1 and Ω_2 are two regions in the finite complex plane and $h: \Omega_1 \to \Omega_2$ is any continuous map, we can induce a homomorphism $h^*: \pi(\Omega_1, a) \to \pi(\Omega_2, h(a))$ as follows:

$$h^*([f]) = [h \circ f]$$

where $h \circ f$ is the composition of f and h. We mention the following facts which can be easily verified:

(1) h^* is well defined: $f \approx f'$ with homotopy F then $h \circ f \approx h \circ f'$ with homotopy $h \circ F$.

(2) $h([f * g])\ (s) = \begin{cases} h(f(2s)) & s \in \left[0, \dfrac{1}{2}\right] \\ h(g(2s-1)) & s \in \left[\dfrac{1}{2}, 1\right] \end{cases}$

implies

$$h \circ (f * g) = (h \circ f) * (h \circ g) \Rightarrow h^*\ ([f] * [g])$$
$$= h^*\ ([f]) * h^*\ ([g])$$

That is, h^* is a homomorphism.

Theorem 4.3.12

If $h : \Omega_1 \to \Omega_2$ and $k : \Omega_2 \to \Omega_3$ are continuous maps between regions and $a \in \Omega_1$ then

(i) $(k \circ h)^* = k^* \circ h^*$

(ii) If $i : \Omega \to \Omega$ is the identity map on a region Ω, then $i^* : \pi(\Omega, a) \to \pi(\Omega, a)$ is the identity isomorphism.

Proof

(i) $(k \circ h)^*\ [f] = [(k \circ h) \circ f] = [k \circ (h \circ f)] = k^*\ ([h \circ f]) = (k^* \circ h^*)$
     $[f]$

(ii) $i^*\ ([f]) = [i \circ f] = [f]$.

Theorem 4.3.13

If $h : \Omega_1 \to \Omega_2$ is a topological homeomorphism between two regions Ω_1 and Ω_2 then

$$h^*: \pi(\Omega_1, a) \to \pi(\Omega_2, h(a))$$

is an isomorphism.

Proof

If $k : \Omega_2 \to \Omega_1$ is the inverse of $h : \Omega_1 \to \Omega_2$ then $k^* \circ h^* = (k \circ h)^*$ which is the identity isomorphism on $\pi(\Omega_1, a)$. Similarly $h^* \circ k^* = (h \circ k)^*$ is the identity isomorphism on $\pi(\Omega_2, h(a))$. Thus $\pi(\Omega_1, a)$ and $\pi(\Omega_2, h(a))$ are isomorphic.

Theorem 4.3.14
Simply connectedness property is a topological property (invariant under homeomorphisms).

Proof
If Ω is simply connected and $h : \Omega \to \Omega_0$ is a topological homeomorphism then for $a \in \Omega$, $\pi(\Omega, a) = \{[\rho_a]\}$ by Theorem 4.3.11. Now by Theorem 4.3.13 $\pi(\Omega_0, h(a))$ is isomorphic to $\pi(\Omega, a)$, the group consisting of only the identity element and again by Theorem 4.3.11, Ω_0 is also simply connected.

Remark 4.3.15
Having characterized simply connected regions as those for which the fundamental group reduces to the identity, we remark that there are other equally interesting characterizations of simply connectedness. However to understand them we need more techniques and hence we postpone it until next chapter.

4.4 Cauchy's Integral Formula and Its Applications

Introduction 4.4.1
In this section we shall use a few preliminary versions of Cauchy's theorem and study the local behaviour of analytic functions. Surprisingly, many interesting developments in complex analysis depend on these results.

Theorem 4.4.2 (Cauchy's integral formula for a disc)
Let $f(z)$ be analytic in an open disc D and $\gamma \subset D$ be a closed curve. Then

$$n(\gamma, z)\, f(z) = \frac{1}{2\pi i} \int_\gamma \frac{f(\zeta)}{\zeta - z}\, d\zeta$$

is valid for all $z \in D \backslash \gamma$ (Here $n(\gamma, z)$ denotes the index of γ with respect to z).

Proof

First note that $z \notin \gamma$ is essential for defining $\int_\gamma \frac{f(\zeta)}{\zeta - z}\, d\zeta$. Put $F(\zeta)$

$$= \frac{f(\zeta) - f(z)}{\zeta - z} \text{ for } \zeta \neq z \ (\zeta \in D). \ F(\zeta) \text{ is analytic at all points of } D \text{ except}$$

at $\zeta = z$ and further we have $\lim_{\zeta \to z} (\zeta - z)\, F(\zeta) = \lim_{\zeta \to z} (f(\zeta) - f(z)) = 0$.
Thus $F(\zeta)$ has $\zeta = z$ as an exceptional point in the sense of Theorem 4.2.11. An application of that theorem now gives $\int_\gamma F(\zeta)\, d\zeta = 0$. But this

is equivalent to $n(\gamma, z)\, f(z) = \frac{1}{2\pi i} \int_\gamma \frac{f(\zeta)}{\zeta - z}\, d\zeta$.

Theorem 4.4.3

Suppose $\phi\,(\zeta)$ is continuous on a curve γ (may or may not be closed) in the finite complex plane. Then

$$F_n(z) = \int_\gamma \frac{\phi(\zeta)}{(\zeta - z)^n}\, d\zeta \quad (n = 1, 2, \ldots)$$

is defined and analytic in each of the regions determined by γ (That is, the components of the complement of γ) and we also have $F'_n\,(z) = n\, F_{n+1}(z)$.

Proof

We shall prove this theorem by induction on "n". Take $n = 1$ and $z_0 \in \gamma^c$ (complement of γ). Choose $\delta > 0$ so that the disc $|z - z_0| < \delta$ does not meet γ (Note that γ^c is open). Restrict z to the neighbourhood $|z - z_0| < \dfrac{\delta}{2}$ so that $|\zeta - z_0| \geq \delta$ and $|\zeta - z| \geq |\zeta - z_0| - |z - z_0| > \delta - \dfrac{\delta}{2} = \dfrac{\delta}{2}\,\forall$ $\zeta \in \gamma$. Now,

$$F_1(z) - F_1(z_0) = \int_\gamma \frac{\phi(\zeta)}{(\zeta - z)}\, d\zeta - \int_\gamma \frac{\phi(\zeta)}{(\zeta - z_0)}\, d\zeta = \int_\gamma \frac{\phi(\zeta)(z - z_0)}{(\zeta - z)(\zeta - z_0)}\, d\zeta.$$

Hence

$$|F_1(z) - F_1\,(z_0)| \leq |z - z_0|\, 2 \int_\gamma \frac{|\phi(\zeta)|}{\delta^2}\, |d\zeta|.$$

Since ϕ is continuous on the compact set γ, $|\phi(\zeta)| \leq M$ for a suitable constant M. Thus

$$|F_1(z) - F_1(z_0)| \leq |z - z_0|\, \frac{2M}{\delta^2}\, l(\gamma)$$

where $l(\gamma)$ is the length of γ given $l(\gamma) = \int_\gamma |d\zeta|$. So given $\varepsilon > 0$ if we choose $\eta < \text{Min}\left\{ \dfrac{\delta}{2}, \dfrac{\varepsilon\delta^2}{2Ml(\gamma)} \right\}$ then $|z - z_0| < \eta$ implies $|F_1(z) - F_1(z_0)|$ $< \varepsilon$ i.e., $F_1(z)$ is continuous at $z_0 \in \gamma^c$. Since z_0 is arbitrary in γ^c, $F_1(z)$ is continuous in each of the regions determined by γ. Note that in the above context ϕ can be replaced by any arbitrary continuous function on γ and the corresponding $F_1(z)$ will be continuous.

We shall now use the continuity of $F_1(z)$ (Of course for various ϕ's) to obtain $F'_1(z) = F_2(z)$. Take $\phi_1(\zeta) = \dfrac{\phi(\zeta)}{(\zeta - z_0)}$ $(z_0 \in \gamma^c)\cdot \phi_1(\zeta)$ is still continuous on γ and by what we have already proved $G_1(z) = \displaystyle\int_\gamma \frac{\phi(\zeta)}{(\zeta - z_0)(\zeta - z)}\,d\zeta$

is continuous at z_0 and so $G_1(z) \to G_1(z_0)$ as $z \to z_0$. Now $\dfrac{F_1(z) - F_1(z_0)}{z - z_0}$

$= G_1(z)$. Thus $F_1'(z_0) = \displaystyle\lim_{z \to z_0} \frac{F_1(z) - F_1(z_0)}{z - z_0} = \lim_{z \to z_0} G_1(z) = G_1(z_0) = F_2(z_0)$.

This completes the proof for the basis of induction (i.e., for $n = 1$).

Let us now assume that $F_{n-1}(z)$ (and hence all such functions obtained by replacing ϕ by any other continuous function on γ) is analytic and $F'_{n-1}(z) = (n - 1)\,F_n(z)$. Consider

$$F_n(z) - F_n(z_0) = \int_\gamma \phi(\zeta)\left(\frac{1}{(\zeta - z)^n} - \frac{1}{(\zeta - z_0)^n}\right)d\zeta.$$

Now

$$\left(\frac{1}{(\zeta - z)^n} - \frac{1}{(\zeta - z_0)^n}\right) = \frac{1}{(\zeta - z)^{n-1}(\zeta - z_0)} + \frac{z - z_0}{(\zeta - z)^n(\zeta - z_0)} - \frac{1}{(\zeta - z_0)^n},$$

$$F_n(z) - F_n(z_0) =$$

$$\int_\gamma \frac{\phi(\zeta)\,d\zeta}{(\zeta - z)^{n-1}(\zeta - z_0)} - \int_\gamma \frac{\phi(\zeta)\,d\zeta}{(\zeta - z_0)^n} + (z - z_0)\int_\gamma \frac{\phi(\zeta)\,d\zeta}{(\zeta - z)^n(\zeta - z_0)}.$$

Our induction hypothesis implies that $G_{n-1}(z) = \displaystyle\int_\gamma \frac{\phi(\zeta)\,d\zeta}{(\zeta - z_0)(\zeta - z)^{n-1}}$ (Re-

placing $\phi(\zeta)$ by $\dfrac{\phi(\zeta)}{(\zeta - z_0)}$ which is also continuous on γ), is continuous

and $G'_{n-1}(z_0) = (n - 1)\,G_n(z_0)$ where $G_n(z) = \displaystyle\int_\gamma \frac{\phi(\zeta)\,d\zeta}{(\zeta - z)^n(\zeta - z_0)}.$

$$F_n(z) - F_n(z_0) = G_{n-1}(z) - G_{n-1}(z_0) + (z - z_0)\int_\gamma \frac{\phi(\zeta)\,d\zeta}{(\zeta - z)^n(\zeta - z_0)}. \quad (1)$$

As before, using $|\zeta - z_0| > \dfrac{\delta}{2}$ and $|\zeta - z| \geq \delta$, $\forall\,\zeta \in \gamma$, we get

$$|F_n(z) - F_n(z_0)| \leq |G_{n-1}(z) - G_{n-1}(z_0)| + |z - z_0|\,\frac{2}{\delta^{n+1}}\,Ml(\gamma).$$

Both the terms on the right side can be made as small as we please by choosing $|z - z_0|$ sufficiently small. (Note that $G_{n-1}(z)$ is analytic at $z_0 \Rightarrow$

$G_{n-1}(z)$ is continuous at z_0 and $\dfrac{2}{\delta^{n+1}} Ml(\gamma)$ is a constant). Thus, $F_n(z)$ (and all others obtained by replacing ϕ by any other continuous function on γ) is continuous. Since

$$G_n(z) = \int_\gamma \frac{\phi(\zeta)\, d\zeta}{(\zeta - z)^n (\zeta - z_0)}$$

(1) can be written as

$$\frac{F_n(z) - F_n(z_0)}{z - z_0} = \frac{G_{n-1}(z) - G_{n-1}(z_0)}{z - z_0} + G_n(z). \qquad (2)$$

Taking limit as $z \to z_0$ in (2) we get

$$F_n'(z_0) = G'_{n-1}(z_0) + G_n(z_0)$$

($G_n(z)$ is continuous at z_0 by replacing ϕ by $\dfrac{\phi(\zeta)}{(\zeta - z_0)}$ in the definition of F_n)

$$= (n - 1)\, G_n(z_0) + G_n(z_0)$$
$$= n\, G_n(z_0)$$
$$= n\, F_{n+1}(z_0).$$

Hence the theorem.

Theorem 4.4.4
Analytic functions in regions have derivatives of all orders which are all analytic.

Proof
Let $f(z)$ be analytic in a region Ω, $z_0 \in \Omega$. Take a closed disc $|z - z_0| \le r$ which is completely contained in Ω and denote by γ its boundary considered as a simple closed curve with $n(\gamma, z) = 1$ or $0 \ \forall\, z \notin \gamma$. Then

$$f(z) = \frac{1}{2\pi i} \int_\gamma \frac{f(\zeta)}{\zeta - z}\, d\zeta \ \forall\, z \text{ with } |z - z_0| < r.$$

Since $f(\zeta)$ is continuous on γ, applying Theorem 4.4.3 repeatedly we get that $f(z)$ is analytic in the interior of γ and

$$f'(z) = \frac{1}{2\pi i} \int_\gamma \frac{f(\zeta)}{(\zeta - z)^2}\, d\zeta; \quad f''(z) = \frac{2}{2\pi i} \int_\gamma \frac{f(\zeta)}{(\zeta - z)^3}\, d\zeta$$

and in general

$$f^{(k)}(z) = \frac{k!}{2\pi i} \int_\gamma \frac{f(\zeta)}{(\zeta - z)^{k+1}}\, d\zeta.$$

All these are analytic in $|z - z_0| < r$ and so at z_0 by Theorem 4.4.3. Since z_0 is arbitrary this completes the proof.

The above formulas enable us to justify "differentiation under the integral sign".

Corollary 4.4.5

If $f(z)$ is analytic in a region Ω then its partial derivatives of all orders exist and are continuous in Ω.

Proof

If $f = u + iv$ we know that u_x, v_x, u_y, v_y exist at each point of Ω and $f'(z)$ $= u_x + iv_x = v_y - iu_y$ by Cauchy-Riemann equations. Now by Theorem 4.4.4 $f'(z)$ is also analytic in Ω and so is continuous in Ω. Thus its real and imaginary parts namely u_x, v_x, u_y, v_y are all continuous. A repeated application of Theorem 4.4.4 for $f'(z)$, $f''(z)$ etc gives our result.

Corollary 4.4.6 (Morera's Theorem)

If $f(z)$ is continuous in a region Ω and for all closed curves γ in Ω, $\int_\gamma f(z) \, dz = 0$, then $f(z)$ is analytic in Ω.

Proof

By Theorem 4.2.4, $f(z) \, dz$ is exact in Ω. This means by Theorem 4.2.3 that $f(z) = g'(z)$ where $g(z)$ is analytic in Ω. But then by Theorem 4.4.4, $g'(z)$ $= f(z)$ is analytic in Ω.

Theorem 4.4.7

If $f(z)$ is analytic in a simply connected region Ω and if $f \neq 0$ any where in Ω then we can define a single-valued analytic branch for $\log f(z)$ in Ω.

Proof

Note that for each $z \in \Omega$ there are infinitely many values for $\log f(z)$ in general. The theorem tells us that at each $z \in \Omega$ we can choose exactly one particular value of $\log f(z)$ say $F_1(z)$ so that $z \mapsto F_1(z)$ is one-to-one and analytic in Ω. We also recall that $F_1(z)$ is a single-valued branch for $\log f(z)$ if and only if $F_1(z)$ is single-valued and $e^{F_1(z)} = f(z)$.

We shall prove more specifically that if $z_0 \in \Omega$ and w_0 is any one of the infinitely many values of $\log f(z_0)$ then there exists a single-valued analytic function $F_1(z)$ such that $e^{F_1(z)} = f(z) \; \forall \; z \in \Omega$ and $F_1(z_0) = w_0$. (So that the branch that we choose for $\log f(z)$ has a predetermined value at a fixed point $z_0 \in \Omega$). Consider $\dfrac{f'(z)}{f(z)}$ which is analytic in Ω as $f(z) \neq 0$ and $f'(z)$ is analytic. By Theorem 4.2.19 we can find an analytic function $F(z)$ such that $F'(z) = \dfrac{f'(z)}{f(z)}$. Consider $e^{-F(z)} f(z)$ which is analytic in Ω. If we let

$$h(z) = e^{-F(z)} f(z) \text{ then}$$

$$h'(z) = e^{-F(z)} f'(z) - F'(z) e^{-F(z)} f(z)$$

$$= e^{-F(z)} \left[f'(z) + f(z) \left(-\frac{f'(z)}{f(z)} \right) \right]$$

$$= 0.$$

Since Ω is connected we see that $h(z)$ is a constant function. To evaluate this constant we shall write $h(z) = h(z_0)$. i.e., $e^{-F(z)} f(z) = e^{-F(z_0)} f(z_0)$. But since w_0 is a chosen value of $\log f(z_0)$ we have $e^{w_0} = f(z_0)$. Thus

$$f(z) = e^{F(z) - F(z_0) + w_0}.$$

Putting $F_1(z) = F(z) - F(z_0) + w_0$ we see that $F_1(z)$ is single-valued, analytic in Ω, $e^{F_1(z)} = f(z)$ and $F_1(z_0) = w_0$.

Theorem 4.4.8 (Liouville's Theorem)
A function $f(z)$ which is analytic and bounded in the entire finite complex plane is a constant.

Proof
Let $|f(\zeta)| \le M \; \forall \; \zeta \in \mathbb{C}$. Fix $a \in \mathbb{C}$, take a circle γ, centre "a" and radius $R > 0$. By what we have proved,

$$f'(a) = \frac{1}{2\pi i} \int_\gamma \frac{f(\zeta)}{(\zeta - a)^2} \, d\zeta$$

$$|f'(a)| \le \frac{1}{2\pi} \frac{M}{R^2} 2\pi R = \frac{M}{R}.$$

Since R can be chosen as large as we want, we can let $R \to \infty$ to get $f'(a) = 0$. Since 'a' is arbitrary, $f(z)$ is a constant (Note that $\mathbb{C}$ is connected).

Corollary 4.4.9 (Fundamental theorem of algebra)
Every non-constant polynomial $p(z)$ with complex coefficients has at least one zero. (By division algorithm a polynomial of degree "n" has exactly "n" zeros including multiplicities).

Proof
If possible let $p(z) \ne 0$ anywhere in $\mathbb{C}$. Then $p(z)$ is analytic and $p(z) \ne 0$ and thus $\frac{1}{p(z)}$ is analytic in the entire plane. If

$$p(z) = a_0 + a_1 z + a_2 z^2 + \ldots + a_n z^n \; (a_n \ne 0)$$

then by triangle inequality

$$|p(z)| \ge |z|^n \left(|a_n| - \frac{|a_{n-1}|}{|z|} - \frac{|a_{n-2}|}{|z|^2} - \ldots - \frac{|a_0|}{|z|^n} \right) \; (z \ne 0).$$

As the right side tends to ∞ as $|z| \to \infty$, $\dfrac{1}{|p(z)|} \to 0$ as $|z| \to \infty$ and

so is bounded for $|z| \geq M$ (for a suitable large M). In $|z| \leq M$, $\dfrac{1}{|p(z)|}$ being

a continuous function is still bounded and so $\dfrac{1}{p(z)}$ is a bounded analytic

function in the entire plane. By Liouville's theorem $\dfrac{1}{p(z)}$ and hence $p(z)$

reduces to a constant, a contradiction to our hypothesis. This proves our corollary.

Notes 4.4.10
(a) We shall now prove that Cauchy's integral formula for a circular disc enables us to write

$$f(a) = \frac{1}{2\pi i} \int_\gamma \frac{f(\zeta)}{\zeta - a} \, d\zeta \quad (|z - a| < r)$$

whenever $f(z)$ is analytic in a region Ω, $a \in \Omega$ and γ is the circle $|z - a| = r$ such that the disc $|z - a| \leq r$ is completely contained in Ω. Indeed there is a bigger concentric circle centered at "a" which is completely contained in Ω (For a proof of this we observe that $r < d\,(a, \Omega^c)$ and the circle with radius "s" where $r < s < d\,(a, \Omega^c)$ will serve our purpose) and $f(z)$ is analytic in this larger disc D and γ is a simple closed curve in D. (Now we can apply Cauchy's integral formula for the disc D with this γ). On the other hand ζ on γ can be parameterized by $\zeta = a + re^{i\theta}$ $(0 \leq \theta \leq 2\pi)$ and

so $f(a) = \int_0^{2\pi} f(a + re^{i\theta}) \, d\theta$. Thus the value of f at the centre is equal to

the mean of its values on the circumference of the circle. This is true of any circle centered at "a" provided the above condition is satisfied.

(b)   We first note that Cauchy's integral formula for a circular disc as stated in Theorem 4.4.2 can be easily extended when there are finitely many exceptional points and γ does not pass through any of them and z is not any of these exceptional points. This is because all the exceptional points of $f(\zeta)$ continue to be exceptional points for the new function $F(\zeta)$ constructed in that proof and one more exceptional point $\zeta = z$ will be added to this list.

(c)   Cauchy's theorem for a circular disc in the presence of exceptional points gave rise to Cauchy's integral formula for a circular disc, i.e., if $f(z)$ is analytic in an open disc D and γ is a closed curve completely contained

in D then $f(z)\,n(\gamma, z) = \dfrac{1}{2\pi i} \int_\gamma \dfrac{f(\zeta)}{\zeta - z} \, d\zeta$ $(z \notin \gamma)$. On the other hand if we assume Cauchy's integral formula to be valid for a circular disc, we can also prove that Cauchy's theorem for a circular disc is equally valid. The proof is as follows.

Let $f(z)$ be analytic in an open disc D and γ be a closed curve in D. Take $z_0 \notin \gamma$ and $z_0 \in D$. Then $F(z) = (z - z_0)f(z)$ is also analytic in D. By integral formula

$$n(\gamma, z_0)\, F(z_0) = \frac{1}{2\pi i} \int_\gamma \frac{F(\zeta)}{\zeta - z_0}\, d\zeta = \frac{1}{2\pi i} \int_\gamma f(\zeta)\, d\zeta.$$

But $F(z_0) = 0 \Rightarrow \int_\gamma f(\zeta)\, d\zeta = 0$.

(d) Similarly the proof of Cauchy's theorem for a circular disc in the presence of exceptional points can be obtained (as above) from the corresponding integral formula for circular regions in the presence of exceptional points.

Theorem 4.4.11

Let $f(z)$ be analytic in a region Ω and $a \in \Omega$. There exists an analytic function $f_1(z)$ in Ω which is equal to $\dfrac{f(z) - f(a)}{z - a}$ for $z \ne a$ and equal to $f'(a)$ for $z = a$.

Proof

Let $F(z) = \dfrac{f(z) - f(a)}{z - a}$ $(a \ne z \in \Omega)$. As we already saw, we can find r, ρ with $\rho > r$ such that $|z - a| \le r$ is contained in $|z - a| < \rho$ which in turn is contained in Ω. Call the disc $|z - a| < \rho$ as D and the simple closed curve $|z - a| = r$ as C. We know that $F(z)$ is analytic in $\Omega \backslash \{a\}$ and hence in $D \backslash \{a\}$. Also $z = a$ is an exceptional point for $F(z)$ since $\lim_{z \to a}(z - a)\, F(z)$ $= \lim_{z \to a}(f(z) - f(a)) = 0$. Thus as observed in Note 4.4.10 (b) Cauchy's integral formula for $F(z)$ is valid in D. Since $a \notin C$ and $n(C, z) = 1$ for $|z - a| < r$ we can easily conclude that

$$F(z) = \frac{1}{2\pi i} \int_C \frac{F(\zeta)}{\zeta - z}\, d\zeta \text{ for } |z - a| < r.$$

On the other hand the right side of the above equality is an analytic function of z in $|z - a| < r$ (See Theorem 4.4.3) with its value at '$z = a$' being

$$\frac{1}{2\pi i} \int_C \frac{F(\zeta)}{\zeta - a}\, d\zeta = \frac{1}{2\pi i} \int_C \frac{f(\zeta)}{(\zeta - a)^2}\, d\zeta - \frac{f(a)}{2\pi i} \int_C \frac{1}{(\zeta - a)^2}\, d\zeta = f'(a)$$

Note that Cauchy's integral formula for the constant function $g(z) \equiv 1$ in D gives

$$1 = g(a) = \frac{1}{2\pi i} \int_C \frac{g(\zeta)}{\zeta - a} \, d\zeta = \frac{1}{2\pi i} \int_C \frac{1}{\zeta - a} \, d\zeta$$

and also (see the proof of Theorem 4.4.4)

$$0 = g'(a) = \frac{1}{2\pi i} \int_C \frac{g(\zeta)}{(\zeta - a)^2} \, d\zeta = \frac{1}{2\pi i} \int_C \frac{1}{(\zeta - a)^2} \, d\zeta$$

Hence there exists an analytic function $f_1(z)$ in Ω which is equal to $F(z)$

$$= \frac{f(z) - f(a)}{z - a} \quad \text{for } z \neq a \text{ and is equal to } f'(a) \text{ at } z = a.$$

Theorem 4.4.12 (Taylor's Theorem)
Let $f(z)$ be analytic in a region Ω, $a \in \Omega$. Then $f(z)$ can be expanded in the following form

$$f(z) = f(a) + \frac{f'(a)}{1!}(z - a) + \frac{f''(a)}{2!}(z - a)^2 + \ldots + \frac{f^{(n-1)}(a)}{(n-1)!}(z - a)^{n-1}$$

$$+ f_n(z)(z - a)^n$$

where $f_n(z)$ is analytic in Ω with $f_n(a) = \dfrac{f^{(n)}(a)}{n!}$. Further $f_n(z)$ is given by

$$f_n(z) = \frac{1}{2\pi i} \int_C \frac{f(\zeta) \, d\zeta}{(\zeta - a)^n (\zeta - z)} \quad \text{where } C \text{ is any circle centre ``}a\text{'' and radius}$$

r such that the disc $|z - a| \leq r$ is contained in Ω. This expansion is valid for $z \in \Omega$ and for $n = 1, 2, \ldots$. Further this line integral representation for $f_n(z)$ is also valid for all z inside C.

Proof
By Theorem 4.4.11 we have an analytic function $f_1(z)$ in Ω such that $f(z) = f(a) + (z - a) f_1(z)$ holds for all z in Ω. Repeating the above argument we can find an analytic function $f_2(z)$ such that $f_1(z) = f_1(a) + (z - a) f_2(z)$. That is $f(z) = f(a) + (z - a) f_1(a) + (z - a)^2 f_2(z)$ and so on. Thus repeating this process n times we find that

$$f(z) = f(a) + (z - a) f_1(a) + (z - a)^2 f_2(a) + \ldots + (z - a)^{n-1} f_{n-1}(a) + (z - a)^n f_n(z)$$

where $f_n(z)$ is analytic in Ω. Differentiating k times ($1 \leq k \leq n$) and putting $z = a$ we find that $f^{(k)}(a) = k! f_k(a)$. Thus

$$f(z) = f(a) + \frac{f'(a)}{1!}(z - a) + \frac{f''(a)}{2!}(z - a)^2 + \ldots + \frac{f^{(n-1)}(a)}{(n-1)!}(z - a)^{n-1}$$

$$+ f_n(z)(z - a)^n$$

where $f_n(z)$ is analytic in Ω and $f_n(a) = \dfrac{f^{(n)}(a)}{n!}$. This completes the first part of our proof. Now for $z \neq a$ we write

$$f_n(z) = \frac{1}{(z-a)^n}\left[f(z) - f(a) - \frac{f'(a)}{1!}(z-a) - \frac{f''(a)}{2!}(z-a)^2 + \dots \right.$$

$$\left. - \frac{f^{(n-1)}(a)}{(n-1)!}(z-a)^{n-1}\right].$$

Since $f_n(z)$ is analytic in Ω using Cauchy's integral formula with the same circle C described in this theorem

$$f_n(z) = \frac{1}{2\pi i}\int_C \frac{f_n(\zeta)\,d\zeta}{(\zeta - z)}.$$

Now for our integration $\zeta \in C$ and so $\zeta \neq a$. Thus

$$f_n(z) = \frac{1}{2\pi i}\left[\int_C \frac{f(\zeta)\,d\zeta}{(\zeta-a)^n(\zeta-z)} - f(a)\int_C \frac{d\zeta}{(\zeta-a)^n(\zeta-z)} - \right.$$

$$\left. - \frac{f'(a)}{1!}\int_C \frac{d\zeta}{(\zeta-a)^{n-1}(\zeta-z)} - \dots - \frac{f^{(n-1)}(a)}{(n-1)!}\int_C \frac{d\zeta}{(\zeta-a)(\zeta-z)}\right].$$

Fix z such that $|z - a| < r$. Let us now define a new function

$$G(w) = \int_C \frac{d\zeta}{(\zeta-w)(\zeta-z)} \quad \text{for } |w-a| < r.$$

For $w = z$, $G(w) = G(z) = \displaystyle\int_C \frac{d\zeta}{(\zeta-z)^2} = 0$ as proved earlier (i.e., applying Cauchy's integral formula for $g(z) \equiv 1$ in $|z - a| < r$ and differentiating under the integral sign). Let $w \neq z$ then

$$G(w) = \frac{1}{z-w}\int_C \left[\frac{1}{\zeta-z} - \frac{1}{\zeta-w}\right]d\zeta$$

$$= \frac{1}{z-w}2\pi i\,[n(C,\,z) - n(C,\,w)] = 0.$$

Thus if we take $\phi(\zeta) = \dfrac{1}{\zeta-z}$ which is continuous on C then $G(w)$

$$= \int_C \frac{\phi(\zeta)\,d\zeta}{\zeta-w} \text{ is identically zero and hence for all integers } k \geq 0,\, G^{(k)}(w) \equiv 0.$$

But by Theorem 4.4.3

$$G^{(k)}(w) = k! \int_C \frac{\phi(\zeta)\,d\zeta}{(\zeta - w)^{k+1}}.$$

Hence $\displaystyle\int_C \frac{d\zeta}{(\zeta - w)^{k+1}(\zeta - z)} \equiv 0$ for all w with $|w - a| < r$ and for all

integers $k \geq 0$. In particular $\displaystyle\int_C \frac{d\zeta}{(\zeta - a)^{k+1}(\zeta - z)} = 0$ $(0 \leq k \leq n - 1)$ and

we find that $f_n(z) = \dfrac{1}{2\pi i}\displaystyle\int_C \dfrac{f(\zeta)\,d\zeta}{(\zeta - a)^n(\zeta - z)}$ is valid for $|z - a| < r$. This

completes the proof of our theorem.

The finite Taylor development obtained above for any analytic function in a region Ω around a point $a \in \Omega$ is powerful enough to discuss some interesting local properties of f. For example if f is analytic in a region Ω and for some $a \in \Omega$, f and all its derivatives vanished at "a" then $f(z) \equiv 0$ in Ω. This is called the principle of analytic continuation (The significance of this name will be clear, a little later) and we proceed to prove this.

Theorem 4.4.13
If $f(z)$ is analytic in a region Ω and for some $a \in \Omega$, $f^{(k)}(a) = 0$ for $k = 0, 1, \ldots$ (Here $f^{(0)}(a)$ means $f(a)$) then $f(z) \equiv 0$ in Ω.

Proof
By Taylor's (finite development) theorem around the given point $a \in \Omega$ we can always write $f(z) = (z - a)^n f_n(z)$ where $f_n(z)$ is analytic in Ω (for any positive integer n). Let $|z - a| \leq r$ be contained in Ω. Then we also

have $f_n(z) = \dfrac{1}{2\pi i}\displaystyle\int_C \dfrac{f(\zeta)\,d\zeta}{(\zeta - a)^n(\zeta - z)}$ where C is the simple closed curve

$|\zeta - a| = r$. Also

$$|f_n(z)| \leq \frac{1}{2\pi}\int_C \frac{|f(\zeta)|\,|d\zeta|}{|\zeta - a|^n|\zeta - z|}.$$

Now $|f(\zeta)|$ is bounded on C (a compact set) by say M. Also $|\zeta - z| \geq |\zeta - a| - |z - a| \geq r - |z - a|$. Hence

$$|f_n(z)| \leq \frac{M}{2\pi}\int_C \frac{|d\zeta|}{|\zeta - a|^n|\zeta - z|} \leq \frac{M}{2\pi}\int_C \frac{|d\zeta|}{r^n(r - |z - a|)} \leq \frac{Mr}{r^n(r - |z - a|)}.$$

Hence $|f(z)| \leq \left(\dfrac{|z-a|}{r}\right)^n \dfrac{Mr}{r - |z - a|}$. But $|z - a| < r \Rightarrow \dfrac{|z-a|}{r} < 1$ and

so $\left(\dfrac{|z-a|}{r}\right)^n \to 0$ as $n \to \infty$. Thus for each fixed z with $|z - a| < r$ we

see that $|f(z)|$ can be made arbitrarily small and in turn this implies $f(z) =$

0 for z with $|z - a| < r$. Thus $f(z)$ identically vanishes in a neighbourhood of "a" namely $|z - a| < r$. Summing up we get that if f and all its derivatives vanish at $z = a$ then there exists a neighbourhood of "a" in which the function and all its derivatives vanish. Let A be the set of all points $z \in \Omega$ at which the function f and all its derivatives vanish. Let $B = \Omega \backslash A$. We also know that f and all its derivatives are continuous in Ω (Theorem 4.4.4) and so if $w_0 \in B$ (with $f^{(k)}(w_0) \neq 0$ for some $k \geq 0$) then there exists a neighbourhood of w_0 through out of which $f^{(k)}(w) \neq 0$. Thus B is also open and in as much as $\Omega = A \cup B$ and Ω is a region, either $A = \phi$ or $B = \phi$. But we are given that $a \in A$ and so $B = \phi$. That is $\Omega \equiv A$. That is $f(z) \equiv 0$ for every $z \in \Omega$.

Definition 4.4.14

If $f(z) \not\equiv 0$ is analytic in a region Ω and $f(z_0) = 0$ then by Theorem 4.4.13 there exists a least positive integer k such that $f^{(k)}(z_0) \neq 0$ and under these conditions we call z_0, a zero of order k (If $k = 1$ we say z_0 is a simple zero) for f.

Note 4.4.15

Let $f(z) \not\equiv 0$ be analytic in a region Ω. If z_0 is a zero of order k for f then applying Theorem 4.4.12 with $a = z_0$ we see that $f(z) = (z - z_0)^k f_k(z)$ where $f_k(z)$ is analytic in Ω and $f_k(z_0) = \dfrac{f^{(k)}(z_0)}{k!} \neq 0.$

Theorem 4.4.16

If $f(z)$ is analytic in Ω and $f(z)$ is equal to zero over a set of points A in Ω such that A has a limit point in Ω then $f(z) \equiv 0$ in Ω.

Proof

Let z_0 be a limit point of A in Ω and let $z_n \to z_0$, $z_n \in A$. Since f is continuous, $f(z_n) \to f(z_0)$ and as $f(z_n) = 0$, $f(z_0) = 0$. If $f \not\equiv 0$, by Note 4.4.15, z_0 is a zero of order k for f and $f(z) = (z - z_0)^k f_k(z)$ with $f_k(z)$ analytic in Ω and $f_k(z_0) \neq 0$. So we can find a neighbourhood of z_0 (say N) in which $f_k(z) \neq 0$ and so $f(z) \neq 0$ except at z_0. i.e., z_0 is an isolated zero. But N contains point of A, other than z_0 a contradiction. This demonstrates that $f(z) \equiv 0$ in Ω.

Corollary 4.4.17

If $f(z)$ and $g(z)$ are analytic in a region Ω and $f(z) = g(z)$ over a set of points having a limit point in Ω then $f(z) \equiv g(z)$ for all $z \in \Omega$.

Proof

Apply Theorem 4.4.16 to $f - g$.

Note 4.4.18

In view of Corollary 4.4.17 there is atmost a unique way of extending a given analytic function from a region Ω to a larger region Ω_1 (in Ω we can find a set of points having a limit point in Ω) and for this reason Theorem 4.4.13 or its equivalent formulations Theorem 4.4.16 and Corollary 4.4.17 are all called "Principle of analytic continuation".

The Theorems 4.4.13 and 4.4.16 imply a very important statement which we record here for our use later.

Theorem 4.4.19

Let $f(z) \not\equiv 0$ be analytic in a region Ω. Then every zero of f is isolated and is of finite order.

Theorem 4.4.20

Let $f(z)$ be analytic in a region Ω, $z_0 \in \Omega$, $w_0 = f(z_0)$ and $f(z) - w_0$ has a zero of order m at z_0 with $m \geq 1$. Let γ_1 and γ_2 be two smooth curves through z_0 making an angle θ. Let $\Gamma_1 = f(\gamma_1)$ and $\Gamma_2 = f(\gamma_2)$ intersect at w_0 making angle ϕ (Both θ and ϕ are measured in the same sense). Then $\phi = m\theta$.

Proof

Let γ_1 be parameterized by $z = z_1(t)$ with $z_1(t_0) = z_0$ and γ_2 be parameterized by $z = z_2(t)$ with $z_2(t_0) = z_0$. Then Γ_1 and Γ_2 are parameterized by $w_1(t) = f(z_1(t))$ and $w_2(t) = f(z_2(t))$ with $w_1(t_0) = w_0 = w_2(t_0)$. Now by definition the angle θ between γ_1 and γ_2 is given by

$$\lim_{t \to t_0} \arg\left(\frac{z_1(t) - z_1(t_0)}{z_2(t) - z_2(t_0)}\right) = \arg\left(\frac{z_1'(t_0)}{z_2'(t_0)}\right) \qquad (3)$$

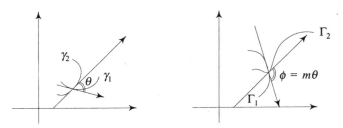

Fig. 4.4 (i)

for a single-valued continuous branch of the argument function. Similarly the angle ϕ between Γ_1 and Γ_2 is given by

$$\lim_{t \to t_0} \arg\left(\frac{w_1(t) - w_1(t_0)}{w_2(t) - w_2(t_0)}\right). \qquad (4)$$

As $f(z) - w_0$ has a zero of order m at z_0 we have by Note 4.4.15,

$$f(z) - w_0 = (z - z_0)^m g_m(z) \qquad (5)$$

where $g_m(z)$ is analytic in a neighbourhood of z_0 and $g_m(z_0) \neq 0$. From (5) we have

$$w_1(t) - w_1(t_0) = (z_1(t) - z_1(t_0))^m \, g_m(z_1(t)) \tag{6}$$
$$w_2(t) - w_2(t_0) = (z_2(t) - z_2(t_0))^m \, g_m(z_2(t)). \tag{7}$$

Hence (6) and (7) give

$$\arg\left(\frac{w_1(t) - w_1(t_0)}{w_2(t) - w_2(t_0)}\right) = m \arg\left(\frac{z_1(t) - z_1(t_0)}{z_2(t) - z_2(t_0)}\right)$$

$$+ \arg g_m(z_1(t)) - \arg g_m(z_2(t)).$$

Taking the limit as $t \to t_0$ we have $\phi = m\theta$ (up to integral multiples of 2π) from (3) and (4). This proves the theorem.

Theorem 4.4.21

Let $f(z) \neq 0$ be analytic in an open disc D. Let $\{z_j\}$ be the zeros of f, each zero being counted as many times as its order indicates. Then for each closed curve γ in D which does not pass through any of the zeros of f we have

$$\frac{1}{2\pi i} \int_\gamma \frac{f'(z)}{f(z)} \, dz = \sum_j n(\gamma, z_j)$$

where the sum on the right side is always finite.

Proof

First we observe that the zeros of f in D can not have limit points in D by Theorem 4.4.16. Thus in the closure of the region enclosed by γ (which is a compact set) say A there can be only finitely many zeros of f. (Recall that every infinite subset of a compact set has a limit point). Hence $n(\gamma, z_i)$ $\neq 0$ only for finitely many z_i. Now γ is contained in an open disc C concentric with D such that $\bar{C} \subset D$ (To get one such C we proceed as follows: Let D be the disc centre a and radius r and observe that $d = \sup_{z \in \gamma}$ $|z - a| < r$ and choose one ρ such that $d < \rho < r$. Then $|z - a| < \rho$ is the required C). By repeating the argument with $\bar{C}$ in the place of A we can again see that $\bar{C}$ and hence the interior of C contains only finitely many zeros of f. Let these finite number of zeros be written as $z_1, z_2, \ldots, z_n$ where we repeat each z_j as many times as the order indicates. Then by repeated use of Note 4.4.15 we can write

$$f(z) = a_0(z - z_1)(z - z_2) \ldots (z - z_n) \, h(z)$$

where $h(z)$ is analytic in the interior of C and has no zeros there. Thus

$$\frac{f'(z)}{f(z)} = \sum_{j=1}^{n} \frac{1}{z - z_j} + \frac{h'(z)}{h(z)} \quad \text{(Take formal logarithmic derivatives). Hence}$$

$$\frac{1}{2\pi i} \int_\gamma \frac{f'(z)}{f(z)} \, dz = \sum_j n(\gamma, z_j). \left[\text{Note that } h(z) \neq 0 \text{ and } h'(z) \text{ is analytic} \Rightarrow \frac{h'}{h}\right.$$

is analytic in C and an application of Cauchy's theorem for a disc gives

$$\int_\gamma \frac{h'(z)}{h(z)} \, dz = 0 \Big].$$

In the above sum we are free to add all other zeros of f that lie in the complement of the interior of C since $n(\gamma, z_j) = 0$ for these zeros. Thus we can rewrite the sum over all zeros of f whether inside C or not and thus our theorem is proved.

Theorem 4.4.22 (Local correspondence Theorem)

Suppose that $f(z)$ is not identically equal to a constant and is analytic in a region Ω. Assume $z_0 \in \Omega$, $f(z_0) = w_0$ and that $f(z) - w_0$ has a zero of order n at z_0. Then for each sufficiently small $\varepsilon > 0$ there exists a corresponding region G containing w_0 such that for every $w \in G$, the equation $f(z) = w$ has exactly n roots (including multiplicities) in $|z - z_0| < \varepsilon$. We can choose ε with the additional property that all the roots of $f(z) = w$ ($w_0 \neq w \in G$) in $|z - z_0| < \varepsilon$ are simple.

Proof

Let $f(z) - w_0$ have a zero of order n at z_0. First choose $\varepsilon > 0$ so that $f(z)$ is defined and analytic in $|z - z_0| \leq \varepsilon$ and also such that z_0 is the only zero of $f(z) - w_0$ in this closed disc. Let γ be the simple closed curve $|z - z_0| = \varepsilon$ and Γ its image under f. Now $w_0 \notin \Gamma$ and so w_0 belongs to the complement of Γ. Thus there exists one and only one region G determined by Γ to which w_0 belongs. Using the properties of the index $n(\Gamma, a)$, it is clear that $n(\Gamma, w_0) = n(\Gamma, w)$ for every $w \in G$.

But
$$n(\Gamma, w_0) = \frac{1}{2\pi i} \int_\Gamma \frac{dw}{w - w_0} = \frac{1}{2\pi i} \int_\gamma \frac{f'(z)}{f(z) - w_0} \, dz$$

(use the definition on both sides with γ as $z = z(t)$). Now using Theorem 4.4.21 with $(f(z) - w_0)$ replacing $f(z)$ we get $n(\Gamma, w_0) = \sum_j n(\gamma, a_j)$ where a_j's are the zeros of $f(z) - w_0$ inside γ. But the number of zeros of $f(z) - w_0$ inside γ is precisely equal to n and since $n(\gamma, a_j) = 1$ or 0 depending on whether a_j lies inside or outside of γ (Prove this separately!) we have $\sum_j n(\gamma, a_j) = n$. Thus $n(\Gamma, w_0) = n$. Further $n(\Gamma, w) = \sum_j n(\gamma, b_j)$ $= n$ for $w \in G$ where b_j's are the zero of $f(z) - w$. i.e., the number of zeros of $f(z)-w$ inside γ (i.e, for which $n(\gamma, b_j)$ is 1) is precisely equal to n.

On the other hand if $\varepsilon > 0$ is so chosen that $|z - z_0| \leq \varepsilon$ is free of zeros of $f'(z)$ also (other than z_0) (This is possible because zeros of $f'(z)$ are also isolated) we can be sure that the solutions of $f(z) - w$ in $|z - z_0| < \varepsilon$ are also simple. (However, $|z - z_0| < \varepsilon$ is connected and so its image under the continuous map f must lie in G only. Thus between $|z - z_0| < \varepsilon$ and the corresponding G the mapping f is just n-to-one).

Corollary 4.4.23 (Open mapping principle)
A non-constant analytic function in a region Ω is an open map. (i.e., the function maps open sets onto open sets).

Proof
Let $f(z)$ be a non-constant analytic function in Ω. Let U be an open subset of Ω. Let $z_0 \in U$ such that $f(z_0) = w_0$. There exists a positive integer $n \geq 1$ such that $f(z) - w_0$ has a zero of order n at z_0. By theorem 4.4.22 there exists $\varepsilon > 0$ and a region G containing w_0 such that $|z - z_0| < \varepsilon$ is a subset of U and each $w \in G$ is assumed by f at n points in $|z - z_0| < \varepsilon$ (Thus there exists at least one point $z \in U$ with $f(z) = w$). Since G is open and $w_0 \in G$ there exists a $\delta > 0$ such that $|w - w_0| < \delta$ is a subset of G. Hence $|w - w_0| < \delta$ is completely contained in $f(U)$. i.e., w_0 is interior to $f(U)$. Since w_0 is arbitrary, $f(U)$ is open.

Corollary 4.4.24
If $f(z)$ is a non-constant analytic function in a region Ω with $f'(z_0) \neq 0$ for some $z_0 \in \Omega$ then the mapping f is a topological homeomorphism and a conformal map between a neighbourhood of z_0 and a region containing $f(z_0) = w_0$.

Proof
By Theorem 4.4.22 there exists an $\varepsilon > 0$ and a corresponding region G containing w_0 such that f maps $|z - z_0| < \varepsilon$, one-to-one, onto G ($f'(z_0) \neq 0$ implies z_0 is a simple zero of $f(z) - w_0$ and $n = 1$ here). This map is obviously continuous and by Corollary 4.4.23 this is also open. Thus both f and f^{-1} (which exists since f is one-to-one) are continuous and so this correspondence between the ε-neighbourhood of z_0 and y is topological. Since we can also make sure that $f'(z) \neq 0$ in $|z - z_0| < \varepsilon$ (for a suitable $\varepsilon > 0$) this correspondence is conformal also.

Theorem 4.4.25 (Maximum principle)
If $f(z)$ is a non-constant analytic function in a region Ω then its absolute value function $|f(z)|$ has no maximum in Ω.

Proof
Let $z_0 \in \Omega$ and $w_0 = f(z_0)$ its image. By Theorem 4.4.23, there exists $\varepsilon > 0$ and $\delta > 0$ such that every point w with $|w - w_0| < \delta$ is the image of some point z in $|z - z_0| < \varepsilon$. i.e., $f(z) = w$. Since in this neighbourhood $|w - w_0| < \delta$ of w_0, we can always find points $w = f(z)$ with $|w| > |w_0|$, $|w_0|$ can not be the maximum of all $|f(z)|$ as z varies over Ω. Since z_0 is arbitrary there can not be any $z_0 \in \Omega$ with $|f(z_0)| = \max_{z \in \Omega} |f(z)|$ and this completes the proof.

Lemma 4.4.26
Let K be a non-empty compact subset of the plane. Let A be one of the components of the interior of K. Then each boundary point of A is also a boundary point of K.

Proof

Let us denote the interior of K by K^0. If 'a' is a boundary point of one of the components A of K^0 then every neighbourhood of 'a' say $\{z : |z - a| < r\}$ intersects both A and its complement. The claim is that, each such neighbourhood intersects both K and K^c. Let us fix one such neighbourhood N. $N \cap A \neq \phi \Rightarrow N \cap K^0 \neq \phi \Rightarrow N \cap K \neq \phi$. On the other hand if $N \cap K^c = \phi$ then $N \subset K$ i.e., 'a' is an interior point of K. But in as much as this neighbourhood is connected it must be completely contained in one component of the interior of K. But one of its points $a \in A$ and A is a component. Thus $N \subset A$ and N can not intersect A^c a contradiction to our assumption. Thus $N \cap K^c \neq \phi$ and 'a' is indeed a boundary point of K.

Lemma 4.4.27

Let E be any non-empty bounded subset of the complex plane. Defining the boundary of E by $\partial E = \overline{E} \cap \overline{(E^c)}$ we have $\partial E \neq \phi$.

Proof

If $\partial E = \phi$ the complex plane is equal to $E^0 \cup (E^c)^0$ where A^0 for any A is the interior of A. (Note that $\mathbb{C} = E^0 \cup \partial E \cup (E^c)^0 \cup (\partial E^c)$ and $\partial E^c = \partial E = \phi$). Now since $\mathbb{C}$ is connected, one of them should be empty. If E^0 is empty, $\overline{E} = E^0 \cup \partial E = \phi \Rightarrow E = \phi$ which is not possible. Thus $(E^c)^0$ must be empty. Thus $E^c \subset (E^c)^0 \cup \partial E^c = \phi$ or $E = \mathbb{C}$ which is unbounded. This is also not possible. Thus $\partial E \neq \phi$.

Theorem 4.4.28 (Maximum modulus theorem)

Let $f(z)$ be a continuous function on a compact set K whose interior is non-empty. If $f(z)$ is analytic in the interior of K, then there exists at least one point z_0 on the boundary of K such that $|f(z_0)| = \sup_{z \in K} |f(z)|$. Moreover if either the interior of K is connected or $f(z)$ is analytic in a region which contains the compact set K then z_0 cannot belong to the interior of K unless of course $f(z)$ is identically a constant function. (Thus in case $f(z)$ is not identically a constant function, z_0 must be a boundary point of K).

Proof

Let K be compact and $K^0 \neq \phi$ and $f(z)$ analytic in K^0 and continuous on K. Then since K is compact and $|f|$ is continuous there exists a point $z_0 \in K$ such that $|f(z_0)| = \sup_{z \in K} |f(z)|$. Let us concentrate on the first part of the above theorem. If z_0 is boundary point of K there is nothing to prove. Let $z_0 \in K^0$. Let A be one of the components of K^0 such that $z_0 \in A$. Since A is open (Components of open sets are open and K^0 is open) we can find $r > 0$ such that $N = \{z: |z - z_0| < r\} \subset A$. A is open, connected and non-empty and so is a region. Now $|f(z_0)|$ is also the maximum of $|f(z)|$ in A and as such it attains its maximum at one of its interior points, namely z_0. Applying Theorem 4.4.25 we see that $f(z)$ reduces to a constant

$(= f(z_0))$ in A and by continuity, $f(z)$ is also the same constant on the boundary of A. Let 'a' be a point on the boundary of A. (Note that $\partial A \neq \phi$ by Lemma 4.4.27). Then $f(z) = f(z_0)$ for every $z \in \overline{A}$. $|f(a)| = |f(z_0)| = \sup_{z \in K} |f(z)|$. Now '$a$' is also a boundary point of K by Lemma 4.4.26 and so our first part is established. Let us now consider the second part of the Theorem. If K^0 is connected $A = K^0$ and if z_0 is an interior point of K then $f(z) \equiv f(z_o)$ in K^0 and hence in K. On the other hand if f is analytic in a region Ω, $K \subset \Omega$ and z_0 is interior to K then over a subregion A of Ω, $f(z) \equiv f(z_0)$ and by principle of analytic continuation we have $f(z) \equiv f(z_0)$ in Ω (and so $f(z) \equiv f(z_0)$ in K) and our second assertion follows.

Lemma 4.4.29 (Minimum Principle)
Let G be a region and f be analytic in G. Assume there is a point c in G and a neighbourhood U of c in G such that $|f(c)| = \inf_{z \in U} |f(z)|$. Then either $f(c) = 0$ or else f is identically equal to a constant in G.

Proof
If $f(c) \neq 0$ then $\dfrac{1}{f}$ is analytic in U and satisfies $\dfrac{1}{|f(z)|} \leq \dfrac{1}{|f(c)|}$. Thus the analytic function $\dfrac{1}{f}$ attains its maximum modulus at an interior point of U. Thus $\dfrac{1}{f}$ and hence f is identically constant in U and by principle of analytic continuation it is the same constant throughout G.

Theorem 4.4.30 (Minimum modulus theorem)
Let G be a bounded region and f a non-constant continuous function in $\overline{G}$ which is analytic in G. Then either f has a point $z_0 \in G$ at which $f(z_0) = 0$ or the minimum of $|f|$ over $\overline{G}$ is assumed only on the boundary of G.

Proof
If f has zero in $\overline{G}$ (which is compact) the conclusion easily follows. If f has no zeros in $\overline{G}$, then $\dfrac{1}{f}$ is analytic in G and by our Theorem 4.4.25, $\dfrac{1}{|f|}$ attains its maximum (note also that $\dfrac{1}{f}$ is continuous on $\overline{G}$) on the boundary of G. i.e., there exists $z_0 \in \partial G$ and not in G such that $\dfrac{1}{|f(z_0)|} \geq \dfrac{1}{|f(z)|}$ for every $z \in \overline{G}$. Therefore $|f(z)| \geq |f(z_0)|$ where $z_0 \in \partial G$.

The maximum modulus theorem can be effectively used in characterizing the class of all mappings which are one-to-one and onto and analytic between certain specified regions in the complex plane. Many of these results depend on, one small but interesting result called, Schwarz lemma which we now proceed to obtain:

Theorem 4.4.31 (Schwarz lemma)
If $f(z)$ is analytic in $|z| < 1$ with $f(0) = 0$ and $|f(z)| \leq 1$ for $|z| < 1$, then
(i) $|f(z)| \leq |z|$ in $|z| < 1$
(ii) $|f'(0)| \leq 1$

If equality holds in (i) for some $z \neq 0$ or if equality holds in (ii) then $f(z) = cz$ with $|c| = 1$.

Proof

Let U denote the open unit disc in the plane. Define $f_1(z) = \dfrac{f(z)}{z}$ for $z \neq 0$. $f_1(0) = f'(0)$. Now clearly f_1 is analytic for $z \neq 0$. On the other hand using finite Taylor development at origin we get

$$f(z) = f(0) + zf'(0) + z^2 h(z)$$

where $h(z)$ is analytic in $|z| < 1$. Since $f(0) = 0$ we have

$$\lim_{z \to 0} \frac{f_1(z) - f_1(0)}{z} = \lim_{z \to 0} \frac{f(z) - zf'(0)}{z^2} = \lim_{z \to 0} h(z) = h(0).$$

Therefore f_1 is analytic at origin also. Now let us take the circle $|z| = r < 1$ and observe that $|f_1(z)| \leq \dfrac{1}{r}$ on $|z| = r$ and hence by maximum modulus theorem $|f_1(z)| \leq \dfrac{1}{r}$ for $|z| \leq r$. Therefore letting r to tend to 1 we see that $|f_1(z)| \leq 1$ for all z in U and this proves (i) and (ii).

On the other hand if equality in $|f_1(z)| \leq 1$ is held at any single point in U then $f_1(z)$ must reduce to a constant say c. Clearly $|c| = 1$ and $f(z) = cz$ with $|c| = 1$.

Corollary 4.4.32 (Generalized Schwarz Lemma)
Let $f(z)$ be analytic in $|z| < R$ and satisfy $|f(z)| < M$ for $|z| < R$. Let $f(z_0) = w_0$ with $|z_0| < R$ and $|w_0| < M$.
Then

$$\left| \frac{M(f(z) - w_0)}{M^2 - \overline{w}_0 f(z)} \right| \leq \left| \frac{R(z - z_0)}{R^2 - \overline{z}_0 z} \right|$$

Proof

Let $\zeta = T(z)$ be a linear fractional transformation which maps $|z| < R$ onto $|\zeta| < 1$ with z_0 going to origin (i.e., $T(z_0) = 0$). Let $\xi = S(w)$ be a linear fractional transformation which maps $|w| < M$ onto $|\xi| < 1$ with $S(w_0) = 0$. Clearly $S \circ f \circ T^{-1}$ maps $|\zeta| < 1$ into $|\xi| < 1$ with 0 going to 0 under this map. We also have explicit expressions for $\zeta = T(z) = \dfrac{R(z - z_0)}{R^2 - \overline{z}_0 z}$ and

$\xi = S(w) = \dfrac{M(\omega) - w_0)}{M^2 - \overline{w}_0}$ (One can obtain these expressions on lines similar to Example 3.2.7). Using Schwarz lemma we obtain

$$|S \circ f \circ T^{-1}) \, (\zeta)| \le |\zeta| \text{ or } |(S \circ f) \, (z)| \le |T(z)|$$

which is equivalent to the required inequality, namely,

$$\left| \frac{M(f(z) - w_0)}{M^2 - \overline{w}_0 f(z)} \right| \le \left| \frac{R(z - z_0)}{R^2 - \overline{z}_0 z} \right|.$$

Using Schwarz lemma we shall now characterize all (analytic automorphisms of *U*) one-to-one conformal mappings of the open unit disc onto itself. (See Example 3.2.7).

Theorem 4.4.33
Let *U* denote the open unit disc of the complex plane, described by $\{z \in \mathbb{C} : |z| < 1\}$ and *T* its boundary. For any $\alpha \in U$, define $\phi_\alpha(z) = \dfrac{z - \alpha}{1 - \overline{\alpha}z}$. Then

(i) ϕ_α is analytic, one-to-one and onto *U*, (ii) $\phi_\alpha(T) = T$, (iii) $\phi_\alpha^{-1} = \phi_{-\alpha}$,

(iv) $\phi_\alpha'(0) = 1 - |\alpha|^2$, (v) $\phi_\alpha'(\alpha) = \dfrac{1}{1 - |\alpha|^2}$.

Proof
The proofs of (i), (ii) and (iii) are provided already while discussing Example 3.2.7. (iv) and (v) can be easily verified.

Theorem 4.4.34
Every one-to-one analytic map *f* of *U* onto itself is given by $f(z) = \lambda \phi_\alpha(z)$ for some $\lambda \in \mathbb{C}$ with $|\lambda| = 1$ and for some $\alpha \in U$.

Proof
Let $f : U \to U$ be one-to-one, onto, analytic. Let *g* be its inverse. Choose $\alpha \in U$ such that $f(\alpha) = 0$ (Note that *f* is onto). Now by local correspondence theorem 4.4.22 $f'(z) \ne 0$ anywhere in *U* (Otherwise in a neighbourhood of a point z_0 with $f'(z_0) = 0$ the mapping *f* is *n* to 1 for $n \ge 2$ which is a contradiction). By open mapping principle *f* is an open map or that *g* is continuous. Let $f(z) = w$ with $g(w) = z$ and $f(z_0) = w_0$. Now

$$g'(w_0) = \lim_{w \to w_0} \frac{g(w) - g(w_0)}{w - w_0} = \lim_{z \to z_0} \frac{z - z_0}{f(z) - f(z_0)} = \frac{1}{f'(z_0)}. \text{ Note that } w$$

$\to w_0 \Leftrightarrow z \to z_0$ by continuity of both *f* and *g*. Thus *g* is also analytic in *U* with

$$g'(w_0) = \frac{1}{f'(z_0)} \text{ whenever } f(z_0) = w_0. \tag{8}$$

Now in general if $F : U \to U$ is one-to-one and analytic with $F(\alpha) = \beta$ where $\alpha, \beta \in U$, then $G = \phi_\beta \circ F \circ \phi_{-\alpha}$ satisfies the conditions of Schwarz's lemma. So $|G'(0)| \le 1$ with equality if and only if $G(z) = \lambda z$ with $|\lambda| = 1$. By chain rule

$$G'(0) = \phi'_\beta \left[(F \circ \phi_{-\alpha})(0)\right](F \circ \phi_{-\alpha})'(0) \quad = \phi'_\beta (\beta) F'(\phi_{-\alpha}(0)) \phi'_{-\alpha}(0)$$

$$= \frac{F'(\alpha)(1 - |\alpha|^2)}{(1 - |\beta|^2)}$$

Thus $|F'(\alpha)| \le \dfrac{(1 - |\beta|^2)}{(1 - |\alpha|^2)}$ with equality if and only if $F = \phi_{-\beta} \circ (\lambda\phi_\alpha)$

with $|\lambda| = 1$. Using the above result with $F = f$, $\beta = 0$ and retaining α we

have $|f'(\alpha)| \le \dfrac{1}{1 - |\alpha|^2}$ and if we take $F = g$ with $\alpha = 0$ and replacing

β by α we also have $|g'(0)| \le 1 - |\alpha|^2$. But $g'(0) = \dfrac{1}{f'(\alpha)}$ by (8) and

so $|f'(\alpha)| \ge \dfrac{1}{1 - |\alpha|^2}$. Thus $|f'(\alpha)| = \dfrac{1}{1 - |\alpha|^2}$ and so $f(z) = \lambda\phi_\alpha(z)$ with

$|\lambda| = 1$. Hence the result.

Corollary 4.4.35
If f is one-to-one, analytic from U onto U with $f(0) = 0$ then $f(z) = \lambda z$
with $|\lambda| = 1$.

Proof
Following from Theorem 4.4.34 and the fact that $f(0) = 0$ implies $\alpha = 0$.

Remarks 4.4.36
Let H denote the upper half plane given by $\{z : \operatorname{Im} z > 0\}$. It is easy to check

that if $w = f(z) = \dfrac{az + b}{cz + d}$ with a, b, c, d real and $(ad - bc)$ positive then

$$\operatorname{Im} w = \frac{(ad - bc)\operatorname{Im} z}{|cz + b|^2}.$$ Further f is one-to-one (being a linear fractional

transformation) and analytic in H (the only pole $z = -\dfrac{b}{c}$ is either real or

∞ and so does not belong to H). Thus $w = f(z)$ is analytic, one-to-one and

maps H into H. On the other hand $w = \dfrac{az + b}{cz + d} \Rightarrow z = \dfrac{-dw + b}{cw - a} =$

$\dfrac{a'w + b'}{c'w + d'}$ with a', b', c', d' are real and $a'd' - b'c' = (-d)(-a) - bc$

$= ad - bc > 0$. Thus w^{-1} is also analytic, one-to-one, from H into itself. i.e.,
$w : H \to H$ is bijective and analytic. Using these observations it is possible
to characterize all possible analytic automorphisms of H (one-to-one, analytic maps of H onto itself).

Lemma 4.4.37

Let $\phi(z) = \dfrac{z - z_0}{z - \bar{z}_0}$, $z_0 \in H$. Then ϕ is one-to-one from H onto U and is analytic in H.

Proof

Since ϕ is a linear fractional transformation with the only pole at $z = \bar{z}_0 \notin H$, ϕ is a one-to-one, analytic in H. Since ϕ maps real axis onto $|w| = 1$ (as is easily verified) it maps H into either $|w| < 1$ or $|w| > 1$ (Note that $\phi(H)$ is connected). Also $\phi(z_0) = 0$ and hence $\phi(H) \subset U$. Similar considerations show that $\phi^{-1}(U) \subset H$ or that $U \subset \phi(H)$. Thus $\phi(H) = U$ completing the proof.

Theorem 4.4.38

Every analytic automorphism f of H is given by $f(z) = \dfrac{az + b}{cz + d}$ where a, b, c, d are real and $ad - bc$ positive.

Proof

Let f be an analytic automorphism of H. Choose $z_0 \in H$ such that $f(z_0) = i$ and define for $z \in H$, $w \in H$

$$F(w) = \frac{w - i}{w + i}; \quad \phi(z) = \frac{z - z_0}{z - \bar{z}_0} \quad \text{so that} \quad F^{-1}(z) = \frac{-iz - i}{z - 1} \ (z \in U).$$

Using Lemma 4.4.37 we have $F \circ f \circ \phi^{-1} : U \to U$ and $F \circ f \circ \phi^{-1}(0) = 0$. Applying Schwarz lemma we have for $w \in U$

$$|F \circ f \circ \phi^{-1}(w)| \leq |w| \tag{9}$$

which implies that $|(F \circ f)(z)| \leq |\phi(z)|$ for every $z \in H$. If equality holds at $z \in H$, $(z \neq z_0)$ then equality holds in (9) for $w \neq 0$ in U. Thus $F \circ f \circ \phi^{-1} = e^{i\theta} I$ where I is the identity map on U i.e.,

$$(F \circ f)(z) = e^{i\theta} \phi(z). \text{ i.e., } f(z) = \frac{-i\lambda\phi(z) - i}{\lambda\phi(z) - 1} \text{ where } \lambda = e^{i\theta}.$$

Summing up we see that for every $z \in H$,

$$\left| \frac{f(z) - i}{f(z) + i} \right| \leq \left| \frac{z - z_0}{z - \bar{z}_0} \right| \tag{10}$$

with equality if and only if $f(z) = \dfrac{-i\lambda\phi(z) - i}{\lambda\phi(z) - 1}$ with $\lambda = e^{i\theta}$. We now observe that F is a bijective analytic map of H onto U, ϕ^{-1} is a bijective analytic map of U onto H and thus $F \circ f \circ \phi^{-1} : U \to U$ is an analytic automorphism of U with $F \circ f \circ \phi^{-1}(0) = 0$. By Theorem 4.4.35, $F \circ f \circ \phi^{-1} = \mu I$ for some $|\mu| = 1$. Hence equality holds in (10). Therefore

$$f(z) = \frac{-i\mu\phi(z) - i}{\mu\phi(z) - 1} \text{ with } |\mu| = 1. \text{ A simple computation shows that } f(z)$$

$$= \frac{az + b}{cz + d} \text{ where } a = i(1 + \mu),\ b = -i(\mu z_0 + \bar{z}_0),\ c = 1 - \mu \text{ and } d = \mu z_0$$

$-\bar{z}_0$ so that $ad - bc = -4\mu\, Im\, z_0$. However $f(z)$ is not altered by multiplying all the constants a, b, c and d by a simple complex number. Hence $f(z)$ can also be written in the form $f(z) = \dfrac{Az + B}{Cz + D}$ where $A = \lambda a,\ B = \lambda b,\ C =$

λc and $D = \lambda d$ so that $AD - BC = \lambda^2(ad - bc)$. If we choose $\lambda^2 = -\dfrac{1}{4\mu}$

or $\lambda = \dfrac{-i}{2\sqrt{\mu}}$, we see that A, B, C, D are real and $AD - BC = Im\, z_0 > 0$.

Thus $f(z)$ can be written in the desired form.

4.5 Singularities

We have so far seen how analytic functions behave in a neighbourhood of a point of analyticity. These are the so called local properties of analytic functions. Equally interesting is the behavior of functions which are analytic in a deleted neighbourhood of points in the plane. To start with let us assume that $f(z)$ is analytic in $0 < |z - a| < \delta$ for a fixed $a \in \mathbb{C}$. Generally speaking we would like to call all such points 'a' as singularities. But to avoid trivialities (for example if f is known to be analytic in $|z - a| < \delta$ and hence also analytic in $0 < |z - a| < \delta$ we need not regard 'a' as a singularity) we explicitly assume that f is not analytic at $z = a$.

Definition 4.5.1
Let $f(z)$ be analytic in $0 < |z - a| < \delta$ for some $\delta > 0$ and be not analytic at $z = a$. Then $z = a$ is called an isolated singularity for f.

Definition 4.5.2
Let $f(z)$ be analytic in $0 < |z - a| < \delta$ for some $\delta > 0$. Let $z = a$ be an isolated singularity for f.
  (i) If there exists an analytic function $g(z)$ in $|z - a| < \delta$ such that $g(z) = f(z)$ for $z \neq a$, we call $z = a$, a removable singularity for f.
  (ii) If $\lim\limits_{z \to a} f(z) = \infty$ then we say $z = a$ is a pole for f.
  (iii) If $z = a$ is neither a removable singularity nor a pole for f then $z = a$ is called an isolated essential singularity for f.

Remark 4.5.3
There are several functions for which $z = a$ will be a non-isolated singu-

larity. For example $f(z) = \dfrac{1}{e^{1/z} + 1}$ at $z = 0$ has a non-isolated (essential) singularity because $z = \dfrac{1}{(2n+1)\,\pi i}$ are all poles for $f(z)$ and $z = 0$ is a limit

point of these singularities and hence in every neighbourhood of $z = 0$ we

will have singularities. Similar is the case with $g(z) = \dfrac{1}{\sin(1/z)}$ wherein z

$= \dfrac{1}{n\pi}$ are poles. Several other examples of this type can be given but the

point to be noted here is that we do not have much to say about the behavior of such functions at this stage and hence we will hereafter be concerned with singularities that are isolated.

Definition 4.5.4

For a function f which is analytic in $|z| > R$ (for some $R > 0$), we always

say $z = \infty$ is an isolated singularity. In this case we let $F(z) = f\!\left(\dfrac{1}{z}\right)$ so that

$F(z)$ becomes an analytic function in $0 < |z| < \dfrac{1}{R}$. We define $z = \infty$ to be

a removable singularity or a pole or an isolated essential singularity for $f(z)$ if $z = 0$ is respectively a removable singularity (or analytic at $z = 0$) or a pole or an isolated essential singularity for $F(z)$.

Example 4.5.5

(i) $F(z) = \dfrac{\sin z}{z}$ at $z = 0$ has a removable singularity. By Theorem 4.4.11

(Take $f(z) = \sin z$, $\Omega = \mathbb{C}$, $a = 0$) there exists an analytic function

$g(z) = \dfrac{\sin z}{z}$ for $z \neq 0$ and $g(0) = 1$. Thus $z = 0$ is a removable singularity

for $f(z)$. From the definition it is also clear that $f(z) = \dfrac{1}{z}$ has a removable

singularity at $z = \infty$.

(ii) $f(z) = \dfrac{1}{(z-a)^{m}}$ (m, a positive integer) has a pole at $z = a$ since $|f(z)|$

$\to \infty$ as $z \to a$. Every polynomial (say) $f(z) = a_0 + a_1 z + \ldots + a_n z^n$ has a pole

at $z = \infty$ since $f\!\left(\dfrac{1}{z}\right) = a_0 + \dfrac{a_1}{z} + \ldots + \dfrac{a_n}{z^n}$ has a pole at $z = 0$.

(iii) $f(z) = e^{1/z}$ has an isolated essential singularity at $z = 0$. Since

$\lim\limits_{z=x\to 0+} e^{1/x} = \infty$ and $\lim\limits_{z=x\to 0-} e^{1/x} = 0$ and so $\lim\limits_{z\to 0} f(z)$ does not exist (Note

that for a removable singularity at $z = a$ for f, $\lim\limits_{z\to a} f(z)$ exists and is

finite). Similarly $f(z) = e^z$ has an isolated essential singularity at $z = \infty$, by definition.

To understand the behavior of functions analytic in a neighbourhood (deleted or otherwise) of $z = a$, we shall first obtain two kinds of series developments which are often called Taylor series and Laurent series. In fact we can even characterize singularities using these series expansions.

Theorem 4.5.6 (Taylor series development)
Let $f(z)$ be analytic in a region Ω, $a \in \Omega$. Then

$$f(z) = f(a) + \frac{f'(a)}{1!}(z - a) + \frac{f''(a)}{2!}(z - a)^2 + \ldots$$

$$+ \frac{f^{(n)}(a)}{n!}(z - a)^n + \ldots$$

where the expansion is valid inside the circle centre 'a' and radius $R = \text{Sup } r$ (Here the supremum is taken over all positive numbers r such that the disc $|z - a| \leq r \subset \Omega$). In other words the radius of convergence of the power series in $z - a$ is at least R. Further the above Taylor development is unique.

Proof
Fix $z \in \Omega$ such that $\rho = |z - a| < R$. From the definition of supremum it is clear that we can find an $r > 0$ such that the closed disc centre 'a' and radius r is completely contained in Ω and $\rho < r$. Under this hypothesis we know by Theorem 4.4.12,

$$f(z) = f(a) + \frac{f'(a)}{1!}(z - a) + \frac{f''(a)}{2!}(z - a)^2 + \ldots$$

$$+ \frac{f^{(n-1)}(a)}{(n-1)!}(z - a)^{n-1} + f_n(z)(z - a)^n$$

where $f_n(z) = \dfrac{1}{2\pi i} \displaystyle\int_C \dfrac{f(\zeta)\,d\zeta}{(\zeta - a)^n\,(\zeta - z)}$, C is the simple boundary of the

disc $|z - a| \leq r \subset \Omega$. To prove our theorem it suffices to show that $(z - a)^n f_n(z) \to 0$ as $n \to \infty$. (Note that if this happens then the partial

sums of $\displaystyle\sum_{k=0}^{\infty} \frac{f^{(k)}(a)}{k!}(z - a)^k$ converge to $f(z)$).

$$|(z - a)^n f_n(z)| \leq \frac{|z - a|^n}{2\pi} \int_C \frac{M|d\zeta|}{r^n(r - \rho)}$$

where $M = \displaystyle\sup_{|z-a| \leq r} |f(z)|$, $|\zeta - a| = r$ and $|\zeta - z| \geq |\zeta - a| - |z - a| = r - $

ρ. i.e., $|(z - a)^n f_n(z)| \leq \left(\dfrac{\rho}{r}\right)^n \dfrac{Mr}{(r - \rho)}$. Since $\rho < r$, we see that $\left(\dfrac{\rho}{r}\right)^n \to$

0 as $n \to \infty$ and so $|(z - a)^n f_n(z)|$ tends to 0 as $n \to \infty$. Since z is arbitrary in $|z - a| < R$, the first part of our theorem follows.

By uniqueness of Taylor development we mean that if $f(z) = \sum_{n=0}^{\infty} a_n (z - a)^n$, the series converging uniformly in every disc $|z - a|$ $\leq r$ contained in Ω then $a_n = \dfrac{f^{(n)}(a)}{n!}$. This follows at once by differentiating the new series term by term (which is valid because of uniform convergence) and putting $z = a$.

Lemma 4.5.7
Let $f(z)$ be analytic in an annulus $r < |z - a| < R$ around $z = a$. For $r < \rho < \sigma < R$ let γ and η denote the circles $|z - a| = \rho$ and $|z - a| = \sigma$. Then

$$\int_{\gamma} f(\zeta)\, d\zeta = \int_{\eta} f(\zeta)\, d\zeta.$$

Proof
We take $\Omega = \{z \in \mathbb{C} : r < |z - a| < R\}$. We first claim that γ and η are homologous to each other. Indeed $n(\gamma, b) = n(\eta, b) = 1$ for all b with $|z - b| \leq r$ and $n(\gamma, b) = n(\eta, b) = 0$ for all b with $|z - b| \geq R$. Hence $\gamma \sim \eta$ and the proof follows from Theorem 4.3.8 with γ replaced by $\gamma - \eta$.

Lemma 4.5.8 (Cauchy's Integral formula for an annulus)
Let f be analytic in $\Omega = \{z \in \mathbb{C} : r < |z - a| < R\}$. Choose $r < \rho < \sigma < R$, γ and η as circles described by $|z - a| = \rho$ and $|z - a| = \sigma$. Then

$$f(z) = \frac{1}{2\pi i} \int_{\eta - \gamma} \frac{f(\zeta)}{\zeta - z}\, d\zeta = \frac{1}{2\pi i} \int_{\eta} \frac{f(\zeta)}{\zeta - z}\, d\zeta - \frac{1}{2\pi i} \int_{\gamma} \frac{f(\zeta)}{\zeta - z}\, d\zeta$$

for $\rho < |z - a| < \sigma$.

Proof
Fix z such that $\rho < |z - a| < \sigma$. By Theorem 4.4.11, $g(\zeta)$ defined by $g(\zeta) = \dfrac{f(\zeta) - f(z)}{\zeta - z}$ for $\zeta \neq z$ and $g(\zeta) = f'(z)$ for $\zeta = z$ is analytic in Ω. Apply Lemma 4.5.7 to g to get

$$\int_{\gamma} g(\zeta)\, d\zeta = \int_{\eta} g(\zeta)\, d\zeta.$$

i.e.,

$$\int_{\gamma} \frac{f(\zeta)}{\zeta - z}\, d\zeta - f(z) \int_{\gamma} \frac{d\zeta}{\zeta - z} = \int_{\eta} \frac{f(\zeta)}{\zeta - z}\, d\zeta - f(z) \int_{\eta} \frac{d\zeta}{\zeta - z}.$$

Since z lies outside of γ and inside of η, $\displaystyle\int_{\gamma} \frac{d\zeta}{\zeta - z} = 0$ and $\displaystyle\int_{\eta} \frac{d\zeta}{\zeta - z} = 2\pi i$.

This completes the proof.

Notes 4.5.9
Let us now consider a series of the form

$$f(z) = \sum_{k=-\infty}^{\infty} a_k(z - a)^k.$$

We can consider this as a sum of two series $f_1(z) = \sum_{k=-\infty}^{-1} a_k(z - a)^k$ and $f_2(z) = \sum_{k=0}^{\infty} a_k(z - a)^k$. One is a power series in $\dfrac{1}{z - a}$ and another is a power series in $z - a$. Thus by our radius of convergence theorem of Abel we can find $\dfrac{1}{r} \geq 0$ and $R \geq 0$ such that $f_1(z)$ represents an analytic function in $\dfrac{1}{|z - a|} < \dfrac{1}{r}$ and $f_2(z)$ represents an analytic function for $|z - a| < R$. That is $f_1(z)$ is analytic in $|z - a| > r$ and f_2 (z) is analytic in $|z - a| < R$. If their intersection is non-empty i.e., if $r < R$ we have a non-empty annular region $r < |z - a| < R$ in which both $f_1(z)$ and $f_2(z)$ are analytic and so $f(z) = f_1(z) + f_2(z)$ becomes analytic in $r < |z - a| < R$. Thus under certain conditions $(r < R)$ a series like $f(z)$ represents an analytic function in an annulus. The interesting point to observe is that the converse is also true i.e., given any analytic function $f(z)$ in an annulus $r < |z - a| < R$ we can always expand $f(z)$ in a series of the above type with the radius of convergence of the positive power series being at least R and the radius of convergence of the negative power series being at least $\dfrac{1}{r}$. This we call as Laurent series development and we proceed to prove this.

Theorem 4.5.10 (Laurent Series Development)
Let f be analytic in an annulus $r < |z - a| < R$. There are complex constants A_n $(-\infty < n < \infty$, n integer) such that $f(z) = \sum_{n=-\infty}^{\infty} A_n(z - a)^n$. Further the coefficients A_n's are given by $A_n = \dfrac{1}{2\pi i} \int\limits_{|\zeta - a| = s} \dfrac{f(\zeta)\, d\zeta}{(\zeta - a)^{n+1}}$ where $r < s < R$ $(-\infty < n < \infty$, n integer). The Laurent series development is unique.

Proof
For every z with $|z - a| < R$ choose $\sigma > r$ such that $|z - a| < \sigma < R$ and define

$$f_1(z) = \dfrac{1}{2\pi i} \int\limits_{|\zeta - a| = \sigma} \dfrac{f(\zeta)\, d\zeta}{(\zeta - z)}.$$

Similarly for each z with $|z - a| > r$ choose $\rho < R$ such that $r < \rho < |z - a|$ and define

$$f_2(z) = \frac{1}{2\pi i} \int\limits_{|\zeta - a| = \rho} \frac{f(\zeta)\,d\zeta}{(\zeta - z)}.$$

We first note that the choice of σ and ρ is immaterial as long as the inequalities are true by Lemma 4.5.7. Also using Theorem 4.4.3 we can conclude that $f_1(z)$ and $f_2(z)$ are well defined and analytic in $|z - a| < R$ and $|z - a| > r$ respectively. Further by Lemma 4.5.8 we also have $f(z) = f_1(z) - f_2(z)$ for every z such that $r < |z - a| < R$. We now develop $f_1(z)$ in a Taylor series as follows:

$$\frac{f(\zeta)}{(\zeta - z)} = \frac{f(\zeta)}{(\zeta - a)} \left(1 - \frac{(z - a)}{(\zeta - a)} \right)^{-1}$$

$$= \frac{f(\zeta)}{(\zeta - a)} \left(1 + \left(\frac{z - a}{\zeta - a} \right) + \left(\frac{z - a}{\zeta - a} \right)^2 + \dots \right)$$

where the series in the right side converges uniformly for $|z - a| \leq \sigma' < \sigma < R$ as $\left| \dfrac{z - a}{\zeta - a} \right| \leq \dfrac{\sigma'}{\sigma} < 1$. Thus term by term integration can be carried

over $|\zeta - a| = \sigma$ for all $|z - a| \leq \sigma'$. Hence $f_1(z) = \dfrac{1}{2\pi i} \displaystyle\int\limits_{|\zeta - a| = \sigma} \dfrac{f(\zeta)}{(\zeta - z)}\,d\zeta$

$$= \sum_{n=0}^{\infty} A_n\, (z - a)^n \text{ where for } n \geq 0,\ A_n = \frac{1}{2\pi i} \int\limits_{|\zeta - a| = \sigma} \frac{f(\zeta)\,d\zeta}{(\zeta - a)^{n+1}}. \text{ Note that}$$

the radius of convergence of the power series in $(z - a)$ is at least R by Theorem 4.5.6. To develop $f_2(z)$ we consider the following manipulations.

$$\frac{f(\zeta)}{(\zeta - z)} = -\frac{f(\zeta)}{(z - a)} \left(1 - \frac{(\zeta - a)}{(z - a)} \right)^{-1}$$

$$= -\frac{f(\zeta)}{(z - a)} \left(1 + \left(\frac{\zeta - a}{z - a} \right) + \left(\frac{\zeta - a}{z - a} \right)^2 + \dots \right)$$

where the series on the right side converge uniformly for $|z - a| \geq \rho' > \rho$ since $\left| \dfrac{\zeta - a}{z - a} \right| \leq \dfrac{\rho}{\rho'} < 1$ and $r < \rho < \rho' \leq |z - a| < R$. Thus $-f_2(z)$

$$= \frac{-1}{2\pi i} \int\limits_{|\zeta - a| = \rho} \frac{f(\zeta)\,d\zeta}{(\zeta - z)} = \sum_{n=1}^{\infty} B_n\, (z - a)^{-n} \text{ where for } n \geq 1,$$

$$B_n = \frac{1}{2\pi i} \int_{|\zeta - a| = \rho} \frac{f(\zeta)\, d\zeta}{(\zeta - a)^{-n+1}}$$ (Note that the radius of convergence of the

above power series in $\dfrac{1}{z-a}$ is at least $\dfrac{1}{r}$ by Theorem 4.5.6). Thus

$$f(z) = f_1(z) - f_2(z) = \sum_{n=0}^{\infty} A_n (z-a)^n + \sum_{n=1}^{\infty} B_n (z-a)^{-n}$$

where $A_n = \dfrac{1}{2\pi i} \displaystyle\int_{|\zeta - a| = \sigma} \dfrac{f(\zeta)\, d\zeta}{(\zeta - a)^{n+1}}$; $B_n = \dfrac{1}{2\pi i} \displaystyle\int_{|\zeta - a| = \rho} \dfrac{f(\zeta)\, d\zeta}{(\zeta - a)^{-n+1}}$.

Now choose $r < s < R$ and let $C = \{\zeta \in \mathbb{C} : |\zeta - a| = s\}$. By Lemma 4.5.7 we see that

$$A_n = \frac{1}{2\pi i} \int_C \frac{f(\zeta)\, d\zeta}{(\zeta - a)^{n+1}} \quad \text{and} \quad B_n = \frac{1}{2\pi i} \int_C \frac{f(\zeta)\, d\zeta}{(\zeta - a)^{-n+1}}$$

$$\left(\text{Note that } \frac{f(\zeta)}{(\zeta - a)^{n+1}} \text{ and } \frac{f(\zeta)}{(\zeta - a)^{-n+1}} \text{ are analytic in } r < |\zeta - a| < R\right)$$

Now changing n to $-m$ in B_n and letting $A_{-m} = B_n$ we see that A_n's

$(-\infty < n < \infty)$ are given by $A_n = \dfrac{1}{2\pi i} \displaystyle\int_C \dfrac{f(\zeta)\, d\zeta}{(\zeta - a)^{n+1}}$. By uniqueness of the

Laurent development we mean that if

$$f(z) = \sum_{n=-\infty}^{\infty} a_n (z-a)^n$$

where the right side converges uniformly for all z with $r < \rho' \le |z - a|$ $\le \sigma' < R$ (ρ' and σ' otherwise arbitrary) then $a_n = A_n$ for $(-\infty < n < \infty)$.

To prove this we let $f(z) = \displaystyle\sum_{m=-\infty}^{\infty} a_m (z - a)^m$ for some $a_m \in \mathbb{C}$ and observe that

$$\frac{f(\zeta)}{(\zeta - a)^{n+1}} = \sum_{k=-\infty}^{-1} a_{k+n+1}\, (\zeta - a)^k + \sum_{k=0}^{\infty} a_{k+n+1}\, (\zeta - a)^k.$$

The right side converge uniformly on $|\zeta - a| = s$ for $r < s < R$ (Take $\rho' = s = \sigma'$). Thus we can integrate the above series term by term (by Corollary 4.1.10) and get

$$A_n = \frac{1}{2\pi i} \int_{|\zeta - a| = s} \frac{f(\zeta)\, d\zeta}{(\zeta - a)^{n+1}} = \frac{a_n}{2\pi i} \int_{|\zeta - a| = s} \frac{d\zeta}{(\zeta - a)} = a_n.$$

Note that all other terms corresponding to $k \neq -1$ vanish after integration

as $(\zeta - a)^k$ for $k \neq -1$ has a primitive namely $\dfrac{(\zeta - a)^{k+1}}{k+1}$ in this annulus

(Theorems 4.2.3 and 4.2.4). Thus the coefficients of the Laurent expansion are uniquely determined.

Note: 4.5.11

In case $f(z)$ is analytic in a neighbourhood of 'a' say $|z - a| < \delta$ we can take $r = 0$ and $R = \delta$ in the above Theorem 4.5.10 and conclude that $f(z)$ has a Laurent expansion in $0 < |z - a| < \delta$.

$$f(z) = \sum_{n=-\infty}^{\infty} A_n(z - a)^n \text{ with } A_n = \frac{1}{2\pi i} \int_{|\zeta - a| = s} \frac{f(\zeta)\, d\zeta}{(\zeta - a)^{n+1}}$$

where $-\infty < n < \infty$ and $0 < s < \delta$. On the other hand since $f(z)$ is analytic

at $z = a$ also, we have $A_0 = \dfrac{1}{2\pi i} \displaystyle\int_{|\zeta - a| = s} \dfrac{f(\zeta)\, d\zeta}{(\zeta - a)} = f(a)$ by Cauchy's integral

formula. For 'n' negative say $n = -m$ $(m \geq 1)$ we have $A_n = A_{-m}$

$$= \frac{1}{2\pi i} \int_{|\zeta - a| = s} \frac{f(\zeta)\, d\zeta}{(\zeta - a)^{-m+1}} = 0 \text{ by Cauchy's theorem for a circular disc.}$$

Thus

$$f(z) = \sum_{n=0}^{\infty} A_n(z - a)^n \text{ with } A_n = \frac{1}{2\pi i} \int_{|\zeta - a| = s} \frac{f(\zeta)\, d\zeta}{(\zeta - a)^{n+1}} = \frac{f^{(n)}(a)}{n!}.$$

(See proof of Theorem 4.4.4) which are the Taylor coefficients of f at 'a'. Hence the Laurent expansion and the Taylor expansion coincide and in as much as the Laurent coefficients are unique the Taylor coefficients also turn out to be unique. This gives yet another proof of the uniqueness of the Taylor development.

We shall now characterize the singularities of a function f analytic in $0 < |z - a| < \delta$ using the concept of algebraic order of f at $z = a$.

Theorem 4.5.12

Let $f(z)$ be analytic in $0 < |z - a| < \delta$. Then $z = a$ is a removable singularity for f if and only if $\lim_{z \to a} (z - a) f(z) = 0$. Further the extended function which is also denoted by $f(z)$ is uniquely determined by the condition $f(a) = \lim_{z \to a} f(z)$.

Proof

If $f(z)$ has a removable singularity at $z = a$ then by definition there exists an analytic function $g(z)$ in $|z - a| < \delta$ such that $g(z) = f(z)$ for $z \neq a$. Then

$$g(a) = \lim_{z \to a} g(z) = \lim_{z \to a} f(z) \qquad (z \to a \Rightarrow z \neq a)$$

and $$\lim_{z \to a}(z - a) f(z) = \lim_{z \to a}(z - a) g(z) = 0 \cdot g(a) = 0.$$

Conversely if $\lim_{z \to a}(z - a) f(z) = 0$ then 'a' is an exceptional point of $f(z)$.

By Remark 4.4.10 (b), Cauchy's integral formula is valid inside $0 < |z - a| < \delta$ and we have

$$f(z) = \frac{1}{2\pi i} \int_C \frac{f(\zeta)}{\zeta - z} \, d\zeta \quad (z \neq a)$$

where C is a circle centre 'a' and completely contained in $|z - a| < \delta$ with its boundary. Since $f(\zeta)$ is continuous on C, by Theorem 4.4.3, we also know that

$$g(z) = \frac{1}{2\pi i} \int_C \frac{f(\zeta)}{\zeta - z} \, d\zeta.$$

is analytic in the interior of C with its value at $z = a$ being

$$g(a) = \frac{1}{2\pi i} \int_C \frac{f(\zeta)}{\zeta - a} \, d\zeta.$$

Thus $g(z)$ is analytic in $|z - a| < \delta$ and $g(z) = f(z)$ for $z \neq a$ and $g(a)$

$= \frac{1}{2\pi i} \int_C \frac{f(\zeta)}{\zeta - a} \, d\zeta$. This proves $z = a$ is a removable singularity for $f(z)$.

Further $f(z) = g(z)$ for $z \neq a$ implies $\lim_{z \to a} f(z) = \lim_{z \to a} g(z) = g(a)$. This also implies that the extended g is unique (If g_1 and g_2 are two extensions then $g_1(a) = \lim_{z \to a} f(z) = g(a) = g_2(a)$).

Theorem 4.5.13
Let $f(z)$ be analytic in $0 < |z - a| < \delta$ with a pole at $z = a$. Then there exists a unique integer $m \geq 1$ such that $f(z) = (z - a)^{-m} f_m(z)$ where $f_m(z)$ is analytic in $|z - a| < \delta$ and $f_m(a) \neq 0$.

Proof
Since $z = a$ is a pole for $f(z)$, $\lim_{z \to a} f(z) = \infty$. Thus in a smaller neighbourhood

of 'a' (if necessary) we can assume that $f(z) \neq 0$. Thus $\dfrac{1}{f(z)}$ is analytic

in this deleted neighbourhood of 'a' and $\lim_{z \to a} \dfrac{(z - a)}{f(z)} = 0$. By

Theorem 4.5.12, $\dfrac{1}{f(z)}$ has a removable singularity and we have the extended

value for $\dfrac{1}{f(z)}$ as $\lim\limits_{z \to a} \dfrac{1}{f(z)} = 0$. Thus the extended function $g(z) = \dfrac{1}{f(z)}$

for $z \neq a$ and $g(a) = 0$ is analytic and in as much as $g(z) \neq 0$, (i.e., $f(z)$ $\neq \infty$) $g(z)$ has a definite unique order for its zero at $z = a$. Hence by Note

4.4.15, $g(z) = \dfrac{1}{f(z)} = (z - a)^m \, g_m(z)$ where $g_m(z)$ is analytic and $g_m(a)$

$\neq 0$ $\left(\text{and so } \dfrac{1}{g_m(z)} \neq 0 \text{ in a small neighbourhood of } z = a\right)$. Thus $f(z) =$

$(z - a)^{-m} f_m(z)$ with $f_m(z) = \dfrac{1}{g_m(z)}$ is analytic at a and $f_m(a) = \dfrac{1}{g_m(a)} \neq 0$.

   From what we have proved so far, the above representation $f(z) =$ $(z - a)^{-m} f_m(z)$ is valid only in a small neighbourhood of 'a' which may be

smaller than the given neighbourhood $|z - a| < \delta$. In fact $f_m(z) = \dfrac{1}{g_m(z)}$ can

be defined only in that small neighbourhood of 'a' wherein $g_m(z) \neq 0$. On the other hand $f_m(z) = (z - a)^m f(z)$ in this smaller neighbourhood. However $(z - a)^m f(z)$ is always analytic throughout $0 < |z - a| < \delta$. Thus we can define $f_m(z)$ throughout of $|z - a| < \delta$ and write $f(z) = (z - a)^{-m} f_m(z)$ for $0 < |z - a| < \delta$ where $f_m(z)$ is analytic in $|z - a| < \delta$, $f_m(a) \neq 0$.

Remark 4.5.14
It is easy to see that the above theorem is equally valid if $0 < |z - a| < \delta$ is replaced by an arbitrary region Ω in which $f(z)$ is analytic except for a pole at $z = a \in \Omega$. We can still write $f(z) = (z - a)^{-m} f_m(z)$ where $f_m(z)$ is analytic in Ω with $f_m(a) \neq 0$.

Definition 4.5.15
If $f(z)$ is analytic in $0 < |z - a| < \delta$ and has a pole at $z = a$ then there exists a unique positive integer 'm' such that (See Theorem 4.5.13) $f(z) = (z - a)^{-m} f_m(z)$ where $f_m(z)$ is analytic throughout $|z - a| < \delta$ and $f_m(a) \neq 0$. This 'm' is called the order of the pole for f at $z = a$.

Theorem 4.5.16
If $f(z)$ is analytic in $0 < |z - a| < \delta$ then $f(z)$ has a pole of order 'm' at $z = a$ if and only if

$$\lim_{z \to a}(z - a)^m f(z) \neq 0 \quad \text{and} \quad \lim_{z \to a}(z - a)^{m+1} f(z) = 0.$$

Proof
Suppose $f(z)$ has a pole of order m at 'a'. Then $f(z) = (z - a)^{-m} f_m(z)$ where $f_m(z)$ is analytic in $|z - a| < \delta$ and $f_m(a) \neq 0$. Thus

$$\lim_{z \to a}(z - a)^m f(z) = f_m(a) \neq 0 \text{ and } \lim_{z \to a}(z - a)^{m+1} f(z) = \lim_{z \to a}(z - a) f_m(a) = 0.$$

Conversely if 'm' satisfies the above conditions then for $z \neq a$, $h(z)$ = $(z - a)^m f(z)$ has a removable singularity at $z = a$ with $h(a) \neq 0$. Now $f(z)$ = $(z - a)^{-m} h(z)$ where $h(z)$ is analytic in $|z - a| < \delta$ and $h(a) \neq 0$. Thus $f(z)$ has a pole of order 'm' at $z = a$.

Corollary 4.5.17

$f(z)$ has a pole of order 'm' at $z = a$ if and only if

$$m = \min_{k}\Big\{k \in \mathbb{N} : (z - a)^k f(z) \text{ is bounded in a neighbourhood of '}a\text{'}\Big\}.$$

Proof

If $f(z)$ has a pole of order 'm' by Theorem 4.5.16 we have

$$\lim_{z \to a}(z - a)^m f(z) \neq 0 \text{ and } \lim_{z \to a}(z - a)^{m+1} f(z) = 0.$$

For $k \geq m$, $(z - a)^k f(z)$ (being analytic) is bounded near 'a'. If $k < m$

$$(z - a)^k f(z) = \frac{(z - a)^m f(z)}{(z - a)^{m-k}} \to \infty \text{ as } z \to a \text{ and so is unbounded near } z = a.$$

Conversely if 'm' is defined as above then $(z - a)^m f(z)$ is bounded near 'a' and so $(z - a)^{m+1} f(z) \to 0$ as $z \to a$. We also claim that $\lim_{z \to a}(z - a)^m f(z) \neq 0$. Otherwise $(z - a)^{m-1} f(z)$ has a removable singularity at $z = a$ and thus $\lim_{z \to a}(z - a)^{m-1} f(z)$ has a finite value which is a contradiction to the minimality of 'm'. Thus by the previous Theorem 'm' is the order of the pole for $f(z)$ at $z = a$.

Theorem 4.5.18

Let $f(z)$ be analytic in $0 < |z - a| < \delta$. Consider the following conditions on $f(z)$ as $z \to a$.

(i) $\lim_{z \to a}|z - a|^\alpha | f(z)| = 0$ (α real)

(ii) $\lim_{z \to a}|z - a|^\beta | f(z)| = \infty$ (β real)

Then one of the following conditions hold:

(a) (i) holds for all real α and $f(z) \equiv 0$

(b) (i) holds for some α and (ii) holds for some β and there is a unique integer 'm' such that (i) holds for $\alpha > m$ and (ii) holds for $\beta < m$.

(c) Neither (i) nor (ii) holds for any α and β real.

Proof

First we observe that (i) holds if $f(z) \equiv 0$ for every α and conversely if (i) holds for all real α then we claim that $f(z) \equiv 0$. If not, by taking $\alpha = 1$, $f(z)$ has a removable singularity at $z = a$ with its value

$$f(a) = \lim_{z \to a} f(z) = \lim_{z \to a} (z - a)^0 f(z) = 0.$$

Thus 'a' is a zero of some finite order 'k' for f and $f(z) = (z - a)^k h(z)$ with $h(z)$ analytic at 'a' and $h(a) \neq 0$. Thus $\lim_{z \to a} (z - a)^{-k} f(z) = h(a) \neq 0$ a contradiction. We also remark that "(ii) holds for all β" cannot happen because in this case taking $\beta = 0$, $\dfrac{1}{f(z)}$ is defined in a small neighbourhood of 'a' and (i) holds for all α for $\dfrac{1}{f(z)}$. Thus $f(z) \equiv \infty$ which cannot happen.

Next we assume that $f(z) \neq 0$. Let (i) hold i.e., $\lim_{z \to a} (z - a)^\alpha f(z) = 0$ for some α. Then it holds for all greater α's. Hence there exists an integer $m > 0$ such that $\lim_{z \to a} (z - a)^m f(z) = 0$ and $\lim_{z \to a} (z - a)^{m+1} f(z) = 0$. [In fact we are free to choose any positive integer greater than α]. Now $(z - a)^m f(z)$ thus has a removable singularity at $z = a$ and its extended value is also zero. Thus $(z - a)^m f(z)$ has a zero of finite order k at $z = a$. So $(z - a)^m f(z) = (z - a)^k h(z)$ where $h(z)$ is analytic at 'a' and $h(a) \neq 0$. Hence

$$\lim_{z \to a} (z - a)^\alpha f(z) = \lim_{z \to a} (z - a)^{\alpha - m + k} h(z)$$

$$= 0 \text{ if } \alpha > m - k$$

$$= \infty \text{ if } \alpha < m - k$$

$$= \mu \text{ if } \alpha = m - k$$

where $\mu = h(a)$ is a non-zero complex number.

Now let (ii) hold. i.e., $\lim_{z \to a} (z - a)^\beta f(z) = \infty$ for some β. Then there exists an integer $m < 0$ so that $\lim_{z \to a} (z - a)^m f(z) = \infty$. [In fact we are free to choose any negative integer less than β]. Now $(z - a)^m f(z)$ thus has a pole at $z = a$ and $(z - a)^m f(z)$ has a pole of finite order l at $z = a$ and so $(z - a)^m f(z) = (z - a)^{-l} k(z)$, where $k(z)$ is analytic at 'a' and $k(a) \neq 0$. Hence

$$\lim_{z \to a} (z - a)^\alpha f(z) = \lim_{z \to a} (z - a)^{\alpha - m - l} k(z)$$

$$= 0 \text{ if } \alpha > m + l$$

$$= \infty \text{ if } \alpha < m + l$$

$$= \lambda \text{ if } \alpha = m + l$$

where $\lambda = k(a)$ is a non-zero complex number.

The integer obtained in (i) or (ii) is unique because if we have two different integers n_1 and n_2 with the same properties choosing a real number p in between them we have that $\lim_{z \to a} (z - a)^p f(z)$ exists and is equal to both 0 and ∞ which is impossible. Summing up we see that if either (i) or (ii) holds for some α or β then there exists a unique integer m such that (i) holds for $\alpha > m$ and (ii) holds for $\beta < m$. This is the case (b). The only other possibility is (c) and our proof is complete.

Definition 4.5.19
In case (i) or (ii) holds for real α or β the unique integer m obtained in the above theorem is called the algebraic order of f at $z = a$. It can be positive, zero or negative but is always an integer. (The reason we call the order algebraic is that it has a sign attached to it). Note also that if $f(z)$ has the algebraic order 'm' then $\lim_{z \to a} (z - a)^m f(z)$ exists as a complex number but is neither 0 nor ∞.

In the following we shall characterize the isolated singularities first using the algebraic order and later using the Laurent series development.

Theorem 4.5.20
Let $f(z)$ be analytic in $0 < |z - a| < \delta$ so that $z = a$ is an isolated singularity for f. Then
   (i) The algebraic order is defined and is equal to 0 if and only if $f(z)$ has a removable singularity at $z = a$ and the extended function has a non-zero value at $z = a$.
  (ii) The algebraic order is defined and equals $-m$ where m is a positive integer if and only if $f(z)$ has a removable singularity at $z = a$ and the extended function has a zero of order m at $z = a$.
 (iii) The algebraic order is defined and equal to $k > 0$ if and only if $f(z)$ has a pole of order k at $z = a$.
 (iv) The algebraic order is not defined at $z = a$ if and only if $z = a$ is an isolated essential singularity for f.

Proof
(i) Let the algebraic order of f at $z = a$ be defined and be equal to 0. Therefore $\lim_{z \to a} (z - a)^\alpha f(z) = 0$ for $\alpha > 0$ and $\lim_{z \to a} f(z) \neq 0$. Thus using $\alpha = 1$, $f(z)$ has a removable singularity with the extended value equal to $\lim_{z \to a} f(z) \neq 0$.

Conversely if $f(z)$ has a removable singularity at $z = a$ with $\lim_{z \to a} f(z) \neq 0$ then as $z \to a$ $(z - a)^\alpha f(z) \to 0$ for $\alpha > 0$ and $(z - a)^\alpha f(z) \to \infty$ for $\alpha < 0$. Hence the algebraic order is defined and is equal to 0.

(ii) Let the algebraic order be defined at $z = a$ and be equal to $-m$ $(m > 0)$. Therefore $\lim_{z \to a} f(z) = 0$ and $\lim_{z \to a} (z - a) f(z) = 0$. Thus $f(z)$ has a removable singularity at $z = a$ and also the extended function has the value $\lim_{z \to a} f(z) = 0$ at $z = a$. Now $h(z) = (z - a)^{-m} f(z)$ also has a removable singularity at $z = a$ (Since $\lim\limits_{z \to a}(z - a)^{-m+1} f(z) = 0$) but $h(z)$ has its extended

value $h(a) = \lim\limits_{z \to a}(z - a)^{-m} f(z) \neq 0$. Therefore $f(z) = (z - a)^m h(z)$ has a

zero of order m at $z = a$.

Conversely let $f(z)$ have a removable singularity at $z = a$ with its extended

value $f(a) = 0$ and assume $f(z)$ (the extended function) has a zero of order

m at $z = a$. Then $f(z) = (z - a)^m h(z)$ where $h(z)$ is analytic in $|z - a| <$

δ and $h(a) \neq 0$. It is now clear that $\lim\limits_{z \to a}(z - a)^\alpha f(z) = 0$ for $\alpha > -m$ and

$\lim\limits_{z \to a}(z - a)^\alpha f(z) = \infty$ for $\alpha < -m$. Hence the algebraic order f at $z = a$ is

defined and is equal to $-m$.

(iii) Let the algebraic order of f at $z = a$ be defined and equal to $k > 0$.
By definition in a neighbourhood of 'a', $(z - a)^\alpha f(z)$ is bounded for $\alpha \geq k$
and is unbounded for $\alpha < k$. Thus Min $\{n \in \mathbb{N}: (z - a)^n f(z)$ is bounded near
'a'$\} = k$ and so by Corollary 4.5.17 $f(z)$ has a pole of order k at $z = a$.
Conversely let $f(z)$ have a pole of order k at $z = a$. Then $k = $ Min $\{n$

$\in \mathbb{N} : (z - a)^n f(z)$ is bounded near 'a'$\}$. Hence for $\alpha > k$, $\lim\limits_{z \to a}(z - a)^\alpha f(z)$

$= \lim\limits_{z \to a}(z - a)^{\alpha-k} (z - a)^k f(z) = 0$. Also for $\alpha < k$ $\lim\limits_{z \to a}(z - a)^\alpha f(z) =$

$\lim\limits_{z \to a}\dfrac{(z - a)^k f(z)}{(z - a)^{k - \alpha}} = \infty$. Therefore k is the algebraic order of f at $z = a$.

(iv) If the algebraic order is not defined at $z = a$ then from (i), (ii), and
(iii) $z = a$ is neither a removable singularity nor a pole for $f(z)$ and so
$z = a$ is an isolated essential singularity. Conversely if $f(z)$ has an essential
singularity at $z = a$ then $z = a$ is neither a pole nor a removable singularity
for $f(z)$ and by (i), (ii) and (iii) its algebraic order is not defined at $z = a$.

We shall now characterize the isolated singularities of a function $f(z)$
analytic in $0 < |z - a| < \delta$ using its Laurent series development. First we
shall fix some terminologies.

Let $f(z)$ be analytic in $r < |z - a| < R$ and have the Laurent expansion

$f(z) = \sum\limits_{k=-\infty}^{\infty} a_k(z - a)^k$. The series $\sum\limits_{k=-\infty}^{-1} a_k(z - a)^k$ containing only negative

powers of $z - a$ will be called the singular part of f and $\sum\limits_{k=0}^{\infty} a_k(z - a)^k$

consisting only non-negative powers of $z - a$ will be called its regular part.

Theorem 4.5.21
Let $f(z)$ be analytic in $0 < |z - a| < \delta$ and have a Laurent series expansion
(as applicable to the annulus $r < |z - a| < R$ with $r = 0$ and $R = \delta$) $f(z)$

$= \sum\limits_{k=-\infty}^{\infty} a_k (z - a)^k$. Then

(i) $f(z)$ has a removable singularity at $z = a$ if and only if $a_k = 0, k < 0$ i.e., if and only if its singular part is zero.

(ii) $f(z)$ has a pole at $z = a$ of order m if and only if $a_{-m} \neq 0$ and $a_k = 0$ for $k < -m$.

(iii) $f(z)$ has an essential singularity at $z = a$ if and only if $a_m \neq 0$ for infinitely many negative integers m.

Proof

(i) If f has a removable singularity at $z = a$, then $\lim\limits_{z \to a} f(z)$ exists and so $f(z)$ is bounded in a deleted neighbourhood of a. i.e., $|f(z)| \leq M$ for $0 < |z - a| < \varepsilon < \delta$ for some $\varepsilon > 0$. Now $a_k = \dfrac{1}{2\pi i} \displaystyle\int\limits_{|\zeta - a| = s} \dfrac{f(\zeta)\, d\zeta}{(\zeta - a)^{k+1}}$ for

any s with $0 < s < \varepsilon$ and so $|a_k| \leq \dfrac{M}{2\pi} \displaystyle\int\limits_{|\zeta - a| = s} s^{-k-1} |d\zeta| = \dfrac{M}{s^k}$. If $k < 0$ the right hand side tends to 0 as $s \to 0$. So $a_k = 0$ for every $k < 0$.

Conversely if $f(z) = \displaystyle\sum_{k=-\infty}^{\infty} a_k (z - a)^k$ with $a_k = 0$ for $k < 0$ then

$\displaystyle\sum_{n=0}^{\infty} a_n (z - a)^n$ is a power series convergent at each point in $0 < |z - a| < \delta$. Thus this power series has a positive radius of convergence around 'a' and

hence $f(z) = \displaystyle\sum_{n=0}^{\infty} a_n (z - a)^n$ represents an analytic function inside this circle of convergence which includes 'a'. Thus 'a' is a removable singularity for $f(z)$.

(ii) Let $f(z)$ have a pole at $z = a$ of order m. Then $f(z) = (z - a)^{-m} g(z)$, $g(a) \neq 0$. Then $g(z) = (z - a)^m f(z)$ has a removable singularity at $z = a$ (Since $m + 1 > m$, $\lim\limits_{z \to a}(z - a)^{m+1} f(z) = 0$) and so $g(z)$ can be defined suitably at $z = a$ so that $g(z)$ is analytic in $|z - a| < \delta$. Now

$$f(z) = \sum_{k=-\infty}^{\infty} a_k(z-a)^k, \quad a_k = \dfrac{1}{2\pi i} \int_C \dfrac{f(\zeta)\, d\zeta}{(\zeta - a)^{k+1}}$$

where C is $|\zeta - a| = s < \delta$. For $k < -m$, $k = -l$ with $l > m$ and $l = m + \alpha$, $\alpha > 0$,

$$a_k = \dfrac{1}{2\pi i} \int_C f(\zeta)\, (\zeta - a)^{l-1}\, d\zeta = \dfrac{1}{2\pi i} \int_C g(\zeta)\, (\zeta - a)^{\alpha - 1}\, d\zeta = 0$$

since $\alpha \geq 1$ and $g(\zeta)$ is analytic in $|z - a| < \delta$. Further

$$a_{-m} = \dfrac{1}{2\pi i} \int_C \dfrac{g(\zeta)\, d\zeta}{(\zeta - a)} = g(a) \neq 0.$$

Conversely let $f(z) = \sum\limits_{k=-\infty}^{\infty} a_k (z - a)^k$ with $a_k = 0$ for $k < -m$ and $a_{-m} \neq 0$
then $f(z) = (z - a)^{-m} \{a_{-m} + a_{-m+1} (z-a) + \ldots\}$. Take $g(z) = a_{-m} + a_{-m+1}$
$(z - a) + \ldots$ so that $g(z) = (z - a)^m f(z)$ is analytic in $0 < |z - a| < \delta$
and has a positive radius of convergence and so is analytic in a
neighbourhood of 'a' including 'a'. Thus $g(z)$ is extended to $z = a$ and
represents an analytic function in $|z - a| < \delta$ with $g(a) = a_{-m} \neq 0$. Thus
$f(z) = (z - a)^{-m} g(z)$, $g(z)$ analytic at 'a' and $g(a) \neq 0$. So $f(z)$ has a pole
of order m at $z = a$.

(iii) Combining (i) and (ii) and using the definition of essential singularity
we see that $f(z)$ has an essential singularity at $z = a$ if and only if
$a_m \neq 0$ for infinitely many negative integers m.

Theorem 4.5.22 (Weierstrass theorem on essential singularity)

Let $f(z)$ be analytic in $0 < |z - a| < \delta$ with $z = a$ as an essential singularity.
Then given any complex number A, $f(z)$ comes as close to A as desired
in any arbitrary neighbourhood of 'a'. i.e., given $A \in \mathbb{C}$, $\varepsilon > 0$, $\delta > 0$ there
exists z_0 with $|z_0 - a| < \delta$ and $|f(z_0) - A| < \varepsilon$. Equivalently if N is any
arbitrary neighbourhood of 'a' in its domain of definition then $f(N \setminus \{a\})$ is
dense in $\mathbb{C}$.

Proof

If this is not true then there exists at least one $A \in \mathbb{C}$, an $\varepsilon > 0$ and a δ
> 0 such that $|z - a| < \delta \Rightarrow |f(z) - A| \geq \varepsilon$. Thus $\lim\limits_{z \to a} |z - a|^{-1} |f(z) - A|$
$= \infty$ so that the condition (ii) of Theorem 4.5.18 is satisfied for $(f(z) - A)$
with $\alpha = -1$. Thus $\lim\limits_{z \to a} |z - a|^{\alpha} |f(z) - A| = 0$ for some $\alpha > 0$. Now

$$\lim_{z \to a} |(z - a)^{\alpha} f(z)| \leq \lim_{z \to a} |(z - a)^{\alpha} [f(z) - A]| + \lim_{z \to a} |(z - a)^{\alpha} A|$$

by triangle inequality. Thus $\lim\limits_{z \to a} |(z - a)^{\alpha} f(z)| = 0$ holds for some α and
so $z = a$ is not an essential singularity for $f(z)$. This contradiction establishes
our theorem.

As an application of the above theorem we shall now characterize all
analytic automorphisms of the entire finite complex plane. (Recall that an
analytic automorphism of a region Ω is an analytic one-to-one mapping of
Ω onto Ω).

Lemma 4.5.23

If $f(z)$ is one-to-one, analytic, from a region Ω onto another region D then
its inverse f^{-1} exists, one-to-one, from D onto Ω and is also analytic.

Proof

Since f is one-to-one and onto, f^{-1} is defined. By open mapping principle
f is open and so f^{-1} is continuous. Thus $z \to z_0 \Leftrightarrow w \to w_0$ where $w = f(z)$

and $w_0 = f(z_0)$. Since f is one-to-one, $f'(z_0) \neq 0$ (Otherwise $f(z) - w_0$ has a zero of order $n \geq 2$ at z_0 and by Theorem 4.4.22, f is not one-to-one in a neighbourhood of z_0). If $g = f^{-1}$ then $\lim\limits_{w \to w_0} \dfrac{g(w) - g(w_0)}{w - w_0} = \lim\limits_{z \to z_0} \dfrac{z - z_0}{f(z) - f(z_0)}$

$= \dfrac{1}{f'(z_0)}$. Hence g is analytic at z_0 and as z_0 is arbitrary in D we get that g is analytic in D.

Theorem 4.5.24
f is an analytic automorphism of $\mathbb{C}$ if and only if $f(z) = az + b$ where $a, b \in \mathbb{C}$ and $a \neq 0$.

Proof
Let $f(z)$ be an analytic automorphism of $\mathbb{C}$. Then by Taylor series development (Theorem 4.5.6) $f(z) = \sum\limits_{n=0}^{\infty} a_n z^n$ where the series has ∞ as its radius of convergence. We now claim that 0 can not be an essential singularity of $h(z) = f\left(\dfrac{1}{z}\right)$. In fact if $z = 0$ were an essential singularity for h then the values of h in any deleted neighbourhood of 0 must be dense in $\mathbb{C}$ by theorem 4.5.22. Let U be the open unit disc. Now $S = \mathbb{C} \backslash \overline{U}$ is open and is non-empty. Thus by open mapping principle $h(S)$ is open and non-empty. Also, $f(S) \cap f(U \backslash \{0\}) = \phi$ as otherwise there exists $w \in S$ and $0 < |\zeta| < 1$ such that $f(w) = f(\zeta)$. But $w \neq \zeta$ (as $|w| > 1$ and $|\zeta| < 1$). This contradicts the injectivity of f. Now $f(S) \cap f(U \backslash \{0\}) = \phi \Rightarrow h(S) \cap h(U \backslash \{0\}) = \phi \Rightarrow h(U \backslash \{0\})$ is not dense in $\mathbb{C}$ (Recall that any dense set in $\mathbb{C}$ will intersect all non-empty open sets in $\mathbb{C}$). Thus the Laurent series for h contains only finitely many negative powers of z. But $h(z) = f\left(\dfrac{1}{z}\right) = \sum\limits_{n=0}^{\infty} a_n z^{-n}$ contains only negative powers of z. This being finite we have $h(z) = a_0$

$+ \dfrac{a_1}{z} + \ldots + \dfrac{a_m}{z^m}$ for some $m \in \mathbb{N}$ and so $f(z) = a_m z^m + a_{m-1} z^{m-1} + \ldots + a_0$ is a polynomial. But f is injective in $\mathbb{C}$ and by fundamental theorem of algebra no polynomial is injective in $\mathbb{C}$ unless it is of the form $a(z - z_0)^N$ for some integer $N > 0$, and $a \neq 0$ (i.e. it should not have more than one 'zero' to start with). Further if $N > 1$ by local correspondence theorem (Theorem 4.4.22), f is a locally N-to-1 map near z_0. This is also impossible as f is one-to-one. Thus $N = 1$ and we have $f(z) = a(z - z_0) = az + b$ for $a, b \in \mathbb{C}$ $(a \neq 0)$. On the other hand, $f(z) = az + b$ is an analytic automorphism of $\mathbb{C}$ as is easily verified. Hence the result.

4.6 Calculus of Residues

Introduction 4.6.1

Cauchy's investigations mark the beginning of the calculus of residues. As an application of his theory, Cauchy derived almost all integral formulas of his time and added new ones also. In fact his theory enables us to compute the values of integrals which are otherwise difficult to workout. In the following we shall develop the residue calculus and apply it to solve some theoretical questions. The actual computation of integrals using residue calculus will be postponed to the next section.

Definition 4.6.2

Let f be holomorphic in $0 < |z - a| < \delta$ with Laurent series expansion $f(z)$

$= \sum\limits_{k=-\infty}^{\infty} a_k(z - a)^k$. Then for any circle $C_\rho = \{z \in \mathbb{C} : |z - a| = \rho\}$ $(0 < \rho$

$< \delta)$, $a_{-1} = \dfrac{1}{2\pi i} \int\limits_{C_\rho} f(\zeta)\, d\zeta$ is independent of ρ and is called the residue of

f at $z = a$, written as $\operatorname*{Res}\limits_{z=a} f(z) = a_{-1}$.

Theorem 4.6.3

Let f be analytic in $0 < |z - a| < \delta$. Then $\operatorname*{Res}\limits_{z=a} f(z)$ is the unique complex number A for which $f(z) - A(z - a)^{-1}$ has a primitive in $0 < |z - a| < \delta$.

Proof

Let $f(z) = \sum\limits_{k=-\infty}^{\infty} a_k(z - a)^k$ be its Laurent expansion, valid for $0 < |z - a| <$

δ. As observed in the proof of Theorem 4.5.10, $\sum\limits_{k=0}^{\infty} a_k(z - a)^k$ converges

uniformly in every $|z - a| \le \rho < \delta$ (its radius of convergence is at least

δ) and $\sum\limits_{k=-\infty}^{-1} a_k (z - a)^k$ converges uniformly in $|z - a| \ge \delta > 0$ with its radius

of convergence in $\dfrac{1}{z - a}$ being ∞.

We now consider $F_1(z) = \sum\limits_{k=0}^{\infty} \dfrac{a_k}{k + 1} (z - a)^{k+1}$ as a power series in

$(z - a)$. This has the same radius of convergence as $\sum\limits_{k=0}^{\infty} a_k (z - a)^k$ which

is at least δ. Similarly $F_2(z) = \sum\limits_{k=-\infty}^{-2} \dfrac{a_k}{k + 1} (z - a)^{k + 1}$ as a power series in

$\dfrac{1}{z-a}$ has the same radius of convergence as $\displaystyle\sum_{k=-\infty}^{-1} a_k(z-a)^k$ which is ∞.

Thus both these series representing F_1 and F_2 define analytic functions in $0 < |z-a| < \delta$.

Thus $F(z) = F_1(z) + F_2(z)$ is analytic in $0 < |z-a| < \delta$ and its derivative is

$$F'(z) = F_1'(z) + F_2'(z) = \sum_{k\neq -1} a_k(z-a)^k = f(z) - a_{-1}(z-a)^{-1}.$$

(For justifying term by term differentiation see Theorem 2.2.1). This proves that the residue a_{-1} satisfies the required conditions of the Theorem. Now if A is any complex number such that $f(z) - A(z-a)^{-1}$ has a primitive $H(z)$

(say) in $0 < |z-a| < \delta$ then $(F(z) - H(z))' = \dfrac{A-a_{-1}}{z-a}$. Since $\dfrac{A-a_{-1}}{z-a}$ has

a primitive, namely $F(z) - H(z)$, by Theorem 4.2.3 and Theorem 4.2.4

$\displaystyle\int_C \dfrac{A-a_{-1}}{z-a}\,dz = 0$ for all closed curves C in $0 < |z-a| < \delta$. On the other

hand if $C = \{z\,|\,|z-a| = \rho < \delta\}$, $\displaystyle\int_C \dfrac{1}{z-a}\,dz = 2\pi i$. This is impossible unless

$A - a_{-1} = 0$ or $A = a_{-1}$. This proves the uniqueness of A with the required properties and the proof is complete.

Example 4.6.4

If $f(z)$ is analytic in $0 < |z-a| < \delta$ and has a removable singularity at $z = a$ then $\displaystyle\operatorname*{Res}_{z=a} f(z) = 0$ (The reason being that $a_{-1} = \dfrac{1}{2\pi i}\int_C f(\zeta)\,d\zeta$ where C

is a circle centre 'a' and radius $r < \delta$ and if $g(z)$ is the extended analytic function in $|z-a| < \delta$, then $\displaystyle\int_C f(\zeta)\,d\zeta = \int_C g(\zeta)\,d\zeta = 0$ by Cauchy's

theorem for a circular disc)

(ii) Let $f(z) = \dfrac{1}{(z-a)^n}$, $n \geq 2$. Then $\displaystyle\operatorname*{Res}_{z=a} f(z) = 0$ since the Laurent

expansion of f is $\displaystyle\sum_{k=-\infty}^{\infty} a_k(z-a)^k$ with $a_k = 0$ for every $k \neq -n$ and $a_{-n} = 1$.

Thus $a_{-1} = \displaystyle\operatorname*{Res}_{z=a} f(z) = 0$.

(iii) If f is analytic in $0 < |z-a| < \delta$ then $\displaystyle\operatorname*{Res}_{z=a} f'(z) = 0$ because the

Laurent expansion of $f'(z)$ at $z = a$ is obtained by term-wise differentiating that of $f(z)$ and in doing so, the term containing $\dfrac{1}{(z-a)}$ does not appear

at all.

Theorem 4.6.5

If f and g are analytic in $0 < |z - a| < \delta$ then for complex numbers α, β,

$$\operatorname*{Res}_{z=a}(\alpha f + \beta g)\,(z) = \alpha \operatorname*{Res}_{z=a} f(z) + \beta \operatorname*{Res}_{z=a} g(z).$$

Proof

Let C be any circle centre a and radius $r < \delta$. Now

$$\operatorname*{Res}_{z=a}(\alpha f + \beta g)\,(z) = \frac{1}{2\pi i}\int_C (\alpha f + \beta g)\,(\zeta)\,\mathrm{d}\zeta$$

$$= \frac{\alpha}{2\pi i}\int_C f(\zeta)\,\mathrm{d}\zeta + \frac{\beta}{2\pi i}\int_C g(\zeta)\,\mathrm{d}\zeta$$

$$= \alpha \operatorname*{Res}_{z=a} f(z) + \beta \operatorname*{Res}_{z=a} g(z).$$

Theorem 4.6.6

If f is analytic in $0 < |z - a| < \delta$ and has a simple pole at $z = a$ then

$$\operatorname*{Res}_{z=a} f(z) = \lim_{z \to a}(z - a)\,f(z).$$

Proof

Let the Laurent series expansion of f at $z = a$ be

$$f(z) = a_{-1}(z - a)^{-1} + a_0 + a_1(z - a) + \dots$$

$$= a_{-1}(z - a)^{-1} + h(z) \text{ (say)}.$$

Here $h(z)$ is analytic in $|z - a| < \delta$ with $h(a) = a_0$. Thus

$$\lim_{z \to a}(z - a)\,f(z) = a_{-1} + \lim_{z \to a}(z - a)\,h(z) = a_{-1} = \operatorname*{Res}_{z=a} f(z).$$

Theorem 4.6.7

If f is analytic in $0 < |z - a| < \delta$ and has a pole of order m at $z = a$ so

that $f(z) = \dfrac{g(z)}{(z - a)^m}$ where $g(z)$ is analytic in $|z - a| < \delta$, then $\operatorname*{Res}_{z=a} f(z)$

$= \dfrac{g^{m-1}(a)}{(m - 1)!}$ where $g^{(m-1)}(a)$ denotes the $(m - 1)$th derivative of g evaluated

at $z = a$.

Proof

$g(z)$ has a Taylor expansion at $z = a$ given by

$$g(z) = c_0 + c_1(z - a) + \dots + c_{m-1}(z - a)^{m-1} + c_m(z - a)^m + \dots$$

Using the uniqueness of Laurent expansion of f at $z = a$ we can say that it is given by

$$f(z) = \frac{c_0}{(z - a)^m} + \dots + \frac{c_{m-1}}{(z - a)} + c_m + \dots$$

By a simple computation $c_{m-1} = \dfrac{g^{(m-1)}(a)}{(m-1)!}$ and by definition $\operatorname*{Res}_{z=a} f(z) =$ c_{m-1}. Hence the result.

Theorem 4.6.8
Let g, h be analytic in $|z - a| < \delta$, $g(a) \neq 0$; $h(a) = 0$; $h'(a) \neq 0$. Then f $= g/h$ has a simple pole at $z = a$ with $\operatorname*{Res}_{z=a} f(z) = \dfrac{g(a)}{h'(a)}$.

Proof
As $h(z)$ has a simple zero at $z = a$ we can write $h(z) = (z - a) h_1(z)$ with $h_1(a) = h'(a) \neq 0$. Therefore

$$f(z) = (z - a)^{-1} \frac{g(z)}{h_1(z)}$$

with $\dfrac{g(z)}{h_1(z)}$ analytic in a neighbourhood of 'a' and

$$\frac{g(a)}{h_1(a)} = \frac{g(a)}{h'(a)} \neq 0.$$

Therefore $z = a$ is a simple pole for $f(z)$ and by Theorem 4.6.6,

$$\operatorname*{Res}_{z=a} f(z) = \lim_{z \to a} (z - a)\, f(z) = \frac{g(a)}{h'(a)}.$$

Theorem 4.6.9 (Residue theorem)
Let Ω be a region and A, a finite subset of Ω. Let γ be a curve which is homologous to 0 in Ω and does not contain points of A. Let f be analytic in $\Omega \backslash A$ then

$$\frac{1}{2\pi i} \int_\gamma f(\zeta)\, d\zeta = \sum_{a \in A} n(\gamma, a) \operatorname*{Res}_{z=a} f(z).$$

Proof
$f(z)$ has a Laurent expansion around each of the finite number of points z_i of A. We shall assume that these neighbourhoods around each of the points of A are mutually disjoint. We shall denote the singular part of $f(z)$ at each of these points by $P_1\left(\dfrac{1}{z - z_1}\right)$, $P_2\left(\dfrac{1}{z - z_2}\right)$, ..., $P_m\left(\dfrac{1}{z - z_m}\right)$ where m denotes the number of points in A. Note that each $P_i\left(\dfrac{1}{z - z_i}\right)$ is analytic outside a small circle around $z = z_i$ ($i = 1, 2, \ldots m$) (See proof of Theorem

4.5.10 and the note 4.5.11). Consider $f(z) - \sum\limits_{i=1}^{m} P_i\left(\dfrac{1}{z-z_i}\right)$. From the

definition of singular parts it follows that $f(z) - P_j\left(\dfrac{1}{z-z_j}\right)$ has a remov-

able singularity at $z = z_j$ and $P_i\left(\dfrac{1}{z-z_i}\right)$ is analytic for $z \neq z_i$ (and hence

analytic at z_j for $j \neq i$). Thus $f(z) - \sum\limits_{i=1}^{m} P_i\left(\dfrac{1}{z-z_i}\right)$ can be extended as an

analytic function throughout of Ω. Hence applying Cauchy's general theorem (Theorem 4.3.8) we get

$$\int_\gamma f(\zeta)\, d\zeta = \sum_{i=1}^{m} \int_\gamma P_i\left(\frac{1}{\zeta-z_i}\right) d\zeta$$

Since the series representing $P_i\left(\dfrac{1}{\zeta-z_i}\right)$ is uniformly convergent on

every compact set in $\{z_i\}^c$ the integration can be done term by term on γ.

Now $\int_\gamma \dfrac{1}{(\zeta-z_i)^m}\, d\zeta = 0$ for $m \neq 1$ as $\dfrac{1}{(\zeta-z_i)^m}$ with $m \neq 1$ has the

primitive $(\zeta - z_i)^{1-m}/(1 - m)$ in the region obtained by omitting 'z_i'. Thus for each i

$$\frac{1}{2\pi i}\int_\gamma P_i\left(\frac{1}{\zeta-z_i}\right)d\zeta = \frac{1}{2\pi i}\operatorname*{Res}_{z=z_i} f(z)\int_\gamma \frac{d\zeta}{(\zeta-z_i)} = n(\gamma, z_i)\operatorname*{Res}_{z=z_i} f(z)$$

Hence $\dfrac{1}{2\pi i}\int_\gamma f(\zeta)\, d\zeta = \sum\limits_{a\in A} n(\gamma, a)\operatorname*{Res}_{z=a} f(z)$ and the proof of our theorem

is complete.

Remark 4.6.10
If in the above theorem $\Omega = \mathbb{C}$ it is sometimes advantageous to evaluate the above integral not using all the residues of f at various points but instead using a single residue of a function related to f as follows:

Theorem 4.6.11
If f is analytic in $\mathbb{C}$ except for a finite number of points and γ is a simple closed curve such that all these finite points are interior to γ. Then

$$\int_\gamma f(\zeta)\, d\zeta = 2\pi i \operatorname*{Res}_{z=0}\left(1/z^2\, f(1/z)\right)$$

Proof

Let $|z| = r$ be large enough so that γ lies inside $|z| < r < \infty$. We have the Laurent expansion for $f(z)$ as $f(z) = \sum_{n=-\infty}^{\infty} c_n z^n$ $(r < |z| < \infty)$ with

$$c_n = \frac{1}{2\pi i} \int_{|z|=R} \frac{f(\zeta)\, d\zeta}{\zeta^{n+1}}, \quad n \in \mathbb{Z} \text{ and } r < R < \infty. \text{ Consider}$$

$$1/z^2\, f(1/z) = \sum_{k=-\infty}^{\infty} \frac{c_n}{z^{n+2}} = \sum_{k=-\infty}^{\infty} \frac{c_{n-2}}{z^n} \quad \left(0 < |z| < \frac{1}{r}\right).$$

Thus $c_{-1} = \operatorname{Res}_{z=0}\left(1/z^2 f(1/z)\right)$. But by taking $n = -1$ in the expression for c_n

we have $c_{-1} = \dfrac{1}{2\pi i} \int_{|z|=R} f(\zeta) d\zeta$. Now if the finite number of points in the

plane at which f is not analytic are denoted by z_i and if η is the circle $|z| = R > r$ then $n(\gamma, z_i) = 1 = n(\eta, z_i)$ $\forall i$. Thus $\gamma \sim \eta$ in $\Omega = \mathbb{C}\backslash\{z_i\}$ and as f is analytic in Ω, $\int_\eta f(\zeta)\, d\zeta = \int_\gamma f(\zeta)\, d\zeta$ (See Theorem 4.3.8).

Thus $\dfrac{1}{2\pi i}\int_\gamma f(\zeta)\, d\zeta = \dfrac{1}{2\pi i}\int_\eta f(\zeta) d\zeta = c_{-1} = \operatorname{Res}_{z=0}\left(1/z^2\, f(1/z)\right)$. Hence

$\int_\gamma f(\zeta)\, d\zeta = 2\pi i \operatorname{Res}_{z=0}\left(1/z^2\, f(1/z)\right).$

Remark 4.6.12

If we want to define residue of f at ∞ as in the case of finite points we have to assume ∞ is an isolated singularity. Then there exists a circle $|z| = r$ outside of which $f(z)$ is analytic (except at ∞) and all the finite singularities of $f(z)$ lie inside $|z| < r$. But a circle with clockwise direction alone is positively oriented with respect to ∞ in the sense that ∞ always lies to the left as the circle is described. Thus if we want the residue theorem to be true, also at ∞, we need to define $\operatorname{Res}_{z=\infty} f(z) = \dfrac{1}{2\pi i}\int_\gamma f(\zeta)d\zeta$ where γ is the circle $|z| = R > r$ described clockwise. In other words if we take $|z| = R$ in the usual anti-clockwise sense $\operatorname{Res}_{z=\infty} f(z) = -\dfrac{1}{2\pi i}\int_{|z|=R} f(\zeta)\, d\zeta$. On the other hand if $f(z)$ has the Laurent expansion $f(z) = \sum_{n=-\infty}^{\infty} c_n z^n$ $(r < |z| < \infty)$ then we have to define $\operatorname{Res}_{z=\infty} f(z) = -c_{-1}$. This indeed is the definition of $\operatorname{Res}_{z=\infty} f(z)$ and we observe the following facts: (i) $\operatorname{Res}_{z=\infty} f(z) = -\dfrac{1}{2\pi i}\int_{|z|=R} f(\zeta)\, d\zeta$ is independent of R so long as $R > r$. This is in fact a consequence of homologous version of Cauchy's theorem and the fact that any two such concentric circles are homologous in $|z| > r$. (ii) Let f be analytic except for isolated singularities and let the set of all finite singularities be bounded (say lies inside $|z| < r$). Then the algebraic sum of all residues of f including that at ∞ equals to 0. This is because $|z| = R > r$ is a simple closed curve and by residue theorem 4.6.9, $\operatorname*{Res}_{z=\infty} f(z) = -\dfrac{1}{2\pi i} \int_{|z|=R} f(\zeta)\, d\zeta = -\text{sum of}$ residues of f at all the finite isolated singularities.

Theorem 4.6.13 (Generalized argument principle)

Let $f(z) \neq 0$ be defined in a region Ω with zeros a_j and poles b_k with no other singularities in Ω. Let $g(z)$ be analytic in Ω. Let γ be homologous to 0 in Ω not passing through a_j's and b_k's.
Then

$$\frac{1}{2\pi i}\int_\gamma g(z)\,\frac{f'(z)}{f(z)}\,dz = \sum_j n(\gamma,\, a_j)\, g(a_j) - \sum_k n(\gamma,\, b_k)\, g(b_k) \qquad (1)$$

where the terms on the right hand side are to be repeated as many times as the orders indicate.

Proof

In general there can be infinitely many zeros and poles of f in Ω. However it should be first noted that the sums on the right hand side of (1) are always finite because inside γ (for which $n(\gamma, a) \neq 0$) there can be at most finitely many a_j's and b_k's as otherwise there will be either a limit point of zeros or a limit point of poles inside or on γ both of which are excluded by our assumptions ($f(z) \neq 0$ and $f(z)$ has no other singularities).

If a_j is a zero of order h for f then, in a neighbourhood of a_j, $f(z) = (z - a_j)^h f_1(z)$ where $f_1(z)$ is analytic and $f_1(a_j) \neq 0$. It follows from the continuity of $f_1(z)$ at a_j that there is a neighbourhood of a_j in which $f_1(z) \neq 0$. Thus in this neighbourhood

$$\frac{f'(z)}{f(z)} = \frac{h}{z - a_j} + \frac{f_1'(z)}{f_1(z)}.$$

Using Taylor expansion for $g(z)$ at $z = a_j$ we get

$$g(z) = g(a_j) + g'(a_j)\,(z - a_j) + \dots$$

and hence in the Laurent expansion for $g(z)\,\dfrac{f'(z)}{f(z)}$ the coefficient of $\dfrac{1}{z - a_j}$

which is nothing but $\operatorname*{Res}_{z=a_j}\left[g(z)\,\dfrac{f'(z)}{f(z)} \right]$ is precisely $hg(a_j)$. Thus $g(z)\,\dfrac{f'(z)}{f(z)}$ has a pole at $z = a_j$ with residue $hg(a_j)$. Similarly at a pole $z = b_k$ of order k for $f(z)$, $g(z)\,\dfrac{f'(z)}{f(z)}$ also has a pole with residue $-kg(b_k)$. Thus using residue theorem 4.6.9 we get

$$\frac{1}{2\pi i}\int_\gamma g(z)\,\frac{f'(z)}{f(z)}\,dz = \sum_j n(\gamma, a_j)\,g(a_j) - \sum_k n(\gamma, b_k)\,g(b_k) \qquad (1)$$

where the terms on the right hand side are to be repeated as many times as the orders indicate.

Theorem 4.6.14 (Argument principle)

If $f(z)$ is analytic in a region Ω except for poles b_k and a_j's are its zeros and $\gamma \sim 0$ in Ω not passing through any of the zeros and poles then

$$\frac{1}{2\pi i}\int_\gamma \frac{f'(z)}{f(z)}\,dz = \sum_j n(\gamma, a_j) - \sum_k n(\gamma, b_k)$$

where the terms on the right hand side are to be repeated as many times as the orders indicate.

Proof

Take $g(z) \equiv 1$ in Theorem 4.6.13.

Corollary 4.6.15

If in the above theorem each $n(\gamma, a)$ is either 0 or 1 then

$$\frac{1}{2\pi i}\int_\gamma \frac{f'(z)}{f(z)}\,dz = N - P$$

where N denotes the number of zeros of f inside γ and P denotes the number of pole of f inside γ including multiplicities.

Theorem 4.6.16 (Rouche's theorem)

Let $\gamma \sim 0$ in Ω be such that $n(\gamma, a)$ is either 0 or 1 for points $a \notin \gamma$. Let $f(z)$ and $g(z)$ be analytic in Ω with $|f(z)| > |g(z)|$ on γ. Then $f(z)$ and $f(z) + g(z)$ have the same number of zeros enclosed by γ including multiplicities.

Proof

First note that if $h(z) = A(z)\,B(z)$ where A and B are analytic then

$$\frac{h'(z)}{h(z)} = \frac{A'(z)}{A(z)} + \frac{B'(z)}{B(z)}.$$

We now write $f + g = f\left(1 + \dfrac{g}{f}\right)$ on γ and so $\dfrac{(f+g)'}{(f+g)} = \dfrac{f'}{f} + \dfrac{\left(1 + \dfrac{g}{f}\right)'}{\left(1 + \dfrac{g}{f}\right)}$

on γ (Note that the inequality $|f(z)| > |g(z)|$ on γ implies in particular that f has no zeros on γ). By the argument principle 4.6.14 (and our assumption that $n(\gamma, a)$ is either 0 or 1 only) we have,

$$\frac{1}{2\pi i} \int_{\gamma} \frac{(f+g)'(z)}{(f+g)(z)} \, dz = \text{No. of zeros of } (f+g) \text{ enclosed by } \gamma$$

and

$$\frac{1}{2\pi i} \int_{\gamma} \frac{f'(z)}{f(z)} \, dz = \text{No. of zeros of } f \text{ enclosed by } \gamma.$$

Now put $\Gamma = (1 + (g/f))(\gamma)$. We know that $|g/f| < 1$ on γ if and only if $|(1 + (g/f)) - 1| < 1$ on γ. That is Γ lies inside the open disc D of centre 1 and radius 1 and hence does not wind around 0. Thus

$$n(\Gamma, 0) = \frac{1}{2\pi i} \int_{\Gamma} \frac{dw}{w} = 0.$$

$\left(\dfrac{1}{w}\right.$ is analytic in D and so we can apply Cauchy's theorem for a circular

disc$\bigg)$. Also

$$0 = n(\Gamma, 0) = \frac{1}{2\pi i} \int_{\Gamma} \frac{dw}{w} = \frac{1}{2\pi i} \int_{\gamma} \frac{(1 + (g/f))'(z)}{(1 + (g/f))(z)} \, dz.$$

Hence the number of zeros of $(f + g)$ enclosed by γ equals the number of zeros of f enclosed by γ including multiplicities.

Theorem 4.6.17
Let $\gamma \sim 0$ in Ω and f, g be analytic in Ω and $n(\gamma, a) = 0$ or 1 for points $a \notin \gamma$. Then $|f(z) - g(z)| < |g(z)|$ on γ implies f and g have the same number of zeros enclosed by γ.

Proof
Replace g by $f - g$ and f by g in Theorem 4.6.16.

Theorem 4.6.18
Let f and g be analytic in a region Ω. $\gamma \sim 0$ in Ω. Let $|f(z) - g(z)| < |f(z)| + |g(z)|$ on γ (Note the strict inequality). Then f and g have the same number of zeros enclosed by γ.

Proof
The condition ensures that both f and g are free from zeros on γ. By continuity of $|f - g|$ and $|f| + |g|$, there exists a neighbourhood $U \subset \Omega$ of γ (U an open set containing γ and contained in Ω) in which f and g are analytic and satisfy $|f(z) - g(z)| < |f(z)| + |g(z)|$ in U. Again in this open set U, f and g are zero free and we consider the analytic function $h(z) = \dfrac{f(z)}{g(z)}$ in U. $h(z)$ is never real and less than or equal to zero in U. In fact

if for some $z \in U$, $h(z) = r < 0$ the inequality $|f(z) - g(z)| < |f(z)| + |g(z)|$ in U implies $|r - 1| < |r| + 1$ whereas $|r - 1| = 1 - r = 1 + |r|$. Thus the principle branch of log h is well defined in U and provides a primitive for $h'/h = (f'/f) - (g'/g)$. It now follows (Theorems 4.2.3 and 4.2.4) that

$$0 = \int_\gamma \frac{h'(z)}{h(z)} dz = \int_\gamma \frac{f'(z)}{f(z)} dz - \int_\gamma \frac{g'(z)}{g(z)} dz.$$

Thus the number of zeros of f enclosed by γ and the number of zeros of g enclosed by γ are equal. This completes the proof of the theorem.

Theorem 4.6.19

Let Ω be a region containing a simple closed curve C and its interior D. Let C be given the positive orientation with respect to D. Let f be analytic in Ω and one-to-one on C. Let C^* be the image of C under f and let C^* take the orientation from $f(z)$ as z describes C. Denote the interior of C^* by D^*. Then C^* is also positively oriented with respect to D^* and f provides a bijective analytic correspondence between D and D^*.

Proof

Since f is continuous and one-to-one on C, C^* is also a simple closed curve with a definite orientation which may be either positive or negative. We show that it cannot be negative. Let $w_0 \in D^*$. If C^* were negatively oriented then $n(C^*, w_0) = \dfrac{1}{2\pi i} \displaystyle\int_{C^*} \frac{dw}{w - w_0} = -1$. Since C^* takes its orientation from $f(z)$ as z describes C and C is positively oriented,

$$-1 = \frac{1}{2\pi i} \int_C \frac{f'(z)}{f(z) - w_0} dz = N$$ where N is the number of zeros of $f(z) - w_0$ inside C. Since f is analytic in D (which is the region enclosed by C) we must have $N \geq 0$. This contradiction proves that C^* is positively oriented and in this case

$$1 = \frac{1}{2\pi i} \int_C \frac{f'(z)}{f(z) - w_0} dz = N$$

That is $f(z) - w_0$ has exactly one solution for each $w_0 \in D^*$. That is f provides a bijective analytic correspondence between D and D^*. (Indeed the above argument gives $D^* \subseteq f(D)$, but f is injective, f is continuous and D is connected $\Rightarrow$ either $f(D) \subseteq D^*$ or $f(D) \cap D^* = \phi$).

Theorem 4.6.20

Let Ω be a region. Let C be a simple closed curve positively oriented with respect to its interior D. Let $C \cup D \subset \Omega$. Let $f(z)$ be analytic except for poles in Ω and C does not pass through any of these poles. Let $f(z)$ map

$C \cup D$ in a one-to-one fashion. Let $f(C) = C^*$ take its orientation from $f(z)$ as z traverses C. Let C^* enclose a region D^*. Then one of the following holds:

    (i) C^* is positively oriented, f is analytic in D and f defines a one-to-one correspondence between D and D^*.

    (ii) C^* is negatively oriented, f is meromorphic (i.e., analytic except for poles) in Ω with just one pole inside of C and f defines a one-to-one correspondence between D and the exterior of D^*.

Proof

(i) Suppose C^* is positively oriented with respect to D^*. Let $w_0 \in C^*$. Now consider the function $f(z) - w_0$. By the argument principle (Theorem 4.6.14)

$$\frac{1}{2\pi i} \int_C \frac{f'(z)}{f(z) - w_0} \, dz = N - P$$

where N is the number of zeros of $f(z) - w_0$ in D and P is the number of poles of $f(z) - w_0$ in D. On the other hand changing the variable in the above integrand by substituting $w = f(z)$ (observe that orientation of $f(C) = C^*$ is assumed to be positive) we get $N - P = \dfrac{1}{2\pi i} \displaystyle\int_{C^*} \dfrac{dw}{w - w_0} =$ $n(C^*, w_0)$ since C^* is the one-to-one image of C). Let $w_0 \in D^*$. We have $n(C^*, w_0) = 1$. Now $N \neq 0$ (otherwise $P = -1 < 0$). However since f is one-to-one in D, $N \leq 1$. Thus $N = 1$ which implies that $P = 0$. From the fact that $N = 1$ for each $w_0 \in D^*$ we get that $f(z)$ provides a one-to-one mapping of D onto D^*. From the fact that $P = 0$ we get that $f(z)$ is analytic in D. The proof of (i) is complete.

(ii) Suppose C^* is negatively oriented with respect to D^*. Then for each $w_0 \in D^*$ we have $n(C^*, w_0) = -1$. Thus $N - P = -1$ as before. But we also have $N \leq 1$. Now there are two possibilities, viz. (a) $N = 0$; $P = 1$ (b) $N = 1$; $P = 2$. We rule out (b) by observing that in this case $\dfrac{1}{f(z) - w_0}$ has a double zero at some point $z_0 \in D$ or two simple zeros in D. Thus by local correspondence theorem for w near 0 there are points $z_1 \neq z_2$ in D such that

$$\frac{1}{f(z_1) - w_0} = \frac{1}{f(z_2) - w_0} = w$$

which contradicts the one-to-one nature of $f(z)$ in D. Thus $N = 0$ and $P = 1$. From the fact that $P = 1$ it follows that $f(z) - w_0$ for $w_0 \in D$ (and therefore $f(z)$ also) has precisely one pole in D. Now let us take w_0 in the exterior of C^*. We now have, $N - P = 0$ (Since $n(C^*, w_0) = 0$). Therefore $N = P = 1$ (as already noted P is exactly 1). Hence $f(z)$ provides a one-to-one correspondence between D and the exterior of C^*.

Remark 4.6.21

A preliminary version of the intermediate value theorem for continuous functions defined on closed intervals can be given as follows: If f is continuous on the closed interval $[a, b]$ with $f(a)\,f(b) < 0$ then there exists a point $x_0 \in (a, b)$ with $f(x_0) = 0$. Actually there is no loss of generality in this theorem if we assume $a = -1$ and $b = 1$ and $f(-1) < 0$ and $f(1) > 0$ (By changing the sign of f if necessary we can assume $f(a) < 0 < f(b)$

and by considering $g(x) = f\left(\dfrac{b-a}{2} x + \dfrac{a+b}{2}\right)$ in $[-1, 1]$ we have, $g(-1)$

$= f(a) < 0 < f(b) = g(1)$ and a zero of g in $(-1, 1)$ gives a zero of f in (a, b)). Now in this context the condition $f(-1) < 0 < f(1)$ can be replaced by $xf(x) > 0$ for $x \in \{-1, 1\}$. Thus the preliminary version can be reworded as follows: Let f be a continuous real function defined on $[-1, 1]$ with $xf(x) > 0$ for $x = 1$ or -1. Then f has at least one zero in $(-1, 1)$. This small but interesting result which is true for real-valued functions of a real-variable has the following generalization in the complex plane.

Theorem 4.6.22

Let Ω be a region in the plane containing the origin. Let γ be a closed curve homologous to zero in Ω such that 0 belongs to the interior of γ. Let $n(\gamma, a)$ be 0 or 1, for $a \notin \gamma$. Suppose that f is analytic in Ω with $\text{Re}\,[\bar{z}\,f(z)] > 0$ for $z \in \gamma$. Then f has exactly one zero in the interior of γ.

Proof

Let $g(z) = cz - f(z)$ where $c = \dfrac{\inf \text{Re}\,[\bar{z}\,f(z)]}{\sup |z|^2}$ $(z \in \gamma)$.

Now for $z \in \gamma$ we have.

$$|f(z)|^2 - |g(z)|^2 = |f(z)|^2 - |cz - f(z)|^2$$
$$= c\,[2\text{Re}\,[\bar{z}f(z)] - c|z|^2] > 0$$

as can be easily verified. Therefore $|f(z)| > |g(z)|$ on γ and applying Rouche's theorem we see that $f(z)$ and $(f + g)\,(z)$ have the same number of zeros. But $(f + g)\,(z) = cz$ has exactly one zero inside γ and hence $f(z)$ has exactly one zero inside γ.

Note that the hypothesis here are stronger in the sense that we assume analyticity of f inside and on γ, and γ is homologous to 0 in Ω. However the conclusion also is stronger because we get that f has exactly one zero inside γ in contrast to the real case wherein there can be more than one zero in (a, b).

4.7 Computation of Integrals

Introduction 4.7.1

In this section we shall demonstrate the power of residue theorem by evaluating several integrals involving real and complex functions. In several of these examples, the residue calculus gives us shorter methods of computing the integrals. On the other hand a few more examples are given in which the evaluation becomes simple and elegant. There are also examples wherein residue calculus is the only known easy way of computing the integrals. Whenever necessary we give general methods with which problems of similar types could be solved and in these cases we illustrate the method by computing just one or two integrals. Even though at the outset the importance of these integrals are not specifically stated we have taken care to compute integrals which are useful in other areas of analysis.

Theorem 4.7.2 (Error integral)

$$\int_{-\infty}^{\infty} e^{-t^2}\, dt = \sqrt{\pi}.$$

Proof

Consider $g(z) = \dfrac{e^{-z^2}}{1 + e^{-2az}}$, $a = (1 + i)\sqrt{\dfrac{\pi}{2}}$. The poles of g are given by those z for which $e^{-2az} = -1$. That is $z = \dfrac{(2n - 1)\pi i}{2a}$ $(n \in \mathbb{Z})$. Since $a^2 = \pi i$ the above simple poles (Note that the poles are simple as $\dfrac{d}{dz}(e^{-2az} + 1) \neq 0$) can be described by $-\dfrac{a}{2} + na$, $n \in \mathbb{Z}$.

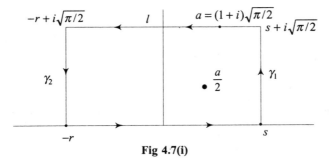

Fig 4.7(i)

Let ∂R be the boundary of the positively oriented rectangle whose vertices are $-r$, s, $s + i\sqrt{\dfrac{\pi}{2}}$, $-r + i\sqrt{\dfrac{\pi}{2}}$ with γ_1, γ_2 as vertical sides and l as the horizontal line through a. The only pole for $g(z)$ available inside this rectangle is $z = \dfrac{a}{2}$ which is got by taking $n = 1$. By residue theorem we have

$$\int_{\partial R} g(z)\, dz = 2\pi i \operatorname*{Res}_{z=\frac{a}{2}} [g(z)]$$

$$\operatorname*{Res}_{z=\frac{a}{2}}[g(z)] = \lim_{z \to \frac{a}{2}}\left(z - \frac{a}{2}\right)\frac{e^{-z^2}}{1 + e^{-2az}}$$

$$= \lim_{z \to \frac{a}{2}}\left[\frac{e^{-z^2}(z - (a/2))}{e^{-2az} - e^{-2a(a/2)}}\right] \quad (\text{Since } e^{-a^2} = -1)$$

$$= \frac{e^{-a^2/4}}{\dfrac{d}{dz}(e^{-2az})\Big|z = \dfrac{a}{2}} = \frac{e^{-a^2/4}}{2a}$$

$$= \frac{1 - i}{2\sqrt{\pi}(1 + i)} \quad \left(\text{Since } a = (1 + i)\sqrt{\frac{\pi}{2}} \text{ and } a^2 = \pi i\right).$$

Therefore

$$\int_{\partial R} g(z)\ dz = \frac{2\pi i(1 - i)}{2\sqrt{\pi}(1 + i)} = \sqrt{\pi}. \tag{1}$$

We observe that

$$g(z) - g(z + a) = \frac{e^{-z^2}}{1 + e^{-2az}} - \frac{e^{-(z+a)^2}}{1 + e^{-2a(z+a)}}$$

$$= \frac{e^{-z^2}}{1 + e^{-2az}} - \frac{e^{-(z^2 + a^2 + 2az)}}{1 + e^{-2az}}$$

$$(\text{Since } e^{-2az} \text{ has period } a).$$

$$= \frac{e^{-z^2}(1 + e^{-2az})}{1 + e^{-2az}} = e^{-z^2}.$$

Thus

$$g(z) - g(z + a) = e^{-z^2}. \tag{2}$$

Now $\displaystyle \int_{\partial R} g(z)\ dz = \int_{-r}^{s} g(z)\ dz + \int_{\gamma_1} g(z)\ dz + \int_{\gamma_2} g(z)\ dz + \int_{l} g(z)\ dz.$ (3)

Further $\displaystyle \left|\int_{\gamma_1} g(z)\ dz\right| \leq \int_{0}^{\sqrt{\pi/2}} \frac{e^{-(s^2 - t^2)}}{1 - e^{-\sqrt{2\pi}(s - t)}}\ dt$

$$\leq e^{-s^2} \int_{0}^{\sqrt{\pi/2}} \frac{e^{t^2}}{1 - e^{-\sqrt{2\pi}}}\ dt \quad \left(\text{if } s > 1 + \sqrt{\frac{\pi}{2}}\right)$$

$$\leq \frac{e^{-s^2}\ e^{\pi/2}}{1 - e^{-\sqrt{2\pi}}}\sqrt{\frac{\pi}{2}} \to 0 \text{ as } s \to \infty.$$

Thus

$$\int_{\gamma_1} g(z)dz \to 0 \text{ as } s \to \infty. \tag{4}$$

Similarly
$$\int_{\gamma_2} g(z)dz \to 0 \text{ as } r \to \infty \tag{5}$$

Further
$$\int_l g(z) \; dz = -\int_{-r}^{s} g\left(t + i\sqrt{\frac{\pi}{2}}\right)dt. \text{ Put } z = t + i\sqrt{\frac{\pi}{2}}$$

$$\int_l g(z) \; dz = -\int_{-r-\sqrt{\pi/2}}^{s-\sqrt{\pi/2}} g(u + a) \; du$$

$$= -\int_{-r-\sqrt{\pi/2}}^{s-\sqrt{\pi/2}} (g(u) - e^{-u^2}) \; du \text{ from (2).} \tag{6}$$

Using all these informations from (1) to (6) and allowing r and s to tend to ∞ in (3) we get

$$\sqrt{\pi} = \int_{-\infty}^{\infty} g(t) \; dt - \int_{-\infty}^{\infty} (g(t) - e^{-t^2}) \; dt = \int_{-\infty}^{\infty} e^{-t^2} \; dt.$$

Here we have used the fact that $\int_{-\infty}^{\infty} g(t) \; dt$ exists finitely.

This can be proved as follows. First we claim that
$$|1 + e^{-2at}| \geq \eta > 0, \; \forall \; t \in \mathbb{R}.$$

Indeed if $t = 0$ the value of $|1 + e^{-2at}| = 2$. Therefore by continuity of the function $1 + e^{-2at}$ we have a $\delta > 0$ such that $|t| < \delta$ implies $|1 + e^{-2at}| \geq 1$. Now let $|t| \geq \delta$. Therefore

$$|1 + e^{-2at}| \geq e^{-2t \, \text{Rea}} -1 \text{ for } t < -\delta$$

and
$$\geq 1 - e^{-2t \, \text{Rea}} \text{ for } t > \delta.$$

That is
$$|1 + e^{-2at}| \geq e^{\delta\sqrt{2\pi}} -1 \text{ for } t < -\delta \text{ and}$$

$$\geq 1 - e^{-\delta\sqrt{2\pi}} \text{ for } t > \delta.$$

Hence if we take $\eta = \min\left\{1, e^{\delta\sqrt{2\pi}} -1, 1 - e^{-\delta\sqrt{2\pi}}\right\}$ we have $|1 + e^{-2at}| \geq \eta$ for every $t \in \mathbb{R}$. Therefore

$$\left|\int_{\mathbb{R}} g(t) \; dt\right| \leq \int_{\mathbb{R}} |g(t)| \; dt \leq \int_{\mathbb{R}} \frac{e^{-t^2}}{\eta} \; dt$$

$$= \frac{2}{\eta} \int_0^{\infty} e^{-t^2} \; dt$$

$$\leq \frac{2}{\eta} \int_0^{1} e^{-t^2} \; dt + \frac{2}{\eta} \int_1^{\infty} e^{-t^2} \; dt$$

$$\leq \frac{2}{\eta} + \frac{2}{\eta} \int_1^\infty e^{-t}\, dt \quad (t \leq t^2)$$

$$= \frac{2}{\eta} + \frac{2}{\eta}\left(\frac{1}{e}\right) < \infty.$$

Theorem 4.7.3 (Translation invariance of error integral)
For any complex number $a = p + iq$ and any positive real number b

$$\int_{-\infty}^{\infty} e^{-b(x+a)^2}\, dx = \frac{1}{\sqrt{b}} \int_{-\infty}^{\infty} e^{-x^2}\, dx = \sqrt{\frac{\pi}{b}}.$$

Proof
We first note that by changing the variable $\sqrt{b}x = y$ and using Theorem
4.7.2, $\int_{-\infty}^{\infty} e^{-bx^2}\, dx$ exists and is equal to $\sqrt{\frac{\pi}{b}}$. We now consider the function

$g(z) = e^{-bz^2}$ which is analytic in the entire complex plane. Take $\gamma = \gamma_0 + \gamma_1 - \gamma_2 + \gamma_3$ where γ_i's are as indicated below.

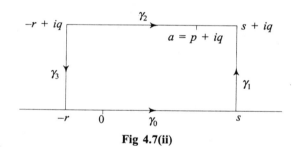

Fig 4.7(ii)

Applying Cauchy's theorem we get that

$$\int_\gamma g(z)\, dz = 0. \qquad (1)$$

On γ_0, $z = x$ with $-r \leq x \leq s$ so

$$\int_{\gamma_0} g(z)\, dz = \int_{-r}^{s} g(x)\, dx. \qquad (2)$$

On γ_1, $z = s + it$ with $0 \leq t \leq q$ and so

$$\left| \int_{\gamma_1} g(z)\, dz \right| \leq M_1 q$$

where $M_1 = \sup\limits_{0 \le t \le q} |g(s+it)| \le \sup\limits_{0 \le t \le q} e^{-b(s^2-t^2)} \le e^{bq^2} e^{-bs^2}$.

Thus $\left| \int\limits_{\gamma_1} g(z)\, dz \right| \le q e^{bq^2} e^{-bs^2} \to 0 \quad \text{as} \quad s \to \infty.$ (3)

Similarly $\left| \int\limits_{\gamma_3} g(z)\, dz \right| \to 0 \quad \text{as } r \to \infty.$ (4)

On γ_2, $z = t + iq$ with $-r \le t \le s$ and so

$$\int\limits_{\gamma_2} g(z)\, dz = \int\limits_{-r}^{s} e^{-b(t+iq)^2}\, dt.$$

Put $x = t - p$ so that $-r-p \le x \le s-p$. We have

$$\int\limits_{\gamma_2} g(z)\, dz = \int\limits_{-r-p}^{s-p} e^{-b(x+p+iq)^2}\, dx = \int\limits_{-r-p}^{s-p} e^{-b(x+a)^2}\, dx.$$

Thus $\int\limits_{\gamma_2} g(z)\, dz \to \int\limits_{-\infty}^{\infty} e^{-b(x+a)^2}\, dx$ as $r, s \to \infty.$ (5)

Using all the informations available from (1) to (5) we see that

$$\int\limits_{-\infty}^{\infty} e^{-bx^2}\, dx - \int\limits_{-\infty}^{\infty} e^{-b(x+a)^2}\, dx = 0.$$

i.e., $\int\limits_{-\infty}^{\infty} e^{-b(x+a)^2}\, dx$ exists and is equal to $\int\limits_{-\infty}^{\infty} e^{-bx^2}\, dx = \sqrt{\dfrac{\pi}{b}}$. This completes

the proof of our theorem.

Theorem 4.7.4
For "a" real with $|a| \le 1$

$$\int\limits_{0}^{\infty} e^{-(1+ia)^2 t^2}\, dt = \frac{1}{2}\, \frac{1-ia}{1+a^2}\, \sqrt{\pi}.$$

Proof
The case $a = 0$ is already dealt with in Theorem 4.7.2.

Case 1 Let $0 < a \le 1$. Let $f(z) = e^{-z^2}$. Take $r > 0$ and integrate $f(z)$ along the closed path which consists of a line from 0 to r along the positive real axis (say γ_1) and the vertical line at r ending at $r(1 + ai)$ (say γ_2) followed by the straight line joining $r(1 + ai)$ to 0 in the complex plane (say γ_3).

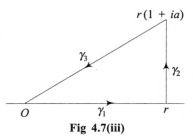

Fig 4.7(iii)

Since e^{-z^2} is analytic every where, in the plane, Cauchy's theorem tells us that

$$\int_{\gamma_1} e^{-z^2} \, dz + \int_{\gamma_2} e^{-z^2} \, dz + \int_{\gamma_3} e^{-z^2} \, dz = 0. \tag{1}$$

We have

$$\int_{\gamma_1} e^{-z^2} \, dz = \int_0^r e^{-t^2} \, dt.$$

Consider γ_2. This line is parameterized by $z = r + it$ where $0 \le t \le ar$.

$$\left| \int_{\gamma_2} f(z) \, dz \right| \le \int_{\gamma_2} |f(z)| \, |dz|$$

$$\le \int_0^{ar} \left| e^{-(r+it)^2} \right| \, dt$$

$$\le e^{-r^2} \int_0^r e^{t^2} \, dt \quad \text{(since } 0 < a \le 1\text{)}$$

$$\le e^{-r^2} \int_0^r e^{rt} \, dt \quad (t^2 = t \cdot t \le tr)$$

$$= \frac{1}{r} - \frac{e^{-r^2}}{r} \le \frac{1}{r}.$$

Thus

$$\int_{\gamma_2} f(z) \, dz \to 0 \ as \ r \to \infty.$$

On γ_3, $z = (1 + ia)t$, $0 \le t \le r$.

$$\int_{\gamma_3} f(z) \, dz = -\int_0^r e^{-(1+ia)^2 t^2} (1 + ia) \, dt$$

$$= -(1 + ia) \int_0^r e^{-(1+ia)^2 t^2} \, dt.$$

From (1)

$$\int_0^r e^{-t^2} \, dt + \int_{\gamma_2} f(z) \, dz - (1 + ia) \int_0^r e^{-(1+ia)^2 t^2} \, dt = 0.$$

Letting $r \to \infty$,

$$\int_0^\infty e^{-t^2} \, dt - (1 + ia) \int_0^\infty e^{-(1+ia)^2 t^2} \, dt = 0.$$

or that

$$\frac{1}{2}\int_{-\infty}^{\infty}e^{-t^2}\,dt = (1+ia)\int_{0}^{\infty}e^{-(1+ia)^2\,t^2}\,dt.$$

Using theorem 4.7.2, we have

$$\int_{0}^{\infty}e^{-(1+ia)^2\,t^2}\,dt = \frac{\sqrt{\pi}}{2}\frac{1}{1+ia} = \frac{\sqrt{\pi}}{2}\frac{1-ia}{1+a^2}.$$

Case 2 $-1 \le a < 0$. Let now $a = -b$ where $0 < b \le 1$. By case 1

$$\int_{0}^{\infty}e^{-(1+ib)^2\,t^2}\,dt = \frac{\sqrt{\pi}}{2}\frac{1-ib}{1+b^2}.$$

Take conjugates on both sides and observe that $\overline{e^z} = e^{\bar{z}}$ and $\overline{\int f\,dt} = \int \bar{f}\,dt$ for complex f (from the definition). We get

$$\int_{0}^{\infty}e^{-(1-ib)^2\,t^2}\,dt = \frac{\sqrt{\pi}}{2}\frac{1+ib}{1+b^2}$$

or that

$$\int_{0}^{\infty}e^{-(1+ia)^2\,t^2}\,dt = \frac{\sqrt{\pi}}{2}\frac{1-ia}{1+a^2}$$

Corollary 4.7.5 (Fresnel Integrals)

$$\int_{0}^{\infty}\cos t^2\,dt = \int_{0}^{\infty}\sin t^2\,dt = \frac{1}{2}\sqrt{\frac{\pi}{2}}.$$

Proof

Taking real and imaginary parts in Theorem 4.7.4 we get

$$\int_{0}^{\infty}e^{(a^2-1)t^2}\cos 2at^2\,dt = \frac{\sqrt{\pi}}{2}\frac{1}{1+a^2}$$

and

$$\int_{0}^{\infty}e^{(a^2-1)t^2}\sin 2at^2\,dt = \frac{\sqrt{\pi}}{2}\frac{a}{1+a^2}$$

for all 'a' with $-1 \le a \le 1$. Substituting $a = 1$ in the above integrals we have

$$\int_{0}^{\infty}\cos 2t^2\,dt = \int_{0}^{\infty}\sin 2t^2\,dt = \frac{\sqrt{\pi}}{4}.$$

Putting $t\sqrt{2} = u$ and changing the variable, we get

$$\int_0^\infty \cos t^2 \, dt = \int_0^\infty \sin t^2 \, dt = \frac{1}{2}\sqrt{\frac{\pi}{2}}.$$

Theorem 4.7.6 (Trigonometric integrals)
Let U denote the open unit disc of the complex plane and ∂U its boundary. Let $R(x, y)$ be a complex-valued rational function. If $z = x + iy$, and R has no poles on ∂U then

$$\int_0^{2\pi} R(\cos \phi, \sin \phi) \, d\phi = 2\pi \sum_{\substack{w \in U \\ w \text{ pole}}} \operatorname*{Res}_{z=w} \tilde{R}(z)$$

where $\tilde{R}(z) = z^{-1} R\left(\dfrac{z + z^{-1}}{2}, \dfrac{z - z^{-1}}{2i}\right)$ and the summation is taken over all

poles w of $\tilde{R}(z)$ inside U.

Proof

For $0 \le \phi \le 2\pi$ and $\zeta = e^{i\phi}$, $\cos\phi = \dfrac{\zeta + \zeta^{-1}}{2}$, $\sin\phi = \dfrac{\zeta - \zeta^{-1}}{2i}$ and $d\phi = \dfrac{d\zeta}{i\zeta}$.

Therefore

$$\int_0^{2\pi} R(\cos\phi, \sin\phi) \, d\phi = \frac{1}{i} \int_{\partial U} R\left(\frac{\zeta + \zeta^{-1}}{2}, \frac{\zeta - \zeta^{-1}}{2i}\right) \zeta^{-1} d\zeta$$

$$= \frac{1}{i} \int_{\partial U} \tilde{R}(\zeta) \, d\zeta$$

$$= 2\pi \sum_{\substack{w \in U \\ w \text{ pole}}} \operatorname*{Res}_{\zeta=w} \tilde{R}(\zeta) \quad \text{(By residue theorem).}$$

Problem 4.7.7

Evaluate $\displaystyle\int_0^{2\pi} \frac{d\phi}{1 - 2a \cos \phi + a^2}$ for $|a| \ne 1$.

Solution

Put $R(x, y) = \dfrac{1}{1 - 2ax + a^2}$ and let U be the open unit disc. Now

$$\tilde{R}(z) = z^{-1} R\left(\frac{z + z^{-1}}{2}, \frac{z - z^{-1}}{2i}\right)$$

$$= z^{-1}\left(\cfrac{1}{1-2a\,\dfrac{1}{2}\left(z+\dfrac{1}{z}\right)+a^2}\right)$$

$$= z^{-1}\left(\cfrac{1}{1-a\left(z+\dfrac{1}{z}\right)+a^2}\right)$$

$$= \frac{1}{(1+a^2)z-az^2-a}$$

$$= \frac{-1}{az^2-(1+a^2)z+a}.$$

The poles are given by $z = a$ or $1/a$. If $|a| < 1$ then a is the only pole inside U and if $|a| > 1$ then $1/a$ is the only pole inside U. The corresponding residues are $\dfrac{1}{(1-a^2)}, \dfrac{-1}{(1-a^2)}$. By theorem 4.7.6 we have

$$\int_0^{2\pi} \frac{d\phi}{1-2a\cos\phi+a^2} = \begin{cases} \dfrac{2\pi}{(1-a^2)} & \text{if } |a|<1 \\[2ex] \dfrac{-2\pi}{(1-a^2)} & \text{if } |a|>1 \end{cases}$$

Remark 4.7.8 (Improper Integrals of First Kind)
Let 'a' be a real number. If $f : [a, \infty) \to \mathbb{C}$ is continuous we define

$$\int_a^\infty f(x)\,dx = \lim_{s\to\infty} \int_a^s f(x)\,dx$$

whenever the limit exists. It is easy to see that

$$\int_a^\infty (f \pm g)\,(x)\,dx = \int_a^\infty f(x)\,dx \pm \int_a^\infty g(x)\,dx$$

and for c complex,

$$\int_a^\infty cf(x)\,dx = c\int_a^\infty f(x)\,dx.$$

Also
$$\int_a^\infty f(x)\,dx = \int_a^b f(x)\,dx + \int_b^\infty f(x)\,dx$$

for every $b > a$. Similarly integrals of the form $\int\limits_{-\infty}^{a} f(x)\ dx$ can be defined and

$$\int\limits_{-\infty}^{\infty} f(x)\ dx = \lim_{r,s\to\infty} \int\limits_{-r}^{s} f(x)\ dx = \int\limits_{-\infty}^{a} f(x)\ dx + \int\limits_{a}^{\infty} f(x)\ dx.$$

It should be noted that r and s should be allowed to tend to ∞ independent of each other. For example $\lim\limits_{r\to\infty} \int\limits_{-r}^{r} f(x)\ dx$ may exist but $\lim\limits_{r,s\to\infty} \int\limits_{-r}^{s} f(x)\ dx$ may not exist as the example $f(x) = x$ demonstrates.

Theorem 4.7.9

If $f : [a,\ \infty) \to \mathbb{C}$ is continuous and if there exists a real $k > 1$ such that $|f(x)\ x^k|$ is bounded then $\int\limits_{a}^{\infty} f(x)\ dx$ exists.

Proof

We first claim that $\int\limits_{a}^{\infty} |f(x)|\ dx$ exists. That is we have to show that $\lim\limits_{s\to\infty} \int\limits_{a}^{s} |f(x)|\ dx$ exists. Since the limits are considered in the real axis it is sufficient to show that for every sequence $s_n \to \infty$, $\int\limits_{a}^{s_n} |f(x)|\ dx$ converges to the same limit. But this happens if and only if each of these sequences is Cauchy and their limits are same. i.e, as $m,\ n \to \infty$ we have to prove that $\int\limits_{s_n}^{s_m} |f(x)|\ dx \to 0$ and if $s_n \to \infty$ and $t_n \to \infty$, $\int\limits_{a}^{s_n} |f(x)|\ dx$ and $\int\limits_{a}^{t_n} |f(x)|\ dx$ converge to the same limit. Let $s_m > s_n$.

Now
$$\left| \int\limits_{s_n}^{s_m} |f(x)|\ dx \right| \le \int\limits_{s_n}^{s_m} \frac{M}{x^k}\ dx \quad \text{(by hypothesis)}$$

$$= \frac{M}{k-1} \left[\frac{1}{s_n^{k-1}} - \frac{1}{s_m^{k-1}} \right] \to 0 \text{ as } s_n,\ s_m \to \infty.$$

Similarly as $n \to \infty$

$$\left| \int\limits_{a}^{s_n} |f(x)|\ dx - \int\limits_{a}^{t_n} |f(x)|\ dx \right| \le \left| \int\limits_{t_n}^{s_n} |f(x)|\ dx \right|$$

$$\le \frac{M}{k-1}\left[\frac{1}{t_n^{k-1}} + \frac{1}{s_n^{k-1}}\right] \to 0. \text{ Thus } \int_a^{s_n}|f(x)|\,dx \text{ and } \int_a^{t_n}|f(x)|\,dx \text{ tend to the}$$

same limit as $n \to \infty$. Using similar arguments it is also easy to see that if

$0 \le f \le g$ and $\int_a^\infty g(x)\,dx$ exists then $\int_a^\infty f(x)\,dx$ also exists. Next $0 \le f(x) +$

$|f(x)| \le 2\,|f(x)|$ $(a \le x < \infty)$. By what we have proved $\int_a^\infty 2\,|f(x)|\,dx$ exists

and therefore $\int_a^\infty (f(x) + |f(x)|)\,dx$ exists. Hence

$$\int_a^\infty (f(x) + |f(x)|)\,dx - \int_a^\infty |f(x)|\,dx = \int_a^\infty f(x)\,dx$$

also exists.

Theorem 4.7.10
Let $p(z)$, $q(z)$ be complex polynomials of degree m, n respectively. Then there are positive real numbers K, L, R such that

$$K\,|z|^{m-n} \le \left|\frac{p(z)}{q(z)}\right| \le L\,|z|^{m-n}, \ \forall\ z \in \mathbb{C} \text{ with } |z| \ge R.$$

Proof

Let $p(z) = \sum_{k=0}^m a_k z^k$. Set $r(z) = \sum_{k=0}^{m-1} |a_k|\,|z^k|$. Then by triangle inequality

$$|a_m|\,|z^m| - r(z) \le |p(z)| \le |a_m|\,|z^m| + r(z). \tag{1}$$

If $|z| \ge 1$ and $k < m$ we have $|z|^k \le |z|^{m-1}$ and so $r(z) \le M|z|^{m-1}$ where $M = \sum_{k=0}^{m-1}|a_k|$. Therefore (1) implies

$$|a_m|\,|z^m| - M|z|^{m-1} \le |p(z)| \le |a_m|\,|z^m| + M|z|^{m-1}.$$

Therefore $\frac{1}{2}|a_m|\,|z^m| \le |p(z)| \le 2|a_m|\,|z^m|$ for all z such that $|z| \ge R_1 =$

max $\{1, 2M/|a_m|\}$. i.e., There exist constants $K_1 = \frac{1}{2}\,|a_m|$ and $L_1 = 2\,|a_m|$

such that $K_1\,|z^m| \le |p(z)| \le L_1\,|z^m|$ for $|z| \ge R_1$. In the same manner there are constants K_2, L_2 and R_2 such that $K_2\,|z^n| \le |q(z)| \le L_2\,|z^n|$ for $|z| \ge R_2$. Choosing $R = $ max $\{R_1, R_2\}$ and $K = K_1 L_2^{-1}$ and $L = L_1 K_2^{-1}$ we have

$$K|z|^{m-n} \le \left|\frac{p(z)}{q(z)}\right| \le L|z|^{m-n}, \ \forall \ z \in \mathbb{C} \text{ with } |z| \ge R.$$

Corollary 4.7.11

If $f(z) = \dfrac{p(z)}{q(z)}$ with deg $p(z) = m$, deg $q(z) = n$ and $n = m + l$ then

$$\lim_{z \to \infty} z^k f(z) = 0 \text{ for every real } k \text{ with } 0 \le k < l.$$

Proof
$|z|^k |f(z)| \le L |z|^{k-l}$ (by Theorem 4.7.10).

Theorem 4.7.12

Let H denote the upper half plane. Let D be a region containing $\overline{H} = H \cup \mathbb{R}$. Let $\Gamma(r) : [0, \pi] \to \overline{H}$ be the curve $\theta \to re^{i\theta}$ (i.e., the closed upper semi-circle of the disc $|z| = r$). Let f be analytic in D except possibly for finitely many poles none of which lies on the real axis. Suppose that $\displaystyle\int_{-\infty}^{\infty} f(x) \ dx$ exists and $\lim_{z \to \infty} zf(z) = 0$ then

$$\int_{-\infty}^{\infty} f(x) \ dx = 2\pi i \sum_{\substack{w \in H \\ w \text{ pole}}} \operatorname{Res}_{z=w} f(z)$$

where the sum is taken over all poles $w \in H$.

Proof
Since the number of poles of f is finite in H, we can choose r large enough so that all the poles of f lie inside the semi-circular region bounded by $\Gamma(r)$. For any such r it follows from the Residue theorem that

Fig 4.7(iv)

$$\int_{-r}^{r} f(x) \ dx + \int_{\Gamma(r)} f(\zeta) \ dz = 2\pi i \sum_{\substack{w \in H \\ w \text{ pole}}} \operatorname{Res}_{z=w} f(z). \qquad (1)$$

Put $\sup_{\zeta \in \Gamma(r)} |f(\zeta)| = M(r)$. Therefore $\left|\displaystyle\int_{\Gamma(r)} f(\zeta) \, d\zeta\right| \le M(r) \, \pi r$. By assumption

$\lim_{z \to \infty} zf(z) = 0$. Thus given $\varepsilon > 0$, there exists r such that $|zf(z)| < \varepsilon$ for $|z| \ge r$. Therefore for r large enough $M(r) \le \dfrac{\varepsilon}{r}$. i.e., $rM(r) < \varepsilon$. Therefore

$$\left| \int\limits_{\Gamma(r)} f(\zeta) \, d\zeta \right| < \pi\varepsilon. \text{ Since } \varepsilon \text{ is arbitrary it follows that } \int\limits_{\Gamma(r)} f(\zeta) \, d\zeta \to 0 \text{ as}$$

$r \to \infty$. Allowing r to tend to ∞ in (1) our theorem follows.

Corollary 4.7.13

If $f(z) = \dfrac{p(z)}{q(z)}$ is a rational function in one complex variable z with the

degree of q exceeding that of p by at least two units and if q has no real zeros then

$$\int\limits_{-\infty}^{\infty} f(x) \, dx = 2\pi i \sum_{\substack{w \in H \\ w \text{ pole}}} \operatorname*{Res}_{z=w} f(z)$$

where the sum is taken over all the finite poles of f in the upper half plane.

Proof

$f(z)$ satisfies the hypothesis of Theorem 4.7.12 in view of Thorem 4.7.9 and Corollary 4.7.11 (We can take $1 < k < 2$ and $k = 1$ successively).

Problem 4.7.14

$\int\limits_{0}^{\infty} \dfrac{x^{m-1}}{1+x^{n}} \, dx = \dfrac{\pi}{n} \left[\sin \dfrac{m\pi}{n} \right]^{-1}$ where m and n are integers, $0 < m < n$; m is

odd and n is even.

Solution

$$\int\limits_{0}^{\infty} \frac{x^{m-1}}{1+x^{n}} \, dx = \frac{1}{2} \int\limits_{-\infty}^{\infty} \frac{x^{m-1}}{1+x^{n}} \, dx. \text{ Let } f(z) = \frac{z^{m-1}}{1+z^{n}}.$$

$$\int\limits_{-\infty}^{\infty} |f(x)| \, dx = 2\int\limits_{0}^{\infty} |f(x)| \, dx \le 2\left(\int\limits_{0}^{1} dx + \int\limits_{1}^{\infty} x^{m-n-1} \, dx \right)$$

$$\left(\text{Note that } 0 < x \le 1 \Rightarrow \frac{x^{m-1}}{1+x^{n}} \le 1, \text{ and for } x \ge 1 \; \frac{x^{m-1}}{1+x^{n}} \le \frac{x^{m-1}}{x^{n}} \right.$$

$= x^{m-n-1} \Bigg)$. Thus $\int\limits_{-\infty}^{\infty} |f(x)| \, dx \le 2 \left(1 + \dfrac{1}{n-m} \right)$ which is finite.

$$\lim_{z\to\infty} z f(z) = \lim_{z\to\infty} \frac{z^{m}}{1+z^{n}} = 0.$$

Thus the hypotheses of Theorem 4.7.12 are satisfied. We now observe that

all the poles of f in H are given by $z_{k} = c^{2k+1}$ $(c = e^{i\pi/n})$ and $0 \le k \le \left(\dfrac{n}{2} - 1 \right)$.

This is because all the n-th roots of (-1) are given by $e^{i(\pi+2k\pi)/n}$, $k = 0$, 1, 2, . . . , $n - 1$. Since the successive roots are equally spaced making an angle $\dfrac{2\pi}{n}$ at the origin, the poles in the upper half plane are given by those $k \geq 0$ for which $\left(\dfrac{2k+1}{n}\right)\pi < \pi$, i.e., $k = 0$, 1, ..., $\left(\dfrac{n}{2}-1\right)$. Further each of these poles are simple and the residues are given by

$$\operatorname*{Res}_{z=z_k} f(z) = \frac{(z_k)^{m-1}}{n(z_k)^{n-1}} = \frac{1}{n}z_k^{m-n} = -\frac{1}{n}z_k^{m} \text{ (since } z_k^{-n} = -1)$$

$$\left(\text{Note that } (1 + z^n) = \prod_1^n (z - z_i) = (z - z_k) \prod_{i \neq k} (z - z_i) \text{ and so } \prod_{i \neq k} (z_k - z_i)\right.$$

$$\left. = \frac{d}{dz}(1 + z^n)|_{z=z_k}\right).$$

[For calculating residues of rational functions with simple poles we shall hereafter use this technique without any further mention]. Now using Theorem 4.7.12 we get

$$\int_0^\infty \frac{x^{m-1}}{1+x^n}\,dx = \frac{1}{2}\int_{-\infty}^\infty \frac{x^{m-1}}{1+x^n}\,dx = \pi i \sum_{k=0}^{\frac{n}{2}-1}\left(-\frac{1}{n}z_k^{m}\right) = -\frac{\pi i}{n}c^m \sum_{k=0}^{\frac{n}{2}-1}c^{2mk}$$

$$= -\frac{\pi i}{n}c^m \frac{c^{mn}-1}{c^{2m}-1} = -\frac{\pi i}{n}\left(\frac{(-1)^m-1}{c^m-c^{-m}}\right)$$

$$= \frac{\pi}{n}\left(\frac{e^{im\pi/n}-e^{-im\pi/n}}{2i}\right)^{-1} = \frac{\pi}{n}\left(\sin\frac{m\pi}{n}\right)^{-1}$$

(Here we have used the fact that m is odd and $c = e^{i\pi/n}$).

Note 4.7.15

We shall now consider improper integrals of the form $\int\limits_{-\infty}^{\infty} g(x)\, e^{iax}\, dx$ where a is real and g satisfies certain conditions. Integrals of this type are called Fourier transforms of g, when viewed as functions of a. In case $g(x)$ is a rational function having a zero of at least order 2 at ∞ (degree of the denominator is greater than that of the numerator by at least 2 units), the method similar to the proof of Corollary 4.7.11 can be easily adapted. On the other hand it is also true that similar result holds where $g(z) = R(z)$, a rational function, has only a simple zero at ∞. However in this case it is not convenient to use semicircles. For one thing it is not easy to estimate the integral over the semicircle and secondly, even if we were successful, we would only have proved that the $\lim\limits_{\rho\to\infty} \int\limits_{-\rho}^{\rho} R(x)\, e^{iax}\, dx$ exists. In case of 'a double zero' at ∞ it actually does not matter because the integral in question is known to be convergent and this limit is the same for all ways of approach to ∞. On the other hand, in the case of simple zero at ∞ we are not assured of the convergence of the integral and so we have to prove that

$$\lim_{r,s\to\infty} \int_{-r}^{s} R(x)\, dx \text{ exists and evaluate it also.}$$

Theorem 4.7.16
Let g be analytic in $\mathbb{C}$ except for finitely many poles none of which is real. Assume that $\lim\limits_{z\to\infty} g(z) = 0$ ($g(z)$ can have a simple zero at ∞). If H denotes the upper half plane and H^- denotes the lower half plane then

$$\int_{-\infty}^{\infty} g(x)\, e^{iax}\, dx = \begin{cases} 2\pi i \sum\limits_{\substack{w\in H \\ w,\, \text{pole}}} \operatorname*{Res}_{z=w} g(z) e^{iaz} & (a>0) \\[2em] -2\pi i \sum\limits_{\substack{w\in H^- \\ w,\, \text{pole}}} \operatorname*{Res}_{z=w} g(z) e^{iaz} & (a<0) \end{cases}$$

Proof
Let $a > 0$. Choose r, s large enough so that all the poles of g in H lie inside the square Q with vertices $-r$, s, $s + iq$, $-r + iq$ where r and s are positive and $q = r + s$. We claim that $I_k = \int\limits_{\gamma_k} g(\zeta)\, e^{ia\zeta}$ $d\zeta$ satisfies $\lim\limits_{r,s\to\infty} I_k = 0$ for $k = 1, 2, 3,$

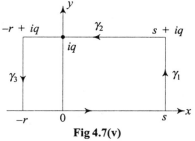

Fig 4.7(v)

The above figure is self-explanatory. If this claim is proved our theorem follows (for the case $a > 0$) as an immediate application of residue theorem. On γ_1 we have $z = s + it$, $0 \leq t \leq q$. Therefore

$$|I_1| \leq \int_0^q |g(s + it)|\, e^{-at}\, dt \leq \alpha \int_0^q e^{-at}\, dt = \alpha\left(\frac{1 - e^{-aq}}{a}\right) \leq \frac{\alpha}{a} \quad \text{where}$$

$\alpha = \sup\limits_{\zeta\in\gamma_1} |g(\zeta)|$. On γ_2, $\zeta = t + iq$, $-r \leq t \leq s$ and $|e^{ia\zeta}| = e^{-aq}$. Thus we have

$$|I_2| \leq \int_{-r}^{s} |g(t + iq)|\, e^{-aq}\, dt \leq \beta e^{-aq}\, (r + s) = \beta e^{-aq}\frac{aq}{a} \leq \frac{\beta}{a} \quad \text{where}$$

$\beta = \sup\limits_{\zeta \in \gamma_2} |g(\zeta)|$ (since $ye^{-y} \leq 1$ for $y > 0$). Similarly $|I_3| \leq \dfrac{\sigma}{a}$ where

$\sigma = \sup\limits_{\zeta \in \gamma_3} |g(\zeta)|$. Now by our hypothesis $\lim\limits_{z \to \infty} g(z) = 0$. Therefore α, β, σ

$\to 0$ as r, $s \to \infty$. Hence the claim is proved, and the proof of the case $a > 0$ is over.

In case $a < 0$ we consider a similar square in the lower half plane and estimate the integrals in the same way and again note that $aq > 0$ since $a < 0$ and $q = -(r + s) < 0$. In this case the negative sign appears in the answer because of the following reason. The chosen rectangle should be given the anti-clockwise orientation but then the real integral in the limit

becomes $\displaystyle\int\limits_{\infty}^{-\infty} g(x)\, e^{iax} dx = -\int\limits_{-\infty}^{\infty} g(x)\, e^{iax} \, dx.$

Corollary 4.7.17

In Theorem 4.7.16 assume that g is real on the real axis with the other hypotheses unchanged. Then

$$\int\limits_{-\infty}^{\infty} g(x)\cos ax\, dx = \left[\begin{array}{l} -2\pi\, \mathrm{Im}\!\left(\sum\limits_{\substack{w \in H \\ w\ \text{pole}}} \operatorname*{Res}\limits_{z=w} g(z)e^{iaz} \right)(a > 0) \\[2em] 2\pi\, \mathrm{Im}\!\left(\sum\limits_{\substack{w \in H^- \\ w\ \text{pole}}} \operatorname*{Res}\limits_{z=w} g(z)e^{iaz} \right)(a < 0) \end{array}\right.$$

$$\int\limits_{-\infty}^{\infty} g(x)\sin ax\, dx = \left[\begin{array}{l} 2\pi\, \mathrm{Re}\!\left(\sum\limits_{\substack{w \in H \\ w\ \text{pole}}} \operatorname*{Res}\limits_{z=w} g(z)e^{iaz} \right)(a > 0) \\[2em] -2\pi\, \mathrm{Re}\!\left(\sum\limits_{\substack{w \in H^- \\ w\ \text{pole}}} \operatorname*{Res}\limits_{z=w} g(z)e^{iaz} \right)(a < 0) \end{array}\right.$$

Proof
Take real and imaginary parts in Theorem 4.7.16.

Problem 4.7.18
Show that

$$\int\limits_{-\infty}^{\infty} \frac{e^{iax}}{x - ib}\, dx = 2\pi i\, e^{-ab}, \quad \int\limits_{-\infty}^{\infty} \frac{e^{iax}}{x + ib}\, dx = 0$$

where $a > 0$, $b \in \mathbb{C}$ with $\mathrm{Re}\, b > 0$.

Proof

Take $g(z) = \dfrac{1}{z \pm ib}$ and apply Theorem 4.7.16. In the first integral the only

simple pole in H is $z = ib$ and the corresponding residue is $e^{(ia)\,(ib)} = e^{-ab}$.

In the second integral the only pole is in lower half plane and no pole in the upper half plane. The result follows.

Problem 4.7.19

Prove that

$$\int_0^\infty \frac{b \cos ax}{x^2 + b^2}\, dx = \int_0^\infty \frac{x \sin ax}{x^2 + b^2}\, dx = \frac{\pi}{2}e^{-ab}$$

if a, b are real and positive.

Solution

Note that $b\cos ax$, $x\sin ax$ and $x^2 + b^2$ are even and hence

$$\int_0^\infty \frac{b \cos ax}{x^2 + b^2}\, dx = \frac{1}{2}\int_{-\infty}^\infty \frac{b \cos ax}{x^2 + b^2}\, dx$$

and

$$\int_0^\infty \frac{x \sin ax}{x^2 + b^2}\, dx = \frac{1}{2}\int_{-\infty}^\infty \frac{x \sin ax}{x^2 + b^2}\, dx .$$

We now apply Corollary 4.7.17 with $g(x) = \dfrac{b}{x^2 + b^2}$ and $\dfrac{x}{x^2 + b^2}$ successively. In both the cases the corresponding functions $g(z)$ have only one simple pole at $z = ib \in H$ with residues $\dfrac{-i}{2}e^{-ab}$ and $\dfrac{1}{2}e^{-ab}$ respectively.

Note 4.7.20

In Theorem 4.7.16 we have assumed that $\lim\limits_{z\to\infty} g(z) = 0$ and $g(z)$ has no poles on the real axis. However if $g(z)$ has a simple pole on the real axis (say $z = \alpha$), then

$\int_{-\infty}^\infty g(x)\, e^{iax}\, dx$ can be defined as follows:

$$\int_{-\infty}^\infty g(x)\, e^{iax}\, dx = \lim_{\delta\to 0}\left\{\int_{-\infty}^{\alpha-\delta} g(x)e^{iax}\, dx + \int_{\alpha+\delta}^\infty g(x)e^{iax}\, dx\right\}.$$

This way of defining the integral, provided the limit exists, is called Cauchy principal value (denoted by P.V.) and calls for evaluation. This method can be generalized when the number of poles on the real axis is finite and the principal values exist at all these points. We shall illustrate this procedure by proving the following theorem.

Theorem 4.7.21

Let $q(z)$ be a rational function with $\lim\limits_{z\to\infty} q(z) = 0$. Let $q(z)$ have a simple pole at $z = 0$ and no other pole on the real axis. Then

$$P.V. \int_0^\infty q(x)\, e^{iax}\, dx = 2\pi i \sum_{\substack{w \text{ pole} \\ w\in H}} \operatorname*{Res}_{z=w}\, (q(z)\, e^{iaz}) + \pi i \operatorname*{Res}_{z=0}(q(z)\, e^{iaz}).$$

Proof

Consider the contour described here where x_1, x_2, $y > 0$ and are sufficiently large to include all the poles of $q(z)$ in the upper half plane. We avoid the origin by using a semicircle γ around the origin of radius δ as shown in the figure. Let the residue of $q(z)$ e^{iaz} at the origin be equal to B. Then by residue theorem (as before the integrals over γ_i's are zero in the limit).

Fig 4.7(vi)

$$\int_{-\infty}^{-\delta} q(x) \, e^{iax} \, dx + \int_{\gamma} q(z) \, e^{iaz} \, dz + \int_{\delta}^{\infty} q(x) \, e^{iax} \, dx$$

$$= 2\pi i \left(\sum_{\substack{w \text{ pole} \\ w \in H}} \operatorname*{Res}_{z=w} (q(z)e^{iaz}) + B \right). \qquad (1)$$

We shall now estimate $\int_{\gamma} q(z) \, e^{iaz} \, dz$. Since the residue of $q(z)$ e^{iaz} at $z = 0$

is B, we can write $q(z) \, e^{iaz} = \dfrac{B}{z} + R_0\,(z)$ where $R_0(z)$ is analytic in a neighbourhood of the origin. Choosing $\delta > 0$ small enough, this representation can be made available in the closed δ-neighbourhood enclosed by γ. Thus

$$\int_{\gamma} q(z) \, e^{iaz} \, dz = \int_{\gamma} \frac{B}{z} dz + \int_{\gamma} R_0(z) \, dz,$$

$$\int_{\gamma} \frac{B}{z} \, dz = i \int_{-\pi}^{0} \frac{B}{\delta e^{i\theta}} \delta e^{i\theta} \, d\theta = \pi Bi.$$

Note that $\left| \int_{\gamma} R_0(z)\, dz \right| \leq M\pi\delta$ for a suitable fixed M since $R_0(z)$

is bounded in a fixed neighbourhood of 0 and as $\delta \to 0$, $\int_{\gamma} R_0(z) \, dz \to 0$.

Thus allowing δ to tend to zero in (1) we get $P.V.\displaystyle\int_{-\infty}^{\infty} q(x) \, e^{iax} \, dx =$

$2\pi i \displaystyle\sum_{\substack{w \text{ pole} \\ w \in H}} \operatorname*{Res}_{z=w} (q(z) \, e^{iaz}) + \pi i B.$

Corollary 4.7.22

Let $q(z)$ be a rational function such that $\lim\limits_{z \to \infty} q(z) = 0$. Let $q(z)$ have finitely many simple poles on the real axis. Then

$$P.V. \int\limits_{-\infty}^{\infty} q(x)\, e^{iax}\, dx = 2\pi i \sum\limits_{\substack{w\ \text{pole} \\ w \in H}} \operatorname*{Res}_{z=w} q(z)\, e^{iaz} + \pi i \sum\limits_{\substack{w\ \text{real} \\ w\ \text{pole}}} \operatorname*{Res}_{z=w} q(z)\, e^{iaz}.$$

Proof
The same method of proof as in Theorem 4.7.20. (But now applied to all the finite number of poles on the real axis). Note that allowing simple poles on the real axis amounts to including one half of the residues.

Problem 4.7.23
Show that

$$P.V. \int\limits_{-\infty}^{\infty} \frac{e^{ix}}{x}\, dx = \pi i, \quad \int\limits_{0}^{\infty} \frac{\sin x}{x}\, dx = \frac{\pi}{2}.$$

Solution
Apply Theorem 4.7.21 and note that origin is the only pole on the real axis with residue 1 and there are no poles in H for $q(z)e^{iz} = \dfrac{e^{iz}}{z}$. The other hypothesis is easily checked. Further we observe that $\dfrac{\sin x}{x}$ is an even function and therefore

$$\int\limits_{0}^{\infty} \frac{\sin x}{x}\, dx = P.V. \frac{1}{2} \int\limits_{-\infty}^{\infty} \frac{\sin x}{x}\, dx = \operatorname{Im}\left(P.V. \frac{1}{2} \int\limits_{-\infty}^{\infty} \frac{e^{ix}}{x}\, dx \right) = \frac{\pi}{2}.$$

Definition 4.7.24
Let $a \in \mathbb{C}$. $z = |z|e^{i\phi}$ be a non-zero complex number where $0 \le \phi < 2\pi$. We define $\log z = \log |z| + i\phi$ and $z^a = e^{a \log z}$. In what follows we shall denote this unique value of ϕ as arg z.

Lemma 4.7.25
With the notations of definition 4.7.24 we have $\lim\limits_{\varepsilon \to 0} (x + i\varepsilon)^a = x^a$ and $\lim\limits_{\varepsilon \to 0}$ $(x - i\varepsilon)^a = x^a\, e^{2\pi i a}$ for every $a \in \mathbb{C}$ with the limit being uniform for x varying over a compact interval on the positive real axis.

Proof
As x varies over the compact interval I on the positive real axis and $\varepsilon > 0$ is small, $(x + i\varepsilon)$ lies in the first quadrant i.e., $\phi = \arg (x + i\varepsilon)$ satisfies

$0 < \phi < \dfrac{\pi}{2}$. In $\left(0, \dfrac{\pi}{2}\right)$, $\theta \to \tan \theta$ is one-to-one, onto and continuous with

inverse $g(x)$ defined from $(0, \infty)$ onto $\left(0, \dfrac{\pi}{2}\right)$ so that $\tan (g(x)) = x$, $\forall \, x$

$\in (0, \infty)$. Now $\tan (\arg (x + i\varepsilon)) = \dfrac{\varepsilon}{x} = \tan \left(g\left(\dfrac{\varepsilon}{x}\right)\right)$. Hence the unique

argument $\phi = \arg (x + i\varepsilon) = g\left(\dfrac{\varepsilon}{x}\right)$ (Since $\theta \to \tan \theta$ is one-to-one). Since

I is a compact interval in $\mathbb{R}^+ = (0, \infty)$ we have $x \geq \delta > 0 \; \forall \, x \in I$. Further
g is also strictly increasing, being the inverse of the strictly increasing

function $\theta \to \tan \theta$. Therefore $\left|g\left(\dfrac{\varepsilon}{x}\right)\right| = g\left(\dfrac{\varepsilon}{x}\right) \leq g\left(\dfrac{\varepsilon}{\delta}\right)$. We now claim that

$g\left(\dfrac{\varepsilon}{\delta}\right) \to 0$ as $\varepsilon \to 0$. This is equivalent to saying that whenever $\varepsilon_n \to 0$,

$g(\varepsilon_n/\delta) \to 0$. If possible let $u \neq 0$ be a limit point of $\{g(\varepsilon_n/\delta)\}$. Necessarily

$u \in \left(0, \dfrac{\pi}{2}\right]$. Then there exists a subsequence $\left(\varepsilon_{n_k}\right)$ of $\{\varepsilon_n\}$ such that

$g\left(\varepsilon_{n_k}/\delta\right) \to u$. Thus by continuity of $\tan \theta$, $\left(\varepsilon_{n_k}/\delta\right) \to \tan u$ as $k \to \infty$.

But $\left(\varepsilon_{n_k}/\delta\right) \to 0 \Rightarrow \tan u = 0$ a contradiction to $u \neq 0$. Hence $\{g(\varepsilon_{n_k}/\delta)\}$
has no non-zero limit point. But this sequence being bounded $\Big($Note that 0

$< g(x) < \dfrac{\pi}{2}\Big)$ the only limit point must be 0. Therefore $g(\varepsilon_n/\delta) \to 0$ as n

$\to \infty$ and so $g\left(\dfrac{\varepsilon}{\delta}\right) \to 0$ as $\varepsilon \to 0$ or that $g\left(\dfrac{\varepsilon}{x}\right) \to 0$ uniformly on I. Now

$\log (x + i\varepsilon) = \dfrac{1}{2}\log (x^2 + \varepsilon^2) + ig\left(\dfrac{\varepsilon}{x}\right)$. Thus $\lim\limits_{\varepsilon \to 0} \log (x + i\varepsilon) = \log x$ the
limit being uniform. We know that

$$0 < \arg (x + i\varepsilon) < 2\pi, \; 0 < \arg (x - i\varepsilon) < 2\pi. \qquad (1)$$

But inequalities (1) force us to conclude that

$$0 < \arg (x + i\varepsilon) + \arg (x - i\varepsilon) < 4\pi.$$

But $\arg (x + i\varepsilon) + \arg (x - i\varepsilon)$ is equal to one of the values of $\arg (x^2 + \varepsilon^2)$
which are $0, \pm 2\pi, \pm 4\pi$, etc. Thus

$$\arg (x - i\varepsilon) = 2\pi - \arg (x + i\varepsilon).$$

By what we have already proved $\arg (x + i\varepsilon) \to 0$ uniformly on I as
$\varepsilon \to 0$. Hence $\arg (x - i\varepsilon) \to 2\pi$ uniformly on I as $\varepsilon \to 0$. Further

$\log(x - i\varepsilon) = \frac{1}{2}\log(x^2 + \varepsilon^2) + i \arg(x - i\varepsilon)$. Thus $\lim\limits_{\varepsilon \to 0} \log(x - i\varepsilon)$
$= \log x + i2\pi$. Hence

$$\lim_{\varepsilon \to 0}(x + i\varepsilon)^a = \lim_{\varepsilon \to 0} e^{a \log(x + i\varepsilon)} = e^{a \log x} = x^a$$

and

$$\lim_{\varepsilon \to 0}(x - i\varepsilon)^a = \lim_{\varepsilon \to 0} e^{a \log(x - i\varepsilon)} = e^{a (\log x + 2\pi i)} = x^a e^{2\pi ai}.$$

Theorem 4.7.26
Let $q(z)$ be analytic in the plane except for finitely many poles, none of which lies on the non-negative real axis. Let $a \in \mathbb{C}\backslash\mathbb{Z}$. Assume
(i) $\lim\limits_{z \to 0} q(z)\, z^a = 0$ (ii) $\lim\limits_{z \to \infty} q(z)\, z^a = 0$ where z^a is given by Definition 4.7.24. Then

$$\int_0^\infty q(x)\, x^{a-1}\, dx = \frac{2\pi i}{1 - e^{2\pi ia}}\left\{ \sum_{w \text{ pole}} \operatorname*{Res}_{z=w} q(z) z^{a-1} \right\}.$$

Proof
We first observe that z^a as a function of z is analytic in $\tilde{C} = \mathbb{C}\backslash[0, \infty)$ (this is similar to the construction of the principal branch for $\log z$). We consider a contour γ in $\tilde{C}$ described as follows:

$\gamma = \gamma_1 + \gamma_2 + \gamma_3 + \gamma_4$ where γ_i's are as shown in the adjacent diagram. Choose ε, r so small and s so large that the region G bounded by γ contains all the poles of q. Since γ is simple and closed

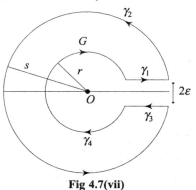

Fig 4.7(vii)

$$\int_\gamma q(\zeta)\, \zeta^{a-1}\, d\zeta = 2\pi i \sum_{\substack{w \in \mathbb{C} \\ w \text{ pole}}} \operatorname*{Res}_{z=w} q(z)\, z^{a-1}$$

$$\int_{\gamma_1} q(\zeta)\, \zeta^{a-1}\, d\zeta = \int_r^s q(x + i\varepsilon)\, (x + i\varepsilon)^{a-1}\, dx.$$

By Lemma 4.7.25, $\lim\limits_{\varepsilon \to 0} \int_{\gamma_1} q(\zeta)\, \zeta^{a-1}\, d\zeta = \int_r^s q(x)\, x^{a-1}\, dx$. Similarly $\lim\limits_{\varepsilon \to 0} \int_{\gamma_3} q(\zeta)$

$\zeta^{a-1}\, d\zeta = -e^{2\pi ia} \int_r^s q(x)\, x^{a-1}\, dx$. Also $\left| \int_{\gamma_2} q(\zeta)\, \zeta^{a-1} d\zeta \right| \leq \alpha 2\pi$ where $\alpha = \sup\limits_{\zeta \in \gamma_2}$

$|q(\zeta)\, \zeta^a|$

Therefore $\lim\limits_{\substack{s\to\infty \\ \varepsilon\to 0}} \int_{\gamma_2} q(\zeta)\ \zeta^{a-1}\ d\zeta = 0$ (Actually $\alpha \to 0$ as $s \to \infty$ by (ii)).

Similarly $\left|\int_{\gamma_4} q(\zeta)\zeta^{a-1}d\zeta\right| \le 2\pi\beta$ where $\beta = \sup\limits_{\zeta\in\gamma_4} |q(\zeta)\ \zeta^a| \to 0$ as $r \to 0$

by hypothesis (i).

Therefore $\lim\limits_{\varepsilon, r\to 0} \int_{\gamma_4} q(\zeta)\ \zeta^{a-1}\ d\zeta = 0$. Putting all these informations to-

gether and allowing $\varepsilon \to 0$, $r \to 0$ and $s \to \infty$, we get the theorem.

Corollary 4.7.27

Let q be a rational function with no pole on the non-negative real axis. Suppose that the degree of the denominator of q is strictly greater than that of the numerator. Then

$$\int_0^\infty q(x)\ x^{a-1}\ dx = \frac{-\pi e^{-\pi i a}}{\sin \pi a}\left\{\sum_{\substack{w\in\mathbb{C} \\ w\ \text{pole}}} \operatorname*{Res}_{z=w} q(z)\ z^{a-1}\right\} \forall\ a \in \mathbb{C}\ (0 < \operatorname{Re} a < 1)$$

Proof

Consider $|z^a|\ |q(z)|$, for the given "a".

$$|z^a| = |e^{a\log z}| \le |e^{a(\log|z| + i\ \arg z)}|\ \text{where } 0 \le \arg z < 2\pi$$

$$\le |z|^{\operatorname{Re} a}\ e^{2\pi|\operatorname{Im} a|}.$$

Thus $|z^a|\ |q(z)| \le e^{2\pi|\operatorname{Im} a|}|z|^{\operatorname{Re}\ a}\ |q(z)|$. Now $q(z)$ is analytic at $z = 0$. Therefore $\lim\limits_{z\to 0}|z^a q(z)| = \lim\limits_{z\to 0}|z^a|\ |q(z)| = 0$. Further the condition that the degree of the denominator polynomial of q is strictly greater than that of the numerator polynomial of q implies $|q(z)| \le \dfrac{M'}{|z|}$ for $|z|$ large.

$$\lim\limits_{z\to\infty} |z^a|\ |q(z)| \le \lim\limits_{z\to\infty}\frac{M}{|z|^{1-\operatorname{Re} a}} = 0$$

with $M = M'\ e^{2\pi|\operatorname{Im} a|}$. Therefore all the conditions of the Theorem 4.7.26 are satisfied. The conclusion follows by observing that

$$\frac{2\pi i}{1 - e^{2\pi i a}} = \frac{-\pi e^{-\pi i a}}{\sin \pi a}.$$

Problem 4.7.28

Compute $\int_0^\infty \dfrac{x^{a-1}}{x + e^{i\phi}}dx$ where 'a' is real with $0 < a < 1$ and $|\phi| < \pi$.

Solution

Consider $q(z) = \dfrac{1}{z + e^{i\phi}}$ which has a pole of order 1 at $z = -e^{i\phi}$ and has

no other poles. This pole is not on the non-negative real axis because $-e^{i\phi}$ $\neq 0$ and is real and positive if and only if $\phi = (2n + 1)\,\pi$ which are excluded. Further

$$\operatorname*{Res}_{z=-e^{i\phi}} q(z)\, z^{a-1} = (-e^{i\phi})^{(a-1)} = e^{i(\phi+\pi)\,(a-1)}$$

$$= -e^{i(a-1)\phi}\, e^{i\pi a}.$$

Hence from our Corollary 4.7.27 we have

$$\int\limits_0^\infty x^{a-1} q(x)\,dx = \frac{-\pi e^{-\pi i a}}{\sin \pi a}\left(-e^{i\phi(a-1)} e^{i\pi a}\right) = \frac{\pi e^{i\phi(a-1)}}{\sin \pi a}.$$

Problem 4.7.29

Compute $\displaystyle\int\limits_0^\infty \frac{x^{a-1}\,dx}{x+1}$ for $0 < a < 1$.

In the solution of problem 4.7.28 putting $\phi = 0$ we get $\displaystyle\int\limits_0^\infty \frac{x^{a-1}}{x+1}\,dx = \frac{\pi}{\sin \pi a}$

$(0 < a < 1)$. On the other hand using the well-known Gamma and Beta functions, the above integral represents $\Gamma(a)\,\Gamma(1-a) = \beta(a, 1-a) = \displaystyle\int\limits_0^1 t^{a-1}$

$(1-t)^{-a}\,dt$. Putting $\dfrac{t}{1-t} = x$ we get $\Gamma(a)\,\Gamma(1, a) = \displaystyle\int\limits_0^\infty \frac{x^{a-1}}{x+1}\,dx = \frac{\pi}{\sin \pi a}$.

Problem 4.7.30

Compute $\displaystyle\int\limits_0^\infty \frac{x^{m-1}}{x^n + e^{i\phi}}\,dx$ where m, n are positive integers with $m < n$ and $|\phi| < \pi$.

Solution

Put $t = x^n$.

we have $\displaystyle\int\limits_0^\infty \frac{x^{m-1}}{x^n + e^{i\phi}}\,dx = \frac{1}{n}\int\limits_0^\infty \frac{t^{\left(\frac{m}{n}-1\right)}}{t + e^{i\phi}}\,dt = \frac{\pi}{n}\left(\sin \frac{m\pi}{n}\right)^{-1} e^{i\left(\frac{m}{n}-1\right)\phi}$.

Problem 4.7.31

$$\int\limits_0^\infty \frac{x^{m-1}}{x^{2n} + 2x^n \cos\phi + 1}\,dx = \frac{\pi}{n}\,\frac{\sin\left(1 - \dfrac{m}{n}\right)\phi}{\sin \dfrac{m\pi}{n}\sin\phi}$$

for $0 < m$, n integers and $|\phi| < \pi$.

Solution

In the Example 4.7.30 multiply the numerator and denominator of the integrand by $x^n + e^{-i\phi}$ and equate the imaginary parts.

Note 4.7.32

The integrals of the type discussed in the Theorem 4.7.26 and its corollary are called Mellin transforms of $q(x)$ when considered as functions of 'a'.

Problem 4.7.33

Prove that $\int\limits_0^{\pi} \log \sin x \, dx = -\pi \log 2$.

Solution

Note that the integrand at $x = 0$, $x = \pi$ becomes $-\infty$ so that the definition

of the integral is $\lim\limits_{\delta \to 0} \int\limits_{\delta}^{\pi-\delta} \log \sin x \, dx$.

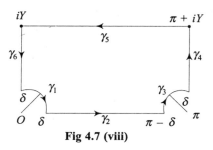

Fig 4.7 (viii)

We now choose the contour Γ which is the boundary of the rectangle whose vertices are 0, π, $\pi + iY$, iY with $Y > 0$. However we avoid the points 0 and π by using small circular quadrants of radius δ.

Consider the function

$$1 - e^{2iz} = -2ie^{iz} \sin z = 1 - e^{-2y} (\cos 2x + i \sin 2x).$$

From the above equality we see that $1 - e^{2iz}$ is real and negative only for $x = n\pi$, $y \leq 0$. Thus in the complement of these regions the principal branch of $\log (1 - e^{2iz})$ is single valued and analytic and our contour lies in this region. Hence by Cauchy's theorem (rather than residue theorem) we have

$$\int\limits_{\Gamma} \log (1 - e^{2iz}) \, dz = 0$$

where $\Gamma = \gamma_1 + \gamma_2 + \gamma_3 + \gamma_4 + \gamma_5 + \gamma_6$ as shown in the figure.

$$\int\limits_{\gamma_4} \log (1 - e^{2iz}) \, dz = i\int\limits_{\delta}^{Y} \log (1 - e^{-2t}) \, dt$$

and

$$\int\limits_{\gamma_6} \log (1 - e^{2iz}) \, dz = i\int\limits_{Y}^{\delta} \log (1 - e^{-2t}) \, dt.$$

Thus integrals over γ_4 and γ_6 cancel each other. Now

$$\left| \int\limits_{\gamma_5} \log (1 - e^{2iz}) \, dz \right| \leq \int\limits_0^{\pi} |\log (1 - e^{-2Y} e^{2it})| \, dt.$$

Now $1 - e^{-2Y} e^{2it}$ varies over the circle centre at 1, with radius e^{-2Y}. Thus as Y becomes larger and larger $1 - e^{2iz}$ as z varies over γ_5, tends to 1 uniformly. Thus $\log (1 - e^{2iz})$ tends to zero as $Y \to \infty$ in the principal branch. Note that the principal value of $\log 1 = 0$. Thus $\int_{\gamma_5} \log (1 - e^{2iz})\, dz$ $\to 0$ as $Y \to \infty$.

$$\left| \int_{\gamma_1} \log (1 - e^{2iz})\, dz \right| \leq \int_0^{\pi/2} |\log (1 - e^{2iz})|\, \delta\, d\theta \qquad (z = \delta e^{i\theta})$$

Now $\left| \dfrac{1 - e^{2iz}}{z} \right| \to 2$ as $z \to 0$ since $\dfrac{d}{dz}(e^{i2z})|_{z=0} = 2i$. Thus in the principal

branch, $\log \left| \dfrac{1 - e^{2iz}}{2z} \right| = \log \left| \dfrac{1 - e^{2iz}}{2\delta} \right| \to 0$ as $\delta \to 0$.

Now with $z = \delta e^{i\theta}$

$$\left| \int_0^{\pi/2} |\log (1 - e^{2iz})|\, \delta\, d\theta \right| \leq \int_0^{\pi/2} |\log |1 - e^{2iz}||\, \delta\, d\theta + \int_0^{\pi/2} \pi \delta\, d\theta$$

(since $|\log w| \leq |\log| w|| + |\arg w| \leq |\log |w|| + \pi$). Now $|\log |(1 - e^{2iz})/2\delta| \leq M$ for sufficiently small δ. Therefore $||\log |1 - e^{2iz}|| \leq M + |\log 2\delta|$. Therefore

$$\int_0^{\pi/2} |\log |1 - e^{2iz}||\, \delta\, d\theta \leq M\delta\, \frac{\pi}{2} + \frac{\pi}{2}\, \delta\, |\log 2\delta| \to 0 \text{ as } \delta \to 0$$

(Use *L' Hospital* rule to prove that $\delta \log 2\delta \to 0$ as $\delta \to 0$).

Thus $\qquad\qquad \int_{\gamma_1} \log (1 - e^{2iz})\, dz \to 0$ as $\delta \to 0$

Similarly $\qquad\qquad \int_{\gamma_3} \log (1 - e^{2iz})\, dz \to 0$ as $\delta \to 0$.

Using all these facts we get $\lim_{\delta \to 0} \int_\delta^{\pi-\delta} \log(-2ie^{ix} \sin x)\, dx = 0$. In the principal

branch $\log (-i) = -i\dfrac{\pi}{2}$ and $\log e^{ix} = ix$.

Hence $\int_0^\pi (\log 2 + \log(-i) + \log e^{ix} + \log \sin x)\, dx = 0$. Therefore

$$\int_0^\pi \log \sin x\, dx = i\frac{\pi^2}{2} - \pi \log 2 - i\frac{\pi^2}{2} = -\pi \log 2.$$

Note 4.7.34

Suppose that a function $F(z)$ of the complex variable z is analytic in the whole plane except for a finite number of poles. Let L_R denote the vertical line segment $z = X + iy$ ($-R \leq y \leq R$, X real) where X is so large that the segment L_R lies to the right of all the singularities of $F(z)$. We now define a new function $f(t)$ on the positive real axis as follows: $f(t) = \dfrac{1}{2\pi i} \lim\limits_{R \to \infty} \int_{L_R} e^{t\zeta} F(\zeta) \, d\zeta$ provided this limit exists. This integral can also be

written as $f(t) = \dfrac{1}{2\pi i}$ P.V. $\int\limits_{-\infty}^{\infty} e^{t\zeta} F(\zeta) \, d\zeta$. For a large class of functions

$F(z)$ the above integral exists and is independent of X and is called the inverse Laplace transform of F. For these functions the following also

holds: $F(z) = \int\limits_{0}^{\infty} e^{-zt} f(t) \, dt$. We also call F as the Laplace transform of f.

Laplace transform and their inverses are very important in the study of ordinary and partial differential equations. Residue theorem is often useful in computing the inverse Laplace transforms. The following theorem demonstrates this fact.

Theorem 4.7.35

Let $F(z)$ be analytic in the plane except for finitely many poles. Let $z_1, z_2, \ldots, z_n$ denote the poles of $F(z)$. Let R_0 be a positive real number such that $|z_i| < R_0$, $\forall i$. For each $R > 2R_0$ let C_R denote the semicircle described by

$z = R_0 + Re^{i\theta} \left(\dfrac{\pi}{2} \leq \theta \leq \dfrac{3\pi}{2} \right)$. Assume that $|F(z)| \leq M_R$ ($z \in C_R$) where

$M_R \to 0$ as $R \to \infty$ then the

Laplace inversion of $F(z) = f(t) = \dfrac{1}{2\pi i}$ P.V. $\int\limits_{-\infty}^{\infty} e^{t\zeta} F(\zeta) \, d\zeta$

exists for $t > 0$, is independent of R_0 (greater than max $|z_i|$)

and
$$f(t) = \sum_{i=1}^{n} \left(\operatorname*{Res}_{z=z_i} e^{zt} F(z) \right).$$

Proof

Consider the contour given by L_R with $x = R_0 > $ max $|z_i|$ and $-R \leq y \leq R$ followed by C_R where $R > 2R_0$.

It is clear that all the poles of $F(z)$ lie inside this contour. So by Residue theorem we have for $t > 0$

$$\int_{L_R} e^{t\zeta} F(\zeta) \, d\zeta + \int_{C_R} e^{t\zeta} F(\zeta) \, d\zeta = 2\pi i \sum_{i=1}^{n} \left(\operatorname*{Res}_{z=z_i} e^{zt} F(z) \right). \qquad (1)$$

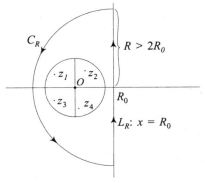

Fig 4.7(ix)

Now
$$\int_{C_R} e^{t\zeta}\, F(\zeta)\, d\zeta = \int_{\pi/2}^{3\pi/2} exp\ (R_0 t + Rte^{i\theta})\ F\ (R_0 + re^{i\theta})\ rie^{i\theta}\ d\theta$$

$$|exp(R_0 t + Rte^{i\theta})| = e^{R_0 t} e^{Rt\ cos\ \theta} \quad \text{and} \quad |F(R_0 + re^{i\theta})| \le M_R$$

so that

$$\left| \int_{C_R} e^{t\zeta}\, F(\zeta)\, d\zeta \right| \le e^{R_0 t} M_R\ R \int_{\pi/2}^{3\pi/2} e^{Rt\ cos\ \theta}\ d\theta. \tag{2}$$

Put $\phi = \theta - \dfrac{\pi}{2}$.

$$\int_{\pi/2}^{3\pi/2} e^{Rt\ cos\ \theta}\, d\theta = \int_0^{\pi} e^{-Rt\ sin\ \phi}\, d\phi$$

$$= \int_0^{\pi/2} e^{-Rt\ sin\ \phi}\, d\phi + \int_{\pi/2}^{\pi} e^{-Rt\ sin\ \phi}\, d\phi$$

$$= 2 \int_0^{\pi/2} e^{-Rt\ sin\ \phi}\, d\phi$$

$\left(\text{Note that in } \displaystyle\int_{\pi/2}^{\pi} e^{-Rt\ sin\ \phi}\, d\phi \text{ we have changed the variable by choosing}\right.$

$\psi = \pi - \phi\Big)$. For $0 \le \phi \le \dfrac{\pi}{2}$, $\dfrac{sin\ \phi}{\phi}$ is decreasing and so $sin\ \phi \ge \dfrac{2\phi}{\pi}$ and

thus

$$2 \int_0^{\pi/2} e^{-Rt\ sin\ \phi}\, d\phi \le 2 \int_0^{\pi/2} e^{-Rt\ 2\phi/\pi}\, d\phi = 2 \left[\frac{e^{\frac{-Rt\ 2\phi}{\pi}}}{\frac{-Rt\ 2}{\pi}} \right]_0^{\pi/2}$$

$$= 2\pi \left(\frac{e^{-Rt}}{-2Rt} - \frac{1}{-2Rt} \right) = \frac{\pi}{Rt}(1 - e^{-Rt}) < \frac{\pi}{Rt}.$$

Thus $\left| \int_{C_R} e^{t\zeta} F(\zeta)\, d\zeta \right| \le \dfrac{e^{R_0 t} M_R \pi}{t} \to 0$ as $R \to \infty$. Hence (1) gives our result (since the right hand side of our result is independent of R_0 the principal value is also independent of R_0).

Problem 4.7.36

Find the Laplace inversion of $F(z) = \dfrac{12}{z^3 + 8}$.

Solution

Consider the function $e^{zt} F(z) = \dfrac{12e^{zt}}{z^3 + 8}$. Its only singularities are simple

poles at $(-8)^{1/3}$ which are given by $z_k = 2e^{i\left(\frac{\pi}{3} + \frac{2k\pi}{3}\right)}$, $k = 0, 1, 2$. More precisely $z_0 = 1 + i\sqrt{3}$, $z_1 = -2$ and $z_2 = 1 - i\sqrt{3}$. The corresponding

residues are $B_k = -\dfrac{z_k}{2} e^{tz_k}$, $k = 0, 1, 2$, (Use $z_k^3 = -8$ and see Example 4.7.14). Let $R_0 = 2$ and $R > 4$. Then for z on the semicircle given by $z = R_0 + Re^{i\theta}$, $\dfrac{\pi}{2} \le \theta \le \dfrac{3\pi}{2}$ we have $|z| \ge R - R_0 > 2$ and $|z^3 + 8| \ge |z|^3 - 8$

$\ge (R - 2)^3 - 8 > 0$. Now it follows that $|F(z)| < \dfrac{12}{(R-2)^3 - 8} \to 0$ as R

$\to \infty$. Thus by previous Theorem 4.7.35 we have,

$$f(t) = B_0 + B_1 + B_2 = B_1 + (B_0 + B_2)$$
$$= e^{-2t} - e^t (\cos \sqrt{3}\, t - \sqrt{3} \sin \sqrt{3}\, t) \qquad (t > 0)$$

Theorem 4.7.37

Let $f(z)$ be an analytic function in the upper half plane except for a finite number of isolated singularities z_k, $k = 1, 2, \ldots$ and continuous on the

real axis with (i) $f(z) = f(-z)$, (ii) $|f(z)| \le \dfrac{M}{|z|^k}$, $k > 1$, for $|z|$ large. Then

$$\int_0^\infty f(x) \log x\, dx = \pi i \sum_{k=1}^n \left(\operatorname*{Res}_{z=z_k}\left[f(z)\left(\log z - i\frac{\pi}{2} \right) \right] \right).$$

Proof

We do not use the principal branch of logarithm but instead define $\log z = \log |z| + i\theta$ where $-\dfrac{\pi}{2} < \theta < \dfrac{3\pi}{2}$ (i.e., $z \in \mathbb{C}\setminus \{z/\operatorname{Re} z = 0 \text{ and } \operatorname{Im} z \le 0\}$).

With this branch take $\phi(z) = f(z) \log z$.
Now for $z = -x$, $x > 0$ we have

$$\phi(z) = f(-x) (\log x + i\pi)$$
$$= f(x) (\log x + i\pi).$$

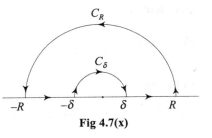

Fig 4.7(x)

Consider the closed contour consisting of real line segments $[-R, -\delta]$, $[\delta, R]$ and the semicircles C_R and C_δ.

$$\int_{-R}^{-\delta} \phi(\zeta) \, \mathrm{d}\zeta = \int_{\delta}^{R} f(x) (\log x + i\pi) \, \mathrm{d}x.$$

$$\left| \int_{C_\delta} \phi(\zeta) \, \mathrm{d}\zeta \right| \le \int_{0}^{\pi} |f(\delta e^{i\theta}) \log (\delta e^{i\theta})| \delta \, \mathrm{d}\theta \le M_1 \int_{0}^{\pi} (|\log \delta| + \pi) \, \delta \, \mathrm{d}\theta$$

(where M_1 is a constant).

It is easy to see that the last integral $\to 0$ as $\delta \to 0$. (Note that since f is continuous at the origin $|f(\delta e^{i\theta})|$ is bounded for small δ).

$$\int_{\delta}^{R} \phi(\zeta) \, \mathrm{d}\zeta = \int_{\delta}^{R} f(x) \log x \, \mathrm{d}x$$

$$\left| \int_{C_R} \phi(\zeta) \, \mathrm{d}\zeta \right| \le \int_{0}^{\pi} |f(Re^{i\theta}) [\log R + \pi]| \, R \, \mathrm{d}\theta$$

$$\le \frac{M}{R^k} \int_{0}^{\pi} [\log R + \pi] \, R \, \mathrm{d}\theta \to 0 \text{ as } R \to \infty.$$

Combining all these and using residue theorem and allowing $R \to \infty$ and $\delta \to 0$ we have

$$\int_{0}^{\infty} [f(x) (\log x + i\pi) + f(x) \log x] \, \mathrm{d}x = 2\pi i \sum_{k=1}^{n} \mathop{\mathrm{Res}}_{z=z_k} f(z) \log z.$$

By Theorem 4.7.12 we have $\int_{0}^{\infty} f(x) \, \mathrm{d}x = \pi i \sum_{k=1}^{n} \mathop{\mathrm{Res}}_{z=z_k} f(z)$. Thus

$$\int_{0}^{\infty} f(x) \log x \, \mathrm{d}x = \pi i \sum_{k=1}^{n} \mathop{\mathrm{Res}}_{z=z_k} f(z) \log z - \frac{\pi}{2} i \pi i \sum_{k=1}^{n} \mathop{\mathrm{Res}}_{z=z_k} f(z)$$

$$= \pi i \sum_{k=1}^{n} \mathop{\mathrm{Res}}_{z=z_k} \left(f(z) \left(\log z - i \frac{\pi}{2} \right) \right).$$

Problem 4.7.38

Prove that $\int_0^\infty \frac{\log x}{1+x^2}\, dx = 0$.

Solution

Take $f(z) = \frac{1}{1+z^2}$. $f(z)$ is even and $|f(z)| \le \frac{1}{|z|^2 - 1} \le \frac{2}{|z|^2}$ for $|z| > \sqrt{2}$.

Hence the conditions of Theorem 4.7.37 are satisfied and the only singularity (a simple pole) in the upper half plane is at $z = i$ and $\underset{z=i}{\text{Res}}\,(f(z)\log$

$z) = \frac{\pi}{4}$ and $\underset{z=i}{\text{Res}}\, f(z) = \frac{1}{2i}$. Hence by Theorem 4.7.37

$$\int_0^\infty \frac{\log x}{1+x^2}\, dx = \pi i\left\{\frac{\pi}{4} - i\frac{\pi}{2}\frac{1}{2i}\right\} = 0.$$

Note 4.7.39
Residue calculus can at times be effectively used to evaluate sums of the

form $\sum_{-\infty}^{\infty} f(n)$ provided the series is convergent. For example if $f(z)$ is

analytic in $\mathbb{C}$ except for finite number of poles none of which coincides with any integer (positive, negative or zero) we consider $g(z) = f(z)\, \pi \cot \pi z$

and use a suitable contour to evaluate $\sum_{-\infty}^{\infty} f(n)$. Similarly using $\pi\, \text{cosec}\, \pi z$

we can evaluate $\sum_{-\infty}^{\infty} (-1)^n f(n)$. We shall illustrate these methods by the
following theorem.

Theorem 4.7.40
Let $f(z)$ be analytic in the finite complex plane except for finitely many poles z_k, $k = 1, 2, \ldots, m$ none of which coincides with any integer. Let $\lim_{z\to\infty} zf(z) = 0$. Then

(i) $\displaystyle\sum_{-\infty}^{\infty} f(n) = -\sum_{k=1}^{n} \underset{z=z_k}{\text{Res}}[f(z)\, \pi \cot \pi z]$

(ii) $\displaystyle\sum_{-\infty}^{\infty} (-1)^n f(n) = -\sum_{k=1}^{n} \underset{z=z_k}{\text{Res}}[f(z)\, \pi \,\text{cosec}\, \pi z]$

Proof
Let C_R be the circle centre origin and radius R so large that it encloses all the poles of f. We shall see to it that this circle does not cut the real axis at integers. By residue theorem

$$\int_{C_R} f(z)\,\pi \cot \pi z\, dz = 2\pi i\left\{\sum_{-N}^{N} f(n) + \sum_{k=1}^{m}\operatorname*{Res}_{z=z_k}[f(z)\,\pi \cot \pi z]\right\} \qquad (1)$$

where $N < R < N + 1$. Since $\sin(\pi z - n\pi) = \dfrac{1}{2i}[e^{i(\pi z - n\pi)} - e^{-i(\pi z - n\pi)}] =$

$(-1)^n \sin \pi z$ we have $\operatorname*{Res}_{z=n} f(z)\,\pi \cot \pi z = \lim_{z \to n} \pi(z - n)\,f(z)\cot \pi z =$

$$\lim_{z \to n}\frac{(z - n)\,\pi\,(-1)^n\,f(z)\cos \pi z}{\sin(\pi z - n\pi)} = (-1)^n f(n)\,\cos \pi n = f(n).\ \text{Given } \varepsilon > 0$$

choose R large so that $|f(z)| < \dfrac{\varepsilon}{|z|}$ $\ (|z| \geq R)$

$$\left|\int_{C_R} f(z)\,\pi \cot \pi z\, dz\right| \leq \frac{\varepsilon}{R}\,\pi \int_{C_R} |\cot \pi z|\,|dz|.$$

Now

$$|\cot \pi z| = \left|\frac{e^{i\pi z} + e^{-i\pi z}}{e^{i\pi z} - e^{-i\pi z}}\right| \leq \left|\frac{e^{-2i\pi z} + 1}{e^{-2i\pi z} - 1}\right|.$$

Observe that if $y > 0$.

$$\left|\frac{1 + e^{-2i\pi z}}{1 - e^{-2i\pi z}}\right| \leq \frac{e^{2\pi y} + 1}{e^{2\pi y} - 1} \to 1 \quad \text{as} \quad y \to \infty.$$

Similarly if $y < 0$.

$$\left|\frac{e^{-2i\pi z} + 1}{e^{-2i\pi z} - 1}\right| \leq \frac{1 + e^{2\pi y}}{1 - e^{2\pi y}} \to 1 \quad \text{as} \quad y \to -\infty.$$

Using the above observations we see that

$$\left|\int_{C_R} f(z)\,\pi \cot \pi z\, dz\right| \leq \frac{\varepsilon}{R}\,\pi M\, 2\pi R = 2\pi^2\, M\varepsilon$$

for a suitable M. Therefore $\int_{C_R} f(z)\,\pi \cot \pi z\, dz \to 0$ as $R \to \infty$.

If $R \to \infty$, N also tends to ∞ and so (1) gives

$$\sum_{-\infty}^{\infty} f(n) = -\sum_{k=1}^{m}\operatorname*{Res}_{z=z_k}[f(z)\,\pi \cot \pi z].$$

Note that in the above proof we have used the fact that $|\cot \pi z|$ is bounded on C_R as $R \to \infty$. Since $\operatorname{cosec}^2 \pi z = 1 + \cot^2 \pi z$, $|\operatorname{cosec} \pi z|$ is also bounded on C_R as $R \to \infty$. Using this result and the fact that $\underset{z=n}{\mathrm{Res}} f(z) \pi \mathrm{cosec}\, \pi z = (-1)^n f(n)$ we get the other result.

Problem 4.7.41

Sum the series $\displaystyle\sum_{-\infty}^{\infty} \frac{1}{n^2 + a^2}$ $(a > 0)$.

Solution

Applying Theorem 4.7.40 to the function $f(z) = \dfrac{1}{z^2 + a^2}$ (it is easily verified that $\lim\limits_{z\to\infty} z f(z) = 0$ and the only simple poles are at $z = \pm ai$).

$$\underset{z=ai}{\mathrm{Res}} \frac{\pi \cot \pi z}{z^2 + a^2} = \lim_{z \to ai} (z - ai) \frac{\pi \cot \pi z}{(z - ai)(z + ai)} = \frac{\pi \cot \pi ai}{2ai}$$

$$\underset{z=-ai}{\mathrm{Res}} \frac{\pi \cot \pi z}{z^2 + a^2} = \lim_{z \to -ai} (z + ai) \frac{\pi \cot \pi z}{(z - ai)(z + ai)} = \frac{-\pi \cot \pi ai}{-2ai}$$

Hence $\displaystyle\sum_{-\infty}^{\infty} \frac{1}{n^2 + a^2} = \frac{\pi}{a} \coth \pi a$.

Problem 4.7.42

Prove that $\displaystyle\sum_{1}^{\infty} \frac{1}{n^2} = \frac{\pi^2}{6}$.

Solution
From the previous example we have

$$2\sum_{1}^{\infty} \frac{1}{n^2 + a^2} + \frac{1}{a^2} = \frac{\pi}{a} \coth \pi a, \quad \sum_{1}^{\infty} \frac{1}{n^2 + a^2} = \frac{1}{2}\left(\frac{\pi}{a} \coth \pi a - \frac{1}{a^2} \right).$$

Allowing a to tend to zero

$$\sum_{1}^{\infty} \frac{1}{n^2} = \lim_{a \to 0} \frac{1}{2}\left(\frac{\pi}{a} \coth \pi a - \frac{1}{a^2} \right) = \lim_{a \to 0} \frac{1}{2}\left(\frac{\pi a \coth \pi a - 1}{a^2} \right)$$

$$= \lim_{a \to 0} \frac{\pi a (e^{\pi a} + e^{-\pi a}) - (e^{\pi a} - e^{-\pi a})}{2a^2 (e^{\pi a} - e^{-\pi a})}.$$

Using L'Hospital's rule twice we get the limit as $\dfrac{\pi^2}{6}$.

4.8 Harmonic Functions

In this section we shall develop the theory of Harmonic functions and obtain the Schwarz reflection principle. We also provide solutions to the so called

Dirichlet's problem for simple domains such as the open unit disc and the upper half plane.

Definition 4.8.1
Let f be a complex function defined on an open set Ω of the plane. Let f be continuous together with its first and second order partial derivatives. If

$$\Delta f = \frac{\partial^2 f}{\partial x^2} + \frac{\partial^2 f}{\partial y^2} = 0$$ we say f is harmonic in Ω. Further Δf will be called

the Laplacian of f.

Theorem 4.8.2
A complex function $f = u + iv$ defined on an open set Ω is harmonic in Ω if and only if u and v are harmonic in Ω.

Proof
Note that $f = u + iv$ is harmonic by definition if and only if u and v are continuous together with their first and second order partial derivatives and $\Delta f = \Delta u + i\Delta v = 0$. i.e. if and only if $\Delta u = 0$ and $\Delta v = 0$.

Theorem 4.8.3
Let f be analytic in a region Ω. Then f is harmonic.

Proof
Since f is analytic, f is continuous together with all its partial derivatives.

(See Theorem 4.4.4). In particular $\dfrac{\partial^2 f}{\partial x \partial y} = \dfrac{\partial^2 f}{\partial y \partial x}$. If $\partial = \dfrac{1}{2}\left(\dfrac{\partial}{\partial x} - i\dfrac{\partial}{\partial y}\right)$

and $\bar{\partial} = \dfrac{1}{2}\left(\dfrac{\partial}{\partial x} + i\dfrac{\partial}{\partial y}\right)$ then $\Delta f = 4\partial\bar{\partial}f$. On the other hand if f is analytic

$\bar{\partial}f = 0$ by Cauchy-Riemann equations (See Theorem 2.1.3). Thus f is harmonic.

Note 4.8.4
The converse of the above theorem is false. For example we can take any non-constant real-valued harmonic function $u(z)$ in a region Ω and consider $f(z) = u(z) + iv(z)$ where $v(z) \equiv 0$. Obviously $f(z)$ is harmonic but being purely real valued, if it is analytic, it must reduce to a constant (See Theorem 2.1.5) contrary to our assumption.

Theorem 4.8.5
The sum of two harmonic functions and the constant multiple of a harmonic function are also harmonic. Further a real valued function u is harmonic if and only if $r\dfrac{\partial}{\partial r}\left(r\dfrac{\partial u}{\partial r}\right) + \dfrac{\partial^2 u}{\partial \theta^2} = 0$ (or equivalently $r^2\dfrac{\partial^2 u}{\partial r^2} + r\dfrac{\partial u}{\partial r}$

$+\dfrac{\partial^2 u}{\partial\theta^2}=0$), where u is considered as a function of r, θ via the polar co-ordinates $x=r\cos\theta$ and $y=r\sin\theta$.

Proof

The linearity of the Laplacian shows that the sum and constant multiple of harmonic functions are harmonic. Now let u be real valued. u is a function of x and y and x and y are functions of r and θ given by $x=r\cos\theta$ and $y=r\sin\theta$.

$$\frac{\partial u}{\partial r}=\frac{\partial u}{\partial x}\frac{\partial x}{\partial r}+\frac{\partial u}{\partial y}\frac{\partial y}{\partial r}=\cos\theta\frac{\partial u}{\partial x}+\sin\theta\frac{\partial u}{\partial y} \qquad (*)$$

$$r\frac{\partial u}{\partial r}=r\cos\theta\frac{\partial u}{\partial x}+r\sin\theta\frac{\partial u}{\partial y}.$$

$$\frac{\partial}{\partial r}\left(r\frac{\partial u}{\partial r}\right)=\frac{\partial}{\partial r}\left(r\cos\theta\frac{\partial u}{\partial x}\right)+\frac{\partial}{\partial r}\left(r\sin\theta\frac{\partial u}{\partial y}\right)$$

$$=\cos\theta\frac{\partial u}{\partial x}+r\cos\theta\left(\frac{\partial^2 u}{\partial x^2}\cos\theta+\frac{\partial^2 u}{\partial y\partial x}\sin\theta\right)$$

$$+\sin\theta\frac{\partial u}{\partial y}+r\sin\theta\left(\frac{\partial^2 u}{\partial y\partial x}\cos\theta+\frac{\partial^2 u}{\partial y^2}\sin\theta\right).$$

$\left(\text{Note that we have to use (*) with }u\text{ replaced by }\dfrac{\partial u}{\partial x}\right).$

$$r\frac{\partial}{\partial r}\left(r\frac{\partial u}{\partial r}\right)=r\cos\theta\frac{\partial u}{\partial x}+r^2\cos^2\theta\frac{\partial^2 u}{\partial x^2}+2r^2\sin\theta\cos\theta\frac{\partial^2 u}{\partial y\partial x}$$

$$+r\sin\theta\frac{\partial u}{\partial y}+r^2\sin^2\theta\frac{\partial^2 u}{\partial y^2}. \qquad (1)$$

$$\frac{\partial u}{\partial\theta}=\frac{\partial u}{\partial x}\frac{\partial x}{\partial\theta}+\frac{\partial u}{\partial y}\frac{\partial y}{\partial\theta}=-\frac{\partial u}{\partial x}r\sin\theta+\frac{\partial u}{\partial y}r\cos\theta$$

$$\frac{\partial^2 u}{\partial\theta^2}=\frac{\partial}{\partial\theta}\left(-\frac{\partial u}{\partial x}r\sin\theta\right)+\frac{\partial}{\partial\theta}\left(\frac{\partial u}{\partial y}r\cos\theta\right)$$

$$=-r\cos\theta\frac{\partial u}{\partial x}+r^2\sin^2\theta\frac{\partial^2 u}{\partial x^2}-\frac{\partial u}{\partial y}r\sin\theta+r^2\cos^2\theta\frac{\partial^2 u}{\partial y^2}$$

$$-2r^2\sin\theta\cos\theta\frac{\partial^2 u}{\partial x\partial y} \qquad (2)$$

From (1) and (2) and the fact that u is harmonic if and only if $\Delta u = 0$, we have

$$r\frac{\partial}{\partial r}\left(r\frac{\partial u}{\partial r}\right) + \frac{\partial^2 u}{\partial\theta^2} = 0 \text{ if and only if } u \text{ is harmonic}$$

Theorem 4.8.6

Let u be a real valued harmonic function in a region Ω. Then,

(i) $f(z) = \dfrac{\partial u}{\partial x} - i\dfrac{\partial u}{\partial y}$ is analytic in Ω $\left(\text{equivalently } \dfrac{\partial u}{\partial x} \text{ is the real part of}\right.$

an analytic function in $\Omega\Big)$.

(ii) If Ω is simply connected then there exists F analytic in Ω such that $u = \operatorname{Re} F$ in Ω (of course this implies that every real-valued harmonic function defined in a region, is at least locally the real part of an analytic function).

Proof

We first observe that since u has continuous second order partial derivatives

$\dfrac{\partial^2 u}{\partial x\partial y} = \dfrac{\partial^2 u}{\partial y\partial x}$. Define $f(z) = \dfrac{\partial u}{\partial x} - i\dfrac{\partial u}{\partial y} = U + iV$ (say). We know that U

and V are continuous together with their first order partial derivatives and

$$\frac{\partial U}{\partial x} = \frac{\partial^2 u}{\partial x^2} = -\frac{\partial^2 u}{\partial y^2} = \frac{\partial V}{\partial y} \text{ and } \frac{\partial U}{\partial y} = \frac{\partial^2 u}{\partial y\partial x} = \frac{\partial^2 u}{\partial x\partial y} = -\frac{\partial V}{\partial x}.$$

Hence U and V satisfy Cauchy-Riemann equations and so $U + iV$ is analytic in Ω.

(ii)   If Ω is simply connected the function $f(z) = \dfrac{\partial u}{\partial x} - i\dfrac{\partial u}{\partial y}$ described

above in (i) is analytic and so by Theorem 4.2.18 has a primitive F(say). Thus if $F(z) = A(z) + iB(z)$ then

$$f(z) = F'(z) = \frac{\partial A}{\partial x} + i\frac{\partial B}{\partial x} = \frac{\partial A}{\partial x} - i\frac{\partial A}{\partial y} = \frac{\partial u}{\partial x} - i\frac{\partial u}{\partial y}$$

so that $A(x, y) = u(x, y) + c$ where c is a real constant (See the proof of Theorem 2.1.5). Thus $u(x, y)$ is the real part of the analytic function $F(z) - c$.

Theorem 4.8.7 (Mean-value property)

Let Ω be a region and u be a real harmonic function defined on Ω. Let $a \in \Omega$ such that the closed disc $\overline{D}(a, r)$ (centre a and radius r) lies in Ω. Then

$$u(a) = \frac{1}{2\pi}\int_0^{2\pi} u(a + re^{i\theta})\, d\theta$$

Proof

Choose r such that the closed disc centre "a" and radius r is contained in Ω (i.e., $\overline{D}(a, r) \subseteq \Omega$). Using (ii) of Theorem 4.8.6 we can choose an analytic function f in D (the corresponding open disc) such that $u = Re\, f$. By Cauchy's integral formula applied to the circle centre "a" and radius r we have $f(a) = \dfrac{1}{2\pi} \int\limits_0^{2\pi} f(a + re^{i\theta})\, d\theta$ (See Note 4.4.10 (a)).

Taking real part on both sides we have

$$u(a) = \frac{1}{2\pi} \int\limits_0^{2\pi} u(a + re^{i\theta})\, d\theta.$$

Theorem 4.8.8 (Maximum and minimum principles for real harmonic functions)

If u is a real-valued non-constant harmonic function in a region Ω then u has neither a maximum nor a minimum in Ω.

Proof

For maximum principle it suffices to show that if there exists $a \in \Omega$ such that $u(z) \leq u(a)$ for all $z \in \Omega$ then $u(z) = u(a)$ in Ω. Let $A = \{z \in \Omega / u(z) = u(a)\} = u^{-1}\{u(a)\}$. This set is closed in Ω, as u is a continuous function and $\{u(a)\}$ is closed in $\mathbb{R}$. We now show that A is open.

Let $z_0 \in A$. Choose $r > 0$ such that $D(z_0, r) \subseteq \Omega$. By mean value property

$$u(z_0) = \frac{1}{2\pi} \int\limits_0^{2\pi} u(z_0 + re^{i\theta})\, d\theta.$$ We first claim that $u(z) = u(a)$ for all points

z with $|z - z_0| = r$. If not there exists some $\theta_0 \in [0, 2\pi]$ such that $u(z_0 + re^{i\theta_0}) < u(a)$. For convenience let us assume $\theta_0 \in (0, 2\pi)$. Choose β such that $u(z_0 + re^{i\theta_0}) < \beta < u(a)$. Using continuity of u we can find a $\delta > 0$ ($0 < \delta < \mathrm{Min}\,(\theta_0, 2\pi - \theta_0)$) such that

$$\theta_0 - \delta \leq \theta \leq \theta_0 + \delta \Rightarrow u(z_0 + re^{i\theta}) < \beta.$$

Now using $u(z) \leq u(a)\ \forall\ z \in \Omega$ we see that

$$u(a) = u(z_0) = \frac{1}{2\pi} \int\limits_0^{2\pi} u(z_0 + re^{i\theta})\, d\theta$$

$$= \frac{1}{2\pi} \left(\int\limits_0^{\theta_0 - \delta} + \int\limits_{\theta_0 - \delta}^{\theta_0 + \delta} + \int\limits_{\theta_0 + \delta}^{2\pi} \right) u(z_0 + re^{i\theta})\, d\theta$$

$$\leq \frac{1}{2\pi} \left[(\theta_0 - \delta)\, u(a) + 2\delta\beta + (2\pi - \theta_0 - \delta)\, u(a) \right]$$

$$< u(a) \text{ a contradiction.}$$

In case $\theta_0 = 0$ or 2π we can easily modify the above proof to get a contradiction.

Thus $u(z) = u(a)$ for every point z with $|z - z_0| = r$. Now since r is arbitrary subject only to the condition $D(z_0, r) \subseteq \Omega$ it follows that $u(z) = u(a)$ throughout a neighbourhood of z_0. This proves that A is open. Since $a \in A$, $A \neq \phi$. By connectedness of Ω it follows that $A = \Omega$. The minimum principle follows from the maximum principle by considering $-u$.

Definition 4.8.9

Let Ω be a region and u a real continuous function on Ω. We say u satisfies

mean-value property in Ω if $\overline{D}(a, r) \subseteq \Omega \Rightarrow u(a) = \dfrac{1}{2\pi} \int\limits_0^{2\pi} u(a + re^{i\theta})\, d\theta$

where $\overline{D}(a, r)$ denotes the closure of the disc centre 'a' radius r.

Corollary 4.8.10

If Ω is a region and u a non-constant real continuous function on Ω with mean-value property, then the maximum and the minimum for u is not attained in Ω.

Proof

Since the only properties of harmonic functions used in the proof of Theorem 4.8.8 are the mean-value property and the continuity, the proof follows.

Corollary 4.8.11

Let Ω be a bounded region and u be a real continuous function on $\overline{\Omega}$ having mean-value property on Ω. If $u(z) = 0$, $\forall\, z \in \partial\Omega$, then $u(z) = 0\ \forall\, z \in \overline{\Omega}$.

Proof

By Corollary 4.8.10, the minimum and the maximum for u are attained only on the boundary. (Here we can assume without loss of generality that u is non-constant. Otherwise u is constant and $u = 0$ on $\overline{\Omega}$ by hypothesis). Since all the boundary values are zero, the theorem follows.

Theorem 4.8.12 (Poisson integral formula)

Suppose $u(z)$ is real and harmonic for $|z| < R$, continuous for $|z| \leq R$ then

$$u(a) = \frac{1}{2\pi} \int\limits_0^{2\pi} \frac{R^2 - |a|^2}{|Re^{i\theta} - a|^2} u(Re^{i\theta})\, d\theta \qquad (|a| < R).$$

Proof

First we assume that $u(z)$ is harmonic in some disc larger than $|z| \leq R$ (say in $|z| < R_1$ with $R_1 > R$). For a given "a" with $|a| < R$ consider the linear fractional transformation $z = S(\zeta) = \dfrac{R(R\zeta + a)}{R + \overline{a}\zeta}$ which maps $|\zeta| \leq 1$ onto

$|z| \leq R$ with $S(0) = a$ (See proof of Corollary 4.4.32). Since the only singularity of $S(\zeta)$ is at $\zeta = -R/\bar{a}$ with $|\zeta| > 1$, $S(\zeta)$ maps the disc $|\zeta| < 1 + \varepsilon$ (for a suitable $\varepsilon > 0$) analytically into $|z| < R_1$. Now $|z| < R_1$ is simply connected and hence from Theorem 4.8.6 (ii) we also know that $u(z) = \text{Re } f(z)$ for some analytic function f in $|z| < R_1$. Thus $u(S(\zeta)) = \text{Re } f(S(\zeta))$ and so $u(S(\zeta))$ is the real part of an analytic function in $|\zeta| < 1 + \varepsilon$. Hence $u(S(\zeta))$ is harmonic in $|\zeta| < 1 + \varepsilon$ (Theorem 4.8.3 and Theorem 4.8.2). By mean-value property of harmonic functions (Theorem 4.8.7) we have

$$u(S(0)) = u(a) = \frac{1}{2\pi} \int_0^{2\pi} u(S(e^{i\phi})) \, d\phi. \tag{1}$$

[Note that the closed disc centre 0 and radius 1 is completely contained in $|\zeta| < 1 + \varepsilon$ which is inside the domain of definition of $u(S(\zeta))$]. Using the mapping properties of $z = S(\zeta)$ we see that $z = S(\zeta) = \dfrac{R(R\zeta + a)}{R + \bar{a}\zeta} \Rightarrow$

$\zeta = \dfrac{R(z - a)}{R^2 - \bar{a}z}$ and as $\zeta = e^{i\phi}$ varies over the unit circle $z = Re^{i\theta}$ varies over $|z| = R$. Thus $\zeta = e^{i\phi} \Rightarrow z = Re^{i\theta}$ and $d\phi = -i\dfrac{d\zeta}{\zeta}$, $dz = iz\,d\theta$. Hence

$$d\phi = -i\frac{d\zeta}{\zeta} = -i\left(\frac{1}{z - a} + \frac{\bar{a}}{R^2 - \bar{a}z}\right) dz = \left(\frac{z}{z - a} + \frac{z\bar{a}}{R^2 - \bar{a}z}\right) d\theta. \tag{2}$$

Substituting $z\bar{z} = R^2$ we get $\dfrac{z}{z - a} + \dfrac{\bar{a}}{\bar{z} - \bar{a}} = \dfrac{R^2 - |a|^2}{|z - a|^2}$. Therefore on

$|z| = R$, $\dfrac{z}{z - a} + \dfrac{z\bar{a}}{R^2 - \bar{a}z} = \dfrac{R^2 - |a|^2}{|z - a|^2}$. Now (2) implies $d\phi = \dfrac{R^2 - |a|^2}{|z - a|^2} d\theta$.

Using (1) we get $u(a) = \dfrac{1}{2\pi} \int_0^{2\pi} \dfrac{R^2 - |a|^2}{|Re^{i\theta} - a|^2} u(Re^{i\theta}) \, d\theta$ which is valid for

$|a| < R$.

In the above derivation we have assume that $u(z)$ is harmonic in a disc larger than $|z| \leq R$. However the result remains true even if we assume $u(z)$ is harmonic in $|z| < R$ and continuous in $|z| \leq R$. Indeed to prove this we need only observe that if $0 < r < 1$ then $u(rz)$ is harmonic in a disc $|z| < \dfrac{R}{r}$ which is larger than $|z| \leq R$. Now applying the above result we get

$$u(ra) = \frac{1}{2\pi} \int_0^{2\pi} \frac{R^2 - |a|^2}{|Re^{i\theta} - a|^2} u(rRe^{i\theta}) \, d\theta \qquad (|a| < R). \tag{3}$$

Since $u(z)$ is continuous on the compact set $|z| \leq R$ and continuous functions on compact sets are uniformly continuous $u(rRe^{i\theta}) \to u(Re^{i\theta})$

uniformly in $0 \leq \theta \leq 2\pi$ as $r \to 1$. Since uniform limit can be taken inside the integral sign we allow $r \to 1$ on both sides of (3) and obtain

$$u(a) = \frac{1}{2\pi} \int_0^{2\pi} \frac{R^2 - |a|^2}{|Re^{i\theta} - a|^2} u(Re^{i\theta}) \, d\theta \qquad (|a| < R).$$

Corollary 4.8.13

Under the hypothesis of the above theorem if $a = re^{i\phi}$ with $r < R$ we have

$$u(re^{i\phi}) = \frac{1}{2\pi} \int_0^{2\pi} \text{Re} \left(\frac{Re^{i\theta} + a}{Re^{i\theta} - a} \right) u(Re^{i\theta}) \, d\theta$$

$$= \frac{1}{2\pi} \int_0^{2\pi} \frac{R^2 - r^2}{R^2 - 2rR \cos(\theta - \phi) + r^2} u(Re^{i\theta}) \, d\theta.$$

Proof

By simple calculation we have

$$\frac{R^2 - |a|^2}{|Re^{i\theta} - a|^2} = \text{Re} \left(\frac{Re^{i\theta} + a}{Re^{i\theta} - a} \right) = \frac{R^2 - r^2}{R^2 - 2rR \cos(\theta - \phi) + r^2}$$

The corollary now follows.

Definition 4.8.14

Let $U(\theta)$ be a piece-wise continuous (continuous except for finitely many points) bounded real-function defined in $0 \leq \theta \leq 2\pi$. Then

$$P_U(z) = \frac{1}{2\pi} \int_0^{2\pi} \text{Re} \left(\frac{e^{i\theta} + z}{e^{i\theta} - z} \right) U(\theta) \, d\theta = \frac{1}{2\pi} \int_0^{2\pi} \frac{1 - |z|^2}{|e^{i\theta} - z|^2} U(\theta) \, d\theta \quad (|z| < 1)$$

is called the Poisson Integral of U.

Notes 4.8.15

(a) We first observe that the integral defining $P_U(z)$ exists for each $z = re^{i\phi}$ with $r < 1$ because $\text{Re} \left(\dfrac{e^{i\theta} + z}{e^{i\theta} - z} \right) U(\theta) = \dfrac{1 - r^2}{1 - 2r \cos(\theta - \phi) + r^2} U(\theta)$ is a bounded function which is continuous except at a finite number of points in $[0, 2\pi]$ and so is Riemann integrable in $[0, 2\pi]$.

(b) Let us define a new function $v(\zeta)$ on $T = (\zeta \in \mathbb{C} / |\zeta| = 1)$ by writing $\zeta = e^{i\theta}$ $(0 \leq \theta < 2\pi)$ and setting $v(\zeta) = U(\theta)$. This function $v(\zeta)$ is also piece-wise continuous on $|\zeta| = 1$. This can be proved as follows. If $\theta_1, \theta_2,$ $\ldots \theta_k$ are the discontinuities of $U(\theta)$ in $[0, 2\pi]$ and if $\zeta_j = e^{i\theta_j}$ $(1 \leq j \leq k)$ then $T \setminus \{-1\} \subset \Omega = \mathbb{C} \setminus (-\infty, 0]$ and in Ω the principal branch of $\log \zeta$

is continuous and so $v(\zeta) = U\left(\frac{1}{i} \log \zeta\right)$ is also continuous at all points of

T except for ζ_j's and probably -1 (if none of these ζ_j's is -1). Thus

$$P_U(z) = \frac{1}{2\pi} \int\limits_0^{2\pi} \mathrm{Re}\left(\frac{e^{i\theta}+z}{e^{i\theta}-z}\right) U(\theta)\, d\theta = \frac{1}{2\pi i} \int\limits_{|\zeta|=1} \frac{1-|z|^2}{|\zeta-z|^2} \frac{v(\zeta)}{\zeta}\, d\zeta.$$

From this observation we can more generally define the Poisson integral of a piece-wise continuous real function $U(\zeta)$ defined on $|\zeta - z_0| = R$ by

$$P_U(z) = \frac{1}{2\pi i} \int\limits_{|\zeta-z_0|=R} \frac{R^2-r^2}{|\zeta-z|^2} \frac{U(\zeta)}{(\zeta-z_0)}\, d\zeta \qquad (r = |z-z_0| < R).$$

(c) The Poisson integral of $U(\theta)$ ($0 \leq \theta \leq 2\pi$) denoted by $P_U(z)$ and defined for $|z| < 1$ is not only a function of z but is also a functional defined on the set of all piece-wise continuous functions defined on $[0, 2\pi]$. The following properties of $P_U(z)$ can be easily verified.
   (i) $P_{U+V}(z) = P_U(z) + P_V(z)$.
   (ii) If c is any complex number $P_{cU}(z) = c\, P_U(z)$
   (iii) If $U(\theta) \geq 0$ then $P_U(z)$ is real and non-negative. In other words P_U is a positive linear functional.
   (iv) If $U(\theta) \equiv c$, a complex constant then $P_U(z) \equiv c$. (Apply Poisson integral formula with $u(z) = 1$ and $u(z) = c$ successively).
   (v) If m, M are real constants such that $m \leq U(\theta) \leq M$ then $m \leq P_U(z) \leq M$. (Use Poisson integral formula, properties of the Riemann integral and (iv)).

Theorem 4.8.16
Given a piece-wise continuous and bounded real valued function $U(\theta)$ for $0 \leq \theta \leq 2\pi$, $P_U(z)$ is harmonic for $|z| < 1$ and $\lim\limits_{z \to e^{i\theta_0}} P_U(z) = U(\theta_0)$ whenever U is continuous at θ_0.

Proof
We first observe that if $\phi(\zeta)$ is bounded and piece-wise continuous on γ

then $\int\limits_\gamma \frac{\phi(\zeta)}{\zeta - z}\, d\zeta$ (This is well defined for z not on γ. See Note 4.8.15 (a))

is analytic in each of the regions determined by γ. The proof is exactly the same as in Theorem 4.4.3 except for the fact that we have to prove that $\int\limits_\gamma |\phi(\zeta)|\, |d\zeta| < \infty$ in our case. To show that this is indeed true in our case we merely observe that $\phi(\zeta)$ is bounded on γ. As observed in Note 4.8.15 (b), define a piece-wise continuous and bounded real function $v(\zeta)$ on T

by writing $\zeta = e^{i\theta}$ and setting $v(e^{i\theta}) = U(\theta)$. (Most often it is convenient to write $v(\zeta)$ as $U(\theta)$ by indentifying $U(\theta)$ with $v(e^{i\theta}) = v(\zeta)$). Now

$$P_U(z) = \frac{1}{2\pi} \int_0^{2\pi} \operatorname{Re}\left(\frac{e^{i\theta} + z}{e^{i\theta} - z}\right) U(\theta)\, d\theta$$

$$= \frac{1}{2\pi}\operatorname{Re} \int_{|\zeta|=1} \frac{\zeta + z}{\zeta - z}\, \frac{v(\zeta)}{i\zeta}\, d\zeta$$

$$= \operatorname{Re}\frac{1}{2\pi i} \int_{|\zeta|=1} \frac{\phi(\zeta)}{\zeta - z}\, d\zeta \quad \text{where } \phi(\zeta) = \frac{(\zeta + z)\, v(\zeta)}{\zeta}.$$

$$= \operatorname{Re} F(z) \text{ (say)}.$$

As $\phi(\zeta)$ is also piece-wise continuous and bounded, $F(z)$ is analytic except on $|z| = 1$ and so its real part is harmonic in $|z| < 1$. Thus $P_U(z)$ is harmonic for $|z| < 1$. We now prove that $\lim_{z \to e^{i\theta_0}} P_U(z) = U(\theta_0)$ whenever U is continuous at θ_0. Without loss of generality we can suppose $U(\theta_0) = 0$ and prove that $\lim_{z \to e^{i\theta_0}} P_U(z) = 0$. In fact if $U(\theta_0) = c \neq 0$ we replace $U(\theta)$ by $U(\theta) - c$ and get $\lim_{z \to e^{i\theta_0}} P_{U-c}(z) = 0$ which in turn implies that $\lim_{z \to e^{i\theta_0}} (P_U(z) - c) = 0$ or $\lim_{z \to e^{i\theta_0}} P_U(z) = c = U(\theta_0)$. Given $\varepsilon > 0$ we choose complementary arcs C_1 (closed) and C_2 (open) of the unit circle such that $e^{i\theta_0}$ is an interior point of C_2 and $|U(\theta)| < \frac{\varepsilon}{2}$ for $e^{i\theta} \in C_2$. This is possible by using just the continuity of U at θ_0. Now consider a function U_1 defined as $U(\theta)$ for $e^{i\theta} \in C_1$ and zero if $e^{i\theta} \in C_2$. Similarly U_2 is defined as $U(\theta)$ for $e^{i\theta} \in C_2$ and zero if $e^{i\theta} \in C_1$. It is clear that for all θ, $U(\theta) = U_1(\theta) + U_2(\theta)$. Thus $P_U = P_{U_1} + P_{U_2}$. Now $P_{U_1}(z)$ is harmonic except on the closed arc C_1 in the same way as P_U was shown to be harmonic except on T. Similarly P_{U_2} is harmonic except on the closure of the arc C_2. Now U_1 is continuous and vanishes at $z = e^{i\theta_0}$. Since

$$\operatorname{Re}\left(\frac{e^{i\theta} + z}{e^{i\theta} - z}\right) = \frac{1 - |z|^2}{|e^{i\theta} - z|^2}$$

it follows that $\quad P_{U_1}(z) = \frac{1}{2\pi} \int_{C_1} \operatorname{Re}\left(\frac{e^{i\theta} + z}{e^{i\theta} - z}\right) U_1(\theta)\, d\theta = 0$

if $z \in C_2$. But $P_{U_1}(z)$ is harmonic and hence continuous every where except on C_1. Hence $\lim\limits_{z \to e^{i\theta_0}} P_{U_1}(z) = 0$. Thus there exists $\delta > 0$ such that $|P_{U_1}(z)|$

$< \dfrac{\varepsilon}{2}$ for $|z - e^{i\theta_0}| < \delta$. We already know $|U(\theta)| < \dfrac{\varepsilon}{2}$ for $e^{i\theta} \in C_2$. Therefore

$|U_2(\theta)| < \dfrac{\varepsilon}{2}$ for every θ and hence $|P_{U_2}(z)| < \dfrac{\varepsilon}{2}$ for $|z| < 1$. Thus

$$|P_U(z)| = |P_{U_1}(z)| + |P_{U_2}(z)| < \frac{\varepsilon}{2} + \frac{\varepsilon}{2} = \varepsilon$$

whenever $|z - e^{i\theta_0}| < \delta$ and $|z| < 1$. Hence the result.

Theorem 4.8.17
Let $a \in \mathbb{C}$, $\rho > 0$ and h, a continuous real function on the circle $|z - a|$ $= \rho$. Let $D(a, \rho)$ denote the open disc center a radius ρ and $\overline{D}(a, \rho)$ its closure. Then there exists a unique continuous function $w(z)$ defined on $\overline{D}(a, \rho)$ with real values such that w is harmonic on $D(a, \rho)$ and $w(z) = h(z)$ for $|z - a| = \rho$.

Proof
Let $f(e^{i\theta}) = h(a + re^{i\theta})$. Then by our hypothesis f is continuous on the unit circle. Thus $P_f(z)$ is harmonic in $|z| < 1$ and has the boundary values f by Theorem 4.8.16. Let $w(z) = P_f\left(\dfrac{z - a}{\rho}\right)$. Clearly w is the required function.

On the other hand if w_1 is also harmonic in $D(a, \rho)$ and continuous on $\overline{D}(a, \rho)$ with the same boundary values h then $w - w_1$ is harmonic in $D(a, \rho)$, continuous on $\overline{D}(a, \rho)$ and is $\equiv 0$ on $|z - a| = \rho$. Corollary 4.8.11 is applicable and we have $w = w_1$ proving the uniqueness of w_1.

Corollary 4.8.18
In addition to the hypothesis of Theorem 4.8.17 if we assume that h is also harmonic in $|z - a| < \rho$ then $P_h = h$.

Proof
Follows from the uniqueness condition given in Theorem 4.8.17.

Note 4.8.19
Given a region Ω with boundary $\partial\Omega$ and a piece-wise continuous bounded real function U on $\partial\Omega$ it is interesting to know whether we can always find a real harmonic function $u(z)$ in Ω whose boundary values agree with U. This problem is called Dirichlet's problem for Ω and our Theorems 4.8.16 and 4.8.17 merely solve this problem in case Ω is an open unit disc. For further information on the solution of the general Dirichlet's problem we refer the reader to L Ahlfor's book on Complex analysis.

Theorem 4.8.20

For a fixed $z \in U = \{z \in \mathbb{C} : |z| < 1\}$ and for every $e^{i\theta} \in T = \partial U$ determine $e^{i\theta^*}$ so that $e^{i\theta}$, z, $e^{i\theta^*}$ are in a straight line. Then

$$P_U(z) = \frac{1}{2\pi} \int_0^{2\pi} U(\theta) d\theta^* = \frac{1}{2\pi} \int_0^{2\pi} U(\theta^*) d\theta$$

(i.e., to find out $P_U(z)$ replace each $U(\theta)$ by the value of U at θ^* and take the average over the interval $[0, 2\pi]$).

Proof

Let θ and θ^* be as above. It is clear that $PA \cdot PA' = PB \cdot PB'$. (See the figure below)

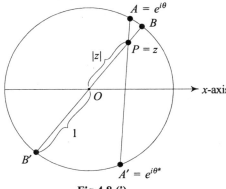

Fig 4.8 (i)

i.e., $|z - e^{i\theta}| \, |z - e^{i\theta^*}| = (1 - |z|)(1 + |z|) = 1 - |z|^2$. But the ratio $\dfrac{z - e^{i\theta}}{z - e^{i\theta^*}}$

is negative as $e^{i\theta}$ and $e^{i\theta^*}$ lie on the same line through z but in different directions. Let $\alpha = \arg(z - e^{i\theta})$ and $\beta = \arg(z - e^{i\theta^*})$ so that $\alpha - \beta = \pm\pi$. We have

$$1 - |z|^2 = (z - e^{i\theta}) \, e^{-i\alpha} \, (z - e^{i\theta^*}) \, e^{-i\beta}$$
$$= (z - e^{i\theta}) \, e^{-i\alpha} \, (\bar{z} - e^{-i\theta^*}) \, e^{2i\beta} \, e^{-i\beta}$$

(Note that $w = \bar{w} e^{2i \arg w}$ for any non-zero complex number w).

Thus

$$1 - |z|^2 = -(z - e^{i\theta})(\bar{z} - e^{-i\theta^*}). \tag{1}$$

In this last equality keep $|z|$ as a constant and differentiate with respect to θ treating θ^* as a function of θ formally so that

$$\frac{e^{i\theta} \, d\theta}{e^{i\theta} - z} = \frac{e^{-i\theta^*} \, d\theta^*}{e^{-i\theta^*} - \bar{z}} \quad or \quad \left| \frac{d\theta^*}{d\theta} \right| = \left| \frac{e^{-i\theta^*} - \bar{z}}{e^{i\theta} - z} \right| = \left| \frac{e^{i\theta^*} - z}{e^{i\theta} - z} \right|$$

But $d\theta^*$ and $d\theta$ are Lebesgue measures and hence

$$\frac{d\theta^*}{d\theta} = \left|\frac{d\theta^*}{d\theta}\right| = \left|\frac{e^{i\theta^*} - z}{e^{i\theta} - z}\right|$$

This combined with (1) gives $d\theta^* = \dfrac{1 - |z|^2}{|e^{i\theta} - z|^2} \, d\theta$. Thus

$$P_U(z) = \frac{1}{2\pi}\int_0^{2\pi} U(\theta)\,d\theta^* = \frac{1}{2\pi}\int_0^{2\pi} U(\theta^*)\,d\theta$$

The last equality is obtained by changing θ to θ^* and observing that $(\theta^*)^* = \theta$.

Theorem 4.8.21

Let Ω be a region and $U : \Omega \to \mathbb{R}$ be a continuous function satisfying the mean value property. Then U is harmonic in Ω.

Proof

Let $a \in \Omega$. Choose an open disc, $D(a, \rho)$ center 'a' and radius ρ so that $\overline{D}(a, \rho) \subset \Omega$. By Theorem 4.8.20, we have a harmonic function W defined on $D(a, \rho)$ such that $W(a + re^{i\theta}) = U(a + re^{i\theta})$. Since W and U both satisfy the mean value property and $(W - U)(z) = 0$ for $|z - a| = \rho$, it follows from corollary 4.8.11 that $U = W$ in $D(a, \rho)$ and hence U must be harmonic in $D(a, \rho)$. Since 'a' is arbitrary the theorem follows.

Theorem 4.8.22 (Harnack's inequality)

Let $D(a, \rho)$ be an open disc, center 'a' and radius ρ and $\overline{D}(a, \rho)$ its closure. If $U : \overline{D}(a, \rho) \to \mathbb{R}$ is conitnuous and harmonic in $D(a, \rho)$ with

$U(z) \geq 0$, $\forall z \in \overline{D}(a, \rho)$ then $\dfrac{\rho - r}{\rho + r} U(a) \leq U(z) \leq \dfrac{\rho + r}{\rho - r} U(a)$, for all $|z - a| \leq r \ (0 \leq r < \rho)$.

Proof

By Poisson integral formula or rather its consequences (Theorem 4.8.17 and Corollary 4.8.18) we have

$$U(a + re^{i\theta}) = \frac{1}{2\pi}\int_0^{2\pi} \frac{\rho^2 - r^2}{\rho^2 - 2r\rho\cos(t - \theta) + r^2} U(a + \rho e^{it})\,dt \qquad (1)$$

Since $\qquad \dfrac{\rho^2 - r^2}{\rho^2 - 2r\rho\cos(t - \theta) + r^2} = \dfrac{\rho^2 - r^2}{|\rho e^{it} - re^{i\theta}|^2}$,

$$\rho - r \leq |\rho e^{it} - re^{i\theta}| \leq \rho + r;$$

we have

$$\frac{\rho - r}{\rho + r} \leq \frac{\rho^2 - r^2}{\rho^2 - 2r\rho\cos(t - \theta) + r^2} \leq \frac{\rho + r}{\rho - r} \qquad (2)$$

Since $U(z) \geq 0$, (1) and (2) give the required inequality for $z = a + re^{i\theta}$. However using the minimum and maximum principles for harmonic functions the required inequality follows for all $|z - a| \leq r$.

Theorem 4.8.23 (Harnack's principle)

Consider a sequence of functions $u_n(z)$, each defined, real-valued and harmonic in a certain region Ω_n. Let Ω be a region such that every point in Ω has a neighbourhood contained in all but a finite number of Ω_n and assume moreover that in this neighbourhood $u_n(z) \leq u_{n+1}(z)$ as soon as n is sufficiently large. Then there are only two possibilities: As $n \to \infty$ either $u_n(z)$ tends uniformly to $+\infty$ on every compact subset of Ω, or $u_n(z)$ tends to a harmonic function $u(z)$ uniformly on compact subsets of Ω.

Proof

Let us first suppose that there exists $z_0 \in \Omega$ with $\lim\limits_{n\to\infty} u_n(z_0) = \infty$. By our assumption there exists 'm', a positive integer and a number $r > 0$ such that for $n \geq m$, $u_n(z)$ is harmonic in $|z - z_0| \leq r$ (contained in Ω_n) and $u_n(z)$ is a non-decreasing sequence in this neighbourhood. Now for $n \geq m$, $u_n(z) - u_m(z) \geq 0$ and is also harmonic. Thus applying Harnack's inequality with 'a' replaced by z_0 and U replaced by $u_n - u_m$, ρ replaced by r and "r" replaced by $\dfrac{r}{2}$, $u_n(z) - u_m(z) \geq \dfrac{1}{3}(u_n - u_m)(z_0)$ for $|z - z_0| \leq \dfrac{r}{2}$. Therefore $u_n(z) \geq u_m(z) + \dfrac{1}{3}(u_n - u_m)(z_0)$. Hence $u_n(z) \geq \min\limits_{|z-z_0|\leq\frac{r}{2}} u_m(z) + \dfrac{1}{3}(u_n - u_m)$

(z_0) for $|z - z_0| \leq \dfrac{r}{2}$. Allowing n to tend to ∞ we see that $u_n(z) \to \infty$ uniformly in $|z - z_0| \leq \dfrac{r}{2}$. On the other hand, if for a certain $z_1 \in \Omega$, $\lim\limits_{n\to\infty} u_n(z_1) < \infty$ then another application of the Harnack's inequality gives the following:

$$u_n(z) - u_m(z) \leq \frac{r + \dfrac{r}{2}}{r - \dfrac{r}{2}}(u_n - u_m)(z_1), \quad |z - z_1| \leq \frac{r}{2}$$

$$= 3(u_n - u_m)(z_1)$$
$$u_n(z) \leq u_m(z) + 3(u_n - u_m)(z_1) \tag{1}$$

$$\leq \max\limits_{|z-z_1|\leq\frac{r}{2}} u_m(z) + 3(u_n - u_m)(z_1) \quad \left(|z - z_1| \leq \frac{r}{2}\right).$$

Hence for $|z - z_1| \leq \dfrac{r}{2}$, $\{u_n(z)\}$ is an increasing and bounded real sequence and so $\lim\limits_{n\to\infty} u_n(z) < \infty$ for every z such that $|z - z_1| \le \dfrac{r}{2}$. From the above observations it follows that the sets A and B on which $\lim\limits_{n\to\infty} u_n(z)$ is respectively finite or infinite are both open and as Ω is connected one of these sets must be empty, i.e., if $A \ne \phi$ then $A = \Omega$ and if $B \ne \phi$ then $B = \Omega$. Thus as soon as the limit is infinite at a single point then this limit is infinite at every point in Ω and $u_n(z) \to \infty$ uniformly in a small neighbourhood around each point in Ω. Since every compact subset of Ω can be covered by finitely many such neighbourhoods $u_n(z) \to \infty$ uniformly on every compact subset of Ω. On the other hand, if the limit is finite at every point then at each $z_1 \in \Omega$ we have using (1)

$$u_n(z) - u_m(z) \le 3(u_n(z_1) - u_m(z_1)) \qquad (2)$$

for $|z - z_1| \le \dfrac{r}{2}$, $\forall\, n \ge m$. Now by our assumption $u_n(z_1)$ converges to a finite number as $n \to \infty$. Hence there exists N such that for all n, $m \ge N$, $|u_n(z_1) - u_m(z_1)| < \dfrac{\varepsilon}{3}$. Therefore $|u_n(z) - u_m(z)| < \varepsilon$ for n, $m \ge N$ by (2). Thus $u_n(z)$ is uniformly Cauchy in a neighbourhood of z_1. Since any compact subset can be covered by finitely many such neighbourhoods it follows that $u_n(z)$ is uniformly Cauchy and so converges uniformly on every compact subset of Ω to a function $u(z)$ (say). In particular $u(z)$ is also continuous being locally a uniform limit of continuous functions. Further

$$u_n(z_0) = \frac{1}{2\pi} \int_0^{2\pi} u_n(z_0 + re^{i\theta})\, \mathrm{d}\theta$$

whenever $|z - z_0| \le r$ is contained in Ω, by mean value property of u_n. Allowing $n \to \infty$ and observing that the convergence is uniform on compact subsets of Ω and in particular on $|z - z_0| = r$, we have

$$u(z_0) = \frac{1}{2\pi} \int_0^{2\pi} u(z_0 + re^{i\theta})\, \mathrm{d}\theta.$$

This proves that u satisfies mean value property and as it is already continuous it follows from Theorem 4.8.21 that $u(z)$ is harmonic in Ω.

Definition 4.8.24

Let Ω be a region in the finite complex plane. We denote by Ω^*, the region obtained by reflecting Ω in the real axis, i.e., $z \in \Omega^* \Leftrightarrow \bar{z} \in \Omega$. We say Ω is symmetric with respect to the real axis if $\Omega = \Omega^*$. We shall also denote the intersection of Ω with the upper half plane as Ω^+, the intersection of Ω with the lower half plane as Ω^- and the intersection of Ω with the real axis as σ.

Note 4.8.25
Let us first denote the upper half plane as H and the lower half plane as H^- (this notation will be assumed throughout this section). If Ω is symmetric with respect to the real axis then σ, Ω^+, Ω^- are all non empty. Indeed if Ω does not intersect the real axis then Ω (being connected) is completely contained in either H or H^-. If $\Omega \subseteq H$ then $z \in \Omega \Rightarrow \bar{z} \in H^-$ and so $\bar{z} \notin \Omega$. But $\Omega = \Omega^* \Rightarrow \bar{z} \in \Omega$.This contradiction proves that $\Omega \subseteq H$ is not possible. A similar reasoning tells us that $\Omega \subseteq H^-$ is not possible. Thus $\Omega \cap \mathbb{R} = \sigma \neq \phi$. If $z \in \sigma$ then there exists a neighbourhood N of z such that $N \subset \Omega$ and this neighbourhood intersects both H and H^-. Hence $\Omega^+ \neq \phi$ and $\Omega^- \neq \phi$.

Theorem 4.8.26 (Schwarz reflection principle)
Let Ω be a region symmetric with respect to the real axis. Let $v(z)$ be a real valued function continuous in $\Omega^+ \cup \sigma$, harmonic in Ω^+, with $v(z) = 0$ on σ. Then v has a harmonic extension V to Ω satisfying $V(\bar{z}) = - V(z)$. Moreover if v is the imaginary part of an analytic function $f(z)$ in Ω^+ then $f(z)$ has an analytic extension satisfying $f(z) = \overline{f(\bar{z})}$ for $z \in \Omega$.

Proof
Define

$$V(z) = \begin{cases} v(z), & z \in \Omega^+ \\ 0, & z \in \sigma \\ -v(\bar{z}), & z \in \Omega^-. \end{cases}$$

We first claim that V is harmonic in Ω. Note that (Solved Exercises: Problem 2 of chapter 2) $V(z)$ is harmonic in Ω^-. Thus it suffices to prove that V is harmonic on σ. Let $x_0 \in \sigma$. Take a disc D centre at x_0 and radius R whose closure is contained in Ω. Denote by P_V the Poission integral of V with respect to this disc D formed with the boundary values of V. More specifically

$$P_V(w) = \frac{1}{2\pi i} \int\limits_{|\zeta - x_0| = R} \frac{R^2 - |w - x_0|^2}{|\zeta - w|^2} \frac{V(\zeta)}{(\zeta - x_0)} d\zeta \qquad (|w - w_0| < R). \quad (1)$$

The difference $V - P_V$ is harmonic in the upper half of the disc D and vanishes on T^+ the upper semi circle of D by Corollary 4.8.17. By definition V vanishes on σ. Also using (1) above if $w = x \in \sigma$ with $|x - x_0| < R$ we can write

$$P_V(x) = \frac{1}{2\pi i} \int\limits_0^{2\pi} \frac{R^2 - |x - x_0|^2}{|x_0 + Re^{i\theta} - x|^2} \frac{V(x_0 + Re^{i\theta})}{Re^{i\theta}} iRe^{i\theta} \, d\theta$$

$$= \frac{1}{2\pi} \int_0^\pi \frac{R^2 - |x - x_0|^2}{|x_0 + Re^{i\theta} - x|^2} \ V(x_0 + Re^{i\theta}) \ d\theta$$

$$+ \frac{1}{2\pi} \int_\pi^{2\pi} \frac{R^2 - |x - x_0|^2}{|x_0 + Re^{i\theta} - x|^2} \ V(x_0 + Re^{i\theta}) \ d\theta.$$

Consider

$$\int_\pi^{2\pi} \frac{R^2 - |x - x_0|^2}{|x_0 + Re^{i\theta} - x|^2} V(x_0 + Re^{i\theta}) \ d\theta = \int_0^\pi \frac{R^2 - |x - x_0|^2}{|x_0 + Re^{-i\theta} - x|^2} V(x_0 + Re^{-i\theta}) \ d\theta$$

(with $\phi = 2\pi - \theta$ and replacing ϕ by θ).

By definition
$V(x_0 + Re^{-i\theta}) = -V(x_0 + Re^{i\theta})$ and $|x_0 + Re^{-i\theta} - x|^2 = |x_0 + Re^{i\theta} - x|^2$.

Hence $P_V(x) = 0$. Hence $V - P_V$ is identically 0 in the upper half of the disc D using maximum and minimum principles for harmonic functions as applied to the region bounded by T^+ and $\sigma \cap D$. The same proof can be repeated for the lower half and we get $V = P_V$ in the entire disc D. Thus V is harmonic in a neighbourhood of x_0 (being equal to the harmonic function P_V in this neighbourhood).

By the defining relations this extension V obviously satisfies $V(\bar{z}) = -V(z)$, $\forall \ z \in \Omega$ (Check this specifically for $z \in \Omega^+$, σ and Ω^-).

For the next part let us assume $v(z) = \text{Im} \ f(z)$ in Ω^+ where $f(z)$ is analytic in Ω^+. Again consider a disc with centre on σ. As before we extend v to the whole of D with $v(\bar{z}) = -v(z)$ and in this disc, which is simply connected, we can also have a conjugate harmonic function which we denote as $-u$. We also normalize u so that $u = \text{Re} \ f$. Now consider $U(z)$
$= u(z) - u(\bar{z})$ defined on D. On the real diameter $\dfrac{\partial U}{\partial y} = 2\dfrac{\partial u}{\partial y} = -2\dfrac{\partial v}{\partial x}$

$= 0 = \dfrac{\partial U}{\partial x}$ (see solution to the Solved Exercise; Problems 2, Chapter 2).

Thus the analytic function $\dfrac{\partial U}{\partial x} - i\dfrac{\partial U}{\partial y}$ (analytic by Theorem 4.8.6 (i)) van-

ishes on the real diameter. Thus the analytic function $\dfrac{\partial U}{\partial x} - i\dfrac{\partial U}{\partial y}$ is zero in

the entire disc D. This implies that $\dfrac{\partial U}{\partial x} = 0$ and $\dfrac{\partial U}{\partial y} = 0$ every where in

D. Therefore U is constant and in as much as it vanishes on $\sigma \cap D$ (for $z \in \sigma \cap D$, $z = \bar{z}$ and so $u(z) = u(\bar{z})$ or that $U(z) = 0$) we get $U = 0$ in D. Hence $u(z) = u(\bar{z})$ in D. Therefore $f(z) = u + iv$ can be extended analytically in D and it satisfies $f(z) = \overline{f(\bar{z})}$.

The construction can be extended to arbitrary discs around various points on σ and they are easily seen to be analytic extensions of each other (Note that $v(z) = \text{Im } f(z)$ throughout Ω^+). Hence f can be extended analytically to the whole of Ω as follows:

$$F(z) = \begin{cases} f(z), & z \in \Omega^+ \\ u(z), & z \in \sigma. \\ \overline{f(\bar{z})}, & z \in \Omega^-. \end{cases}$$

This $F(z)$ is analytic in Ω^+ by hypothesis. It is analytic in Ω^- (See Solved Exercises Problem 1, Chapter 2) and $F(z) = u(z) + iv(z) = u(z)$ on σ is also analytic in a neighbourhood of σ (namely in the union of the discs D). The fact $F(z) = \overline{F(\bar{z})}$ is easy to check.

As far as analytic functions are concerned the above theorem can be reworded as follows.

Theorem 4.8.27
Let Ω be a region symmetric with respect to the real axis. Lel Ω^+, Ω^-, σ be as in definition 4.8.24. If f is analytic in Ω^+, $\text{Im } f(z) \to 0$ as $z \to a$ a point on σ then $f(z)$ can be extended analytically to the whole of Ω and the extended function (which is also denoted by f) satisfies $f(z) = \overline{f(\bar{z})}$ for all $z \in \Omega$.

Theorem 4.8.28 (Solution of the Dirichlet's problem for the upper half plane)
Let f be a real-valued piece-wise continuous and bounded function on the real axis. Let

$$u(x + iy) = \frac{1}{\pi} \int_{-\infty}^{\infty} \frac{yf(t)\,dt}{|t - w|^2} \quad \text{where } w = x + iy.$$

Then u is harmonic in the upper half plane and $\lim_{y \to 0} u(x + iy) = f(x)$ at each point $x(\in \mathbb{R})$ at which f is continuous.

Proof

Let $\phi(z) = i\left(\dfrac{1-z}{1+z}\right)$. Then ϕ maps the unit circle onto the real axis with interior corresponding to the upper half plane. Thus $f(\phi(z))$ is a piece-wise continuous bounded function on the unit circle. Thus

$$P_U(z) = \frac{1}{2\pi} \int_{-\pi}^{\pi} \frac{1 - |z|^2}{|e^{i\theta} - z|^2} U(e^{i\theta})\,d\theta$$

where $U(e^{i\theta}) = f(\phi(e^{i\theta}))$ gives a harmonic function in $|z| < 1$ with boundary values $U(e^{i\theta})$ at the points of continuity for $f \circ \phi$. We shall now change the variable $z = \phi^{-1}(w)$. As w varies over the upper half plane $\phi^{-1}(w)$ varies over the interior of the unit disc and

$$P_U(\phi^{-1}(w)) = \frac{1}{2\pi} \int_{-\pi}^{\pi} \frac{1 - |\phi^{-1}(w)|^2}{|e^{i\theta} - \phi^{-1}(w)|^2} f\left(i\left(\frac{1 - e^{i\theta}}{1 + e^{i\theta}}\right)\right) d\theta$$

gives a function defined in the upper half plane with boundary values $f(t)$ (with $t = \phi(e^{i\theta})$) at all points t at which f is continuous. We shall now show that $P_U(\phi^{-1}(w))$ is harmonic in w in the upper half plane and $P_U(\phi^{-1}(w)) =$

$u(x + iy) = \dfrac{1}{\pi} \int_{-\infty}^{\infty} \dfrac{yf(t)\,dt}{|t - w|^2}$ where $w = x + iy$. Indeed if $F(z)$ is a non-constant

analytic function in a region Ω and g is real valued and harmonic in $F(\Omega)$ then for all $z_0 \in \Omega$, there is a neighbourhood of z_0 (say $N(z_0)$) in which $g(F(z))$ is the real part of $G(F(z))$ where $G(w)$ is an analytic function in $N(w_0) = F(N(z_0))$. (This can be proved as follows. Since F is open (open map) and continuous, for each fixed $w_0 = F(z_0) \in F(\Omega)$ and a circular neighbourhood $N(w_0) \subseteq F(\Omega)$ of w_0 there exists a neighbourhood $N(z_0) \subseteq \Omega$ of z_0 such that F maps $N(z_0)$ into $N(w_0)$. $N(w_0)$ is simply connected and an application of Theorem 4.8.6 (ii) yields $g = \mathrm{Re}\ G$, where G is analytic in $N(w_0)$). Thus for all z in $N(z_0)$, $\mathrm{Re}\ G(F(z)) = g(F(z))$. But $G(F(z))$ is analytic in $N(z_0)$. Therefore $g(F(z))$ is harmonic in $N(z_0)$. This observation enables us to conclude that $u(w) = P_U(\phi^{-1}(w))$ is harmonic as w varies in the upper half plane. Therefore $u(w) = u(x + iy) = P_U(\phi^{-1}(w))$ is harmonic in the upper half plane with boundary values $f(t)$ at points of continuity for f. Now

$$u(w) = u(x + iy) = \frac{1}{2\pi} \int_{-\pi}^{\pi} \frac{1 - |\phi^{-1}(w)|^2}{|e^{i\theta} - \phi^{-1}(w)|^2} f\left(i\left(\frac{1 - e^{i\theta}}{1 + e^{i\theta}}\right)\right) d\theta.$$

(Note that θ can be allowed to vary either from 0 to 2π or from $-\pi$ to π

to cover the unit circle). Let $t = i\left(\dfrac{1 - e^{i\theta}}{1 + e^{i\theta}}\right) = \tan \dfrac{\theta}{2}$. As θ varies from

$-\pi$ to π, t varies from $-\infty$ to ∞. Further $w = \phi(z) = i\left(\dfrac{1 - z}{1 + z}\right) \Rightarrow$

$z = \phi^{-1}(w) = \left(\dfrac{i - w}{i + w}\right)$ and $t = \phi(e^{i\theta}) \Rightarrow e^{i\theta} = \dfrac{i - t}{i + t}$. Therefore

$$u(x + iy) = \frac{1}{2\pi} \int_{-\infty}^{\infty} \frac{1 - \left|\dfrac{i - w}{i + w}\right|^2}{\left|\dfrac{i - t}{i + t} - \dfrac{i - w}{i + w}\right|^2} \frac{2f(t)}{1 + t^2} dt$$

$$= \frac{1}{\pi} \int\limits_{-\infty}^{\infty} \frac{\operatorname{Re} i\bar{w}}{|w-t|^2} \, f(t) \, dt$$

$$= \frac{1}{\pi} \int\limits_{-\infty}^{\infty} \frac{y f(t) \, dt}{|t-w|^2}.$$

SOLVED EXERCISES

1. Compute $\int\limits_{|z|=r}(\operatorname{Re} z) \, dz$.

    Solution $\operatorname{Re} z = \dfrac{z + \bar{z}}{2}$ and on $|z| = r$, $z = re^{i\theta}$, $\bar{z} = re^{-i\theta}$ and $dz = ire^{i\theta} \, d\theta$. Thus

    $$\int\limits_{|z|=r}(\operatorname{Re} z) \, dz = \int\limits_{0}^{2\pi} \frac{1}{2} \, (re^{i\theta} + re^{-i\theta}) \, ire^{i\theta} \, d\theta$$

    $$= i\frac{r^2}{2} \int\limits_{0}^{2\pi} (\cos 2\theta + i \sin 2\theta + 1) \, d\theta$$

    $$= i\frac{r^2}{2} \left[\frac{\sin 2\theta}{2} - i\frac{\cos 2\theta}{2} + \theta \right]_{0}^{2\pi}$$

    $$= i\frac{r^2}{2}(2\pi) = i\pi r^2$$

2. Compute $\int\limits_{|z|=1}|z-1| \, |dz|$.

    Solution On $|z| = 1$ we have $z = e^{i\theta}$ ($0 \le \theta \le 2\pi$) and $dz = ie^{i\theta} \, d\theta$ so that $|z - 1|^2 = 1 + 1 - 2 \cos \theta = 4 \sin^2 (\theta/2)$, $|dz| = d\theta$. Thus

    $$\int\limits_{|z|=1}|z-1| \, |dz| = \int\limits_{0}^{2\pi} 2 \sin (\theta/2) \, d\theta$$

    $$= -4(\cos (\theta/2))_{0}^{2\pi} = -4[-1 -1] = 8.$$

3. Suppose f is analytic in a region containing a closed curve γ. Prove that $\operatorname{Re} \int\limits_{\gamma} \overline{f(z)} \, f'(z) \, dz = 0$.

Solution Write $w = f(z)$, $dw = f'(z) \, dz$, $\Gamma = f(\gamma)$.

$$\int\limits_{\gamma} \overline{f(z)} f'(z) \, dz = \int\limits_{\Gamma} \bar{w} \, dw.$$

Let the closed curve Γ be given by the equation $w = w(t) = u(t) + iv(t)$ ($a \le t \le b$). Since Γ is a closed curve $w(b) = w(a)$ or that $u(b) = u(a)$

and $v(b) = v(a)$. Now by definition

$$\text{Re} \int_\Gamma \overline{w} \, dw = \text{Re} \left\{ \int_a^b (u(t) - iv(t)) \, (u'\,(t) + iv'(t)) \, dt \right\}$$

$$= \int_a^b (u(t) \, u'\,(t) + v\,(t) \, v'\,(t)) \, dt$$

$$= \frac{1}{2} \int_a^b \frac{d}{dt} ((u(t))^2 + (v(t))^2) \, dt$$

$$= \frac{1}{2} [(u(b))^2 + (v(b))^2] - [(u(a))^2 + (v(a))^2]$$

$$= 0.$$

Thus $\int_\Gamma \overline{w} \, dw = \int_\gamma \overline{f(z)} f'(z) \, dz$ has zero real part.

4. If C is a positively oriented simple closed curve in the plane show that the area enclosed by C is given by

$$A = \frac{1}{2i} \int_C \overline{z} \, dz.$$

Solution By elementary calculus the area enclosed by a simple closed curve C in the plane is given by

$$\frac{1}{2} \int_C R^2 \, d\phi$$

where (R, ϕ) are the polar coordinates of a variable point on C. Let $z = z(t) = R(t) \, e^{i\phi(t)}$ for $a \le t \le b$. By taking logarithmic derivatives, it follows that

$$\frac{R'(t)}{R(t)} dt + i\phi'(t) \, dt = \frac{z'(t)}{z(t)} dt. \tag{1}$$

It is clear that

$$d\phi = \phi'(t) \, dt. \tag{2}$$

$$(R(t))^2 = z(t)\overline{z(t)}. \tag{3}$$

Thus using (1), (2) and (3) we get

$$A = \frac{1}{2} \int_a^b (R(t))^2 \, \phi'(t) \, dt$$

$$= \frac{1}{2i} \int_a^b \overline{z(t)} \, z'(t) \, dt - \frac{1}{2i} \int_a^b R(t) \, R'(t) \, dt$$

$$= \frac{1}{2i} \int_C \overline{z} \, dz - \frac{1}{4i} \int_a^b \frac{d}{dt} (R^2(t)) \, dt$$

$$= \frac{1}{2i} \int_C \overline{z} \, dz - 0$$

$\left(\text{since } \dfrac{d}{dt}(R^2\,(t)) \text{ is a total differential and } C \text{ is a closed curve}\right.$

$\displaystyle\int_a^b \frac{d}{dt}(R^2(t))\ dt = R^2\,(b) - R^2\,(a) = 0.$ In fact $z(a) = z(b) \Rightarrow R(a) = $

$\left. R(b)\right).$ Thus $A = \dfrac{1}{2i}\displaystyle\int_C \bar{z}\,dz.$

5. Prove that a function which is analytic in the finite complex plane (also called an entire function) and which satisfies an inequality $|f(z)| \le c\,|z|^n$ ($z \in \mathbb{C}$ or for large $|z|$) for some non-negative integer n must be a polynomial of degree atmost n.

   Solution We shall prove the above proposition by induction on n. If $n = 0$, the function is entire and bounded and consequently must be a constant by Liouville's theorem. On the other hand if $n = 1$ again

   $g(z) = \dfrac{f(z)}{z}$ is analytic for $z \ne 0$ and at $z = 0$ $\left(\text{since } \lim\limits_{z \to 0} z\,g(z) = f(0)\right.$
   $= 0 \text{ as } |f(z)| \le c\,|z|\left.\right)$ it has a removable singularity with $g(0) = f'(0)$, g is entire and bounded and so $g(z) = c$, a constant or that $f(z) = cz$ which is a polynomial of degree at most 1.

   We shall now assume that if $|f(z)| \le c|z|^m$ for some $m \ge 1$ then $f(z)$ must be polynomial of degree atmost m and prove the same statement when m is replaced by $(m + 1)$. Let $f(z)$ be entire and satisfy $|f(z)| \le c\,|z|^{m+1}$. This implies in particular that $f(0) = 0$, $\left|\dfrac{f(z)}{z}\right|$
   $\le c|z|^m$ and thus $f'(0) = \lim\limits_{z \to 0}\dfrac{f(z)}{z} = 0.$ Put

   $$g(z) = \begin{cases} \dfrac{f(z)}{z} & \text{for } z \ne 0 \\ 0 & \text{for } z = 0 \end{cases}$$

   Now $g(z)$ is analytic at all points $z \ne 0$ and at $z = 0$ it has a removable singularity (Since $\lim\limits_{z \to 0} z g(z) = \lim\limits_{z \to 0} f(z) = f(0) = 0$) and its extended value is $\lim\limits_{z \to 0} g(z) = f'(0) = 0.$ Thus $g(z)$ is entire. But now $|g(z)| \le |z|^m$ for all z from the definition. Hence by induction hypothesis $g(z)$ must be a polynomial of degree atmost m since $f(z) = zg(z)$ for all z (including $z = 0$) $f(z)$ must be a polynomial of degree atmost $(m + 1)$. This completes the proof. See solution for problem 20(ii) for modification in case the inequality is valid only for large $|z|$.

6. Let Ω be a region and $f : \Omega \to \mathbb{C}$ analytic. Let $c \in \Omega$ and $f'(c) \ne 0$. If B is an open disc centred at c and $B \subseteq \Omega$ is such that $\sup\limits_{\zeta \in B}|f'(\zeta)$
   $- f'(c)| < |f'(c)|$ then show that f is injective in B. Hence or other-

wise determine the largest open disc around the origin wherein $f(z) = z^2 + z$ is injective.

Solution   The analytic function $f'(\zeta) - f'(c)$ has $f(\zeta) - f'(c)\zeta$ as a primitive in Ω. Hence (See Theorem 4.1.6 viii) if γ is the line joining two points $z, w \in B$, then

$$\int_\gamma [f'(\zeta) - f'(c)] \, d\zeta = (f(w) - f'(c)w) - (f(z) - f'(c)z)$$

$$= (f(w) - f(z)) - f'(c) (w - z). \qquad (1)$$

On the other hand

$$\left| \int_\gamma [f'(\zeta) - f'(c)] \, d\zeta \right| \leq \int_\gamma |f'(\zeta) - f'(c)| \, |d\zeta|$$

$$\leq \left(\sup_{\zeta \in \gamma} |f'(\zeta) - f'(c)| \right) |w - z|. \qquad (2)$$

Since $\gamma \subset B$, our hypothesis now implies (using (1) and (2))

$$\left| \frac{f(w) - f(z)}{w - z} - f'(c) \right| < |f'(c)| \ (w, z \in B \ w \neq z). \qquad (3)$$

We now claim that f is injective in B. If not we can find w, z in B, $w \neq z$ such that $f(w) = f(z)$. For this w and z we have from (3), $|f'(c)| < |f'(c)|$ a contradiction. This contradiction establishes the injectivity of f in B. Consider $f(z) = z^2 + z$. This is analytic throughout of $\Omega = \mathbb{C}$. ($f'(0) = 1 \neq 0$). Take $c = 0$ and B as the open disc centre 0 and radius $r < \dfrac{1}{2}$.

Now

$$\sup_{\zeta \in B} |f'(\zeta) - f'(0)| = \sup_{\zeta \in B} |f'(\zeta) - 1|$$

$$= \sup_{\zeta \in B} |2\zeta + 1 - 1|$$

$$< 1 = |f'(0)|.$$

Since $r < \dfrac{1}{2}$ is arbitrary it follows that f is injective in $|z| < \dfrac{1}{2}$. On the other hand f can not be injective in any larger disc because $f'(-1/2) = 0$ and in every small neighbourhood of $z = -1/2$, f is not one-to-one by local correspondence theorem and every open disc larger than $|z| < \dfrac{1}{2}$ contains the point $z = -1/2$. Thus the largest open disc, around origin in which f is injective, is $|z| < \dfrac{1}{2}$.

7. Let f be analytic in a region containing 0. Let $f'(0) \neq 0$. Prove that there exists an analytic function $g(z)$ defined in a neighbourhood of 0 such that

$$f(z^n) = f(0) + (g(z))^n$$

Solution Consider the analytic function $f(z) - f(0)$ which has a simple zero at $z = 0$. Hence we can write $f(z) - f(0) = zh(z)$ where $h(z)$ is analytic in an open disc around origin with $h(0) \neq 0$. (See note 4.4.15). By continuity of h we will have $h(z) \neq 0$ in another disc around origin. Let B be an open disc of radius $r < 1$ in which

$$f(z) - f(0) = zh(z) \text{ with } h(z) \neq 0.$$

If $z \in B$, $|z| < r < 1$ then $|z|^n < r^n < r$ and hence we have

$$f(z^n) = f(0) + z^n h(z^n) \ (z \in B).$$

Thus $f(z^n) - f(0) = z^n \zeta(z)$ where $\zeta(z) = h(z^n)$ is analytic in B. Since B is simply connected we can choose an analytic branch for the n-th root of $\zeta(z)$. (Use Theorem 4.4.7 to define $\log \zeta(z)$ and define n-th root of $\zeta(z)$ as $e^{(\log \zeta(z))/n}$). Denote this by $\eta(z)$ so that $(\eta(z))^n = \zeta(z)$. From (1) we have

$$f(z^n) = f(0) + z^n \, (\eta(z))^n$$
$$= f(0) + (z\eta(z))^n$$
$$= f(0) + (g(z))^n$$

with $g(z) = z\eta(z)$ analytic in B.

8. Under the hypothesis of local correspondence theorem prove that the n, n-th roots of the equation $f(z) = w$ (as w varies in the prescribed neighbourhood of w_0) are given as the solutions of a polynomial equation

$$z^n + a_1(w) z^{n-1} + a_2(w) z^{n-2} + \ldots + a_{n-1}(w) z + a_n(w) = 0.$$

where the coefficients $a_i(w)$ $(1 \leq i \leq n)$ are analytic functions of w in a neighbourhood of w_0.

Solution By the local correspondence theorem (Theorem 4.4.22) we know that the equation $f(z) = w$ where $|w - w_0| < \delta$ has exactly n roots $z_1(w), z_2(w), \ldots, z_n(w)$ in the neighbourhood $|z - z_0| < \varepsilon$. Applying the generalized argument principle (Theorem 4.6.13) with $g(z) = z^m$ ($m = 1, 2, \ldots$) we have

$$\frac{1}{2\pi i} \int_{|z-z_0|=\varepsilon} \frac{z^m f'(z)}{f(z) - w} \, dz = \sum_{j=1}^n (z_j(w))^m \tag{1}$$

Imitating the proof of Theorem 4.4.3 we can easily prove that the LHS of (1) represents an analytic function of w in $|w - w_0| < \dfrac{\delta}{2}$.

(Note that with the notations of Theorem 4.4.22 for $|z - z_0| = \varepsilon$, $|f(z) - w_0| \geq \delta$ and if $|w - w_0| < \dfrac{\delta}{2}$ then $|f(z) - w| \geq \dfrac{\delta}{2}$ and this is precisely what is required to imitate the proof of Theorem 4.4.3). Thus the power sums of the roots $z_j(w)$ are analytic functions of w in a neighbourhood of w_0. If the polynomial equation whose roots are $z_j(w)$ $(1 \leq j \leq m)$ is given by

$$z^n + a_1(w) z^{n-1} + a_2(w) z^{n-2} + \ldots + a_{n-1}(w) z + a_n(w) = 0.$$

then we know that each $a_j(w)$ (the elementary symmetric functions of the roots $z_j(w)$) can be expressed as polynomials in the power sums of $z_j(w)$ and hence they are also analytic.

9. Prove or disprove the existence of analytic functions in a neighbourhood of 0 satisfying any of the following conditions.

(i) $f\left(\dfrac{1}{n}\right) = (-1)^n \left(\dfrac{1}{n}\right)$ $(n = 1, 2, \ldots)$.

(ii) $f\left(\dfrac{1}{n}\right) = \dfrac{1}{n^2} - 1$ $(n = 1, 2, \ldots)$.

(iii) $|f^{(n)}(0)| \geq (n!)^2$ $(n = 1, 2, \ldots)$.

(iv) $f \neq 0$ and $\left|f\left(\dfrac{1}{n}\right)\right| \leq e^{-n}$ $(n = 1, 2, \ldots)$.

Solution Since $f(z) = z^2 - 1$ is analytic in the whole plane and satisfies $f\left(\dfrac{1}{n}\right) = \dfrac{1}{n^2} - 1$ we need to consider (i), (iii). (iv) only. Indeed we shall prove that no function analytic at 0 can satisfy any of these conditions.

Consider (i). Assume there exists an analytic function $f(z)$ with $f\left(\dfrac{1}{n}\right) = (-1)^n \left(\dfrac{1}{n}\right)$ $(n = 1, 2, \ldots)$. Since for each $n = 1, 2, \ldots$ $(-1)^n$ and $(-1)^{n+1}$ have opposite sign $f\left(\dfrac{1}{n}\right)$ and $f\left(\dfrac{1}{n+1}\right)$ must be real and be of opposite sign. Thus by continuity of f there exists x_n such that $\dfrac{1}{n+1} < x_n < \dfrac{1}{n}$ and $f(x_n) = 0$. Evidently $x_n \to 0$ as $n \to \infty$ and so $f(x_n) \to f(0)$. But $f(0) = \lim_{n \to \infty} f\left(\dfrac{1}{n}\right) = \lim_{n \to \infty} \dfrac{(-1)^n}{n} = 0$. Thus 0 is a limit point of zeros of f. Thus f can not be analytic at $z = 0$ (See Theorem 4.4.16) unless $f = 0$. But $f = 0$ can not satisfy (i).

Now consider (iii). Again if we assume that there is an analytic functions f at 0 with $\dfrac{|f''(0)|}{n!} \geq (n!)$ then

$$f(z) = \sum_0^\infty c_n z^n,$$

with $|c_n| \geq n!$. This implies that the radius of convergence R of $\sum_0^\infty c_n z^n$ satisfies

$$\frac{1}{R} = \limsup_{n \to \infty} |c_n|^{1/n} \geq \limsup_{n \to \infty} (n!)^{1/n} = \infty.$$

Thus $R = 0$ and f can not be analytic in a neighbourhood of the origin.

Consider (iv). If there exists a function $f(z) \neq 0$ analytic at the origin satisfying $\left| f\left(\dfrac{1}{n}\right) \right| \leq e^{-n}$ then $f(0) = 0$. (Note that by continuity $f(0) = \lim_{n \to \infty} f\left(\dfrac{1}{n}\right)$ and so $|f(0)| \leq \lim_{n \to \infty} e^{-n} = 0$.) Thus $f(z)$ should have a zero of finite order at $z = 0$ say m. Now

$$f(z) = z^m \, g(z) \tag{1}$$

where $g(z)$ is analytic at the origin with $g(0) \neq 0$. But for n large we have from (1) above that

$$\left| g\left(\frac{1}{n}\right) \right| \leq n^m \left| f\left(\frac{1}{n}\right) \right| \leq n^m e^{-n}$$

and $n^m e^{-n} \to 0$ as $n \to \infty$. This shows that $g(0) = \lim_{n \to \infty} g\left(\dfrac{1}{n}\right) = 0$ a contradiction. This contradiction shows that no such f can exist.

10. Let U be the open unit disc and $f : U \to U$ analytic with $f(0) = 0$. Let $\zeta = e^{2\pi i/n}$ (n is a fixed positive integer). Prove that for all $z \in U$,

$$|f(\zeta z) + f(\zeta^2 z) + \ldots + f(\zeta^n z)| \leq n \, |z|^n$$

with equality at $z \neq 0$ if and only if $f(z) = az^n$ with $|a| = 1$.

Solution Define $F(z) = f(\zeta z) + f(\zeta^2 z) + \ldots + f(\zeta^n z)$. By hypothesis $F(0) = 0$. But for $1 \leq k \leq n - 1$,

$$F^{(k)}(z) = \zeta^k f^{(k)}(\zeta z) + \zeta^{2k} f^{(k)}(\zeta^2 z) + \ldots + \zeta^{nk} f^{(k)}(\zeta^n z).$$

Thus for $1 \leq k \leq n - 1$,

$$\begin{aligned}
F^{(k)}(0) &= f^{(k)}(0) \left(\zeta^k + \zeta^{2k} + \zeta^{3k} + \ldots + \zeta^{nk} \right) \\
&= f^{(k)}(0) \left(1 + \zeta^k + \zeta^{2k} + \ldots + \zeta^{(n-1)k} \right) \\
&\qquad\qquad\qquad\qquad \text{(since } \zeta^n = 1\text{)}
\end{aligned}$$

$$= f^{(k)}(0)\left(\frac{\zeta^{nk} - 1}{\zeta^k - 1}\right) = 0.$$

(as k is not a multiple of n).
But $F^{(n)}(0) = nf^{(n)}(0)$. (Since each ζ^{nj} in the summation is 1 and there are n terms). Thus F has a zero of order atleast n at $z = 0$. Let the order of F at $z = 0$ be m so that $m \geq n$. Now $F(z) = z^m G(z)$ where G is analytic at 0 and $G(0) \neq 0$. Thus $\dfrac{F(z)}{z^m}$ is analytic in U (whenever $z \neq 0$ but $z \in U$, it is clear that $\dfrac{F(z)}{z^m}$ is analytic at z but $\dfrac{F(z)}{z^m} = G(z)$ is analytic at $z = 0$ also). We first note that $|F(z)| \leq n$ for all $z \in U$ by our hypothesis. Now fix $z_0 \in U$. Choose r such that $1 > r > |z_0|$.

On $|z| = r$, $\left|\dfrac{F(z)}{z^m}\right| \leq \dfrac{n}{r^m}$ and by Maximum modulus theorem the

inequality holds for $|z| \leq r$. Thus $\left|\dfrac{F(z_0)}{z_0^m}\right| \leq \dfrac{n}{r^m}$ and allowing $r \to 1$

we see that $|F(z_0)| \leq n|z_0|^m \leq n|z_0|^n$. Since z_0 is arbitrary in U it follows that $|F(z)| \leq n|z|^n$ and the required inequality follows.

Assume equality holds for some $z \neq 0$. Then $\left|\dfrac{F(z)}{z^n}\right|$ attains its maxi-

mum modulus n at some $z \in U$ and hence by maximum modulus theorem it reduces to a constant. Therefore

$$F(z) = n \, a \, z^n \text{ with } |a| = 1.$$

Now define $k(z) = f(z) - az^n$. Consider $k(\zeta z) + k(\zeta^2 z) + \ldots + k(\zeta^n z)$. By definition this is equal to $F(z) - naz^n$ which is 0. Hence we have

$$k(\zeta z) + k(\zeta^2 z) + \ldots + k(\zeta^n z) = 0. \tag{1}$$

We know by Schwarz Lemma that $|f(z)|^2 < 1, \forall z \in U$. Hence

$$|f(\zeta^j z)|^2 < 1, \forall z \in U \text{ and } j = 0, 1, 2, \ldots (n-1),$$

$$\text{i.e., } |az^n + k(\zeta^j z)|^2 < 1$$

or that

$$|az^n|^2 + 2\text{Re } az^n \overline{k(\zeta^j z)} + |k(\zeta^j z)|^2 < 1. \tag{2}$$

(2) gives in particular that

$$|az^n|^2 + 2\text{Re } az^n \overline{k(\zeta^j z)} < 1. \, j = 0, 1, 2, \ldots (n-1). \tag{3}$$

Using (2) with $j = 0$ we get

$$|k(z)|^2 < 1 - |z|^{2n} - 2\text{Re } az^n \overline{k(z)}. \tag{4}$$

Using (1) in (4) (observe that $k(\zeta^n z) = k(z)$) we get

$$|k(z)|^2 < (1 - |z|^{2n}) + 2\text{Re } az^n \left(\overline{k(\zeta z)} + \overline{k(\zeta^2 z)} + \dots + \overline{k(\zeta^{n-1} z)} \right) \quad (5)$$

Further using (3) in (5)

$$|k(z)|^2 < (1 - |z|^{2n}) + (n - 1)(1 - |z|^{2n}) = n(1 - |z|^{2n}). \quad (6)$$

Fix $z \in U$. Take $r > |z|$. For each such r

$$|k(z)|^2 \le \sup_{|z| \le r} |k(z)|^2 = \sup_{|z| = r} |k(z)|^2 \quad (7)$$

Equality in the above is a consequence of the maximum modulus theorem. Thus (6) and (7) give

$$|k(z)|^2 \le n(1 - r^{2n}). \quad (8)$$

Since we can choose $r > |z|$ arbitrarily we can allow $r \to 1$ in (8). Therefore $k(z) = 0$. Since z is arbitrary it follows that $k(z) \equiv 0$ which implies that $f(z) = az^n$, $|a| = 1$. Hence we have

$$|F(z)| = n|z|^n \Rightarrow f(z) = az^n, \ |a| = 1.$$

On the other hand if $f(z) = az^n$ with $|a| = 1$ it is easy to check that $|F(z)| = n|z|^n$ and we have thus discussed the case of equality completely.

11. Let f be an entire function such that $|f'(z)| \le |z|$, $\forall z$. Show that $f(z) = a + bz^2$ with $|b| \le \dfrac{1}{2}$.

Solution Since

$$|f'(z)| \le |z| \quad (1)$$

by Solved Exercise 5, $f'(z)$ is a polynomial of degree atmost 1. Therefore $f'(z) = c + dz$ and hence $f'(0) = c$. But by (1), $|f'(0)| \le 0$, i.e., $f'(0) = 0$. Hence $c = 0$ and so $f'(z) = dz$ and thus $f(z) = a + bz^2$ for some constants a, b. Now consider the representation $f(z) = a + bz^2$. Then $f'(z) = 2bz$. Putting $z = 1$ we get

$$f'(1) = 2b. \quad (2)$$

But by (1), $|f'(1)| \le 1$ and therefore from (2) we have $|2b| \le 1$ which implies that $|b| \le \dfrac{1}{2}$.

12. Show that an entire function satisfying $f(z + 1) = f(z)$ and $f(z + i) = f(z)$ is a constant.

Solution First observe that the given hypotheses imply that

$$f(z + n) = f(z) \text{ and } f(z + mi) = f(z), \ \forall m, n \in \mathbb{Z}.$$

Take any complex number $a + ib$. There exists integers n, m such that $n \le a < n + 1$ and $m \le b < m + 1$. Thus we can write

$$a = n + x, \ 0 \le x < 1.$$
$$b = m + y, \ 0 \le y < 1.$$

Therefore $f(a + ib) = f(n + x + i(m + y)) = f(x + iy)$ where $0 \le x < 1, \ 0 \le y < 1$.

Thus the set of values of $f(z)$ throughout the plane are the same as the set of its values on $I \times I$ in the plane where $I = [0, 1]$. $I \times I$ being compact, f is bounded on $I \times I$ and hence everywhere. Thus f is a bounded entire function and therefore by Liouville's theorem f reduces to a constant.

13. Suppose f is analytic in $|z| \le 1$ with $|f(z)| \le 2$ for $|z| = 1$, Im $z \ge 0$ and $|f(z)| \le 3$ for $|z| = 1$, Im $z < 0$. Then prove that $|f(0)| \le \sqrt{6}$.

Solution   Consider $g(z) = f(z) f(-z)$ which is analytic in $|z| \le 1$. By integral formula,

$$g(0) = \frac{1}{2\pi i} \int_{|z|=1} \frac{g(\zeta)}{\zeta} d\zeta.$$

We write the set of all z with $|z| = 1$ as $c_1 \cup c_2$, where c_1 is the set of all z with $|z| = 1$, Im $z \ge 0$ and c_2 is the set of all z with $|z| = 1$, Im $z < 0$. Thus

$$(f(0))^2 = g(0) = \frac{1}{2\pi i}\left(\int_{c_1} + \int_{c_2} \right) \frac{g(\zeta)}{\zeta} d\zeta.$$

Now $\left| \dfrac{1}{2\pi i} \displaystyle\int_{c_1} \dfrac{g(\zeta)}{\zeta} d\zeta \right| \le \dfrac{1}{2\pi} \displaystyle\int_{c_1} \dfrac{|g(\zeta)|}{|\zeta|} |d\zeta| \le 3.$

(Note that $\zeta \in c_1 \Rightarrow -\zeta \in c_2$).

Similarly

$$\left| \frac{1}{2\pi i} \int_{c_2} \frac{g(\zeta)}{\zeta} d\zeta \right| \le \frac{1}{2\pi} \int_{c_2} \frac{|g(\zeta)|}{|\zeta|} |d\zeta| \le 3.$$

Hence $|f(0)|^2 \le 6 \Rightarrow |f(0)| \le \sqrt{6}$.

14. Let f be an entire function satisfying $|f(z)| \le \dfrac{1}{|\text{Im } z|}$ for $|z|$ sufficiently large. Then show that $f \equiv 0$.

Solution   Fix $R > 0$ and consider $g(z) = (z^2 - R^2) f(z)$. Let $|z| = R$, Re $z \ge 0$.

$$|(z - R) f(z)| \le \left| \frac{z - R}{\text{Im } z} \right| = \sec \theta \le \sqrt{2}$$

where θ represents the acute angle between the vectors $\overrightarrow{zP}$ and $\overrightarrow{zR}$. Here P represents the foot of the perpendicular from z to the real diameter. $\left(O\hat{R}z = \dfrac{\pi}{2} - \theta = O\hat{z}R \Rightarrow R\hat{O}z = 2\theta. \right.$ But $0 \le R\hat{O}z \le \dfrac{\pi}{2}.$

Hence $0 \le \theta \le \dfrac{\pi}{4}$ and as $\sec \theta$ is increasing in the first quadrant

$\sec \theta \le \sqrt{2}$. Similarly if z is in the fourth quadrant, $\left| \dfrac{z-R}{\text{Im } z} \right| \le \sec \theta$

for the corresponding θ, with $0 \le \theta \le \dfrac{\pi}{4}$). Again for $|z| = R$, Re z ≤ 0, we can show that $|(z + R) f(z)| \le \sqrt{2}$. Thus using $|z| \pm R \le 2R$ and $2\sqrt{2} \le 3$ we get

$$|g(z)| \le |z + R| |z - R| |f(z)| \le 3R, \ \forall \ |z| = R.$$

Since $g(z)$ is analytic we can apply maximum modulus theorem for g and conclude that $|g(z)| \le 3R, \ \forall \ |z| \le R$. Hence

$$|f(z)| \le \frac{3R}{|z^2 - R^2|} \le \frac{3R}{R^2 - |z|^2}.$$

Note that R can be chosen arbitrarily. Keeping z fixed and allowing R to tend to ∞ we see that $|f(z)| = 0$. Since z is arbitrary we see that $f \equiv 0$.

15. Using Morera's theorem show that $f(z) = \int\limits_0^\infty \dfrac{e^{zt}}{1+t} dt$ is analytic in the left half plane $L = \{z \in \mathbb{C} : \text{Re } z < 0\}$.

Solution Since $|e^{zt}| = e^{xt}$ where $z = x + iy$ we have $\left| \dfrac{e^{zt}}{1+t} \right| \le e^{xt}$ with

$x < 0$ for $0 \le t < \infty$. But $\int\limits_0^\infty e^{xt} dt$ exists whenever $x < 0$. Thus $f(z)$ is well defined for $z \in L$. We now show that $f(z)$ is continuous. Fix $z = x + iy \in L$ and let $z_n \to z$ with $z_n \in L$.

$$|f(z_n) - f(z)| \le \int\limits_0^\infty \frac{|e^{z_n t} - e^{zt}|}{1+t} dt \le \int\limits_0^\infty \frac{e^{xt} |e^{(z_n - z)t} - 1|}{1+t} dt.$$

First note that for all z, $|e^z - 1| \le |z| e^{|z|}$ (consider the power series expansion for $e^z - 1$). Choose N large so that $|z_n - z| < \dfrac{|x|}{2}$ for $n \ge N$. We now get

$$|f(z_n) - f(z)| \le \int\limits_0^\infty \frac{|z_n - z| t e^{xt} e^{|z_n - z|t}}{1+t} dt$$

$$\le |z_n - z| \int\limits_0^\infty e^{xt} e^{|x| t/2} dt$$

$$\le M |z_n - z| \ (\text{since } x = -|x|)$$

where $M = \int\limits_0^\infty e^{-|x| t/2} dt$. Thus f is continuous at $z \in L$. Since z is arbitrary we have proved the continuity of $f(z)$ in L. Next we prove that $\int\limits_\gamma f(z) \, dz = 0$ for all closed curves γ in L. By definition, if $z = z(s)$

$(a \le s \le b)$ is the parametric equation of γ, we have

$$\int_{\gamma} f(z)\ dz = \int_{a}^{b}\int_{0}^{\infty} \frac{e^{z(s)t}}{1+t}\ dt\ z'(s)\ ds.$$

But the integral $\int_{0}^{\infty} \frac{e^{z(s)t}}{1+t} z'(s)\ dt$ being uniformly convergent in $a \le s \le$ b (Note that as $z(s)$ varies over γ, Re $z(s) \le -\delta$ for some $\delta > 0$ and $z'(s)$ is bounded on γ). Using the theory of double integrals (For example see Theorem 10. 8 on p 443 of Multiple integrals Field theory and series by B. M. Badak, S. V. Fomin, Mir Publishers, Moscow) we see that

$$\int_{a}^{b} z'(s) ds \int_{0}^{\infty} \frac{e^{z(s)t}}{1+t}\ dt = \int_{0}^{\infty} \frac{dt}{1+t} \int_{a}^{b} e^{z(s)t} z'(s)\ ds$$

$$= \int_{0}^{\infty} \frac{dt}{1+t} \int_{\gamma} e^{zt} dz = 0.$$

(Since e^{zt} for each fixed $t \in (0, \infty)$ is an entire function, Cauchy's theorem is applicable). Thus $f(z)$ is continuous and satisfies $\int_{\gamma} f(z)\ dz$ $= 0$ for all closed curves γ in L and by Morera's theorem $f(z)$ is analytic in L. (If the reader is familiar with Measure Theorey he can use Dominated convergence theorem and Fubini's theorem to get the result more directly).

16. Suppose f is entire and $|f(z)| = 1$ for $|z| = 1$. Prove that $f(z) = cz^N$ with $|c| = 1$.

Solution If $f(z)$ is a constant then $f(z) = c$ with $|c| = 1$. Thus the required equality holds with $n = 0$. Let f be non-constant. Since f is analytic in $|z| \le 1$ it can have only finitely many zeros say $\alpha_1, \alpha_2, \ldots,$ α_N in $|z| < 1$ with α_i's repeated as many times as their orders indicate. (Note that none of these can lie on $|z| = 1$ as $|f(z)| = 1$ for $|z| = 1$). Consider

$$g(z) = \prod_{i=1}^{N}\left(\frac{z - \alpha_i}{1 - \overline{\alpha}_i z}\right).$$

We have $|g(z)| = 1$ for $|z| = 1$ and g has the same zeros with same orders as that of f atleast in $|z| < 1$. Now put $h(z) = \dfrac{f(z)}{g(z)}$. Since f and g have the same set of zeros with the same orders, h is analytic in $|z| \le 1$. However the additional zeros of $h(z)$ because of the factors

$(1 - \overline{\alpha}_i z)$ do not contribute to the set of zeros in $|z| \leq 1$ because these zeros correspond to $1/\overline{\alpha}_i$ which lie outside of $|z| \leq 1$. Thus $h(z)$ is analytic and has no zeros in $|z| \leq 1$. We now apply the maximum and the minimum modulus theorems for $h(z)$ and since $|h(z)| = 1$ for $|z| < 1$, $|h(z)| \equiv 1$ throughout $|z| \leq 1$. Thus $h(z)$ reduces to a constant function of modulus 1, i.e., $h(z) = c$ with $|c| = 1$ or that $f(z) =$

$$c \prod_{i=1}^{N} \left(\frac{z - \alpha_i}{1 - \alpha_i z} \right).$$ with $|c| = 1$. This equality which holds for $|z| \leq 1$

should continue to hold throughout $\mathbb{C} \setminus \{1/\overline{\alpha}_i, \ i = 1, 2, \ \dots \ N\}$ by principle of analytic continuation. Since $f(z)$ is entire it has no poles in $\mathbb{C}$ and thus each $\alpha_i = 0$ $\left(\text{Otherwise} \lim_{z \to 1/\overline{\alpha}_i} f(z) = \infty, \text{ a contradiction} \right).$

Thus $f(z) = cz^N$ with $|c| = 1$. This completes the solution.

17. Prove that if $f(z)$ is analytic and bounded in $|z| < 1$ then for each ζ with $|\zeta| < 1$

$$f(\zeta) = \frac{1}{\pi} \iint_{|z|<1} \frac{f(z) \, dx \, dy}{(1 - \overline{z}\zeta)^2}. \tag{1}$$

Deduce that if $f(z)$ is analytic and bounded in a closed disc $|z| \leq s$ then for each ξ with $|\xi| < s$

$$f(\xi) = \frac{s^2}{\pi} \iint_{|w| \leq s} \frac{f(w) \, du \, dv}{(s^2 - \overline{w}\xi)^2} \quad (w = u + iv).$$

Solution   Note that the double integral $\dfrac{1}{\pi} \displaystyle\iint_{|z|<1} \dfrac{f(z) \, dx \, dy}{(1 - \overline{z}\zeta)^2}$ exists be-

cause $f(z)$ is continuous and bounded in $|\zeta| < 1$ and $|1 - \overline{z}\zeta| \geq 1 -$

$|\zeta| > 0$ in $|z| < 1$ so that $\dfrac{f(z)}{(1 - \overline{z}\zeta)^2}$ is also continuous and bounded.

We now start evaluating this double integral using polar coordinates.

Put $x = r \cos \theta$, $y = r \sin \theta$ so that $dx \, dy = \begin{vmatrix} \dfrac{\partial x}{\partial r} & \dfrac{\partial x}{\partial \theta} \\ \dfrac{\partial y}{\partial r} & \dfrac{\partial y}{\partial \theta} \end{vmatrix} dr \, d\theta = r$

$dr \, d\theta$. Thus the right side of (1) can be written as

$$\frac{1}{\pi} \int_{r=0}^{1} \int_{\theta=0}^{2\pi} \frac{f(r \cos \theta, r \sin \theta)}{(1 - \overline{z}\zeta)^2} r \, dr \, d\theta.$$

$$= \frac{1}{\pi} \int_{r=0}^{1} r \, dr \int_{\theta=0}^{2\pi} \frac{f(r\cos\theta, r\sin\theta)}{(1-\bar{z}\zeta)^2} \, d\theta. \tag{2}$$

Now for r fixed $z = re^{i\theta} \Rightarrow dz = rie^{i\theta} \, d\theta = izd\theta \Rightarrow d\theta = \frac{dz}{iz}$. Hence this expression reduces to

$$\int_{r=0}^{1} 2r \, dr \frac{1}{2\pi i} \int_{|z|=r} \frac{f(z) \, dz}{z \, (1-\bar{z}\zeta)^2} = \int_{r=0}^{1} 2r \, dr \frac{1}{2\pi i} \int_{|z|=r} \frac{zf(z) \, dz}{(z-r^2\zeta)^2}$$

$$= \int_{r=0}^{1} 2r \, dr \, (zf(z))' \, (r^2\zeta).$$

(By Cauchy's integral formula). Put $g(z) = zf(z)$. Then the last integral reduces to $\int_{r=0}^{1} 2r \, g'(r^2\zeta) \, dr$. Let $r^2\zeta = s$ so that $2\zeta r \, dr = ds$. Therefore

$$\int_{r=0}^{1} 2r \, g'(r^2\zeta) \, dr = \int_{0}^{\zeta} \frac{1}{\zeta} g'(s) \, ds$$

$$= \frac{1}{\zeta}(g(\zeta) - g(0))$$

$$= \frac{\zeta f(\zeta)}{\zeta} \quad \text{(since } g(0) = 0\text{)}$$

$$= f(\zeta).$$

This completes the proof of the first part.

In case $f(z)$ is analytic and bounded in a closed disc $|z| \leq s$ we consider $F(z) = f(sz)$. This function is well defined for $|z| \leq 1$ and applying the previous result we can write for $|\zeta| < 1$

$$F(\zeta) = \frac{1}{\pi} \iint_{|z|<1} \frac{F(z) \, dx \, dy}{(1-\bar{z}\zeta)^2} \quad (z = x + iy)$$

i.e.,

$$f(s\zeta) = \frac{1}{\pi} \iint_{|z|<1} \frac{f(sz) \, dx \, dy}{(1-\bar{z}\zeta)^2}.$$

Now put $w = sz$, $\xi = s\zeta$ with $w = u + iv$ and $z = x + iy$ so that $u = sx$, $v = sy$, $du \, dv = s^2 \, dx \, dy$ and $|\xi| < s$. We get

$$f(\xi) = \frac{1}{\pi} \iint_{|w|<s} \frac{f(w) \, du \, dv}{\left(1 - \dfrac{\bar{w}\xi}{s^2}\right)^2 s^2}$$

$$= \frac{s^2}{\pi} \iint\limits_{|w|<s} \frac{f(w)\, du\, dv}{(s^2 - \overline{w}\xi)^2}.$$

We now observe that if $f(w)$ is continuous and bounded by M in $|w| \le s$ then

$$\left| \iint\limits_{|w|=s} \frac{f(w)\, du\, dv}{(s^2 - \overline{w}\xi)^2} \right| \le \frac{M}{(s^2 - s|\xi|)^2} \iint\limits_{|w|=s} du\, dv = 0$$

since the two-dimensional area of the circumference of the circle $|w| = s$ is zero. Thus

$$\iint\limits_{|w|<s} \frac{f(w)\, du\, dv}{(s^2 - \overline{w}\xi)^2} = \iint\limits_{|w|\le s} \frac{f(w)\, du\, dv}{(s^2 - \overline{w}\xi)^2}$$

and we indeed get

$$f(\xi) = \frac{s^2}{\pi} \iint\limits_{|w|\le s} \frac{f(w)\, du\, dv}{(s^2 - \overline{w}\xi)^2} \qquad (w = u + iv).$$

18. Let f be analytic in $0 < |z - a| < \delta$. Prove that f has algebraic order m at $z = a$ if and only if $\lim\limits_{z \to a} (z - a)^m f(z)$ exists but is neither 0 nor ∞.

Solution Let the algebraic order of f at $z = a$ exist and be equal to m. If $m = 0$ from the characterisation of algebraic order (see Theorem 4.5.20) it is clear that $\lim\limits_{z \to a} f(z)$ exists and is neither 0 nor ∞. Let $m \ne 0$. If $m > 0$, f has a pole of order m and if $m < 0$, f has a zero of order $-m$. In both cases

$$f(z) = (z - a)^{-m} f_m(z)$$

where $f_m(z)$ is analytic at $z = a$ and is neither 0 nor ∞ at $z = a$. Thus $\lim\limits_{z \to a} (z - a)^m f(z) = f_m(a)$ exists and is neither 0 nor ∞. Thus the necessary part is proved.

For the converse if $\lim\limits_{z \to a} (z - a)^m f(z)$ exists and is neither 0 nor ∞ it is easy to see that for $\alpha > m$ (write $\alpha = m + \beta$, $\beta > 0$) $\lim\limits_{z \to a} (z - a)^\alpha f(z) = 0$ and for $\alpha < m$ (write $\alpha = m - \beta$, $\beta > 0$) $\lim\limits_{z \to a} (z - a)^\alpha f(z) = \infty$. Thus m is the algebraic order of f at $z = a$.

19. If f and g have algebraic orders h and k respectively at $z = a$, prove that
    (A) The algebraic order l of $f \pm g$ satisfies $l \le \max(h, k)$.
    (B) The algebraic order of fg is $h + k$ and
    (C) The algebraic order of f/g is $h - k$.

Solution

Recall the definition of algebraic order at $z = a$. Consider the following conditions.

(i) $\lim\limits_{z \to a}(z - a)^\alpha \, f(z) = 0$ for $\alpha > h$.

(ii) $\lim\limits_{z \to a}(z - a)^\alpha \, f(z) = \infty$ for $\alpha < h$.

(iii) $\lim\limits_{z \to a}(z - a)^\beta \, g(z) = 0$ for $\beta > k$.

(iv) $\lim\limits_{z \to a}(z - a)^\beta \, g(z) = \infty$ for $\beta < k$.

Now $f \pm g$ is well defined and analytic in a deleted neighbourhood of a. Further $\lim\limits_{z \to a}(z - a)^\alpha \, (f \pm g) \, (z) = 0$ holds for $\alpha > \max (h, k)$ using (i) and (iii). If possible let $l > \max (h, k)$. Choose an integer l' so that $l > l' > \max (h, k)$. Since $l' > \max (h, k)$ by our observation

$$\lim_{z \to a}(z - a)^{l'} \, (f \pm g) \, (z) = 0.$$

But from the definition of the order, $l' < l$ and so

$$\lim_{z \to a}(z - a)^{l'} \, (f \pm g) \, (z) = \infty.$$

This contradiction guarantees that $l \le \max (h, k)$. This proves (A).

Now consider (fg). This function is well defined and analytic in a deleted neighbourhood of 'a'. Consider

$$\lim_{z \to a}(z - a)^{h+k} \, (fg)(z) = \lim_{z \to a}(z - a)^h \, f(z)\lim_{z \to a}(z - a)^k \, g(z)$$

which exists but is neither 0 nor ∞. Thus by the solution of the problem 18 $(h + k)$ is the algebraic order of fg. This proves (B).

In view of B it is sufficient to prove that the algebraic order of $\dfrac{1}{g}$ exists and is $-k$. Since g has a definite algebraic order k at $z = a$, $g(z)$ can not vanish in every neighbourhood of a (otherwise there exists $z_n \to a$ with $g(z_n) = 0$. Now if $\beta < k$, $\lim\limits_{z \to a}(z - a)^\beta \, g(z) = \lim\limits_{z_n \to a} (z_n - a)^\beta g(z_n) = 0$ a contradiction). Thus $\dfrac{1}{g(z)}$ is analytic in a deleted neighbourhood of a. Now (iii) and (iv) together imply that

$$\lim_{z \to a}\frac{1}{g(z)}(z - a)^{-\beta} = \infty \text{ for } \beta > k \qquad \text{i.e., } -\beta < -k.$$

and

$$\lim_{z \to a}\frac{1}{g(z)}(z - a)^{-\beta} = 0 \text{ for } \beta < k \qquad \text{i.e., } -\beta > -k.$$

In other words

$$\lim_{z \to a}\frac{1}{g(z)}(z - a)^\alpha = \infty \text{ for } \alpha < -k$$

and $$\lim_{z \to a} \frac{1}{g(z)} (z - a)^\alpha = 0 \text{ for } \alpha > -k.$$

This shows that $-k$ is the algebraic order of $\frac{1}{g}$ at $z = a$ and this completes the proof of (C) (of course using (B)).

20. By definition any function which is defined and analytic for $|z| > R$ (for any $R > 0$) is said to have an isolated singularity at $z = \infty$. We say that f has a removable singularity at ∞, if $f(1/z)$ has a removable singularity at $z = 0$ or is analytic at $z = 0$. Similarly we say f has a zero of order m at $z = \infty$ if $f(1/z)$ has a removable singularity at $z = 0$ and the extended function has a zero of order m at $z = 0$. f is said to have a pole of order m at $z = \infty$ if $f(1/z)$ has a pole of order m at $z = 0$. Prove that if

   (i) f is entire and has a removable singularity at $z = \infty$ then f is a constant.

   (ii) If f is entire and has a non-essential singularity at $z = \infty$ then f is a polynomial.

Solution (i) Since f has a removable singularity at $z = \infty$, $f(1/z)$ has a removable singularity at $z = 0$ and consequently $f(1/z)$ is bounded near origin, i.e., there exists M and δ (both positive numbers) such that $|f(1/z)| \le M$ if $|z| < \delta$. i.e., $|f(z)| \le M$ if $|z| > \frac{1}{\delta}$. On the other hand the entire function f is certainly bounded inside the compact set $|z| \le \frac{1}{\delta}$ and consequently $|f(z)| \le K$ for all $z \in \mathbb{C}$, i.e., f is a bounded entire function and by Liouville's theorem f reduces to a constant.

   (ii) In view of (i) we can assume f has a pole at $z = \infty$. But then $f(1/z)$ has a pole at $z = 0$ and so $z^m f(1/z)$ is analytic at 0 for some positive integer m. So $|f(1/z)| \le C |1/z|^m$ for all small $|z|$ or that $|f(z)| \le C |z|^m$ for large $|z|$ and by solved exercises 5, f reduces to a polynomial. (Note that by choosing C large enough we can assume that

$$\left| f(z) - \sum_0^{m-1} c_k z^k \right| \le C |z|^m$$

for all z with $c_k = f^{(k)}(0)/k!$).

21. Show that an isolated singularity of f is removable whenever either Re $f(z)$ or Im $f(z)$ is bounded above or below in a deleted neighbourhood of the singularity.

Solution
Case 1 Let Re $f(z)$ be bounded above in a deleted neighbourhood of the singularity 'a'. Assume that f is not a constant and Re $f(z) \le M$

and that $M \leq 1$ (there is no loss of generality in this assumption). Consider $\phi(w) = \dfrac{w - M + 1}{w - M - 1}$ which is a linear fractional transformation with a single simple pole at $w = M + 1$. $\phi(w)$ is analytic at all points in the extended complex plane except at $w = M + 1$. We first observe that

$$\text{Re } w \leq M \Rightarrow |\phi(w)| \leq 1, \ \phi^{-1}(1) = \infty.$$

$$\left(\text{Indeed } |\phi(w)|^2 = \frac{|w - M|^2 + 1 + 2\,\text{Re}(w - M)}{|w - M|^2 + 1 - 2\,\text{Re}(w - M)} \leq 1 \Leftrightarrow \text{Re }(w - M) \leq 0\right).$$

Thus $\phi \circ f$ analytic in a deleted neighbourhood of "a" as long as $f(z)$, does not assume $M + 1$ in this neighbourhood. However $f(z)$ assumes $M + 1$ if and only if Re $f(z)$ assumes $M + 1$ which is impossible by our condition that Re $f(z) \leq M$. Hence $\phi \circ f$ is analytic in a deleted neighbourhood of 'a' and further in this neighbourhood $|\phi \circ f| \leq 1$. Hence by Theorem 4.5.12. $F = \phi \circ f$ has a removable singularity at $z = a$ with $F(a) = \lim_{z \to a} (\phi \circ f)(z)$. Note that $|F(z)| \leq 1$ in a deleted neighbourhood of a. We can now conclude that $f = \phi^{-1} \circ F$ also has a removable singularity at $z = a$ provided $F(a) \neq 1$. (This is because ϕ^{-1} is analytic at all points except at 1 where it has a pole). By what we have already observed F is analytic in a deleted neighbourhood of 'a' and $|F(z)| \leq 1$ holds in a deleted neighbourhood of 'a'. Incase $F(a) = 1$, by Maximum modulus theorem F reduces to a constant and so is f. Thus indeed we have proved that $F(a) \neq 1$.

Case 2 Re $f(z)$ is bounded below.
By case (i) $-f(z)$ has a removable singularity at $z = a$ and so is f.

Case 3 Im $f(z)$ is bounded above or below.
Im $f(z) = \text{Re }(-if)$ and by cases (i) and (ii) $(-if)$ has a removable singularity at $z = a$. Thus f also has a removable singularity at $z = a$.

22. Define Fibonacci numbers c_n, $n = 0, 1, 2, \ldots$ Obtain an explicit expression for these c_n's and prove the recurrence relation $c_n = c_{n-1} + c_{n-2}, \ \forall \ n \geq 2$.

Solution Fibonacci numbers c_n are defined as the Taylor coefficients of the infinite series representation of the rational function $\dfrac{-1}{z^2 + z - 1}$ around origin. We define α and β as the roots of the quadratic equation $z^2 + z - 1 = 0$ and obtain the Taylor development of $\dfrac{-1}{z^2 + z - 1}$ around the origin as follows.

$$\frac{-1}{z^2+z-1} = \frac{-1}{(z-\alpha)(z-\beta)}$$

$$= -(z-\alpha)^{-1}(z-\beta)^{-1}$$

$$= -\alpha^{-1}\left(1-\frac{z}{\alpha}\right)^{-1}\beta^{-1}\left(1-\frac{z}{\beta}\right)^{-1}$$

$$= \frac{-1}{\alpha\beta}\left\{\left(1+(z/\alpha)+(z/\alpha)^2+\ldots\right)\left(1+(z/\beta)+(z/\beta)^2\ldots\right)\right\}$$

$$= \frac{-1}{\alpha\beta}\left\{1+z\left(\frac{1}{\alpha}+\frac{1}{\beta}\right)+\ldots\right.$$

$$\left. + z^n\left(\frac{1}{\beta^n}+\frac{1}{\alpha\beta^{n-1}}+\ldots+\frac{1}{\alpha^n}\right)+\ldots\right\}.$$

Now the coefficient of z^n is given by

$$c_n = \frac{-1}{\alpha\beta}\frac{1}{\beta^n}\frac{(\beta/\alpha)^{n+1}-1}{(\beta/\alpha)-1}$$

$$= \frac{-1}{(\alpha\beta)^{n+1}}\frac{\beta^{n+1}-\alpha^{n+1}}{\beta-\alpha}$$

Since α and β are the roots of the quadratic equation $z^2+z-1=0$

we have $\alpha = \dfrac{\sqrt{5}-1}{2}$; $\beta = -\dfrac{(\sqrt{5}+1)}{2}$. Hence the expression for c_n

reduces to (using $(-1)^{n-1} = (-1)^{n+1}$)

$$c_n = \frac{(-1)^{n+1}}{\sqrt{5}}\left(\beta^{n+1}-\alpha^{n+1}\right)$$

$$c_{n-2}+c_{n-1} = \frac{(-1)^{n+1}}{\sqrt{5}}\left(\beta^{n-1}-\alpha^{n-1}\right)+\frac{(-1)^n}{\sqrt{5}}\left(\beta^n-\alpha^n\right)$$

$$= \frac{(-1)^{n+1}}{\sqrt{5}}\left((\beta^{n-1}-\alpha^{n-1})-(\beta^n-\alpha^n)\right)$$

$$= \frac{(-1)^{n+1}}{\sqrt{5}}\left(\beta^{n+1}\left(\frac{1}{\beta^2}-\frac{1}{\beta}\right)-\alpha^{n+1}\left(\frac{1}{\alpha^2}-\frac{1}{\alpha}\right)\right).$$

Now $\dfrac{1}{\beta^2}-\dfrac{1}{\beta} = \dfrac{1}{\beta}\left(\dfrac{1}{\beta}-1\right) = \dfrac{-2}{1+\sqrt{5}}\left(\dfrac{-2}{1+\sqrt{5}}-1\right) = \dfrac{2}{1+\sqrt{5}}\dfrac{3+\sqrt{5}}{1+\sqrt{5}} = 1$

and $\dfrac{1}{\alpha^2} - \dfrac{1}{\alpha} = \dfrac{1}{\alpha}\left(\dfrac{1}{\alpha} - 1\right) = \dfrac{2}{\sqrt{5}-1}\left(\dfrac{2}{\sqrt{5}-1} - 1\right) = \dfrac{2}{\sqrt{5}-1}\dfrac{3-\sqrt{5}}{\sqrt{5}-1} = 1.$

Hence we have $c_{n-2} + c_{n-1} = \dfrac{(-1)^{n+1}}{\sqrt{5}}(\beta^{n+1} - \alpha^{n+1}) = c_n.$

$$\text{i.e., } c_{n-2} + c_{n-1} = c_n.$$

23. Define Bernoulli numbers. Obtain the recurrence relation satisfied by these numbers and prove that the sequence of Bernoulli numbers is unbounded.

Solution Define $g(z) = \dfrac{z}{e^z - 1}$, $z \neq 0$. Since

$$\lim_{z \to 0} g(z) = \lim_{z \to 0} \dfrac{1}{(e^z - 1)/z} = 1 \text{ we have } \lim_{z \to 0} z\, g(z) = 0.$$

Thus g has a removable singularity at the origin and its extended value at $z = 0$ is given by $g(0) = \lim_{z \to 0} g(z) = 1$. Thus the function $g(z)$ defined by

$$g(z) = \dfrac{z}{e^z - 1}, \; z \neq 0$$

$$g(0) = 1$$

is analytic at origin and we write the Taylor series of $g(z)$ around origin in the form

$$g(z) = \dfrac{z}{e^z - 1} = \sum_{\gamma=0}^{\infty} \dfrac{B_\gamma}{\gamma!} z^\gamma, \quad B_\gamma \in \mathbb{C}. \tag{1}$$

We define $B_{2\gamma}$ ($\gamma = 0, 1, 2, ..$) as Bernoulli numbers. We first observe that

$$\cot z = i\,\dfrac{e^{2iz} + 1}{e^{2iz} - 1}$$

$$= i\left(1 - \dfrac{2}{1 - e^{2iz}}\right)$$

$$= i + z^{-1}\, g(2iz).$$

$\cot z$ from the definition is an odd function and so $z \cot z$ is an even function. Since

$$\cot z = i + z^{-1}\, g(2iz)$$

$$g(2iz) = z \cot z - iz.$$

Thus $g(z) = \dfrac{z}{2i} \cot(z/2i) - i(z/2i)$ or that $g(z) + \dfrac{z}{2} = \dfrac{z}{2i} \cot(z/2i)$.

Therefore $\left(g(z) + \dfrac{z}{2} \right)$ is an even function and hence its odd Taylor coefficients around origin are zero. Using (1) it now follows that

$B_0 = g(0) = 1$, $B_1 = -\dfrac{1}{2}$ and $B_{2\gamma+1} = 0$, $\gamma \ge 1$. Hence

$$g(z) = 1 - \frac{z}{2} + \sum_1^\infty \frac{B_{2\gamma}}{(2\gamma)!} z^{2\gamma}. \tag{2}$$

First observe that

$$1 = \frac{e^z - 1}{z} g(z) = \sum_1^\infty \frac{z^{\gamma-1}}{\gamma!} \sum_0^\infty \frac{B_\gamma}{\gamma!} z^\gamma. \tag{3}$$

Now using the uniqueness of the Taylor series expansion for analytic functions we can multiply the infinite series on the right side of (3) and compare the coefficients of z^{n-1}, for each $n \ge 2$. This gives

$$\frac{1}{n!} \frac{B_0}{0!} + \frac{1}{(n-1)!} \frac{B_1}{1!} + \frac{1}{(n-2)!} \frac{B_2}{2!} + \ldots + \frac{1}{1!} \frac{B_{n-1}}{(n-1)!} = 0.$$

Multiplying throughout by $n!$ and using the relation

$$n^C r = \frac{n!}{r!(n-r)!}$$

we get $n^C_0 B_0 + n^C_1 B_1 + \ldots + n^C_{n-1} B_{n-1} = 0$. This recurrence relation determines all the Bernoulli numbers. A simple computation shows that

$$B_0 = 1, \ B_2 = \frac{1}{6}, \ B_4 = -\frac{1}{30}, \ B_6 = \frac{1}{42}, \ \ldots$$

Further the radius of convergence of the Taylor series around origin representing $g(z)$ is exactly 2π (the distance between origin and the nearest singularity for $g(z)$ which is $\pm 2\pi i$ is precisely equal to 2π). This observation enables us to show that the sequence $B_{2\gamma}$ is unbounded. Infact, if this sequence $B_{2\gamma}$ is bounded (say $|B_{2\gamma}| \le M$) from (2) we see that the radius of convergence R of the series representing $g(z)$ will be equal to ∞. This is because

$$\frac{1}{R} = \overline{\lim_{\gamma \to \infty}} \left(\frac{|B_{2\gamma}|}{(2\gamma)!} \right)^{1/2\gamma}$$

$$\leq \overline{\lim_{\gamma \to \infty}} \left(\frac{M}{(2\gamma)!} \right)^{1/2\gamma} = 0$$

since $\lim_{n \to \infty} (n!)^{1/n} = \infty$. This contradiction proves that $\{B_{2\gamma}\}$ is indeed unbounded.

24. Determine all entire functions $f(z)$ satisfying the differential equation

$$f''(z) + f(z) = 0. \tag{1}$$

Solution   Since $f(z)$ is entire $f(z)$ is given by

$$f(z) = \sum_{n=0}^{\infty} a_n z^n, \text{ with } a_n = \frac{f^{(n)}(0)}{n!}$$

where the infinite series expansion is valid for all z.

$$f''(z) + f(z) = 0 \Rightarrow 2a_2 = -a_0 \text{ or that } a_2 = \frac{-a_0}{2!}.$$

We show that $a_{2n} = (-1)^n \dfrac{a_0}{(2n)!}$ by induction. Since this is verified for $n = 1$, we assume

$$a_{2m} = (-1)^m \frac{a_0}{(2m)!} \text{ and prove } a_{2m+2} = (-1)^{m+1} \frac{a_0}{(2m+2)!}.$$

Differentiating (1), $2m$ times, we have

$$f^{(2m+2)}(z) = -f^{(2m)}(z).$$

Putting $z = 0$ we get $(2m + 2)!\, a_{2m+2} = -(2m)!\, a_{2m}$

$$= -(2m)!\,(-1)^m \frac{a_0}{(2m!)}$$

$$a_{2m+2} = (-1)^{m+1} \frac{a_0}{(2m+2)!}.$$

Similarly we obtain $a_3 = \dfrac{a_1}{3!}$ by differentiating (1) once and putting $z = 0$. Again by induction we get

$$a_{2n+1} = (-1)^n \frac{a_1}{(2n+1)!}.$$

Thus $f(z) = a_0 \left(1 + \sum_{n=1}^{\infty} (-1)^n \frac{z^{2n}}{(2n)!} \right) + a_1 \left(z + \sum_{n=1}^{\infty} (-1)^n \frac{z^{2n+1}}{(2n+1)!} \right)$

$$= a'\cos z + b'\sin z$$

$$= a'\left(\frac{e^{iz} - e^{-iz}}{2}\right) + b'\left(\frac{e^{iz} - e^{-iz}}{2i}\right) = a\, e^{iz} + b\, e^{-iz}.$$

On the other hand functions of the form $ae^{iz} + be^{-iz}$ satisfy (1) as is easily verified. Thus the set of all entire functions satisfying $f''(z) + f(z) = 0$ are nothing but $ae^{iz} + be^{-iz}$ where $a, b \in \mathbb{C}$.

25. Let f and g be analytic in an open set D, $\overline{B_r(0)} \subseteq D$, $(r > 0)$. Let $|a| = r$ be such that $g(a) = 0$, $g'(a) \neq 0$, $f(a) \neq 0$, $g(z) \neq 0$ for $z \neq a$ in $|z| \leq r$. If the power series representation of f/g around origin

is $\sum_0^\infty a_n z^n$ show that $\lim\limits_{n \to \infty} \dfrac{a_n}{a_{n+1}} = a$.

Solution f/g is analytic at all points in $|z| \leq r$ except at 'a'. At 'a', g has a simple zero and f is non-zero. Hence f/g has a simple pole at 'a'. Therefore

$$\frac{f(z)}{g(z)} = (z - a)^{-1} h(z)$$

where $h(z)$ is analytic at $z = a$ and $h(a) \neq 0$. (Theorem 4.5.13). Let us represent $h(z)$ as a power series around origin by

$$h(z) = \sum_0^\infty c_n z^n, \ h(a) \neq 0$$

Therefore

$$\frac{f(z)}{g(z)} = -a^{-1}\left(1 - \frac{z}{a}\right)^{-1} \sum_0^\infty c_n z^n$$

$$= -a^{-1}\left(\sum_0^\infty \frac{z^n}{a^n}\right)\left(\sum_0^\infty c_n z^n\right).$$

But we are given that the power series representing f/g is $\sum_0^\infty a_n z^n$. Using the uniqueness of Taylor series around origin we have

$$a_n = -a^{-1}\left(\frac{c_0}{a^n} + \frac{c_1}{a^{n-1}} + \ldots + c_n\right)$$

Similarly $\quad a_{n+1} = -a^{-1}\left(\frac{c_0}{a^{n+1}} + \frac{c_1}{a^n} + \ldots + c_{n+1}\right)$. Thus

$$\frac{a_n}{a_{n+1}} = \frac{a(c_0 + c_1 a + \ldots + c_n a^n)}{(c_0 + c_1 a + \ldots + c_{n+1} a^{n+1})}$$

Taking limits on both sides as $n \to \infty$,

$$\lim_{n \to \infty} \frac{a_n}{a_{n+1}} = a \frac{h(a)}{h(a)} = a.$$

$\left(\text{Note that } h(a) = \sum_0^\infty c_n a^n \right).$

26. Let D be an open subset of $\mathbb{C}$ and $a \in D$. Let $f : D\backslash\{a\} \to \mathbb{C}$ be analytic. Show that if f' has a removable singularity at 'a' then f also has a removable singularity at 'a'.

Solution Since f' is having a removable singularity at $z = a$, we can redefine f' at $z = a$ in such a way that f' becomes analytic in $|z - a| < \delta$. Therefore f' has a primitive g (say) in $|z - a| < \delta$, i.e., g is analytic in $|z - a| < \delta$ and $g' = f'$ in $|z - a| < \delta$. Consider $h = g - f$ in $0 < |z - a| < \delta$. h is analytic and $h' = 0$ in $0 < |z - a| < \delta$ which is a region. Hence h is a constant in $0 < |z - a| < \delta$ say c. Now $f(z) = g(z) - c$ in $0 < |z - a| < \delta$. Hence

$$\lim_{z \to a} (z - a) f(z) = \lim_{z \to a} (z - a) g(z) - \lim_{z \to a} c \, (z - a) = 0.$$

Hence 'a' is a removable singularity for f.

27. Show that an odd entire function has only odd terms in its power series around origin and an even entire function has only even terms in its power series around the origin.

Solution Let $f(z)$ have the power series representation

$$f(z) = \sum_0^\infty a_n z^n.$$

(i) Let $f(z)$ be an odd function. Therefore

$$f(-z) = -f(z). \tag{1}$$

Differentiating (1), $2k$ times, we get

$$f^{(2k)} (-z) = -f^{(2k)} (z), \quad k = 1, 2, \ldots .$$

Putting $z = 0$ we get $f^{(2k)} (0) = -f^{(2k)} (0)$ or that $f^{(2k)} (0) = 0$.

This in turn gives $a_{2k} = \dfrac{f^{(2k)} (0)}{(2k)!} = 0$. Hence $f(z)$ has only odd terms in its power series representation around the origin.

(ii) Let $f(z)$ be an even function. Therefore

$$f(-z) = f(z). \tag{2}$$

Differentiating (2) $(2k - 1)$ times we get

$$-f^{(2k-1)} (-z) = f^{(2k-1)} (z).$$

Putting $z = 0$ we get

$$-f^{(2k-1)} (0) = f^{(2k-1)} (0) \text{ or that } f^{(2k-1)} (0) = 0.$$

This gives $a_{2k-1} = \dfrac{f^{(2k-1)} (0)}{(2k-1)!} = 0$. Hence $f(z)$ has only even

terms in its power series representation around the origin.

28. Let f be an entire function such that $|f(z)| \le a + b|z|^{3/2}$ where a, b are constants. Show that f is at most a linear polynomial.

Solution Given that $|f(z)| \le a + b |z|^{3/2}$ we get $|f(z^2)| \le a + b|z|^2$ $\le c|z|^3$ for some constant c and for large $|z|$. By solved exercise 5 (see also solution for exercise 20(ii)) we see that $f(z^2)$ is a polynomial in z of degree atmost 3, i.e,

$$f(z^2) = a_0 + a_1 z + a_2 z^2 + a_3 z^3 \tag{1}$$

Since $f(z^2)$ is an even function by the previous problem we have $a_1 = a_3 = 0$. Thus

$$f(z^2) = a_0 + a_2 z^2 \text{ or that } f(z) = a_0 + a_2 z.$$

This shows that f is atmost a linear polynomial (i.e., a polynomial of degree atmost 1).

29. Show that if f is analytic in $|z| \le 1$, there must be some positive

integer n such that $f\left(\dfrac{1}{n}\right) \ne \dfrac{1}{n+1}$.

Solution Suppose the contrary, i.e., f is an analytic function in $|z|$

≤ 1 such that $f\left(\dfrac{1}{n}\right) = \dfrac{1}{n+1}$ for all positive integers n. Now

$$f(0) = \lim_{n\to\infty} f\left(\frac{1}{n}\right) = \lim_{n\to\infty} \frac{1}{n+1} = 0.$$

Consider the function $(z + 1) \dfrac{f(z)}{z}$ which is equal to 1, at $z = \dfrac{1}{n}$ for

all positive integers n. Now $\dfrac{f(z)}{z}$ is analytic for $z \ne 0$ in $|z| \le 1$ and

has a removable singularity at $z = 0$ because

$$\lim_{z\to 0} z\frac{f(z)}{z} = \lim_{z\to 0} f(z) = f(0) = 0.$$

Hence $\dfrac{f(z)}{z}$ is analytic everywhere in $|z| \le 1$. Thus $(z + 1)\dfrac{f(z)}{z}$ is

analytic everywhere in $|z| \le 1$. Therefore by principle of analytic continuation $(z + 1)\dfrac{f(z)}{z} \equiv 1$ in $|z| \leq 1$ (The constant function 1 and

$(z + 1)\dfrac{f(z)}{z}$ agree over the set $\{1/n\}$ with limit point 0 inside $|z| <$

1). Therefore

$$f(z) = \frac{z}{z+1} \text{ in } |z| \leq 1.$$

But this function is no longer analytic in $|z| \leq 1$ because $z = -1$ is a pole. But this contradicts the analytic nature of f on $|z| \leq 1$ and proves

that there exists atleast one n such that $f\left(\dfrac{1}{n}\right) \neq \dfrac{1}{n+1}$.

30. If $f(z)$ is entire and satisfies $|f(z)| \geq |z|^N$ for sufficiently large $|z|$ and $N \in \mathbb{N}$, then prove that $f(z)$ is a polynomial of degree atleast N.

Solution   Consider $f\left(\dfrac{1}{z}\right)$. We know that $z \to \dfrac{1}{z}$ and $\dfrac{1}{z} \to f\left(\dfrac{1}{z}\right)$

are analytic functions in $\mathbb{C}^* = \mathbb{C} \backslash \{0\}$. Thus $f\left(\dfrac{1}{z}\right)$ is analytic in $\mathbb{C}^*$. The

given hypothesis implies in particular that $\lim\limits_{z \to 0} f\left(\dfrac{1}{z}\right) = \infty$ and so

$f\left(\dfrac{1}{z}\right)$ has a pole at $z = 0$. We shall first assume that $f\left(\dfrac{1}{z}\right)$ has a pole

of order m at $z = 0$ and write

$$f\left(\frac{1}{z}\right) = z^{-m} \, g(z)$$

where $g(z)$ is analytic at 0 with $g(0) \neq 0$. Thus $g(z) = z^m f\left(\dfrac{1}{z}\right)$ is well

defined and analytic in a neighbourhood of 0. But for z outside this

neighbourhood we can easily define $g(z)$ as $z^m f\left(\dfrac{1}{z}\right)$ to conclude that

$g(z)$ is entire. Further more if $|z|$ is sufficiently large

$$\left|\frac{g(z)}{z^m}\right| = |f(1/z)| \leq c$$

since $f\left(\dfrac{1}{z}\right) \to f(0)$ as $|z| \to \infty$ by continuity. Thus $g(z)$ is entire and

satisfies $|g(z)| \leq c|z|^m$, for large $|z|$. This gives that $g(z)$ is a polynomial of degree at most m. Again using the fact that $g(z)$ is analytic at origin we see that for small $|z|$, $|g(z)| \geq |z|^m \dfrac{1}{|z|^N} = |z|^{m-N} \to \infty$ as $|z| \to 0$ if $m < N$. Thus $m \geq N$. Hence $g(z) = a_0 + a_1 z + \ldots + a_m z^m$ with $m \geq N$ and $a_0 \neq 0$ or that

$$f\left(\frac{1}{z}\right) = \frac{a_0}{z^m} + \frac{a_1}{z^{m-1}} + \ldots + a_m.$$

Thus

$$f(z) = a_0 z^m + a_1 z^{m-1} + \ldots + a_m$$

with $a_0 \neq 0$, i.e., f is a polynomial of degree $m \geq N$.

31. If f is analytic in $|z| > 0$ and $|f(z)| \leq |z|^{1/2} + |z|^{-1/2}$. Prove that f is a constant.

Solution We claim that $f(z)$ has a removable singularity at the origin. Indeed,

$$\lim_{z \to 0} |zf(z)| \leq \lim_{z \to 0} (|z|^{3/2} + |z|^{1/2}) = 0.$$

Thus the extended function is entire. Consider $zf(z)$. For large $|z|$,

$$|z\,f(z)| \leq |z|^{3/2} + |z|^{1/2}$$
$$< 2\,|z|^{3/2} \qquad \text{(if } |z| > 1\text{)}.$$

Hence by solved exercise 28, $zf(z)$ is atmost a linear polynomial. Therefore

$$zf(z) = a + bz, \ a, \ b \in \mathbb{C}.$$

But $\lim_{z \to 0} zf(z) = 0 \Rightarrow a = 0$. Hence $zf(z) = bz$ or $f(z) = b$, a constant.

32. How many roots of the equation $z^7 - 2z^5 + 6z^3 - z + 1 = 0$ lie in $|z| < 1$.

Solution Let $f(z) = 6z^3$, $g(z) = z^7 - 2z^5 - z + 1$. On $|z| = 1$, $|f(z)| = 6$, $|g(z)| < 5 < 6 = |f(z)|$. Hence by Rouche's Theorem (Theorem 4.6.16) $f(z)$ and $g(z) + f(z)$ have the same number of zeros inside $|z| < 1$. However $f(z)$ has exactly three zeros at $z = 0$ in $|z| < 1$ and so the required equation has exactly three zeros in $|z| < 1$.

33. How many roots of $z^4 - 6z + 3 = 0$ have their modulii between 1 and 2.

Solution Let $f(z) = -6z$, $g(z) = z^4 + 3$. On $|z| = 1$, $|g(z)| \leq 4 < 6 = |f(z)|$. Thus by Rouche's Theorem (Theorem 4.6.16) $f(z) + g(z) = z^4 - 6z + 3$ also has the same number of zeros as $f(z)$ in $|z| < 1$. But f has a single zero in $|z| < 1$. Hence $z^4 - 6z + 3$ has a single zero in $|z| < 1$. Now let $h(z) = z^4$, $k(z) = -6z + 3$. On $|z| = 2$, $|k(z)| \leq 15 < 16 = |h(z)|$ and $h(z)$ has exactly four zeros in $|z| < 2$. Thus $h(z) + k(z) = z^4 - 6z + 3$ also has exactly four zeros in $|z| < 2$. This same polynomial has just one root in $|z| < 1$. Further on $|z| = 1$,

$|z^4 - 6z + 3| \geq 6 - 1 - 3 = 2 > 0$ and so $z^4 - 6z + 3$ has no zeros in $|z| = 1$. Thus the number of roots of $z^4 - 6z + 3$ lying on $1 < |z| < 2$ is precisely $4 - 1 = 3$.

34. Let f be a function analytic inside and on the unit circle. Suppose that $|f(z) - z| < |z|$ on the unit circle.
   (a) Show that $|f'(1/2)| \leq 8$.
   (b) Show that f has precisely one zero inside the unit circle.

   Solution (a) By Cauchy's integral formula,

$$f'(z) = \frac{1}{2\pi i} \int_{|\zeta|=1} \frac{f(\zeta)}{(\zeta - z)^2} \, d\zeta$$

Therefore

$$f'(1/2) = \frac{1}{2\pi i} \int_{|\zeta|=1} \frac{f(\zeta)}{(\zeta - (1/2))^2} \, d\zeta.$$

This in turn implies that

$$|f'(1/2)| \leq \frac{1}{2\pi} \int_{|\zeta|=1} \frac{|f(\zeta)|}{|\zeta - (1/2)|^2} \, |d\zeta|.$$

Using $|f(z) - z| < |z|$ on $|z| = 1$ and the fact that $\left| \zeta - \frac{1}{2} \right| \geq \frac{1}{2}$ for $|\zeta| = 1$ we get

$$|f'(1/2)| \leq \frac{1}{2\pi} 8 \int_{|\zeta|=1} |d\zeta| = 8.$$

   Therefore $|f'(1/2)| \leq 8$.
   (b) Let $F(z) = f(z) - z$ and $G(z) = z$. It is given that $|F(z)| < |G(z)|$. By Rouche's theorem G and $F + G$ have the same number of zeros inside $|z| < 1$, i.e., $f(z)$ and z have the same number of zeros inside $|z| < 1$ which implies that $f(z)$ has precisely one zero inside $|z| < 1$.

35. Let a be a real number greater than 1. Prove that the equation $ze^{a-z} = 1$ has only one solution in $|z| < 1$ and that this solution is real and positive.

   Solution Consider $|ze^{a-z}|$ on $|z| = 1$. We write $z = x + iy$ with $-1 \leq x \leq 1$, $-1 \leq y \leq 1$.

$$|ze^{a-z}| = |z| \, e^{Re(a-z)} = e^{a-x} \geq e^{a-1} > 1$$

   since $a > 1$. Thus the constant function $G(z) = 1$ and $F(z) = ze^{a-z}$ satisfy $|F(z)| > |G(z)|$ on $|z| = 1$. Hence by Rouche's Theorem $F(z)$ and $F(z) - G(z) = ze^{a-z} - 1$ have the same number of zeros inside $|z| < 1$. But

$$F(z) = 0 \Leftrightarrow ze^{a-z} = 0 \Leftrightarrow z = 0.$$

(Note that e^{a-z} never vanishes). Thus F has exactly one zero inside $|z| = 1$. Hence the same is true of $ze^{a-z} - 1$ and this is the first part of our result. We have to prove that this solution is real and positive. Consider the function $h(x) = xe^{a-x}$ defined for $0 \leq x \leq 1$. Since $h(0) = 0$ and $h(1) > 1$, $h(0) < 1 < h(1)$ and by intermediate value property of continuous functions, $h(x) = 1$ for some $0 < x < 1$, i.e., $xe^{a-x} = 1$. Thus the unique solution of $ze^{a-z} = 1$ in $|z| < 1$ is real and positive.

36. Show that every analytic function in the closed unit disc which is real on the unit circle is a constant.

Solution Let $a = \alpha + i\beta$ be a non-real complex number (i.e., $\beta \neq 0$). We claim that $f(z) \neq a$ for any z in $|z| < 1$. By argument principle, N, the number of zeros of $(f(z) - a)$ in $|z| < 1$ is given by

$$N = \frac{1}{2\pi i} \int_{|z|=1} \frac{f'(z)}{f(z) - a} \, dz.$$

Case 1 $\beta > 0$. Consider

$$\text{Im } (f(z) - a) = \text{Im } (f(z) - (\alpha + i\beta)).$$
$$= -\beta < 0 \text{ on } |z| = 1.$$

Thus $(f(z) - a)$ describes a closed curve $\Gamma = (f - a) (\gamma)$ (with γ as the unit circle) in the lower half plane. Hence $n(\Gamma, 0) = 0 \Rightarrow N = 0$. Thus $f(z) \neq a$ in $|z| < 1$.

Case 2 $\beta < 0$.
A similar argument as in case 1 shows that Γ lies in the upper half plane and again $N = 0$. Thus $f(z)$ never assumes non-real values and hence $f(z)$ is always real in $|z| < 1$. Since $\text{Im } f(z) \equiv 0$, $f(z)$ reduces to a constant.

37. Evaluate $\displaystyle\int_0^{2\pi} \frac{d\phi}{(a + \cos \phi)^2}$ $(a \in \mathbb{R}, a > 1)$.

Solution We shall use Theorem 4.7.6. Put $R(x, y) = (a + x)^{-2}$.

Now $$\tilde{R}(z) = z^{-1} R\left(\frac{z + z^{-1}}{2}, \frac{z - z^{-1}}{2i}\right)$$

$$= z^{-1} \left(\frac{1}{a + \frac{1}{2}\left(z + \frac{1}{z}\right)}\right)^2 = \frac{4z}{(z^2 + 2az + 1)^2}.$$

The poles are given by $-a \pm \sqrt{(a^2 - 1)}$. Since

$$\left(-a + \sqrt{(a^2 - 1)}\right)\left(-a - \sqrt{(a^2 - 1)}\right) = 1 \quad \text{and} \quad \left|-a - \sqrt{(a^2 - 1)}\right| > a > 1,$$

the only pole inside D (the open unit disc) is $-a + \sqrt{(a^2 - 1)} = \alpha$ (say) and is of order 2. Thus

$$\operatorname*{Res}_{z=\alpha} \tilde{R}(z) = \frac{1}{1!}\left\{\frac{d}{dz}\{(z - \alpha)^2 \, \tilde{R}(z)\}\right\}_{z=\alpha}$$

$$= \left\{\frac{d}{dz}\left\{\frac{4z}{(z - \beta)^2}\right\}\right\}_{z=\alpha} \quad \left(\text{with } \beta = -a - \sqrt{(a^2 - 1)}\right)$$

$$= \frac{-4(\alpha + \beta)}{(\alpha - \beta)^3}$$

$$= \frac{a}{(a^2 - 1)^{3/2}} \cdot$$

Hence by Theorem 4.7.6 we have

$$\int_0^{2\pi} \frac{d\phi}{(a + \cos \phi)^2} = \frac{2\pi a}{(a^2 - 1)^{3/2}} \cdot$$

38. Compute $h_n = \sum_{k=0}^{n-1} e^{\frac{2\pi i k^2}{n}}$, $n = 1, 2, \ldots$ using residue calculus.

Solution Let $h_n (z) = \sum_{k=0}^{n-1} e^{\frac{2\pi i(z+k)^2}{n}}$, $n = 1, 2, \ldots$ Note that h_n

$= h_n(0)$. Consider $M_n (z) = \dfrac{h_n (z)}{e^{2\pi i z} - 1}$. $M_n(z)$ for each n is meromor-

phic with poles at all integers. Consider the parallelogram P with

vertices $-\dfrac{1}{2} - cr, \dfrac{1}{2} - cr, \dfrac{1}{2} + cr, -\dfrac{1}{2} + cr$ where $c = e^{i\pi/4} = \dfrac{1+i}{\sqrt{2}}$

(Therefore $c^2 = i$) and $r > 0$. See the figure below for the contours γ_i of P. Inside P, $M_n(z)$ has only one simple pole at $z = 0$ with residue

equal to $\lim_{z \to 0} z \, M_n (z) = \lim_{z \to 0} \dfrac{h_n (z)}{(e^{2\pi i z} - 1)/z} = \dfrac{h_n(0)}{2\pi i} = \dfrac{h_n}{2\pi i}$. Let the

boundary of the parallelogram be given by $\partial P = \gamma_1 + \gamma_2 + \gamma_3 - \gamma_4$. By

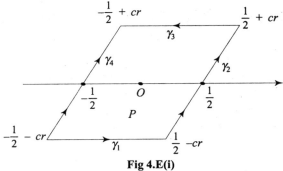

Fig 4.E(i)

residue theorem $\int_{\partial P} M_n (\zeta) \, d\zeta = h_n$. Put

$$I(r) = \int_{\gamma_2} M_n(\zeta) \, d\zeta - \int_{\gamma_4} M_n(\zeta) \, d\zeta.$$

We first observe that

$$h_n(z + 1) - h_n (z) = e^{\frac{2\pi i z^2}{n}} (e^{4\pi i z} - 1) \qquad (2)$$

$$\left(\text{Since } \sum_{k=0}^{n-1} e^{\frac{2\pi i (z+1+k)^2}{n}} - \sum_{k=0}^{n-1} e^{\frac{2\pi i (z+k)^2}{n}} = \sum_{k=1}^{n} e^{\frac{2\pi i (z+k)^2}{n}} - \sum_{k=0}^{n-1} e^{\frac{2\pi i (z+k)^2}{n}} \right.$$

$$\left. = e^{\frac{2\pi i (z+n)^2}{n}} - e^{\frac{2\pi i z^2}{n}} = e^{\frac{2\pi i z^2}{n}} \left(e^{\frac{2\pi i (n^2 + 2nz)}{n}} - 1 \right) = e^{\frac{2\pi i z^2}{n}} \left(e^{4\pi i z} - 1 \right) \right).$$

Thus

$$M_n(z + 1) - M_n (z) = \frac{h_n(z+1)}{\left(e^{2\pi i (z+1)} - 1 \right)} - \frac{h_n(z)}{\left(e^{2\pi i z} - 1 \right)}$$

$$= \frac{h_n(z+1) - h_n(z)}{\left(e^{2\pi i z} - 1 \right)} = e^{\frac{2\pi i z^2}{n}} \left(e^{2\pi i z} + 1 \right) \text{ by (2)}.$$

Now consider $\int_{\gamma_2} M_n (\zeta) \, d\zeta$. Put $\zeta = z + 1$. As ζ varies over γ_2, z

varies over γ_4. Thus $\int_{\gamma_2} M_n (\zeta) \, d\zeta = \int_{\gamma_4} M_n (\zeta + 1) \, d\zeta$. Hence

$$I(r) = \int_{\gamma_4} (M_n(\zeta + 1) - M_n (\zeta)) \, d\zeta = \int_{\gamma_4} e^{\frac{2\pi i \zeta^2}{n}} \left(e^{2\pi i \zeta} + 1 \right) d\zeta. \text{ We shall}$$

use the following identities:

$$\frac{2\pi i \zeta^2}{n} + 2\pi i \zeta = \frac{2\pi i}{n}\left(\zeta + \frac{n}{2}\right)^2 - \frac{1}{2}\pi i n \quad \text{and} \quad e^{-\frac{1}{2}\pi i n} = (-i)^n.$$

Now $I(r) = \int_{\gamma_4} e^{\frac{2\pi i \zeta^2}{n}} d\zeta + (-i)^n \int_{\gamma_4} e^{\frac{2\pi i}{n}\left(\zeta + \frac{n}{2}\right)^2} d\zeta$

On γ_4, $\zeta = -\frac{1}{2} + ct$, $-r < t < r$. $c^2 = i$. Therefore

$$I(r) = c\int_{-r}^{r} e^{\frac{2\pi i}{n}\left(-\frac{1}{2} + ct\right)^2} dt + c(-i)^n \int_{-r}^{r} e^{\frac{2\pi i}{n}\left(\frac{n-1}{2} + ct\right)^2} dt$$

$$= c\int_{-r}^{r} e^{-\frac{2\pi}{n}\left(t - \frac{1}{2c}\right)^2} dt + c(-i)^n \int_{-r}^{r} e^{-\frac{2\pi}{n}\left(\frac{n-1}{2c} + t\right)^2} dt.$$

By Theorem 4.7.3 it follows that

$$\lim_{r \to \infty} I(r) = c(1 + (-i)^n) \int_{-\infty}^{\infty} e^{-\frac{2\pi}{n} s^2} ds = c(1 + (-i)^n) \sqrt{\frac{n}{2\pi}} \int_{-\infty}^{\infty} e^{-t^2} dt$$

$$= (1 + (-i)^n) \, c \sqrt{\frac{n}{2}}.$$

We now show that $I_1(r) = \int_{\gamma_1} M_n(\zeta) \, d\zeta$ and $I_2(r) = \int_{\gamma_2} M_n(\zeta) \, d\zeta$ tend

to zero as $r \to \infty$, γ_1 is parameterized by $\zeta = t - cr$, $-\frac{1}{2} \le t \le \frac{1}{2}$.

Thus $\left| \int_{\gamma_1} M_n(\zeta) \, d\zeta \right| \le \frac{1}{4}\sup_{\zeta \in \gamma_1} |M_n(\zeta)|$. Further for $\zeta \in \gamma_1$,

$$|M_n(\zeta)| = |M_n(t - cr)| = \left| \frac{h_n(t - cr)}{(e^{2\pi i(t - cr)} - 1)} \right| \le \left| \frac{h_n(t - cr)}{(e^{\pi\sqrt{2}} - 1)} \right| \quad (r \ge 1).$$

On γ_1, $h_n(t - cr) = \sum_{k=0}^{n-1} e^{\frac{2\pi i(t - cr + k)^2}{n}}$ and for each k and large r

$$\left| e^{\frac{2\pi i(t - cr + k)^2}{n}} \right| \le \left| e^{\frac{2\pi i}{n}\left((t + k)^2 + (cr)^2 - 2(t + k)cr\right)} \right|$$

$$= \left| e^{\dfrac{2\pi i}{n}\left((t+k)^2 + ir^2 - \dfrac{2(t+k)r}{\sqrt{2}}(1+i)\right)} \right|$$

$$= e^{-\dfrac{2\pi}{n}\left(r^2 - \sqrt{2}(t+k)r\right)}$$

$$= e^{-\dfrac{2\pi}{n}\left(\left(r - \dfrac{1}{\sqrt{2}}(t+k)\right)^2 - \dfrac{(t+k)^2}{2}\right)}$$

$$\le M e^{-\dfrac{2\pi}{n}\left(r - \dfrac{1}{\sqrt{2}} - \dfrac{k}{\sqrt{2}}\right)^2} \quad \text{(use } t \le 1 \text{ and } M = e^{\pi n}\text{)}.$$

Since this is true for each k, $\displaystyle\sup_{\zeta \in \gamma_1} |M_n(\zeta)| \to 0$ as $r \to \infty$.

(Note that n is fixed). Similarly $\displaystyle\int_{\gamma_3} M_n(\zeta)\, d\zeta \to 0$ as $r \to \infty$.

Using all these we get that

$$h_n = (1 + (-i)^n)\left(\frac{1+i}{\sqrt{2}}\right)\sqrt{\frac{n}{2}} = \frac{(1+(-i)^n)}{(1-i)}\sqrt{n}.$$

39. Prove using contour integration $\displaystyle\sum_{k=0}^{n}({}^n c_k)^2 = {}^{2n}c_n$.

Solution ${}^n c_k$ is the coefficient of z^k in the binomial expansion of $(1 + z)^n$. Let C be any simple closed circle, centre origin. Since $\dfrac{(1+z)^n}{z^{k+1}}$ is having a pole at origin with residue ${}^n c_k$, by residue theorem we have

$$^n c_k = \frac{1}{2\pi i}\int_C \frac{(1+z)^n}{z^{k+1}}\, dz. \tag{1}$$

We now observe that the constant term in the binomial expansion of $(1 + z)^n\left(1 + \dfrac{1}{z}\right)^n$ is $\displaystyle\sum_{k=0}^{n}({}^n c_k)^2$. Therefore the coefficient of $\dfrac{1}{z}$ in the

expansion of $(1+z)^n\left(1 + \dfrac{1}{z}\right)^n \dfrac{1}{z}$ is $\displaystyle\sum_{k=0}^{n}({}^n c_k)^2$. Thus again by residue theorem

$$\sum_{k=0}^{n}({}^n c_k)^2 = \frac{1}{2\pi i}\int_C (1+z)^n\left(1 + \frac{1}{z}\right)^n \frac{1}{z}\, dz$$

$$= \frac{1}{2\pi i} \int_C \frac{(1+z)^{2n}}{z^{n+1}} \, dz$$

$$= 2^n c_n \text{ by (1)}.$$

40. Suppose f is entire and is of the form

$$f(x, y) = u(x) + iv(y)$$

Show that f is at most a linear polynomial.

Solution   If f is analytic then we know that its real and imaginary parts are harmonic. Hence $u(x, y) \equiv u(x)$, $v(x, y) \equiv v(y)$ are both

harmonic. Thus $\dfrac{\partial^2 u}{\partial x^2} + \dfrac{\partial^2 u}{\partial y^2} = 0 \Rightarrow u''(x) = 0$ or that $u'(x) = a$ or

that $u(x) = ax + b$ (where $'$ denotes the derivative with respect to x)

with $a, b \in \mathbb{R}$. Similarly $\dfrac{\partial^2 v}{\partial x^2} + \dfrac{\partial^2 v}{\partial y^2} = 0 \Rightarrow v''(y) = 0$ or that $v'(y)$

$= c$ (Here $'$ denotes the derivative with respect to y) or that $v(y) = cy$

$+ d$ with c and d real. But the $C - R$ equation for f gives $\dfrac{\partial u}{\partial x} = \dfrac{\partial v}{\partial y}$

or that $u'(x) = v'(y)$ which is the same as $a = c$. Thus $f(z) = u(x)$ $+ iv(y) = ax + b + i(cy + d) = ax + icy + b + id = az + \alpha$ with $b + id = \alpha$. Thus $f(z)$ is at most a linear polynomial.

41. Let the function $u(z)$ be harmonic in the disc $|z| \leq \rho$. Prove that

$$\frac{d}{dr} \int_0^{2\pi} u(re^{i\phi}) \, d\phi = \int_0^{2\pi} \frac{\partial}{\partial r} u(re^{i\phi}) \, d\phi \quad (r < \rho) \text{ and deduce Gauss's mean}$$

value theorem from this.

Solution   By definition

$$\frac{d}{dr} \int_0^{2\pi} u(re^{i\phi}) \, d\phi = \lim_{\Delta r \to 0} \left(\int_0^{2\pi} \left[\frac{u((r + \Delta r) e^{i\phi}) - u(re^{i\phi})}{\Delta r} \right] \right) d\phi.$$

Also for every fixed $\phi \in [0, 2\pi]$

$$\lim_{\Delta r \to 0} \left[\frac{u((r + \Delta r) e^{i\phi}) - u(re^{i\phi})}{\Delta r} \right] = \frac{\partial u}{\partial r} (re^{i\phi}).$$

Note that by mean value theorem for real variables

$$\frac{u((r + \Delta r) e^{i\phi}) - u(re^{i\phi})}{\Delta r} = \frac{\partial u}{\partial r} (se^{i\phi})$$

for some s between r and $r + \Delta r$. Now using uniform continuity of $\dfrac{\partial u}{\partial r}(z)$ for $|z| \le r'$ where $r < r' < \rho$, given $\varepsilon > 0$ we can find a $\delta > 0$ such that whenever $|z_1 - z_2| < \delta$ and $|z_1| \le r'$, $|z_2| \le r'$,

$$\left| \frac{\partial u}{\partial r}(z_1) - \frac{\partial u}{\partial r}(z_2) \right| < \varepsilon.$$ Thus for sufficiently small Δr,

$$\left| \frac{u((r + \Delta r)\, e^{i\phi}) - u(re^{i\phi})}{\Delta r} - \frac{\partial u}{\partial r}(re^{i\phi}) \right| < \varepsilon \text{ uniformly for all } \phi \text{ (Note}$$

that δ is independent of ϕ, $|se^{i\phi} - re^{i\phi}| < \delta$ as soon as $|s - r| < \delta$). Thus we see that as $\Delta r \to 0$,

$$\frac{u((r + \Delta r)\, e^{i\phi}) - u(re^{i\phi})}{\Delta r} \to \frac{\partial u}{\partial r}(re^{i\phi})$$

uniformly for $\phi \in [0, 2\pi]$. Thus by Riemann integration theory

$$\frac{d}{dr} \int_0^{2\pi} u(re^{i\phi})\, d\phi = \int_0^{2\pi} \frac{\partial}{\partial r} u(re^{i\phi})\, d\phi \tag{1}$$

(At this stage any reader familiar with measure theory will discover that we are trying to avoid an immediate application of Lebesgue Dominated Convergence theorem).

Since $|z| < \rho$ is simply connected there exists a harmonic conjugate of u say v. Therefore $\dfrac{\partial u}{\partial r} = \dfrac{1}{r}\dfrac{\partial v}{\partial \phi}$ and we have

$$\int_0^{2\pi} \frac{\partial}{\partial r} u(re^{i\phi})\, d\phi = \frac{1}{r} \int_0^{2\pi} \frac{\partial}{\partial \phi} v(re^{i\phi})\, d\phi$$

$$= \frac{1}{r} \left[v(re^{i\phi}) \right]_0^{2\pi}$$

$$= 0. \tag{2}$$

Using (1) and (2) we get that

$$\frac{d}{dr} \int_0^{2\pi} u(re^{i\phi})\, d\phi = 0.$$

We now deduce the Gauss's mean value theorem

$$u(0) = \frac{1}{2\pi} \int_0^{2\pi} u(re^{i\phi})\, d\phi.$$

Since $\dfrac{d}{dr} \displaystyle\int_0^{2\pi} u(re^{i\phi})\, d\phi = 0$, $\displaystyle\int_0^{2\pi} u(re^{i\phi})\, d\phi$ is a constant independent of r.

We evaluate the constant by taking the limit, $\lim\limits_{r\to 0}\int_0^{2\pi} u(re^{i\phi})\,d\phi = u(0)$

$2\pi.$ $\left(u(z) \to u(0) \text{ as } z \to 0 \text{ implies } u(re^{i\phi}) \to u(0) \text{ as } r \to 0 \text{ uniformly} \right.$
for $\phi \in [0, 2\pi]$).

Hence $u(0) = \dfrac{1}{2\pi}\int_0^{2\pi} u(re^{i\phi})\,d\phi.$

42. Prove that the solution of the Dirichlet's problem for the disc $|z| \le R$ is given by

$$P_U(re^{i\phi}) = \sum_{-\infty}^{\infty} c_n \left(\frac{r}{R}\right)^{|n|} e^{in\phi} \quad (r < R)$$

where $U(t)$ is a piece-wise continuous function of $|z| = R$ and

$c_n = \dfrac{1}{2\pi}\int_0^{2\pi} U(Re^{i\phi}) e^{-in\phi}\,d\phi.$ (These coefficients c_n are called the

Fourier coefficients of $U(Re^{i\phi})$).

Solution   By Corollary 4.8.17 the required unique solution of the given Dirichlet's problem is given by

$$P_U(z) = \frac{1}{2\pi}\int_0^{2\pi} \frac{R^2 - |z|^2}{|Re^{i\phi} - z|^2} U(Re^{i\phi})\,d\phi \quad (z = re^{i\theta}).$$

We write

$$\frac{R^2 - |z|^2}{|Re^{i\phi} - z|^2} = \frac{t}{t-z} + \frac{\bar{z}}{\bar{t}-\bar{z}} \quad \text{with } t = Re^{i\phi}.$$

$$= \frac{1}{1-(z/t)} + \frac{\overline{(z/t)}}{1-\overline{(z/t)}}$$

$$= 1 + \sum_1^{\infty}(z/t)^n + \sum_1^{\infty}\left(\overline{(z/t)}\right)^n$$

$$= 1 + \sum_1^{\infty}\left((z/t)^n + \overline{(z/t)^n}\right)$$

$$= 1 + \sum_1^{\infty}(r/R)^n e^{in(\theta-\phi)} + \sum_1^{\infty}(r/R)^n e^{-in(\theta-\phi)}$$

$$= \sum_0^{\infty}(r/R)^n e^{in(\theta-\phi)} + \sum_{-1}^{-\infty}(r/R)^{-n} e^{in(\theta-\phi)}$$

$$= \sum_{-\infty}^{\infty}(r/R)^{|n|} e^{in(\theta-\phi)}.$$

Since the series in question converges uniformly in $[0, 2\pi]$ and $U(Re^{i\phi})$ is a bounded function we can write

$$P_U(re^{i\theta}) = \sum_{-\infty}^{\infty} c_n \left(\frac{r}{R}\right)^{|n|} e^{in\theta}$$

where $c_n = \dfrac{1}{2\pi} \int\limits_0^{2\pi} U(Re^{i\phi}) \, e^{-in\phi} \, d\phi.$

43. Use the Solved Exercises: No. 42 to compute the series solution of the Dirichlet's problem for

$$U(Re^{i\phi}) = \begin{cases} 0 & \text{for } \phi \in (0, \pi) \\ 1 & \text{for } \phi \in (\pi, 2\pi) \end{cases}$$

Solution   For $n \in \mathbb{Z}, n \neq 0$ we have

$$c_n = \frac{1}{2\pi} \int\limits_0^{2\pi} U(Re^{i\phi}) \, e^{-in\phi} \, d\phi$$

$$= \frac{1}{2\pi} \int\limits_\pi^{2\pi} e^{-in\phi} \, d\phi$$

$$= -\frac{1}{2\pi in} \left[e^{-in\phi} \right]_\pi^{2\pi}$$

$$= -\frac{1}{2\pi in} \, [1 - (-1)^n].$$

Thus $c_n = \begin{cases} -\dfrac{1}{\pi in} & \text{if } n \text{ is odd.} \\ 0 & \text{if } n \text{ is even and } n \neq 0. \end{cases}$

whereas $c_0 = \dfrac{1}{2\pi} \int\limits_\pi^{2\pi} d\phi = \dfrac{1}{2}.$

We shall write all odd numbers as $(2k - 1)$ $(-\infty < k < \infty)$ and get

$$P_U(re^{i\theta}) = \frac{1}{2} - \frac{1}{\pi i} \sum_{-\infty}^{\infty} \left(\frac{r}{R}\right)^{|2k-1|} \frac{e^{i(2k-1)\theta}}{(2k-1)}$$

$$= \frac{1}{2} - \frac{2}{\pi} \sum_{k=1}^{\infty} \left(\frac{r}{R}\right)^{2k-1} \frac{\sin(2k-1)\theta}{(2k-1)}$$

$$\left\{\text{Note that } \sum_{-\infty}^{0}\left(\frac{r}{R}\right)^{|2k-1|}\frac{e^{i(2k-1)\theta}}{(2k-1)} = -\sum_{l=0}^{\infty}\left(\frac{r}{R}\right)^{2l+1}\frac{e^{-(2l+1)i\theta}}{(2l+1)}\right.$$

$$= -\sum_{m=1}^{\infty}\left(\frac{r}{R}\right)^{2m-1}\frac{e^{-(2m-1)i\theta}}{(2m-1)}$$

$$\left. = -\sum_{k=1}^{\infty}\left(\frac{r}{R}\right)^{2k-1}\frac{e^{-i(2k-1)\theta}}{(2k-1)}.\right.$$

In the above equalities we have changed the variables k as $-l$ and l as $m - 1$ successively$\Big\}$.

44. Show that $u(0) = \dfrac{1}{2\pi}\int_0^{2\pi}u(e^{i\theta})\,d\theta$ where $u(z) = \log|1+z|$ and hence

compute $\displaystyle\int_0^{\pi}\log\,(\sin\,\theta)\,d\theta$.

Solution Since $u(e^{i\theta}) = \log|1 + e^{i\theta}|$ is undefined at $\theta = \pi$, the definition of $\int_0^{2\pi}u(e^{i\theta})\,d\theta$ must be as follows.

$$\int_0^{2\pi}u(e^{i\theta})\,d\theta = \lim_{\delta\to 0}\left(\int_0^{\pi-\delta}u(e^{i\theta})d\theta + \int_{\pi+\delta}^{2\pi}u(e^{i\theta})\,d\theta\right) \tag{1}$$

Since the principal branch of $\log(1+z)$ is analytic in $|z| < 1$, and its real part is $\log|1+z|$, we conclude that $u(z) = \log|1+z|$ is harmonic in $|z| < 1$. Hence by mean value property of harmonic functions

$$u(0) = \frac{1}{2\pi}\int_0^{2\pi}u(re^{i\theta})\,d\theta \text{ for each } r < 1. \tag{2}$$

Thus $0 = u(0) = \dfrac{1}{2\pi}\int_0^{2\pi}u(re^{i\theta})\,d\theta$. But

$$\int_0^{2\pi}u(re^{i\theta})\,d\theta = \lim_{\delta\to 0}\left(\int_0^{\pi-\delta}u(re^{i\theta})d\theta + \int_{\pi+\delta}^{2\pi}u(re^{i\theta})\,d\theta\right) \tag{3}$$

where the limit is uniform for $0 < r < 1$ (This is because the modulus of the difference of the left and right sides of equality (3) is dominated by $\displaystyle\int_{\pi-\delta}^{\pi+\delta}|u(re^{i\theta})|\,d\theta \le (\log 2)\,(2\delta)$ which tends to 0 uniformly for all r with $0 < r < 1$).

We now claim that $u(re^{i\theta}) \to u(e^{i\theta})$ uniformly as $r \to 1$ in $[0, \pi - \delta]$ and in $[\pi + \delta, 2\pi]$. Infact,

$$\left| \frac{1 + r^2 + 2r\cos\theta}{2 + 2\cos\theta} - 1 \right| = \left| \frac{(1 - r^2) + (1 - r)(2\cos\theta)}{2 + 2\cos\theta} \right|$$

$$\leq \left| \frac{(1 - r)(1 + r + 2\cos\theta)}{2 + 2\cos\theta} \right|$$

$$\leq \left| \frac{(1 - r)(4)}{2(1 + \cos\theta)} \right| \leq \frac{2(1 - r)}{\varepsilon}$$

where $\varepsilon = 1 + \cos(\pi - \delta)$ or $1 + \cos(\pi + \delta)$ depending on $\theta \in$ $[0, \pi - \delta]$ or $\theta \in [\pi + \delta, 2\pi]$. Thus $\log\left| \dfrac{1 + re^{i\theta}}{1 + e^{i\theta}} \right| \to 0$ as $r \to 1$, uniformly for $\theta \in [0, \pi - \delta]$ or for $\theta \in [\pi + \delta, 2\pi]$. Hence

$$\left(\int_0^{\pi - \delta} + \int_{\pi + \delta}^{2\pi} \right) u(re^{i\theta})\, d\theta \to \left(\int_0^{\pi - \delta} + \int_{\pi + \delta}^{2\pi} \right) u(e^{i\theta})\, d\theta \qquad (4)$$

as $r \to 1$. Using (1), (3) and (4) we see that

$$\int_0^{2\pi} u(re^{i\theta})\, d\theta \to \int_0^{2\pi} u(e^{i\theta})\, d\theta. \qquad (5)$$

Using (2) and (5) we get

$$0 = u(0) = \frac{1}{2\pi} \int_0^{2\pi} u(e^{i\theta})\, d\theta.$$

Thus we have verified the mean value property. We now compute $\int_0^{\pi} \log(\sin\theta)\, d\theta$. We have

$$u(e^{i\theta}) = \log|1 + e^{i\theta}|$$

$$= \frac{1}{2}\log|1 + e^{i\theta}|^2$$

$$= \frac{1}{2}(\log 2 + \log(1 + \cos\theta)).$$

Therefore

$$0 = u(0) = \frac{1}{2\pi} \int_0^{2\pi} \frac{1}{2}(\log 2 + \log(1 + \cos\theta))\, d\theta.$$

i.e., $\quad -\log 2 = \dfrac{1}{2\pi} \int_0^{2\pi} \dfrac{1}{2} 2 \log \sin (\theta/2)\, d\theta.$

$$= \dfrac{1}{\pi} \int_0^{\pi} \log \sin \theta\, d\theta$$

which implies $\int_0^{\pi} \log (\sin \theta)\, d\theta = -\pi \log 2.$

EXERCISES

1. Compute $\int_\gamma \operatorname{Re} z\, dz$ where γ is the line segment joining 0 to $1 + i$.

2. Compute $\displaystyle\int_{|z|=\rho} \dfrac{|dz|}{|z-a|^2}.$

3. If $P(z)$ is a polynomial, compute the value of $\displaystyle\int_{|z-a|=R} P(z)\, d\bar z$

   {**Hint:** Write $z = a + Re^{i\theta}$ $(0 \le \theta \le 2\pi)$ and use $\int_0^{2\pi} e^{ik\theta}\, d\theta = 0$ for $k \ne 0$}.

4. If g is continuous on $|z| = 1$, show that
   $$\overline{\int_{|z|=1} g(z)\, dz} = - \int_{|z|=1} \overline{g(z)}\, z^2 dz.$$

5. Let γ be a curve defined by $\gamma(t) = a\cos t + ib \sin t$, $0 \le t \le 2\pi$. Compute $\int_\gamma |\zeta|^2\, d\zeta$.

6. Show that any function which is meromorphic in the extended complex plane must be a rational function (See solved problem 20).

7. How many roots of the equation $z^4 + 8z^3 + 3z^2 + 8z + 3 = 0$ lie in the right half plane. (Hint: Find the change in the argument of $f(y)$ as y varies between $-\infty$ and ∞, where $f(z)$ is the given polynomial).

8. Describe a set of conditions under which the formula $\int_\gamma \log z dz = 0$ is meaningful and valid.

9. Develop the following functions as a power series around the origin.
   (a) $f(z) = \sin^2 z$
   (b) $f(z) = \cos (z^2 - 1)$.

10. Let a, b be complex numbers with $|a| < 1 < |b|$. For positive integers n, m compute $\displaystyle\int_{|z|=1} \dfrac{dz}{(z-a)^m (z-b)^n}.$

11. Let f be analytic in $|z| < r$, $(r > 1)$. Prove that

    (i) $\dfrac{1}{\pi} \displaystyle\int_0^{2\pi} f(e^{it}) \cos^2 (t/2) \, dt = f(0) + \dfrac{1}{2}f'(0)$

    (ii) $\dfrac{1}{\pi} \displaystyle\int_0^{2\pi} f(e^{it}) \sin^2 (t/2) \, dt = f(0) - \dfrac{1}{2}f'(0)$

    (**Hint:** Write $\cos^2 (t/2) = \dfrac{1}{2}(1 + \cos t) = \dfrac{1}{4}(2 + e^{it} + e^{-it})$ and

    $\sin^2 (t/2) = \dfrac{1}{4} (2 - e^{it} - e^{-it})$.)

12. Find the radius of convergence of the series representing

    (i) $f(z) = \dfrac{e^z}{1 - az}$, $a \in \mathbb{C}$.

    (ii) $f(z) = \dfrac{\sin^2 z}{z}$

    around origin.

13. Let f be an analytic function in a region $G \subseteq \mathbb{C}$. Show that the following are equivalent.
    (i) f is a polynomial.
    (ii) There is a point $c \in G$ such that $f^{(n)}(c) = 0$ for almost all $n \in \mathbb{N}$. (This means: for all but a finite number of n's).

14. Let G be a region in $\mathbb{C}$ which is symmetric about $\mathbb{R}$, i.e., $G = \{\bar{z} : z \in G\}$. Let f be an analytic function on G. Show that the following are equivalent.

    (i) $f(G \cap \mathbb{R}) \subseteq \mathbb{R}$.   (ii) $f(\bar{z}) = \overline{f(z)}$, $\forall \, z \in G$.

15. Show that every automorphism of the open unit disc fixing the origin is a rotation.

16. Prove that every automorphism of either the open unit disc or the upper half plane with two distinct fixed points is the identity.

17. Prove the following sharper version of Schwarz lemma. If $f : U \to U$ is analytic (where U is the open unit disc) with a zero of order $n \geq 1$ at $z = 0$ then $|f(z)| \leq |z|^n$ and $|f^{(n)}(0)| \leq n!$. Discuss the cases of equalities in the above inequalities.

18. For $z, w \in U$ the open unit disc put $\Delta(z, w) = \left| \dfrac{z - w}{1 - \bar{w}z} \right|$. Let $f : U \to U$ be analytic. Prove that $\Delta(f(z), f(w)) \leq \Delta(z, w)$ with equality if and only if f is an automorphism of U.

19. Let U denote the open unit disc and H denote the upper half plane. If $P : U \to H$ is analytic with $P(0) = 1$ then prove that

    $$\dfrac{1 - |z|}{1 + |z|} \leq |p(z)| \leq \dfrac{1 + |z|}{1 - |z|} \quad \text{and} \quad |p'(0)| \leq 2.$$

20. If C is any contour from -1 to 1 which, except for its end points, lies in the upper half plane, then prove that $\int_C z^i dz = (1 + e^{-\pi})(1 - i)$ where z^i denotes the principal branch given by $z^i = e^{i \log z}$ with $|\text{Im} \log z| < \pi$.

21. The Legendre's polynomial is defined by
$$P_n(z) = \frac{1}{n! z^n} \frac{d^n}{dz^n}(z^2 - 1)^n.$$
Show that $P_n(1) = 1$ and $P_n(-1) = (-1)^n$.

22. Let $f(z)$ be entire with $\text{Re } f(z)$ bounded above in $\mathbb{C}$. Show that $f(z)$ is a constant.

23. Find the Laurent expansion of $f(z) = \dfrac{1}{z^2(1-z)}$ valid in (a) $0 < |z| < 1$. (b) $1 < |z| < \infty$.

24. Using the Taylor expansion for $\dfrac{1}{w}$ around $w = 1$ prove that
$$\log (1 + z) = \sum_1^\infty \frac{(-1)^{n+1}}{n} z^n \text{ for } |z| < 1 \text{ where } \log (1 + z) \text{ denotes the}$$
principal branch of $\log (1 + z)$.

25. Suppose f and g are analytic at z_0 with $f(z_0) = 0 = g(z_0)$ and $g'(z_0) \neq 0$. Prove that $\displaystyle\lim_{z \to z_0} \frac{f(z)}{g(z)} = \frac{f'(z_0)}{g'(z_0)}$.

26. Show that $\dfrac{1}{\cosh z} = \sum_0^\infty \dfrac{E_n}{n!} z^n$, $|z| < \dfrac{\pi}{2}$ where $E_0 = 1$, $E_2 = -1$, $E_4 = 5$, $E_6 = -61$, $E_{2n+1} = 0$, $n = 0, 1, 2, \ldots$ (These E_n's are called Euler numbers).

27. Let $f(z)$ be analytic in an annular region containing the unit circle $z = e^{i\theta}$, $|\theta| \leq \pi$. Show that
$$f(z) = \frac{1}{2\pi} \int_{-\pi}^\pi f(e^{i\theta}) \, d\theta + \frac{1}{2\pi} \sum_1^\infty \int_{-\pi}^\pi f(e^{i\theta}) \left((ze^{i\theta})^n + (ze^{-i\theta})^{-n} \right) d\theta$$
where z is any point in the annular region. Deduce the Fourier series expansion of $U(\theta) = \text{Re } f(e^{i\theta})$ given by
$$u(\theta) = \frac{1}{2\pi} \int_{-\pi}^\pi u(\phi) \, d\phi + \frac{1}{\pi} \sum_1^\infty \int_{-\pi}^\pi u(\phi) \cos (n\theta - n\phi) \, d\phi.$$

28. Expand $f(z) = \dfrac{z}{1 + z^3}$ in a series of (i) positive powers of z (ii) negative powers of z. In each case specify the region in which the expansion is valid.

29. Find the Laurent expansion of

   (i) $\dfrac{z}{z+2}$, $\quad |z| > 2.$

   (ii) $\cos (1/z)$, $\quad z \neq 0.$

   (iii) $\dfrac{1}{(z+1)^2 \, (z+1)}$, $\quad 1 < |z| < 2.$

   (iv) $\dfrac{e^z}{z(z^2+1)}$, $\quad 0 < |z| < 1.$

30. Prove the following expansions

   (i) $e^z = e + e\sum\limits_{1}^{\infty} \dfrac{1}{n!} \, (z-1)^n$

   (ii) $\dfrac{1}{z} = \sum\limits_{1}^{\infty}(-1)^n \, (z-1)^n,$ $\qquad |z-1| < 1.$

   (iii) $\dfrac{1}{z^2} = 1 + \sum\limits_{1}^{\infty}(n+1) \, (z+1)^n,$ $\qquad |z+1| < 1.$

31. Using Residues prove that $\displaystyle\int\limits_{0}^{\infty}\dfrac{(\log x)^2}{1+x^2} \, dx = \dfrac{\pi^3}{8}$

32. Assume that f is analytic in a region Ω and that at every point of Ω either $f = 0$ or $f' = 0$. Show that f is constant in Ω.
   (**Hint:** First consider f^2).

33. Show that a non-constant analytic function can not map a region onto a straight line or onto a circular arc.
   (**Hint:** Assume the contrary and conclude that either $f(z) = a + bt$, $t \in \mathbb{R}$, $\forall \, z \in \Omega$ or $|f(z) - a| = R$ for all $z \in \Omega$).

34. Let f be entire, real on the real axis and purely imaginary on the imaginary axis. Show that f is odd.

35. Suppose f is a non-constant analytic function in the closure $\bar{S}$ of a region S in the plane and $f(z)$, for some $z \in \bar{S}$, is a boundary point of $\Omega = f(S)$. Prove that z is a boundary point of S.

36. Show that every function f analytic in a symmetric region Ω can be written in the form $f_1 + if_2$ where f_1, f_2 are analytic on Ω and real on the real axis.

   (**Hint:** Put $f_1 = \dfrac{f+g}{2}$ $f_2 = \dfrac{f-g}{2i}$ where $g(z) = \overline{f(\bar{z})}$.

37. If f is analytic in $|z| \leq 1$ and satisfies $|f| = 1$ on $|z| = 1$, show that $f(z)$ is rational (See problem 16).

38. Let Ω be a simply connected region in $\mathbb{C}$. Given a harmonic function $u \, (z)$ in Ω obtain an expression for $v(z)$ where $f(z) = u(z) + iv(z)$ is analytic in Ω.

(**Hint:** Try $v(z) = \text{Im} \int\limits_{z_0}^{z} \left(\dfrac{\partial u}{\partial x}(w) - i \dfrac{\partial u}{\partial y}(w) \right) dw$ for a suitable arc joining z_0 and z).

39. If u and v are conjugate harmonic functions, prove that their product (uv) is harmonic.

40. Prove that the angle subtended at a point z in the upper half plane by a given segment of the real axis is harmonic in the variable z.

41. Suppose f is entire and $|f(z)| \le \dfrac{1}{|\text{Re } z|^2}$ show that f is identically zero. (**Hint:** First observe that f is bounded on the real axis. Apply Poisson representation formula

$$f(\xi + i\eta) = \frac{1}{\pi} \int\limits_{\mathcal{R}} \frac{\eta f(x)\, dx}{(x-\xi)^2 + \eta^2}$$

and split the integral as the sum of two integrals one over $|x| < k$ and another over $|x| > k$ for any $k > 0$ and prove that $|f(z)| \le \dfrac{c}{|\text{Im } z|}$ for all $z \ne 0$ and then use Solved exercise 14).

42. If f is analytic and bounded in $\text{Im } z \ge 0$ and real on the real axis, prove that f is a constant.
    (**Hint:** Use Poisson representation formula).

43. Let Ω be the punctured plane described by $0 < |z| < \infty$. Show that $u(z) = \log |z|$ is harmonic in Ω but does not possess a harmonic conjugate in Ω.

44. Let Ω be a region. Let u be a real valued function of x, y belonging to Ω such that $\dfrac{\partial^2 u}{\partial x^2}, \dfrac{\partial^2 u}{\partial x\, \partial y}, \dfrac{\partial^2 u}{\partial y\, \partial x}, \dfrac{\partial^2 u}{\partial y^2}$ exist, are continuous and satisfy $\dfrac{\partial^2 u}{\partial x^2} + \dfrac{\partial^2 u}{\partial y^2} = 0$ in Ω. If Ω is simply connected, prove that these hypotheses are sufficient to conclude that u is continuous. From this deduce that the same result is true if Ω is any region.

45. Let $f(z)$ be analytic in the closed annulus $r_1 \le |z| \le r_2$. Prove that $M(r) = \max\limits_{|z|=r} |f(z)|$ satisfies

$$\log M\,(r) \le \frac{\log r_2 - \log r}{\log r_2 - \log r_1} \log M\,(r_1) + \frac{\log r - \log r_1}{\log r_2 - \log r_1} \log M\,(r_2)$$

$(r_1 < r < r_2)$, i.e., $\log M\,(r)$ is a convex function of $\log r$.

NOTES

The development of the theory of complex integration has its origin in the 19th century. Historically the mathematicians who can be associated with this development are Gauss (1811), Poisson (1813) and Cauchy (1814), see also [2, 3]. Indeed Cauchy's idea of combining two real integrals of the type $\int (u dx - v dy)$ and $\int (v dx + u dy)$ (which come up in mathematical physics in the study of two dimensional flows of incompressible fluids) into a single integral $\int f dz$ with $f = u + iv$ and $z = x + iy$ can be truly regarded as the beginning of the complex integral calculus. A good exposition of the integral calculus in the complex plane along with other reference materials can be found in [12, 13]. The most important application of the Cauchy theory lies in the local power series expansion of an analytic function.

The concept of an analytic function satisfying the complex differentiability condition is due to Cauchy and Riemann whereas the concept of an analytic function which has a local power series expansion is due to Weierstrass. While the former definitions is a natural extension of the real differentiability condition for real valued functions of a real variable the latter definition is the one which is useful if one wants to develop a function theory over general complete valued fields K other than $\mathbb{R}$ or $\mathbb{C}$ (the so called p-adic function theory). More interesting informations regarding the history of the development of the theory of functions of a complex variable can be had from [10].

The French mathematician Edouard Goursat (1858-1936) was the first to give a proof of Cauchy's theorem for a rectangle (first assuming the continuity of the derivative and later dispensing with this requirement). However the German mathematician Pringsheim (1850-1941) essentially simplified Goursat's proof and gave it the elegant and final form, that it has even today. If the reader so wishes he can take for granted the statement of the "General form of Cauchys' theorem" and skip the proof given in 4.3. However a post-graduate student must be aware of the intricacies in the detailed proof of this topological theorem. The general form of Cauchy's theorem leads to the following general form of Cauchy's integral formula: If γ is a cycle in a non-empty open set G and $\gamma \sim 0$ in G and if f is analytic in G then for $z_0 \in G\backslash\gamma$ we have

$$\frac{1}{2\pi i} \int_\gamma \frac{f(z)}{z - z_0} dz = n(\gamma, z_0) f(z_0).$$

On the other hand taking $F(z) = (z - z_0) f(z)$ it is easy to see that this integral formula also implies Cauchy's theorem $\left(\text{indeed } \frac{1}{2\pi i} \int_\gamma f(z) dz = \frac{1}{2\pi i} \right.$

$\left. \int_\gamma \frac{F(z)}{z - z_0} dz = F(z_0) = 0 \right).$ Using this observation there is yet another proof of the general form of Cauchy's theorem (due to Dixon) which the reader can supplement. (See the list of special topics given towards the end of this "NOTES"). Our proof of theorem 4.3.8 is due to Artin as available in [5].

For a few specific functions and special curves γ it may be easy to calculate $\int_\gamma f(z)\, dz$ directly and prove that this integral reduces to 0 (whenever this is true) without any necessity to invoke Cauchy's theorem in any form. However if we do not know much about the function or the curve the only way is to appeal to Cauchy's theorem in one form or the other to conclude $\int_\gamma f(z)\, dz = 0$ (whenever this is true). Most often Cauchy's theorem for simply connected regions or its homotopy version will be sufficient. Thus it is interesting to ask how useful is the most general form of Cauchy's theorem. At this stage theoretically one has to convince oneself that there are situations in which only the general form of Cauchy's theorem and no other form is applicable. For this conviction it is necessary to exhibit a region Ω and a curve γ such that Ω is not simply connected, $\gamma \sim 0$ in Ω but is not homotopic to a constant path (In this case $\int_\gamma f(z)\, dz = 0$ for all analytic functions $f(z)$ in Ω can be proved only by applying the general form of Cauchy's theorem). That this indeed is the case can be proved with a little more effort (See the list of topics towards the end of this "NOTES").

As an application of Cauchy's theorem for simply connected regions we have proved the existence of a single-valued analytic branch for $\log f(z)$ whenever $f(z)$ is analytic and free from zeros in a simply connected region Ω (See Theorem 4.4.7). On the other hand there is an elegant but purely algebraic proof of this theorem (See the list of topics towards the end of this "NOTES"). The reader is invited to master this.

The famous Jordan curve theorem asserts that every Jordan curve in the plane determines exactly two components. The notion of "winding number of a curve with respect to a point" (See Theorem 4.2.19) leads to a simple proof of one part of the above theorem namely that each Jordan curve determines at least two components. (See [1] p 118).

Using Cauchy's integral formula for discs it is easy to see that every f analytic in a neighbourhood of a point $c \in \mathbb{C}$ satisfies

$$f(c) = \frac{1}{2\pi} \int_0^{2\pi} f(c + re^{i\theta})\, d\theta$$

where the closed disc centre c and radius r lies in the region of analyticity of f. This equality is called mean value property for analytic functions. Using this equality there is an elegant proof of Liouville's theorem given by E. Nelson [9]. Solved Exercise No. 14, is also called extended Liouville's theorem. Further another version of the same theorem states that if $f(z)$ is entire and satisfies any one of the four inequalities

$$-A|z|^n \le \operatorname{Re} f \le A|z|^n$$
$$-A|z|^n \le \operatorname{Im} f \le A|z|^n,$$

then f is a polynomial of degree atmost n. For a proof we refer to [4].

The principle of analytic continuation as stated in Theorem 4.4.16 has the following interesting Corollary. "If Ω is an open set and $H(\Omega)$ denotes the set of all analytic functions in Ω then $H(\Omega)$ is an integral domain (already $H(\Omega)$ is a ring with unit element under the pointwise addition and multiplication) if and only if Ω is a region. Indeed if Ω is a region and $f(z)\,g(z) = 0$ for all $z \in \Omega$ with $f(c) \ne 0$ for some $c \in \Omega$ then by continuity, $f(z) \ne 0$ in a small neighbourhood of c and in this neighbourhood $g(z)$ is identically 0 which implies (by Theorem 4.4.16) $g(z) \equiv 0$ in Ω. Thus $fg = 0$, $f \ne 0$ implies $g = 0$. Conversely if $H(\Omega)$ is an integral domain and Ω is not connected (Ω is given to be open) then we can write $\Omega = \Omega_1 \cup \Omega_2$ where Ω_1 and Ω_2 are non-empty disjoint open sets. If we define $f(z) = 0$ for $z \in \Omega_1$, $f(z) = 1$ for $z \in \Omega_2$, $g(z) = 1$ for $z \in \Omega_1$ and $g(z) = 0$ for $z \in \Omega_2$ then it is easy to see that f and g are in $H(\Omega)$ and that $fg = 0$ with neither $f = 0$ nor $g = 0$ contradicting the given algebraic nature of $H(\Omega)$. Using the principle of analytic continuation there is a "quantitative form of the open mapping theorem" from which also the open mapping theorem can be easily deduced. See [10, p 257].

Recall that in real variable theory the relative maxima and minima for a differentiable function f are found among the critical points of f (those points at which $f' = 0$). In contrast to this situation there is an "anti-calculus proposition" due to Erdos which says that "if f is analytic throughout a closed disc D and assumes its maximum modulus at a boundary points α then $f'(\alpha) \ne 0$ unless f is a constant". A proof of this can be found in [4, p 73].

Schwarz original formulation of his theorem (Schwarz lemma) runs as follows. If $f : U \to \Omega$ is a bijective analytic function (here U is the open unit disc and Ω is a region) with $f(0) = 0$ and ρ_1 and ρ_2 are the least and the greatest values of $|w|$ as w varies over $\partial\Omega$ then for all $z \in U$

$$\rho_1 |z| \le |f(z)| \le \rho_2 |z|.$$

Its reformulation as Schwarz lemma and its elegant proof which we have given in the text are due to Caratheodary. A beautiful application of Schwarz lemma is the following theorem due to "Study". "If $f : U \to \Omega$ is a bijective analytic function with $f(0) = 0$ and $\Omega_r = f(D(0, r))$ ($0 < r < 1$) is the image of the open disc centre 0 and radius r under f then Ω convex implies Ω_r is convex for all r, $0 < r < 1$. Similarly if Ω is star-shaped with respect to the origin then so is Ω_r. A proof of this can be had from [10]. Several variants and generalizations (some of which we have worked out in the solved exercises section) are available in the literature. See [10, p 274-275].

If we define a singular point for a function f as one at which f fails to be analytic we always have atleast one singular point on the boundary of the disc of convergence (explicitly assuming its radius to be positive) of a power series of $f(z) = \sum\limits_{k=0}^{\infty} a_k(z - c)^k$ (See [10, p 234]). Using this interesting observation Pringsheim in 1894 proved the following theorem. "Let the power series $f(z) = \sum\limits_{k=0}^{\infty} a_k z^k$ have a positive radius of convergence R and suppose that all but finitely many of the coefficients are real and nonnegative then $z = R$ is a singularity for f. For a proof see [10, p 235]. This theorem can be considered as a generalization of the property of the geometric series $\sum\limits_{k=0}^{\infty} z^k$ which has the only singularity at $z = 1$ on its circle of convergence. The descriptions of the idea of a pole in terms of the growth condition as well as the series development are due to Riemann [11]. The word "pole" was introduced in 1875 by Briot and Bouquet. They also gave the name "meromorphic" for functions which are analytic except for poles in 1875. Meromorphic functions may not only be added, subtracted and multiplied but can even be divided and here lies their advantage over analytic functions. This makes their algebraic structure simpler in contrast to that of analytic functions. In particular the set of all meromorphic functions in a given region forms a Field.

The Theorem 4.5.22 is more popularly known as "Casorati-Weierstrass theorem" due to its discovery by the Italian Mathematician Casorati. The statement of "Casorati-Weierstrass theorem" is equivalent to the following statement. "Let $f(z)$ be analytic in $0 < |z - a| < \delta$. Then $z = a$ is an isolated essential singularity for f if and only if there exists a sequence $\{z_n\}$ in $0 < |z - a| < \delta$ such that $\lim\limits_{n \to \infty} z_n = a$ and that $\{f(z_n)\}$ has no limit in $\mathbb{C} \cup \{\infty\}$. This is rather easy and we leave its proof to the reader. On the other hand far more than the contents of the "Casorati-Weierstrass theorem" is true. If f is analytic on $\Omega = \{z : 0 < |z - a| < \delta\}$ and has an isolated essential singularity at $z = a$ then $f(\Omega)$ is either the whole of $\mathbb{C}$ (as in the case of $f(z) = \sin (1/z)$ at $z = 0$) or the whole of $\mathbb{C}$ with just one exception (as in the case of $f(z) = \exp (1/z)$ at $z = 0$). This famous theorem is due to Picard and its proof is outside the scope of this book. However there is a simple consequence of "Casorati-Weierstrass theorem" which says the following. "If f is a transcendental entire function (an entire function which is not a polynomial) then for every $a \in \mathbb{C}$ there is sequence $\{z_n\}$ such that $z_n \to \infty$ and $f(z_n) \to a$. For a proof see [10, p 308].

The classical literature on the residue calculus is very extensive. The booklet by the Finnish Mathematician Ernst Lindelof [6] needs to be mentioned here. The most recent monograph by Mitrinovic and Keckic [7] will be a very good supplement. (Incidentally the latter monograph also contains a short biography of Cauchy). Theorem 4.6.22 is due to M.H. Shih. See [14]. At this stage we would like to point out that a beginner interested in "Complex analysis" must develop the habit of reading books other than the text book to which he may be "confined" initially. In fact with the help of the teacher the students can start knowing more about the subject by analysing a given topic from different points of view and also by trying to know more about certain aspects of the theory from other sources (like new books, journals etc.) and presenting them as lectures at appropriate forums. In this connection we would like to bring a few topics of interest (the teachers can supplement this list) to the readers at this juncture which are relevant to the contents of the fourth chapter and also indicate the sources from which they can learn and develop them.

Topic I: There are several ways of proving the fundamental theorem of algebra. Collect as many as you can and present them. You can start with [10, p 267 and 391].

Topic II: A complex number α is said to be algebraic if α satisfies an equation $p(\alpha) = 0$ where p is a polynomial with integer coefficients. It is shown in algebra that the set K of all algebraic numbers is countable and is a field extension of the rationals $\mathbb{Q}$ and is therefore not the whole of $\mathbb{C}$. Prove the following: (1) There exists a transcendental entire function (i.e., a non-polynomial) $f(z) = \sum a_k z^k$ with $a_k \in \mathbb{Q}$ for all k and is such that $f(K) \subseteq K$ and $f(\mathbb{Q}) \subseteq \mathbb{Q}$. (2) If A is a countable set in $\mathbb{C}$ and B is a dense subset of $\mathbb{C}$ prove that there exists a transcendental entire function f with $f(A) \subseteq B$. (3) Construct a transcendental entire function which together with all its derivatives take algebraic values at algebraic numbers. (See [10] p 254-255).

Topic III: Understand and present Dixon's proof of the General form of the Cauchy's theorem. See [5].

Topic IV: Prove the existence of a region Ω which is not simply connected and a cycle γ in Ω which is homologous to zero in Ω but not homotopic to a constant in Ω.

Topic V: Search and find out several versions and extensions of Schwarz lemma and prove as many of these as possible. Start from [10, p 274].

Topic VI: Obtain and prove several versions and variations of Liouville's theorem. Use [4].

Topic VII: Give a pure algebraic proof of Theorem 4.4.7. Start proving the lifting lemma of exercise 12 (a) on p 342 of [8] and interpret the results taking $Y = \Omega$, $E = \mathbb{C}$, $B = \mathbb{C}\backslash\{0\}$, $P(z) = e^z$ and f the given analytic function.

References

1. L. Ahlfors., Complex Analysis, McGraw-Hill (3rd ed. 1979), New York.
2. A.L. Cauchy., Mémoire sur ies intégrales défines, Euvres (1) 1, 1814, 319-506.
3. A.L. Cauchy., Mémorie sur les intégrales défines, prises entre des limites imaginaires, Euvres (2) 15, 1825, 41-89. Also reprinted in Bull. Sci. Math. (1) 7, 1874, 265-304 and 8, 1875, 43-55 and 148-159.
4. Joseph Bak and Donald J. Newman., Complex Analysis, Under graduate text in mathematics, Springer-Verlag, New York-Heidelberg-Berlin. 1982.

5. S. Lang., Complex Analysis, Addison-Wesely, 1977, Reading, 2nd ed., Graduate texts in Mathematics, 103, Springer Verlag, 1985, New York.

6. E. Lindelöf., Le calcul des résidus et ses applications à la théorie des fonctions, Gauthier-Villars, 1905, Paris. Reprinted by Chelsea Publ. Co., 1947, New York.

7. D.S. Mitrinovic and J.D. Keckic., The Cauchy method of Residues: Theory and Applications, D. Reidel, 1984, Dordrecht.

8. J.R. Munkres., Topology A first course, Prentice Hall of India Pvt. Ltd., New Delhi. 1988.

9. E. Nelson., A proof of Liouville's theorem, Proc. Amer. Math. Soc., 12, 1961, p 995.

10. R. Remmert., Theory of Complex functions, Graduate texts in Mathematics, 122, Springer-Verlag, New York, 1991.

11. B. Riemann., Grundlagen für eine allgemeine Theorie der Functionen einer veränderlichen complexen Grösse, Inagural Dissertation, 1851, Göttingen, Werke, 5-43.

12. P, Stäckel., Integration durch imaginäres Gebiet. Ein Beiträg zur Geschichte der Funktionentheorie, Biblio. Math. (3), 1, 1900, 109-128.

13. P, Stäckel., Beiträge zur Geschichte der Funktionen theorie im achtzehnten Jahrhundert, Biblio. Math. (3), 2, 1901, 111-121.

14. M.H. Shih., An analog of Balzano's Theorem for functions of a complex variable, Amer. Math. Monthly, 89, 1982, 210-211.

✦ 5

Riemann Mapping Theorem

5.1 Riemann Mapping Theorem and Boundary Behaviour

Introduction 5.1.1

We recall that topology is the study of properties of topological spaces which are invariant under topological homeomorphisms. Similarly Group theory, Ring theory etc., are also properties of the corresponding groups, rings etc., which are invariant under respective isomorphisms. In the same way function theory is concerned with properties of regions in the complex plane which are invariant under "Conformal equivalence". We say two regions Ω_1, Ω_2 are conformally equivalent if there exists a one-to-one, analytic (and hence conformal) mapping from Ω_1 onto Ω_2. Indeed under these conditions there will also be a one-to-one analytic (and hence conformal) mapping from Ω_2 to Ω_1 (by its inverse mapping). It is also easy to see that if ϕ is one such mapping from Ω_1 onto Ω_2, then there exists a bijection between the ring of analytic functions on Ω_2 denoted by $H(\Omega_2)$ onto the corresponding $H(\Omega_1)$. In fact $f \rightarrow f \circ \phi$ is actually a ring isomorphism between $H(\Omega_2)$ and $H(\Omega_1)$. This observation enables us to transfer problems related to $H(\Omega_2)$ into those related to $H(\Omega_1)$ and if in addition Ω_1 has a simple structure we can even solve these problems and carry the solutions back to $H(\Omega_2)$. It is in this context that the Riemann mapping theorem assumes significance. Essentially this theorem states that every simply connected region which is not the whole plane is conformally equivalent to the open unit disc. Before discussing a proof of this theorem we shall first develop some pre-requisites.

Definition 5.1.2

Let Ω be any region in the complex plane. Let $H(\Omega)$ denote the set of all analytic functions on Ω. Suppose $\mathcal{F} \subseteq H(\Omega)$. We say that $\mathcal{F}$ is a normal family if every sequence of members of $\mathcal{F}$ admits a subsequence which converges uniformly on compact subsets of Ω.

Definition 5.1.3

A family $\mathcal{F} \subseteq H\ (\Omega)$ is said to be equicontinuous on $E \subseteq \Omega$ if given $\varepsilon > 0$ there exists $\delta > 0$, depending only on ε and not on $f \in \mathcal{F}$ such that $|f(z) - f(z')| < \varepsilon$ whenever $|z - z'| < \delta$ for every $f \in \mathcal{F}$ and for every z, z' in E.

Theorem 5.1.4 (Weierstrass theorem for sequence of analytic functions)

Let $\{f_n(z)\}$ be a sequence of analytic functions defined on a common region Ω. Let $f_n(z) \to f(z)$ uniformly on compact subsets of Ω. Then $f(z)$ is analytic in Ω and further $f_n'\ (z) \to f'(z)$ uniformly on compact subsets of Ω. (This also implies $f_n^{(k)}(z) \to f^{(k)}(z)$ uniformly on compact subsets of Ω for each $k = 2, 3, \ldots$).

Proof

$f(z)$, being the uniform limit of $f_n(z)$ on compact subsets of Ω, is continuous in Ω. By Cauchy's integral formula, $f_n(z) = \dfrac{1}{2\pi i}\displaystyle\int_C \dfrac{f_n(\zeta)}{\zeta - z}\,d\zeta$ where C is the circle $|\zeta - a| = r$ such that $|\{\zeta \in \mathbb{C} | \zeta - a| \le r\} \subseteq \Omega$ and $|z - a| < r|$. Since $f_n(\zeta) \to f(\zeta)$ uniformly on C, we have, by taking limits on both sides, $f(z) = \dfrac{1}{2\pi i}\displaystyle\int_C \dfrac{f(\zeta)}{\zeta - z}\,d\zeta$ which proves that $f(z)$ is analytic in $|z - a| < r$. (Note that by continuity of f and Theorem 4.4.3 the right side integral represents an analytic function of z in $|z - a| < r$). Since a is arbitrary, subject only to the condition that the disc $|\zeta - a| \le r$ is contained in Ω, $f(z)$ is analytic throughout of Ω. Similarly using $f_n'(z) = \dfrac{1}{2\pi i}\displaystyle\int_C \dfrac{f_n(\zeta)}{(\zeta - z)^2}\,d\zeta$

we get $\lim\limits_{n \to \infty} f_n'(z) = f'(z)$ (at least point-wise). We now claim that this convergence is uniform for $|z - a| \le \rho < r$. Indeed given $\varepsilon > 0$ we can find N such that for $n \ge N$ and for all z with $|z - a| \le \rho < r$

$$|f_n'\ (z) - f'(z)| \le \frac{1}{2\pi}\int_C \left|\frac{f_n(\zeta) - f(\zeta)}{(\zeta - z)^2}\right||d\zeta| \le \frac{1}{2\pi}\varepsilon\,\frac{2\pi r}{(r - \rho)^2}.$$

Thus f_n' tends to f' uniformly as $n \to \infty$ for $|z - a| \le \rho < r$. However any compact subset of Ω can be covered by finitely many closed discs of the form $\{z\,|z - a_1| \le \rho_1\} \subset \Omega$. Hence the convergence is uniform on any compact subset of Ω. A repeated application gives $f_n^{(k)}(z) \to f^{(k)}(z)$ uniformly on compact subsets of Ω.

Corollary 5.1.5

If $f(z) = \sum\limits_{1}^{\infty} f_n(z)$ converges uniformly on every compact subset of a region Ω then $f(z)$ is analytic and $f'(z) = \sum\limits_{1}^{\infty} f_n'(z)$, the convergence being uniform on every compact subset of Ω.

Proof

Apply Theorem 5.1.4 for the sequence of partial sums of $\sum_{1}^{\infty} f_n(z)$.

Theorem 5.1.6

Let $\mathcal{F} \subseteq H(\Omega)$ and $\mathcal{F}$ be uniformly bounded on compact subsets of the region Ω in the following sense. To each compact set $K \subseteq \Omega$ there is a number $M(K) < \infty$ such that $|f(z)| \leq M(K)$, $\forall f \in \mathcal{F}$ and $\forall z \in K$. Then $\mathcal{F}$ is a normal family.

Proof

Define $V_n = \{z : |z| > n\} \cup \bigcup_{a \in \Omega^c} D\left(a, \dfrac{1}{n}\right)$ where $D(a, r) = \{z : |z - a|$ $< r\}$ and Ω^c is the complement of Ω. Put $K_n = V_n^c$. We first claim that

(i) $\Omega = \bigcup_{n=1}^{\infty} K_n$, with K_n compact for each n. (ii) $K_n \subset K_{n+1}^{\circ}$ $(n = 1, 2,$ . .) where K_{n+1}° denotes the interior of K_{n+1}. (iii) each compact subset of Ω lies in some K_n.

V_n is evidently open. Thus K_n is closed. But $K_n \subset \overline{D(0, n)}$. Hence K_n is compact for each n. It is clear that Ω^c is contained in V_n and hence $K_n \subset \Omega$.

Thus $\bigcup_{n=1}^{\infty} K_n \subseteq \Omega$. On the other hand let $w \in \Omega \subseteq \mathbb{C}$ so that there exists m

such that $|w| \leq m$. Let $A_k = \{z : |z| > k\}$ and $B_k = \bigcup_{a \in \Omega^c} D\left(a, \dfrac{1}{k}\right)$ so that $V_k = A_k \cup B_k$. We claim that there exists at least one k such that $w \notin V_k$. If possible let $w \in V_k$ for all k. This means that for each k, $w \in A_k$ or B_k. Certainly $w \notin A_k$ for $k \geq m_0$ for some m_0 (since $|w| > k$ for infinitely many k implies that $w \notin \mathbb{C}$). Thus $w \in B_k \forall k \geq m$. Therefore for each $k \geq m$ there exists $a_k \in \Omega^c$ such that $|w - a_k| < \dfrac{1}{k}$. Thus w is a limit point of Ω^c and hence $w \in \Omega^c$ (note that is Ω^c closed). Therefore $w \notin V_n$ for some n which is the same thing as saying $w \in K_n$ for some n. It now follows that $\Omega \subseteq \bigcup_{n=1}^{\infty} K_n$ and hence $\Omega = \bigcup_{n=1}^{\infty} K_n$.

To prove (ii) we observe that $z \in K_n \Rightarrow D\left(z, \dfrac{1}{n} - \dfrac{1}{n+1}\right) \subseteq K_{n+1}$. Indeed

$z \in K_n \Rightarrow |z| \leq n$ and $w \in D\left(z, \dfrac{1}{n} - \dfrac{1}{n+1}\right) \Rightarrow |w - z| \leq \dfrac{1}{n} - \dfrac{1}{n+1} \Rightarrow$

$|w| \leq n + 1$. i.e., $w \notin A_{n+1}$. We also observe that $w \in D\left(z, \dfrac{1}{n} - \dfrac{1}{n+1}\right) \Rightarrow$

$w \notin B_{n+1}$. (Otherwise there exists $a \in \Omega^c$ such that

$$|w - a| < \frac{1}{n+1} \Rightarrow |z - a| \leq |z - w| + |w - a|$$

$$\leq \frac{1}{n} - \frac{1}{n+1} + \frac{1}{n+1} = \frac{1}{n}.$$

i.e., $z \in B_n$ contradicting the fact that $z \in K_n$). Therefore $w \in K_{n+1}$.

Hence $D\left(z, \dfrac{1}{n} - \dfrac{1}{n+1}\right) \subseteq K_{n+1}$. Thus (ii) is proved.

Let K be any compact subset of Ω. (ii) implies that $K \subset \Omega \subset \bigcup\limits_{n=1}^{\infty} K_{n+1}^{\circ}$ which implies that K is contained in a finite union of open sets K_{n+1}° which in turn is contained in a single compact subset of the form K_N for some N. This proves (iii). From the proof of (ii) above it follows that

$$2\delta_n = \frac{1}{n} - \frac{1}{n+1}$$

satisfies $D(z, 2\delta_n) \subseteq K_{n+1}$ for all $z \in K_n$. We now prove that for each n the family of all restrictions of $f \in \mathcal{F}$ to K_n is equicontinuous in the sense of definition 5.1.3. Indeed if z', $z'' \in K_n$, $|z' - z''| < \delta_n$ and γ denotes the positively oriented boundary of the circle, centre z' and radius $2\delta_n$ we have (by Cauchy integral formula).

$$f(z') - f(z'') = \frac{1}{2\pi i}\int_{\gamma} f(\zeta)\left(\frac{1}{\zeta - z'} - \frac{1}{\zeta - z''}\right) d\zeta$$

$$= \frac{(z' - z'')}{2\pi i}\int_{\gamma} \frac{f(\zeta)\, d\zeta}{(\zeta - z')(\zeta - z'')}$$

Using $|\zeta - z'| = 2\delta_n$ and $|\zeta - z''| \geq |\zeta - z'| - |z' - z''| \geq 2\delta_n - \delta_n = \delta_n$ we get

$$|f(z') - f(z'')| \leq \frac{M(K_{n+1})|z' - z''|}{2\pi(2\delta_n \delta_n)}\, 2\pi 2\delta_n = \frac{M(K_{n+1})|z' - z''|}{\delta_n}.$$

This proves that for z', $z'' \in K_n$, $|z' - z''| < \delta = \dfrac{\varepsilon \delta_n}{M(K_{n+1})}$ we have $|f(z') - f(z'')| < \varepsilon$, $\forall f \in \mathcal{F}$. Thus the restrictions of members of $\mathcal{F}$ to each K_n is equicontinuous.

Let $\{f_j\}$ be any sequence from $\mathcal{F}$. To prove that $\mathcal{F}$ is normal we have to extract a subsequence which converges uniformly on every compact subset of Ω. In as much as each such compact set is contained in K_n for some n it suffices to prove that there exists a subsequence converging uniformly on each K_n. Now each K_n is a compact subset of the plane and so there exists a countable dense subset $E_n \subseteq K_n$, $\forall\ n$. Let $E = \overset{\infty}{\underset{1}{\bigcup}} E_n$. Since E (being countable union of countable sets) is countable, we enumerate E by $z_1, z_2, z_3, \ldots.$ We are given that $\mathcal{F}$ is uniformly bounded on each compact set and hence on each $\{z_j\}$. Thus $\{f_n(z_1)\}$ is a bounded sequence and so it admits a subsequence say $f_{11}, f_{12}, \ldots$ converging at z_1. Evaluating all these f_{1j}' s at z_2, again we get a bounded sequence which admits a subsequence of $\{f_{1j}\}$ say $f_{21}, f_{22}, \ldots$ converging at z_2. Proceeding like this we get an array of sequences as follows:

$$\begin{pmatrix} f_{11}, f_{12}, f_{13}, \ldots \\ f_{21}, f_{22}, f_{23}, \ldots \\ f_{31}, f_{32}, f_{33}, \ldots \\ \vdots\vdots\vdots\vdots\vdots\vdots\vdots\vdots\vdots\vdots\vdots \end{pmatrix}$$

Note that $\{f_{1j}\}$ converges at z_1 and $\{f_{2j}\}$ converges at z_2 and also at z_1 (Since $\{f_{2j}\}$ is a subsequence of $\{f_{1j}\}$). Thus the k-th sequence $\left\{f_{kj}\right\}_{j=1}^{\infty}$ not only converges at z_k but also at each z_i with $i \le k$. Consider the diagonal sequence $g_1 = f_{11}$, $g_2 = f_{22}$, $\ldots.$ This sequence $\{g_n\}$ being a subsequence of each $\left\{f_{ij}\right\}_{j=1}^{\infty}$ from the stage $n = i$ onwards it follows that g_n converges at every point of E. We claim that $\{g_n\}$ converges uniformly on each K_N. Since uniform convergence of a sequence is equivalent to it being uniformly Cauchy, we shall merely prove that $\{g_n\}$ is uniformly Cauchy. Fix K_N. Note that $\{g_n | K_N\}$ is equicontinuous. Thus given $\varepsilon > 0$ there exists $\delta > 0$ such that $z, w \in K_N$, $|z - w| < \delta \Rightarrow |g_n(z) - g_n(w)| < \dfrac{\varepsilon}{3}$ for all n. Cover K_N by open balls of radius $\dfrac{\delta}{2}$ and extract a finite sub cover say $B_1, B_2, \ldots B_M$ (K_N is compact). Since $E \cap K_N$ is dense in K_N ($E \cap K_N \supseteq E_N$ and E_N is dense in K_N) we can choose $w_i \in E \cap B_i \cap K_N$ for $1 \le i \le M$. (Note that $B_i \cap K_N$ is open in K_N and so must intersect the dense set $E \cap K_N$). Since $w_i \in E$ and $\{g_n\}$ is convergent at every point of E, the sequence $\{g_n(w_i)\}$ is convergent for $1 \le i \le M$. Hence for these finitely many $w_i's$ we can choose a single stage L such that for $n, m \ge L$,

$$|g_m(w_i) - g_n(w_i)| < \frac{\varepsilon}{3} \ (1 \le i \le M).$$

Now let $z \in K_N$ be an arbitrary point. Choose $n, m \geq L$.

$$|g_m(z) - g_n(z)| \leq |g_m(z) - g_m(w_i)| + |g_m(w_i) - g_n(w_i)| + |g_n(w_i) - g_n(z)|$$

where $w_i \in K_N$ is so chosen that $|z - w_i| < \delta$ (this is possible since $z \in B_i$ for some $1 \leq i \leq M$ and in this B_i there is a corresponding $w_i \in K_N$). Using equicontinuity of the sequence $\{g_n | K_N\}$ we have $|g_n(z) - g_n(w_i)| < \frac{\varepsilon}{3}$, $|g_m(w_i) - g_m(z)| < \frac{\varepsilon}{3}$ and by our choice, $m, n \geq L \Rightarrow$

$|g_m(w_i) - g_n(w_i)| < \frac{\varepsilon}{3}$. Putting all these together we get that $\{g_n(z)\}$ is uniformly Cauchy on K_N. This gives us the required result.

Note 5.1.7

As already observed (Chapter 1, Solved exercises No. 12) the finite complex plane and the open unit disc are homeomorphic to each other. However, in view of Liouville theorem there can not be any bijective analytic map between these two regions. On the other hand "given any simply connected region Ω other than the whole plane there does exist a bijective analytic map of Ω onto the open unit disc U" is the most important feature of the Riemann mapping theorem and we proceed to give a proof of this famous theorem.

Theorem 5.1.8 (Riemann mapping theorem)

Let $\Omega \neq \mathbb{C}$ be a simply connected region. $z_0 \in \Omega$, β real. Then there exists a unique bijective analytic function $h : \Omega \to U$ satisfying $h(z_0) = 0$ and with one value of arg $(h'(z_0)) = \beta$.

Proof

We first prove the uniqueness. In fact if h_1 and h_2 are two such maps it easily follows that $h_2 \circ h_1^{-1}$ is a bijective analytic map of U onto U and by Theorem 4.4.34, $h_2 \circ h_1^{-1}$ must be a linear fractional transformation say S. Our hypotheses also imply that $S(0) = 0$ and $S'(0) = h_2'(h_1^{-1}(0))(h_1^{-1})'(0)$

$= \dfrac{h_2'(z_0)}{h_1'(z_0)}$ and so arg $S'(0) = \beta - \beta = 0$. The only linear fractional transformations from U onto U satisfying $S(0) = 0$ are given by $S(z) = e^{i\lambda}z$. (See Theorem 4.4.35). arg $S'(0) = 0$ now gives $\lambda = 0$. Thus $S(z) \equiv z$.

Let Σ denote the class of all $\psi \in H(\Omega)$ which are one-to-one in Ω and which map Ω into U. We shall show that Σ is non-empty and that there exists $h \in \Sigma$ which maps Ω onto U.

Since $\Omega \neq \mathbb{C}$ choose $w_0 \notin \Omega$. Since Ω is simply connected and $z - w_0$ as a function of z never vanishes in Ω, we can get a $\phi \in H(\Omega)$ such that

$\phi^2(z) = z - w_0.$ $\Big[$ In fact using Theorem 4.4.7 choose an analytic branch for

$\log(z - w_0)$ and put $\phi = \exp\left\{\dfrac{1}{2}\log(z - w_0)\right\}\Big].$ We first claim that this ϕ

is one-to-one. Indeed

$$\phi(z_1) = \phi(z_2) \Rightarrow \phi^2(z_1) = \phi^2(z_2) \Rightarrow z_1 - w_0 = z_2 - w_0 \Rightarrow z_1 = z_2.$$

We also claim that there are no two points z_1 and z_2 in Ω such that $\phi(z_1) = -\phi(z_2)$ [$\phi(z_1) = -\phi(z_2) \Rightarrow \phi^2(z_1) = \phi^2(z_2) \Rightarrow z_1 = z_2$ and $\phi(z_1) = 0$]. ϕ being one-to-one is non-constant and its analyticity implies that it is an open map (by the open mapping principle namely Corollary 4.4.23). Thus $\phi(\Omega)$ is open and non-empty. Choose $a \in \phi(\Omega)$ and $r > 0$ such that $D(a, r) \subseteq \phi(\Omega)$. Without loss of generality we suppose that $0 < r < |a|$. Now $D(-a, r) \cap \phi(\Omega)$ is empty (if not there exists $z \in \Omega$ such that $w = \phi(z) \in \phi(\Omega)$ and $|w + a| < r$. But then $-w \in D(a, r) \subseteq \phi(\Omega)$ a contradiction to our earlier observation). Thus for every $z \in \Omega$, $|\phi(z) + a|$

$\geq r$ or that $\psi(z) = \dfrac{r}{\phi(z) + a} \in H(\Omega)$ and maps Ω into $\overline{U}$. But by

maximum principle $\left|\dfrac{r}{\phi(z) + a}\right|$ must be less than 1. Thus $\psi(z) \in \Sigma$. Note

that ϕ is injective $\Rightarrow \psi$ is injective. Therefore Σ is non-empty. This is the first step. Our next step (step II) is to show that if $\psi \in \Sigma$ is not onto then there exists $\psi_1 \in \Sigma$ such that for our given $z_0 \in \Omega$, $|\psi_1'(z_0)| > |\psi'(z_0)|$. Indeed let $\alpha \in U$ be such that $\psi(z) \neq \alpha$ for every $z \in \Omega$. Put $\phi_\alpha(z) =$

$\dfrac{z - \alpha}{1 - \bar{\alpha}z}$ (see Theorem 4.4.33 for the properties of ϕ_α) and $\phi_\alpha \circ \psi$ is never

zero (as $\phi_\alpha(z) = 0 \Leftrightarrow z = \alpha$). Since Ω is simply connected there exists $g \in H(\Omega)$ (with value in U) such that $g^2(z) = (\phi_\alpha \circ \psi)(z)$ (proof as before). Put $s(w) = w^2$ so that $s \circ g = \phi_\alpha \circ \psi$. If $g(z_0) = \beta \in U$ then we consider $\phi_\beta \circ g = \psi_1$, g being an analytic branch of a multi-valued function is always one-to-one and all ϕ_β's are bijections of U onto U. Thus ψ_1 is analytic and injective. Also ψ_1 maps Ω into U since g maps Ω into U (Note that $|g^2(z)|$ $= |(\phi_\alpha \circ \psi)(z)| < 1 \Rightarrow |g(z)| < 1$). Thus we have $\psi_1 \in \Sigma$. We now have

$$\psi = \phi_{-\alpha} \circ s \circ g = \phi_{-\alpha} \circ s \circ \phi_{-\beta} \circ \psi_1 \text{ (Note that } \phi_\beta^{-1} = \phi_{-\beta})$$

$$= F \circ \psi_1 \text{ (say)}$$

with $F : U \to U$ such that $F(z) = (\phi_{-\alpha} \circ s \circ \phi_{-\beta})(z)$. If $F(0) = \gamma$ with $\gamma \in U$, by generalized Schwarz's lemma (See also the proof of Theorem 4.4.34) $|F'(0)| \leq 1 - |\gamma|^2$. Now if $\gamma \neq 0$, $1 - |\gamma|^2 < 1$ and even if $\gamma = 0$, $|F'(0)|$ $\neq 1$ because otherwise $F(z) = e^{i\lambda}z$ and F is one-to-one whereas $F(z) = (\phi_{-\alpha} \circ s \circ \phi_{-\beta})(z)$ is never one-to-one. (For example if $w \neq 0$ and $w \in U$

there exists $z_1 \neq z_2$ both belonging to U such that $\phi_{-\beta}(z_1) = w$, $\phi_{-\beta}(z_2) = -w$. Thus $F(z_1) = F(z_2)$ but $z_1 \neq z_2$. Notice here that $\phi_{-\beta}$ is one-to-one and $s(w) = s(-w)$. Thus $|F'(0)| < 1$ and the chain rule (using $\psi = F \circ \psi_1$) gives

$$\psi'(z_0) = F'(\psi_1(z_0)) \; \psi_1'(z_0) = F'(0) \; \psi_1'(z_0)$$

$(\psi_1(z_0) = (\phi_\beta \circ g)(z_0) = \phi_\beta(\beta) = 0)$. Hence, $|\psi'(z_0)| < |\psi_1'(z_0)|$ and this completes our step II.

Our final step III consists in putting $\eta = \sup\limits_{\psi \in \Sigma} |\psi'(z_0)|$ and showing that there exists $h_1 \in \Sigma$ such that $|h_1'(z_0)| = \eta$. Indeed if this is proved, by step II, h_1 must be onto because there can be no $\psi_1 \in \Sigma$ with $|\psi_1'(z_0)| > |h_1'(z_0)| = \eta$. To start with, there is always a sequence $\psi_n \in \Sigma$ such that $|\psi_n'(z_0)| \to \eta$. (Supremum of a set not belonging to it is always a limit point of the set). Since Σ is uniformly bounded on the whole of Ω in the sense of Theorem 5.1.6, Σ is a normal family and thus every sequence admits a subsequence converging uniformly on compact subsets of Ω. Thus $\{\psi_n(z)\}$ mentioned above admits a subsequence (say) $\{\xi_n(z)\}$ (i.e., if $\{\psi_{n_k}\}$ in the subsequence then we can put $\psi_{n_k}(z) = \xi_k(z)$) which converges uniformly on compact subsets of Ω to (say) $h_1(z)$ so that by Theorem 5.1.4 $h_1(z) \in H(\Omega)$ and $\xi_n'(z) \to h_1'(z)$ uniformly on compact subsets of Ω. In particular

$$|\xi_n'(z_0)| \to |h_1'(z_0)|.$$

We have $|\xi_n(z_0)| < 1$, $\forall\, n$ and so $|h_1(z)| \leq 1$ by taking point-wise limit. But $|h_1'(z_0)| = \eta > 0$. (each $\psi \in \Sigma$ satisfies $\psi'(z_0) \neq 0$ by one-to-one nature and so $\eta > 0$). Thus $h_1(z)$ is a non-constant analytic map and by maximum modulus theorem

$$|h_1(z)| < 1, \; \forall\, z \in \Omega.$$

Thus h_1 maps Ω into U and all that remains, is to prove that h_1 is injective in Ω.

Take $z_1 \neq z_2$, $(z_1, z_2 \in \Omega)$. Our claim is $h_1(z_1) \neq h_1(z_2)$. Draw a circle centre z_2 radius r so that $D = \{z \in \mathbb{C}: |z - z_2| \leq r\} \subset \Omega$ and D does not contain z_1 and $h_1(z) - h_1(z_1)$ has no zeros on $|z - z_2| = r$. This is possible because zeros of $h_1(z) - h_1(z_1)$ are isolated and even if $h_1(z) - h_1(z_1) = 0$ at $z = z_2$, there exists such a neighbourhood of z_2. Now $\delta = \inf\limits_{z \in \partial D} |h_1(z) - h_1(z_1)| > 0$ and so, for large n, we have on ∂D

$$|(\xi_n(z) - \xi_n(z_1)) - (h_1(z) - h_1(z_1))| < \delta \leq |h_1(z) - h_1(z_1)|$$

(In fact using uniform convergence of $\xi_n(z) - \xi_n(z_1)$ to $h_1(z) - h_1(z_1)$ on ∂D we can get for large n

$$|(\xi_n(z) - \xi_n(z_1)) - (h_1(z) - h_1(z_1))| < \delta \leq |h_1(z) - h_1(z_1)|$$

for $z \in \partial D$). An immediate application of Rouches theorem (Theorem 4.6.16) gives that $h_1(z) - h_1(z_1)$ and $\xi_n(z) - \xi_n(z_1)$ have the same number of zeros inside D. However, $\xi_n \in \Sigma \Rightarrow \xi_n(z) - \xi_n(z_1) = 0$ if and only if $z = z_1$ and so there are no zeros for $\xi_n(z) - \xi_n(z_1)$ inside D. Thus $h_1(z) - h_1(z_1) \neq 0$ in D and in particular $h_1(z_2) - h_1(z_1) \neq 0$ or that $h_1(z_1) \neq h_1(z_2)$ proving injectivity of h_1. Thus $h_1 \in \Sigma$. This completes the proof of step III. Using all these steps we get that $h_1 \in \Sigma$ and is such that h_1 maps Ω bijectively onto U.

We now claim that for this h_1, $h_1(z_0) = 0$. Otherwise $h_1(z_0) = \beta \neq 0 \Rightarrow (\phi_\beta \circ h_1) \in \Sigma$ and

$$(\phi_\beta \circ h_1)'(z_0) = \phi_\beta'(\beta) h_1'(z_0).$$

Thus $|(\phi_\beta \circ h_1)'(z_0)| = \dfrac{|h_1'(z_0)|}{1 - |\beta|^2} > |h_1'(z_0)|$ a contradiction. Let λ be equal to one of the values of arg $h_1(z_0)$. Put $h(z) = e^{i(\beta - \lambda)} h_1(z)$. It is clear that $h \in \Sigma$, $h(z_0) = 0$ and one of the values of arg $h'(z_0) = \beta - \lambda + \lambda = \beta$ and so h satisfies all our conditions and the proof of our theorem is complete.

Note 5.1.9

With the notations of the above Riemann mapping theorem $|h'(z_0)| = \sup\limits_{\psi \in \Sigma} |\psi'(z_0)|$. However it is also true that $|h'(z_0)|$ maximises $|\psi'(z_0)|$ as ψ varies over the larger family of all analytic mappings from Ω into U (not necessarily one-to-one). This is an easy consequence of Schwarz's lemma and we state and prove it in the following.

Theorem 5.1.10

Let Ω be a simply connected region other than the whole plane. If f is any analytic function from Ω into U and h is the Riemann mapping function from Ω onto U with $h(z_0) = 0$. Then $|f'(z_0)| \leq |h'(z_0)|$ with equality if and only if $f(z) = \lambda h(z)$ for some constant λ with $|\lambda| = 1$.

Proof

The function $g = f \circ h^{-1}$ maps U into U and is analytic there. Hence, by extended form of Schwarz's lemma

$$|g'(0)| \leq 1 - |g(0)|^2 \leq 1$$

and $|g'(0)| = 1$ if and only if $g(z) = \lambda z$ with $|\lambda| = 1$. This, by chain rule, is the same as saying

$$|f'(z_0)| = |g'(0) h'(z_0)| \leq |h'(z_0)|$$

with equality if and only if $f(z) = \lambda h(z)$ with $|\lambda| = 1$.

Remark 5.1.11
In this section we shall be concerned with the boundary behaviour of the Riemann mapping function from a simply connected region $\Omega \neq \mathbb{C}$ onto U. There are certain very interesting sufficient conditions under which the Riemann mapping function can be extended topologically or analytically to include a part of the boundary of Ω. Instead of giving the most general situation under which this can be done we shall be just content with two simple situations which are not only geometrically interesting but are also frequently encountered. But first we need a few definitions, examples and preliminary results.

Definition 5.1.12
Let Ω be a region and $\{z_n\}$ a sequence in Ω. We say that the sequence $\{z_n\}$ tends to the boundary of Ω or $z_n \to \partial\Omega$ if the points z_n ultimately stay away from each point of Ω, i.e., given $z \in \Omega$ there exists an $\varepsilon > 0$ and an $n_0 \in \mathbb{N}$ such that $|z - z_n| \geq \varepsilon$ for $n \geq n_0$.

Proposition 5.1.13
Let Ω be a region. For a sequence $\{z_n\}$ in Ω, $z_n \to \partial\Omega$ if and only if given any compact subset $K \subset \Omega$ there exists $n_0 \in \mathbb{N}$ such that $z_n \notin K$ for $n \geq n_0$.

Proof
Sufficiency follows at once since given $z \in \Omega$ we can find a $\rho > 0$ such that the disc $\overline{D(z, \rho)} \subset \Omega$ and then we can take $K = \overline{D(z, \rho)}$ and choose $\varepsilon < \rho$.

For the necessity let K be compact and $K \subset \Omega$. For each $z \in K$ there exists ε_z and n_z such that $|z - z_n| \geq \varepsilon_z$ for $n \geq n_z$. The corresponding discs $D(z, \varepsilon_z) = \{\zeta \in \mathbb{C} : |\zeta - z| < \varepsilon_z\}$ is a cover for K and by compactness

$$K \subset \bigcup_{i=1}^{N} D(w_i, \varepsilon_{w_i}).$$

If $n \geq M = \max_{1 \leq i \leq N} \{n_{w_i}\}$ then it is clear that $z_n \notin K$ for $n \geq M$.

Proposition 5.1.14
If f is a one-to-one bicontinuous mapping of a region Ω onto another region Ω_1 and $z_n \to \partial\Omega$ then $f(z_n) \to \partial\Omega_1$.

Proof
Let K be a compact subset of Ω_1. Then $f^{-1}(K)$ is a compact subset of Ω and we can find n_0 such that

$$n \geq n_0 \Rightarrow z_n \notin f^{-1}(K).$$

This is the same thing as saying $f(z_n) \notin K$ for $n \geq n_0$ and by Proposition 5.1.13, $f(z_n) \to \partial\Omega_1$.

Remark 5.1.14a

The Proposition 5.1.14 obtained above must be clearly understood. Our definition of $z_n \to \partial\Omega$ does not mean that $z_n \to a$ point of $\partial\Omega$. However if a sequence $z_n \in \Omega$ is such that $z_n \to z_0 \in \partial\Omega$ certainly $z_n \to \partial\Omega$. Thus, the condition $f(z_n) \to \partial\Omega_1$ does not always imply that $f(z_n) \to a$ point of $\partial\Omega_1$. For example if $\Omega_1 = U$, all that one can say is $|f(z_n)| \to 1$ and not that $f(z_n) \to w$ with $|w| = 1$. To infer such conditions (which imply that f can be extended continuously to $\partial\Omega$) we need other observations and hypothesis.

For example let G be the domain bounded by the non-negative real axis in the $z = x + iy$ plane. There exists a single valued analytic branch for $\zeta = f(z) = \sqrt{z}$ such that $f(-1) = i$ (say $z = re^{i\theta}$, $0 < \theta < 2\pi$ then $\zeta = r^{1/2} e^{i\theta/2}$). This maps G onto the upper half plane Im $\zeta > 0$. Now take a linear fractional transform of Im $\zeta > 0$ onto $|w| < 1$ in such a way that the positive real axis corresponds to $|w| = 1$ and Im $w > 0$ and the negative real axis corresponds to $|w| = 1$ and Im $w < 0$. If now $z_n = x_n + iy_n$ tends to a boundary point $x > 0$, $y = 0$ of G from the half plane $y > 0$ then $w = w(\zeta)$ (the composition of ζ and w described above) actually tends to a point w with $|w| = 1$ and Im $w = v > 0$. On the other hand if z_n tends to the same point $x > 0$ $y = 0$ of ∂G from the lower half plane $y < 0$ then w (ζ) actually tends to $\overline{w}$ ($w = u + iv$ as before with $|w| = 1$ and $v > 0$). Thus if we take a sequence $\{z_n\}$ with Im $z_{2j} > 0$ and Im $z_{2j+1} < 0$ and such that $z_n \to x > 0$ then $z_n \to \partial G$ but w_n does not tend to a unique limit. In fact $\{w_{2j}\} \to w$ and $\{w_{2j+1}\} \to \overline{w}$ (with $w \neq \overline{w}$) so that $|w_n| \to 1$ or that $\{w_n\} \to \partial U$ but w_n does not tend to a boundary point of U. We shall also say "$z \to \partial\Omega$" instead of "for every sequence $\{z_n\} \to \partial\Omega$".

Definition 5.1.15

Let $\gamma \subseteq \partial\Omega$ be an open circular arc or an open straight line segment. We say γ is a free boundary arc if given any point $z \in \gamma$ there exists an open disc D centre z such that $D \cap \gamma = D \cap \partial\Omega$ (i.e., γ stays away from rest of the boundary of Ω).

Note that in the above definition we can further assume that each such disc D determines two open semidiscs D_1 and D_2 each of which lies entirely in Ω° (interior of Ω) or entirely in $(\Omega^c)^\circ$ (exterior of Ω). In as much as $z \in \partial\Omega$, one of these must be in Ω and some times both D_1 and D_2 may be in Ω. If only one of D_1 and D_2 lie in Ω we say z is a one-sided free boundary point and if both lie in Ω we say z is a two-sided free boundary point. But since γ as a point set is connected and the set of one-sided free boundary points and the set of the two-sided free boundary points are both open, all its points must be of the same type and correspondingly we define γ as a one-sided free boundary arc or as a two-sided free boundary arc. Further in the case of a single one-sided free boundary arc γ, the region always lies on the same side of γ.

Definition 5.1.16

Let Ω be a simply connected region and $\beta \in \partial\Omega$. β is said to be a simple boundary point if to each sequence $\{\alpha_n\}$ from Ω with $\alpha_n \to \beta$ there is a curve γ with parametric interval $[0, 1]$ and a sequence of points $\{t_n\}$ with the following properties.

  (i) $0 < t_1 < t_2 < \dots < t_n < \dots$

  (ii) $t_n \to 1$ as $n \to \infty$

  (iii) $\gamma(t_n) = \alpha_n$, $\gamma(1) = \beta$, $\gamma(t) \in \Omega$ for $0 \le t < 1$.

In other words there is a curve in Ω passing through α_n and ending at β.

Example 5.1.17

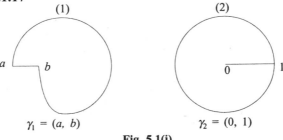

(1)   (2)

a b 0 1

$\gamma_1 = (a, b)$ $\gamma_2 = (0, 1)$

Fig. 5.1(i)

In the above regions γ_1 in (1) is a one-sided free boundary arc and γ_2 in (2) is a two-sided free boundary arc where as $\beta = 0$ in (2) is also a two-sided free boundary point. All points of γ_1 are simple and all points of γ_2 are not simple. $\beta = 1$ in (2) is also not a simple boundary point.

Let Ω be the interior of the square with vertices $0, 1, 1 + i, i$. Remove from Ω the line segments

$$\left[\frac{1}{2n}, \frac{1}{2n} + \frac{n-1}{n}i\right] \text{ and } \left[\frac{1}{2n+1} + \frac{i}{n}, \frac{1}{2n+1} + i\right] \quad n \ge 2$$

Consider the resulting region Ω, which is simply connected.

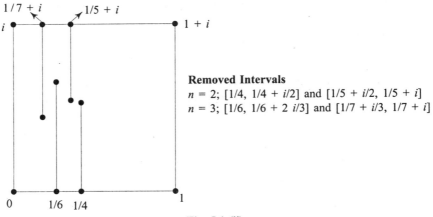

$1/7 + i$ $1/5 + i$

i $1 + i$

Removed Intervals

$n = 2;\ [1/4, 1/4 + i/2]$ and $[1/5 + i/2, 1/5 + i]$

$n = 3;\ [1/6, 1/6 + 2\ i/3]$ and $[1/7 + i/3, 1/7 + i]$

0 $1/6$ $1/4$ 1

Fig. 5.1 (ii)

Here all points iy, $0 \leq y \leq 1$ are boundary points which are not simple and the open line segment $\{iy : 0 < y < 1\}$ is not a free boundary arc either one-sided or two-sided.

These facts are geometrically clear and we invite the reader to supply the rigorous proofs.

Theorem 5.1.18

Let Ω be a proper simply connected region in the plane and γ, a one-sided free boundary arc which is either a line segment or a circular arc. Let f be a Riemann mapping function mapping Ω onto U with $f(z_0) = 0$ for some $z_0 \in \Omega$. Then f can be analytically extended to a region containing $\Omega \cup \gamma$ and the image of γ under this extension is an arc γ' on the unit circle. Further the extended function is one-to-one on γ.

Proof

Using a preliminary analytic linear fractional transformation we can map γ onto an open linear segment on the real axis. Thus there is no loss of generality in assuming that γ is a finite open interval in $\mathbb{R}$. Choose a small open disc around a fixed point $x_0 \in \gamma$ so that the half disc which lies in Ω does not contain the unique point z_0 with $f(z_0) = 0$. In this half disc (which is simply connected) a single valued analytic branch for $\log f(z)$ can be defined and its real part tends to 0 as z approaches a point on the real diameter (since by Proposition 5.1.14 $|f(z)| \to 1$ as z approaches a point on the diameter and hence Re $\log f(z) = \log |f(z)| \to 0$). Thus by reflection principle (Theorem 4.8.26 applied to $i \log f (z)$) $\log f(z)$ has an analytic extension to the whole disc and consequently $f(z) = e^{\log f(z)}$ also has an analytic extension to the whole disc. This process can be made applicable to all points on γ and the extensions to overlapping discs must coincide because in parts of their intersections they are identical with the given $f(z)$. Thus we get an analytic extension of f to a neighbourhood of γ. We also note that by the same reflection principle for $z \notin \Omega \cup \gamma$ but z belonging to the given neighbourhood of γ, $|f(z)| > 1$ and for $z \in \gamma$, $|f(z)| = 1$. For the extended function f (in a neighbourhood of γ) we first start proving $f'(z) \neq 0$ for any point on γ. Indeed, if not, there exists $z_1 \in \gamma$ such that $f'(z_1) = 0$. By local correspondence theorem we can find a neighbourhood of z_1 and a corresponding neighbourhood of $f(z_1)$ in which f is not one-to-one. This contradicts the behaviour of the extended f described above.

Now we prove that the extended f is one-to-one on γ. Let A and B be two points on the open segment γ and consider the closed segment $\overline{AB}$. Choose discs $N(x, r_x)$ centered around each point x of $\overline{AB}$ with radius r_x so that $N(x, r_x)$ has its upper half discs $D(x, r_x) \subseteq \Omega$ (Note that if one such disc lies on one side of γ then all such discs around points on γ also lie on the same side of γ by connectedness of γ and the evident fact that each such point has a neighbourhood in which all boundary points are of the same type). Cover $\overline{AB}$ by a finite number of discs say

$$\overline{AB} \subseteq \bigcup_{0}^{n} N(x_i, r_{x_i}/2)$$

with $x_0 = A$ and $x_n = B$. Take $\delta = \min\left\{\frac{1}{2}r_{x_i}\right\}$. Take points y_x for each

$x \in \overline{AB}$ such that $y_x \in D(x, r_x/2)$, $|y_x - x| = \delta$ and such that $\overline{xy_x}$ is perpendicular to $\overline{AB}$. Join all these y_x to form a parallel segment $\overline{A'B'}$. See figure below.

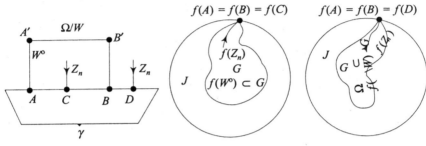

Fig. 5.1(iii)

All points on the closed wedge $W = [A, A', B', B]$ except for the closed segment $\overline{AB}$, belong to Ω and so their images belong to U. ($y \in W \Rightarrow$ $0 < |y - x| \le \delta$ for some $x \in \overline{AB}$ and $|x - x_i| < \frac{1}{2} r_{x_i}$ for some i. Therefore $|y - x_i| < r_{x_i} \Rightarrow y \in D(x, r_{x_i}) \subseteq \Omega$).

Thus f maps W except for $\overline{AB}$ into U and $\overline{AB}$ corresponds to an arc $\overline{f(A)f(B)}$ on the unit circle. Let W°, $\Omega\backslash W$ be two components of Ω determined by $[A, A', B', B]$ and correspondingly let G and J be the components determined by its image in U. If $f(A) = f(B)$ then either $f(W^{\circ}) \subseteq G$ or $f(W^{\circ}) \subseteq J$ by connectedness. If $f(W^{\circ}) \subseteq G$ let $Z_n \in W^{\circ}$ be such that $Z_n \to C$ where C lies between A and B on $\overline{AB}$ so that $|f(Z_n)| \to 1$ and $f(Z) \in \partial f(W^{\circ}) \in \partial G$. Therefore $f(C) = f(A) = f(B)$, i.e., f is constant on $\overline{AB}$, a contradiction to the fact that $f' \ne 0$ on γ. Similarly if $f(W^{\circ}) \subseteq J$ then $f(\Omega\backslash W) \subseteq G$ and as Z_n tends to any connected part of $\gamma\backslash\overline{AB}$, $f(Z_n) \to f(A) = f(B)$, i.e., f is constant on each connected subsegment of $\gamma\backslash\overline{AB}$, again a contradiction.

We shall now obtain yet another result regarding the boundary behaviour of Riemann mapping function. This result extends the Riemann mapping function from a simply connected region $\Omega \subset \mathbb{C}$ to its boundary where the boundary need not be a line segment or a circular arc.

Definition 5.1.19

A complex function $\phi(t)$ of a real variable t defined in $a < t < b$ is said to be real analytic if for every t_0, $a < t_0 < b$,

$$\phi(t) = \phi(t_0) + \phi'(t_0)\ (t - t_0) + \frac{1}{2!}\phi''(t_0)\ (t - t_0)^2 + \ldots$$

where the series converges in some interval $(t_0 - \rho, t_0 + \rho)$, $\rho > 0$.

Definition 5.1.20

γ is said to be a regular simple analytic arc if γ is given by a parametric equation $\phi(t)$ where $\phi(t)$ is real analytic, $\phi'(t) \neq 0$ and ϕ is one-to-one.

Proposition 5.1.21

If ϕ is real analytic complex function in (a, b) then $\phi(t)$ can be extended as an analytic function in a region Δ symmetric to the real axis containing the segment (a, b).

Proof

If $a < t_0 < b$ we have $\phi(t) = \phi(t_0) + \phi'(t_0)\ (t - t_0) + \ldots$ in some interval $(t_0 - \rho, t_0 + \rho)$ $\rho > 0$. By Abel's radius of convergence theorem this power series around t_0 has a certain radius of convergence which is greater than or equal to ρ. Thus the series representing $\phi(t)$ converges for complex values of t also, provided $|t - t_0| < \rho$ and in this disc $\phi(t)$ is analytic. Since t_0 is arbitrary we can extend $\phi(t)$ around each point of (a, b) as an analytic function. However in overlapping intervals these analytic functions also coincide as both are extensions of $\phi(t)$ on the real axis. Thus $\phi(t)$ can be defined as an analytic function in a region Δ symmetric to the real axis containing the segment (a, b).

Definition 5.1.22

γ is said to be a free one-sided regular simple analytic boundary arc of Ω if γ is a regular, simple arc and is given by a parametric equation $\phi(t)$ which is real analytic and which can be extended to a simply connected region Δ symmetric to the interval (a, b) with the property that $\phi(t) \in \Omega$ whenever t lies in the upper half of Δ and $\phi(t)$ lies outside of Ω if t lies in the lower half of Δ.

Theorem 5.1.23

Let Ω be a proper simply connected region in the plane and f is the Riemann Mapping function that maps Ω onto the open unit disc $|z| < 1$ with $f(z_0) = 0$ for some $z_0 \in \Omega$. Let γ be a free one-sided regular simple analytic arc on the boundary of Ω. Then $f(z)$ has an analytic extension to a region containing $\Omega \cup \gamma$ and that γ is mapped onto an arc of the unit circle.

Proof

Let γ be parameterized by $\phi(t)$ with the properties of ϕ listed in the definition of γ as a regular, simple one-sided free boundary arc. Since around each point of γ we can draw a circular disc in such a way that its intersection with Ω does not contain z_0, there is no loss of generality in assuming $\phi(t) \neq z_0$ for $t \in \Delta$. (Otherwise take the union of such discs around each point of γ and take its preimage under ϕ intersect with Δ and take its reflection in the real axis and take the union). Now Δ is a simply connected region, $f(\phi(t))$ is defined on the upper half of Δ and has no point there at which $f(\phi(t)) = 0$. Hence $\log f(\phi(t))$ has a single-valued analytic branch in the upper half of Δ. As $t \to t_0$, $t \in \Delta$, $t_0 \in (a, b)$, $\phi(t) \to \partial\Omega$ and so $f(\phi(t)) \to \partial U$ and in particular Re $\log f(\phi(t)) \to 0$. Hence by reflection principle (Theorem 4.8.26 applied to $i \log f(\phi(t))$) $\log f(\phi(t))$ has an analytic continuation to the whole of Δ.

Define $F(t) = e^{\log f(\phi(t))}$ $(t \in \Delta)$. Thus F is an analytic function defined in Δ with

$$F(t) = f(\phi(t)) \text{ for } t \in \text{ upper half of } \Delta. \tag{1}$$

and $F(t)$ defines a new function on (a, b) as well as on the lower part of Δ. Now define $g(z)$ on $\Omega \cup \gamma$ as follows. For each $z \in \gamma$, pick a unique $t \in (a, b)$ (ϕ is one-to-one on (a, b)) such that $\phi(t) = z$. $\phi'(t) \neq 0$ implies that the complex derivative of the extended ϕ is also non-zero at t and so by local correspondence theorem there exists a neighbourhood of t say L_z and a corresponding neighbourhood of z say N_z in which ϕ gives a one-to-one onto analytic map. Thus ϕ^{-1} exists as an analytic function from N_z to L_z. We can now consider

$$g(z) = \begin{cases} f(z), & z \in \Omega \\ F(\phi^{-1}(z)) & z \in \cup N_z \end{cases}.$$

This definition gives an extension of f on Ω to a region containing $\Omega \cup \gamma$. If for two points z_1 and z_2 in γ the corresponding discs N_{z_1} and N_{z_2} intersect then

$$g(z) = F(\phi^{-1}(z)), \ z \in N_{z_1}$$

$$g(z) = F(\phi^{-1}(z)), \ z \in N_{z_2}.$$

Now wherever z belongs to that part of $N_{z_1} \cap N_{z_2}$ lying in Ω by virtue of (1)

$$f(z) = F(\phi^{-1}(z)), \ z \in \Omega \cap N_{z_1} \cap N_{z_2}.$$

Hence the different definitions of $g(z)$ are analytic continuations of each other and this provides the required extension stated in the theorem. The fact that γ is mapped onto an arc of the unit circle is a consequence of Proposition 5.1.14.

5.2 Schwarz-Christoffel Formula

In the previous section we have seen how the Riemann mapping function of a proper simply connected region can be extended analytically to certain parts of the boundary. However there is no way we can get the explicit expressions for such functions and their inverses from these theorems because the proofs are existential in nature. In some simple cases it is however possible to get the actual expressions for the mapping functions. One such case is the Schwarz's Christoffel formula which gives the mapping function of $\overline{U}$ onto a region bounded by a finite number of linear segments making interior angles $\alpha_k \pi$ for $k = 1, 2, \ldots n$. Before obtaining this formula we shall first prove a theorem that the Riemann mapping function of a polygonal region can actually be extended homeomorphically to include all the boundary points.

Theorem 5.2.1

Let Ω be a bounded simply connected region whose boundary is a closed polygonal path without self-intersections. Let the consecutive vertices be $z_1, z_2 \ldots z_n$ in the positive cyclic order with interior angle at each z_k being equal to $\alpha_k \pi$ with $0 < \alpha_k < 2$ so that $\sum \alpha_k = (n - 2)\pi$. Let U be the open unit disc. Then the Riemann mapping function $f : \Omega \to U$ can be extended as a homeomorphism of $\overline{\Omega}$ onto $\overline{U}$ with z_k corresponding to w_k ($k = 1, 2, \ldots n$) and in which the sides $[z_{k-1}, z_k]$ correspond to the closed arcs $\overline{w_{k-1}, w_k}$ on $T = \partial U$.

Proof

By definition

$$\alpha_k \pi = \arg \left(\frac{z_{k-1} - z_k}{z_{k+1} - z_k} \right)$$

(with $0 < \alpha_k < 2$, $k = 1, 2, \ldots n$, $z_{n+1} = z_1$). Since each open line segment (z_{k-1}, z_k) of the polygonal path is a one sided free boundary arc in the sense of our definition 5.1.15, by Theorem 5.1.18, f can be extended to the open segment (z_{k-1}, z_k) continuously and in a one-to-one fashion in such a way that each open side is mapped onto an open arc of the unit circle. Our aim is to show that these arcs are disjoint and cover ∂U fully. In the following, by f, we shall mean this extended f. Our first aim is to define $f(z_k)$ and establish the continuity of f at z_k for each k.

We first consider a circular sector S_k which is the intersection of Ω with a sufficiently, small disc around z_k. Now $(z - z_k)$ is analytic and non-zero in S_k so that a single-valued analytic branch $\zeta(z) = (z - z_k)^{1/\alpha_k}$ can be constructed. By the properties of the power function we see that ζ maps S_k onto a semicircular region S'_k around origin. The inverse function for this branch thus satisfies

$$z(\zeta) = z_k + \zeta^{\alpha_k}$$

for a suitable branch for ζ^{α_k} and maps S'_k onto S_k. Now set

$$g(\zeta) = f(z_k + \zeta^{\alpha_k}) = f(z(\zeta)).$$

Now as ζ tends to a point on the diameter of S'_k, $z(\zeta)$ tends to a point on the rectilinear segments through z_k and so $f(z(\zeta)) \to \partial U$ by Proposition 5.1.14. In particular $|g(\zeta)| \to 1$. Now we can use the reflection principle and conclude that g can be extended analytically to a region symmetric to S'_k along the diameter. Denoting the extended function also as g we see that as $\zeta \to 0$ in S'_k, $g(\zeta) \to g(0)$ by continuity. If $z \to z_k$ with $z \in S_k$ then z can be written as $z_k + \zeta^{\alpha_k}$ for $\zeta \in S'_k$ with $\zeta \to 0$ in S'_k, i.e., $f(z) \to g(0)$, as $\zeta \to 0$. Therefore $f(z_k + \zeta^{\alpha_k}) = g(\zeta) \to g(0)$ as $\zeta \to 0$. We call $g(0)$ as $f(z_k) = w_k \in T$. Thus $f(z) \to f(z_k)$ as $z \to z_k$ with $z \in S_k$. Now we claim that as $z \to z_k$ along $L_1 = (z_{k-1}, z_k)$ or $L_2 = (z_k, z_{k+1})$ then $f(z) \to f(z_k)$ ($k = 1, 2 \ldots n$). This will complete the continuous extension of f to the corners of the polygon.

Take a sequence $a_1, a_2, \ldots$ tending to z_k along L_1. Choose $b_1, b_2, \ldots \in S_k$ such that $|b_j - a_j| < \dfrac{1}{j}$ and $|f(b_j) - f(a_j)| < \dfrac{1}{j}$ for every j. $\Big($By continuity of the extended f there exists a neighbourhood of a_j say N_j such that for every $z \in N_j$, $|f(z) - f(a_j)| < \dfrac{1}{j}$ and in N_j there exists a point $b_j \in S_k$ such that $|b_j - a_j| < \dfrac{1}{j}\Big)$. This proves that $f(z) \to f(z_k)$ as $z \to z_k$ along L_1 (Note that $b_n \to z_k$ in S_k and so $f(b_n) \to f(z_k)$ but $\lim_n f(b_n) = \lim_n f(a_n)$). Similarly $f(z) \to f(z_k)$ as $z \to z_k$ along L_2. Thus f is continuous at z_k and that the images of L_1 and L_2 under the extended f have a common point $f(z_k) = w_k$.

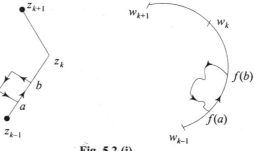

Fig. 5.2 (i)

Next take any two points a, b on the line segment $[z_{k-1}, z_k]$ as in the above figure. Draw a rectangle R with a, b as two of its vertices in such a way that the closed rectangle R except for the boundary $[a, b]$ lies entirely in the polygonal region. The image of R under the extended f is a region Δ in U whose boundary contains the arc of the unit circle corresponding to $[a, b]$. Throughout the interior and the entire boundary of R the extended f is analytic and one-to-one. Hence if we give the positive orientation for ∂R by Theorem 4.6.19(i) its image under the extended f must be positively oriented with respect to Δ. In particular the circular arc must be positively oriented with respect to the interior of the circle $|z| = 1$. This means that as z describes the linear segment $[z_{k-1}, z_k]$ its image under the extended f describes the unit circle in the positive sense with its end points w_{k-1}, w_k. Since this is true for each $k = 1, 2, ... n$, we see that the arcs $[w_{k-1}, w_k]$ for various k are disjoint except for the end points and cover the whole of T. Therefore the extended f maps the interior of the given polygon onto U, maps its boundary points onto points on T with z_k corresponding to w_k for every k in a one-to-one manner and the extended map is a homeomorphism between the closure of the polygon and $\overline{U}$. (Note that continuity of f from a compact space onto $\overline{U}$ implies f is closed and so f^{-1} is continuous)

Theorem 5.2.2 (Schwarz-Christoffel Formula for the Open Unit Disc)
Let Ω be a bounded simply connected region bounded by a polygon with interior angles $\alpha_k \pi$ $(k = 1, 2, ... n)$. If $F(w)$ is a one-to-one analytic map of U onto Ω then

$$F(w) = c\int_0^w \prod_1^n (\zeta - w_k)^{-\beta_k} d\zeta + c' \qquad (w \in U)$$

where $\beta_k = 1 - \alpha_k$ $(k = 1, 2, ... n)$, with $\Sigma \beta_k = 2$, c, c' complex constants and w_k $(k = 1, 2, ... n)$ are distinct points of the unit circle with $F(w_k) = z_k$ the vertices of the polygon, with integration over any path joining 0 and w in U.

Proof
By Theorem 5.2.1 the Riemann mapping function $f = F^{-1}$ from Ω onto U can be extended to $\overline{\Omega}$ and the extended f maps $\overline{\Omega}$ homeomorphically onto $\overline{U}$. Further $F = f^{-1}$. With the same notations as in Theorem 5.2.1, consider $g(\zeta) = f(z_k + \zeta^k)$, $\zeta \in S_k'$, and extend $g(\zeta)$ to the whole of the disc D symmetric to the real diameter using reflection principle and put $g(0) = f(z_k) = w_k$ $(1 \leq k \leq n)$. Thus $g(\zeta)$ is analytic at the origin with $g(0) = w_k$. Further $g'(0) \neq 0$. (If not by local properties of analytic functions there exists a neighbourhood of $\zeta = 0$ and a neighbourhood of $g(0) = w_k$ in which the map is atleast 2-to-1. If w belongs to this neighbourhood of $g(0)$ with $|w| < 1$ there exists two distinct points $a, b \in D$ such that $g(a) = g(b) = w$. But a, b can not both belong to S_k or its boundary because then f will not be one-to-one. On the other hand if $a \in S_k'$ and b belongs to the reflection of

S_k' in the real diameter then $g(a) \in U$ and $g(b)$ belongs to exterior of U by reflection principle. Again $g(a) \neq g(b)$. Thus $g'(0) \neq 0$). Hence g gives a one-to-one analytic map of a neighbourhood of 0 onto a neighbourhood of $g(0) = w_k$ such that in this neighbourhood g has an inverse g^{-1} which is also analytic at $g(0)$ with $(g^{-1})' \, (g(0)) = \dfrac{1}{g'(0)} \neq 0$. Thus in suitable neighbourhoods of 0 and w_k respectively

$$w = g(\zeta) = f(z_k + \zeta^k) = w_k + \sum_1^\infty a_m \zeta^m \qquad (a_1 \neq 0)$$

and

$$\zeta = g^{-1}(w) = \sum_1^\infty b_m (w - w_k)^m. \qquad (b_1 \neq 0)$$

Thus $\zeta^{\alpha_k} = (w - w_k)^{\alpha_k} G_k(w)$ where $G_k(w)$ is analytic and not equal to 0 in a neighbourhood of w_k and the equality being valid in a deleted neighbourhood of w_k. $\left(\text{Note that} \left(\dfrac{g^{-1}(w)}{w - w_k} \right) \text{ is analytic and non-vanishing} \right.$ in a neighbourhood of w_k and so its α_kth power can be defined as an analytic function in this neighbourhood $\Big)$. But since f and F are inverses of each other we also have

$$F(w) = z_k + \zeta^{\alpha_k}, \qquad (\zeta \in S_k', \ w \in S_k)$$

and thus

$$F(w) - z_k = (w - w_k)^{\alpha_k} G_k \, (w) \qquad (1)$$

valid for $w \in S_k$. Thus $F(w)$ can now be extended to a deleted neighbourhood of w_k by (1). Differentiating this in the deleted neighbourhood of w_k we find that

$$F'(w) = \alpha_k(w - w_k)^{\alpha_k - 1} G_k \, (w) + G_k'(w) \, (w - w_k)^{\alpha_k}$$

or

$$F'(w) \, (w - w_k)^{\beta_k} = \alpha_k G_k(w) + G_k'(w) \, (w - w_k).$$

Now the right hand side is analytic in w in the whole of a neighbourhood of w_k including w_k. Thus $F'(w) \, (w - w_k)^{\beta_k}$ is analytic in a neighbourhood of w_k and at $w = w_k$

$$F'(w) \, (w - w_k)^{\beta_k} \neq 0. \qquad (2)$$

We can do this for each k and we conclude that

$$H(w) = F'(w) \prod_{k=1}^n (w - w_k)^{\beta_k}$$

is analytic everywhere in $|w| < 1$ and in suitable neighbourhoods of w_k, one for each k, neither $F'(w)$ nor $\prod_{k=1}^n (w - w_k)^{\beta_k}$ vanish in U or in $T \backslash \{w_1, w_2,$

.... w_n} as F is one-to-one and $w_k \in T$. The only terms which may take the value 0 at each w_k are $F'(w)$ and $(w - w_k)^{\beta_k}$ for that k and their product does not vanish at $w = w_k$ by (2). Hence $F'(w) \prod_{k=1}^{n} (w - w_k)^{\beta_k}$ is analytic in $\overline{U}$ and not equal to 0 in $\overline{U}$. We now claim that $H(w)$ is indeed a constant throughout $\overline{U}$.

First we note that $H(w)$ is analytic in some simply connected region containing $\overline{U}$ and not equal to 0 there. Infact using theorem 5.1.18 and the above discussions, around each $z \in T$ we can choose a disc $D(z, r)$ of radius r such that in $D(z, r)$, $H(w)$ is analytic and not equal to 0 there.

Cover T by $\dfrac{r}{2}$ open discs and extract a finite subcover for T say

$\overset{m}{\underset{i=1}{\bigcup}} D\left(z_i, \dfrac{r_i}{2}\right)$. Choose $\delta = \min\limits_{1 \le i \le m}\left\{\dfrac{r_i}{2}\right\}$ and take any z with $|z| < 1 + \delta$. Choose $|\zeta| = 1$ so that $|\zeta - z| < \delta$. Now there exists z_i such that $|\zeta - z_i| < \dfrac{r_i}{2}$, and $\qquad |z \qquad\qquad - \qquad\qquad z_i|$

$\le |z - \zeta| + |\zeta - z_i| < \dfrac{r_i}{2} + \delta < r_i$. Therefore $z \in D(z_i, r_i)$ and $H(w)$ is analytic in $D(z_i, r_i)$. Hence $|z| < 1 + \delta$ is the required region of analyticity of $H(w)$. Thus $\log H(w)$ has an analytic branch in a region containing $|w| = 1$. We shall now trace the arg $H(w) = \text{Im } (\log H(w))$ which is a continuous function on T. If $w = e^{i\theta}$ varies on the unit circle between $w_k = e^{i\theta_k}$ and $w_{k+1} = e^{i\theta_{k+1}}$ (strictly between the two points, i.e., $\theta \ne \theta_k$ or θ_{k+1}) then

$$\text{arg } H(e^{i\theta}) = \text{arg } F'(e^{i\theta}) + \Sigma \text{arg } (e^{i\theta} - e^{i\theta_k})^{\beta_k}. \qquad (3)$$

Now as $e^{i\theta} = w$ varies between w_k and w_{k+1} as stated above arg $F'(e^{i\theta})$ represents the angle (upto an integral multiple of 2π) between the tangent to the image of the arc $\overline{w_k w_{k+1}}$ (which is a linear segment) at $F(e^{i\theta})$ and the tangent to the arc $\overline{w_k w_{k+1}}$ of the unit circle at $e^{i\theta}$ in that order,

$$\text{i.e., arg } F'(e^{i\theta}) = \phi_1 - \left(\dfrac{\pi}{2} + \theta\right)$$

where ϕ_1 represents the argument of the polygonal line segment $\overline{z_k z_{k+1}}$ which is constant for all such θ. Now

$$w - w_k = e^{i\theta} - e^{i\theta_k}$$

$$= e^{i\left(\frac{\theta+\theta_k}{2}\right)} \left\{ e^{i\left(\frac{\theta-\theta_k}{2}\right)} - e^{-i\left(\frac{\theta-\theta_k}{2}\right)} \right\}$$

$$= 2i \, e^{i\left(\frac{\theta+\theta_k}{2}\right)} \sin\left(\dfrac{\theta-\theta_k}{2}\right).$$

From (3) we see that as $e^{i\theta}$ varies between w_k and w_{k+1}.

$$\arg H(e^{i\theta}) = \left[\phi_1 - \left(\frac{\pi}{2} + \theta\right)\right] + \Sigma \beta_k \left[\frac{\pi}{2} + \frac{\theta + \theta_k}{2}\right]$$

$$= \left[\phi_1 - \left(\frac{\pi}{2} + \theta\right) + \left(\frac{\pi}{2} + \frac{\theta}{2}\right)\Sigma \beta_k + \Sigma \frac{\beta_k \theta_k}{2}\right]$$

$$= \text{constant independent of } \theta.$$

Thus $\arg H(e^{i\theta})$ is constant on the arc $\overline{w_k w_{k+1}}$ or that $\text{Im} \log H(w)$ is constant on each such arc of the unit circle. Since $\text{Im} \log H(w)$ is a continuous function of w throughout the unit circle it must be the same constant on the unit circle. Further $\text{Im} \log H(w)$ being harmonic inside the unit circle, $\text{Im} \log H(w)$ by maximum and minimum principles for harmonic function is constant throughout $\overline{U}$ and hence $\log H(w)$ is constant which in turn implies $H(w)$ is a constant. Therefore,

$$H(w) = F'(w) \prod_1^n (w - w_k)^{\beta_k} = c$$

or
$$F'(w) = c \prod_1^n (w - w_k)^{-\beta_k} \qquad (w \in \overline{U})$$

and an integration gives our result. Note that since $F'(w)$ is analytic in U any path joining 0 and w in U can be taken for the integration and our proof is complete.

Theorem 5.2.3 (Schwarz-Christoffel formula for the upper half plane)
Let Ω be a simply connected region bounded by a polygon with interior angles $\alpha_k \pi$ ($k = 1, 2, \dots n$) and vertices $z_1, z_2, \dots z_n$. If $F(\zeta)$ is a one-to-one analytic map of $\text{Im } \zeta > 0$ onto Ω then

$$F(\zeta) = c \int_0^\zeta \prod_1^{n-1} (w - \xi_k)^{-\beta_k} dw + c'$$

where $\beta_k = 1 - \alpha_k$ ($k = 1, 2, \dots n$) with $\sum_1^n \beta_k = 2$, ξ_k ($k = 1, 2, \dots n - 1$) are reals and c and c' are complex constants and $F(\xi_k) = z_k$, ($1 \le k \le n - 1$), $F(\infty) = z_n$ (or with $\infty = \xi_n$, $F(\xi_n) = z_n$), where the integration is over the line joining 0 and ζ in the upper half plane.

Proof
We first map $|w| < 1$ onto $\text{Im } \zeta > 0$ by

$$\zeta = \phi(w) = i\left(\frac{1 + w}{1 - w}\right).$$

Then $(F \circ \phi) \, (w)$ maps $|w| < 1$ onto Ω. If w_i's are such that $(F \circ \phi) \, (w_i)$ $= z_i$, $(1 \le i \le n)$ or that $\phi(w_i) = \xi_i \, (1 \le i \le n - 1)$ and $\phi(w_n) = \infty$ then by Theorem 5.2.2.

$$(F \circ \phi)' \, (w) = c_1 \prod_1^n (w - w_k)^{-\beta_k},$$

$$\text{i.e., } F' \, (\phi \, (w)) \, \phi'(w) = c_1 \prod_1^n (w - w_k)^{-\beta_k}.$$

Putting $\zeta = \phi(w)$ we see that

$$F'(\zeta) = \frac{c_1}{\phi'(w)} \prod_1^n (\phi^{-1}(\zeta) - \phi^{-1}(\xi_k))^{-\beta_k}$$

$$\phi'(w) = \frac{2i}{(1 - w)^2} = 2i \left(1 - \frac{\zeta - i}{\zeta + i} \right)^{-2} = 2i \left(\frac{2i}{\zeta + i} \right)^{-2} = \frac{(\zeta + i)^2}{2i}$$

and $\qquad \phi^{-1}(\zeta) - \phi^{-1}(\xi_n) = \phi^{-1}(\zeta) - \phi^{-1} \, (\infty) = \left(\dfrac{\zeta - i}{\zeta + i} - 1 \right) = \dfrac{-2i}{\zeta + i}.$

Thus

$$F'(\zeta) = \frac{c_1}{2i} (1 - w)^2 \prod_1^{n-1} (\phi^{-1} \, (\zeta) - \phi^{-1}(\xi_k))^{-\beta_k} \left(\frac{-2i}{\zeta + i} \right)^{-\beta_n}$$

But $\beta_n = 2 - \sum_1^{n-1} \beta_i$ and $1 - w = 1 - \dfrac{\zeta - i}{\zeta + i} = \dfrac{2i}{\zeta + i}$. Thus

$$F' \, (\zeta) = \frac{2ic_1}{(\zeta + i)^2} \prod_1^{n-1} \left(\frac{\zeta - i}{\zeta + i} - \frac{\xi_k - i}{\xi_k + i} \right)^{-\beta_k} (-2i)^{-\beta_n} (\zeta + i)^{2 - \sum_1^{n-1} \beta_k}$$

$$= c \prod_1^{n-1} (\zeta - \xi_k)^{-\beta_k}$$

for a different constant c as can be easily verified. To prove our theorem just integrate both sides on a line segment joining 0 and ζ.

Remark 5.2.4

(1) In the Schwarz-Christoffel formula for a circular disc by applying a linear fractional transformation if necessary we can choose three points w_i $(1 \le i \le n)$ on T at our will, i.e., if we want to choose w_1', w_2', w_3' in the place of w_1, w_2, w_3 in the formula for a one-to-one conformal map of $|\zeta| < 1$ onto the polygon, choose a linear fractional transformation $\phi(\zeta)$ so that $\phi(\zeta)$ maps $|\zeta| < 1$ onto $|w| < 1$ and maps w_1', w_2', w_3' onto the set

$\{w_1, w_2, w_3\}$ in a suitable order and some w_j' to w_j for other j's with all $w_i' \in T$. This way ϕ maps $|\zeta| \le 1$ onto $|w| \le 1$ with $\{w_i'\} \to \{w_i\}$. Consider $F_1(\zeta) = (F \circ \phi)(\zeta)$. $F_1(\zeta)$ is another conformal map of $|\zeta| < 1$ onto the same polygon with $\{w_i'\}$ corresponding to the vertices $\{z_i\}$ and so

$$F_1(\zeta) = c \int_0^\zeta \prod_1^n (w - w_i')^{-\beta_k} dw + c'$$

(2) The polygon with vertices z_i's $(1 \le i \le n)$ need not always be convex. It will be so if and only if $\alpha_k \pi < \pi$ or that $1 - \alpha_k = \beta_k > 0$. In case one particular $\alpha_k = 1$ or $\beta_k = 0$ the corresponding vertex is only a point in an otherwise $(n - 1)$ sided polygon.

(3) In the case of the Schwarz-Christoffel formula for the upper half plane there are only $(n - 1)$ parameters as we have chosen ∞ to correspond to the nth vertex of the polygon. Also β_n does not appear explicitly in the

formula although $\beta_n = 2 - \sum_1^{n-1} \beta_k$. As $0 < \alpha_k \pi < 2\pi$ implies $0 < \alpha_k < 2$,

$\beta_k = 1 - \alpha_k$ satisfies $-1 < \beta_k < 1$. Thus β_n also satisfies $-1 < \beta_n < 1$. Now if we consider the case wherein the given polygon is convex (i.e.,) $\beta_k > 0$ for every k then in the representation

$$F(\zeta) = c \int_0^\zeta \prod_1^{n-1} (w - \zeta_k)^{-\beta_k} dw + c'$$

the integrand is dominated (for $|w|$ large say $|w| > M$ where $M =$

$\sup_{1 \le k \le n-1} |\zeta_k|$) by $\left(\dfrac{1}{|w| - M} \right)^{2-\beta_n}$. Since the exponent $2 - \beta_n > 1$ for all 0

$< \beta_n < 1$ the integral $\int_0^\zeta \prod_1^{n-1} (w - \zeta_k)^{-\beta_k} dw$ converges at $\zeta = \infty$ also, by

comparison test and the point at ∞ corresponds under F to the vertex z_n

with angle $\alpha_n \pi$., i.e., $z_n = c \int_0^\infty \prod_1^{n-1} (w - \zeta_k)^{-\beta_k} dw + c'$. Here also α_n can be

equal to 1 in which case it is actually an $(n - 1)$ sided polygon.

5.3 Conformal Mappings of Doubly Connected Regions and Regions of Finite Connectivity

Introduction 5.3.1
It is a simple consequence of the Riemann Mapping Theorem that any two proper simply connected regions in the plane are conformally equivalent. In fact if Ω_1 and Ω_2 are two such regions and ϕ_1 and ϕ_2 respectively map Ω_1

and Ω_2 one-to-one conformally onto U then $\phi_2^{-1} \circ \phi_1$ is the required conformal equivalence. This is a specific property of simply connected regions. In fact even elementary doubly connected regions (say annuli) need not be conformally equivalent to each other. We will now obtain a necessary and sufficient condition under which any two annular regions around origin can be conformally equivalent to each other.

If only we are interested in the conformal equivalence of annular regions (around origin) of all types one has to classify them for comparison as follows. We first put

$$A\ (r,\ R) = \{z \in \mathbb{C}/r < |z| < R\} \text{ for } r,\ R \in [0,\ \infty],\ r < R.$$

Table I	Table II
(1) $A\ (0,\ r)\ (0 < r < \infty)$	(1) $A\ (0,\ s)\ (0 < s < \infty)$
(2) $A\ (0,\ \infty)$	(2) $A\ (0,\ \infty)$
(3) $A\ (r,\ R)\ (0 < r < R < \infty)$	(3) $A\ (s,\ S)\ (0 < s < S < \infty)$
(4) $A\ (r,\ \infty)\ (0 < r < \infty)$	(4) $A\ (s,\ \infty)\ (0 < s < \infty)$

Any one region listed in Table I can be compared to any one in Table II. As a first observation we remark that conformally $A\ (0,\ r)$ and $A\left(\dfrac{1}{r},\ \infty\right)$ are equivalent by the map $z \to \dfrac{1}{z}$. Hence we shall avoid case (4) in both tables as they are taken care of by (1). Thus we are left with (1), (2), (3) only, on both the tables. Since conformal equivalence is a symmetric relation we have to settle the existence or otherwise of conformal equivalence between the following regions only.

$$A(0,\ r) \quad \text{and} \quad A(0,\ s)$$
$$A(0,\ r) \quad \text{and} \quad A\ (0,\ \infty)$$
$$A(0,\ r) \quad \text{and} \quad A(s,\ S)$$
$$A(0,\ \infty) \quad \text{and} \quad A\ (s,\ S)$$
$$A\ (r,\ R) \quad \text{and} \quad A\ (s,\ S)$$

We shall first observe that $A\ (0,\ r)$ and $A(0,\ s)$ are conformally equivalent and the map is $z \to \dfrac{s}{r}\ z$. On the other hand neither $A(0,\ r)$ and $A\ (0,\ \infty)$ nor $A(0,\ r)$ and $A\ (s,\ S)$ are conformally equivalent. In fact if $A\ (0,\ r)$ and $A\ (0, \infty)$ are equivalent then there exists an analytic map (the inverse)

$$f \colon A(0,\ \infty) \to A(0,\ r)$$

for which as $z \to 0$, $|f(z)|$ must tend to either 0 or r only and in any case $f(z)$ has a removable singularity at $z = 0$ and the extended function is thus entire and bounded. Thus by Liouville's theorem f must be a constant, an absurd conclusion. Similarly the existence of a conformal equivalence between $A(0, r)$ and $A(s, S)$ will lead to an extension at $z = 0$ such that $z = 0$ is interior to $|z| < r$ but $f(0)$ can not be interior to $A(s, S) \cup \{f(0)\}$. Similarly there is no conformal equivalence between $A(0, \infty)$ and $A(s, S)$ (the extension is a bounded entire function). Thus we are left with only one case of determining the conformal equivalence of $A(r, R)$ and $A(s, S)$. In fact the above deliberations suggest that $A(r, R)$ is conformally equivalent to $A(s, S)$ if and only if $Rs = rS$. We shall see below that this indeed is the case. Note that once this theorem is proved not all annular regions are conformal to each other and in particular to each real number there corresponds a family of annular regions which are conformally equivalent to each other and to different real numbers, there correspond different conformal equivalence classes.

Theorem 5.3.2
Let $A(r_1, R_1)$ and $A(r_2, R_2)$ be two annular regions with $0 < r_i < R_i < \infty$ $(i = 1, 2)$. These regions are conformally equivalent if and only if $R_1/r_1 = R_2/r_2$.

Proof
$A(r_1, R_1)$ is conformally equivalent to $A(1, R_1/r_1)$ as we already saw. Thus there is no loss of generality in assuming $r_1 = r_2 = 1$. Assuming this we have only to prove that $A(1, R_1)$ is conformally equivalent to $A(1, R_2)$ if and only if $(R_1 = R_2)$.

Since one part is obvious we shall just suppose $A(1, R_1)$ and $A(1, R_2)$ are conformally equivalent and show that $R_1 = R_2$. Assume f is analytic and maps $A_1 = A(1, R_1)$ conformally onto $A_2 = A(1, R_2)$. Let K be the circle centre 0 and radius $r = \sqrt{R_2}$. Consider $f^{-1} : A_2 \to A_1$ which is also analytic. Observe that $K \subseteq A_2$ because of the inequalities $1 < \sqrt{R_2} < R_2$. Thus $f^{-1}(K)$ is a compact subset contained in A_1. Hence for sufficiently small ε, $A(1, 1 + \varepsilon) \cap f^{-1}(K) = \phi$. See the figure below.

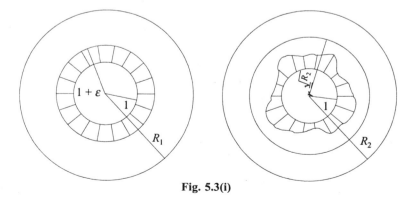

Fig. 5.3(i)

Now $V = f(A(1, 1 + \varepsilon))$ is a connected subset of A_2 which does not intersect K. Therefore V is contained in $A(1, r)$ or $A(r, R_2)$. Again there is no loss of generality in assuming $V \subseteq A(1, r)$. (In case $V \subseteq A(r, R_2)$ verify that $g = R_2/f$ is a conformal equivalence from A_1 onto A_2 with $A(1, 1 + \varepsilon) \cap g^{-1}(K) = \phi$ and $g(A(1, 1 + \varepsilon)), \subseteq A(1, r)$ by converting inequalities for $|f|$ into those involving $|g|$ and we can proceed with g instead of f). Thus our assumption implies that if $1 < |z_n| < 1 + \varepsilon$ $(n = 1.2, \ldots)$ is a sequence of complex numbers with $|z_n| \to 1$ then $f(z_n) \in V$ and further $f(z_n)$ tends to ∂V so that $|f(z_n)| \to 1$ or R_2. (Applying proposition 5.1.14 twice). But in so far as ∂V does not intersect $|w| = R_2$ the only possibility is that $|f(z_n)| \to 1$. Similarly if z_n is a sequence in A_1 with $|z_n| \to R_1$ then $|f(z_n)| \to R_2$ as $n \to \infty$. Now define $\alpha = \dfrac{\log R_2}{\log R_1}$. Put $u(z) = 2 \log |f(z)| - 2\alpha \log |z|$ ($z \in A_1$). Note that $u(z)$ is well defined and is also harmonic (locally both log $|f(z)|$ and $\log |z|$ in A_1 are real parts of analytic branches of $\log f$ and $\log z$ respectively and are therefore harmonic).

Let ∂ be the Cauchy-Riemann differential operator given by $\partial = \dfrac{1}{2}\left(\dfrac{\partial}{\partial x} - i\dfrac{\partial}{\partial y}\right)$. Since f is analytic in A_1, $\overline{\partial} f = 0$ and $\partial f = f' = \partial f/\partial x =$

$(-i)\,\partial f/\partial y$. Thus $\partial \bar{f} = 0$ (verify that $\partial \bar{f} = \overline{(\overline{\partial} f)}$ where $f = u_1 + iv_1$ and so $\bar{f} = u_1 - iv_1$). We now use the chain rule to conclude that

$$\partial(2 \log |f|) = \partial \log |f|^2 = \partial \log \bar{f} f = \partial \log \bar{f} + \partial \log f = f'/f.$$

$$\text{and } \partial\,(2\log |z|) = 1/z.$$

Thus $\partial u = \dfrac{f'(z)}{f(z)} - \dfrac{\alpha}{z}$ for $z \in A_1$. Recalling that $|z_n| \to 1$ implies $|f(z_n)| \to 1$ and $|z_n| \to R_1$ implies $|f(z_n)| \to R_2$, we see that $u(z)$ gets extended continuously to the boundary of A_1 and as $z \to \partial A_1$, $u(z) \to 0$, i.e., the continuously extended values of the harmonic function $u(z)$ on the boundary of A_1 are zero. But since non-constant harmonic functions do not have maximum or minimum inside their regions of definition we see that max $u(z) = $ min $u(z) = 0$, i.e., $u(z) \equiv 0$ in A_1 or that $\dfrac{f'(z)}{f(z)} \equiv \dfrac{\alpha}{z}$. Let us now take $\gamma(t) = \sqrt{R_1}e^{it}$ $(-\pi \le t \le \pi)$ and set $\Gamma = f(\gamma)$ to get

$$n\,(\Gamma, 0) = \dfrac{1}{2\pi i}\int_\Gamma \dfrac{dw}{w} = \dfrac{1}{2\pi i}\int_\gamma \dfrac{f'(z)}{f(z)}\,dz = \dfrac{\alpha}{2\pi i}\int_\gamma \dfrac{dz}{z} = \alpha n(\gamma, 0) = \alpha.$$

(Note that $n(\gamma, 0) = 1$). Therefore, α is an integer. On the other hand $R_1, R_2 > 1$ imply $\alpha > 0$ and so α is a positive integer. Now let us consider

$h(z) = \dfrac{f(z)}{z^\alpha}$ which is an analytic function in A_1.

Now
$$h'(z) = f'(z)\, z^{-\alpha} + f(z)\, (-\alpha)\, z^{-\alpha-1}$$

$$= \frac{f(z)}{z^{\alpha}} \left(\frac{f'(z)}{f(z)} - \frac{\alpha}{z} \right) = 0 .$$

Therefore $h(z)$ is a constant equal to c (say) throughout A_1. Thus $f(z) = cz^{\alpha}$ but as $f(z)$ is one-to-one in A_1 it follows that α is not ≥ 2. (In case $\alpha \geq 2$, take any $z \in A_1$ and take all possible α-th roots of z which are again in A_1 and are getting mapped onto the same z^{α}). Thus $\alpha = 1$ or that $R_1 = R_2$.

Definition 5.3.3
We say Ω is of finite connectivity n if the complement of Ω in the extended plane has exactly n components. We say a region Ω is of infinite connectivity if this complement has infinitely many components. We shall see later that $n \geq 2$ is equivalent to the region being not simply connected.

Example 5.3.4
A circular region (i.e., an open disc) in which $n - 1$ points are removed is of connectivity n. On the other hand, $\mathbb{C}\backslash\mathbb{N}$ where $\mathbb{N}$ is the set of all natural numbers is of infinite connectivity.

In this small section we shall describe how we can associate a "homology basis" consisting of $(n - 1)$ cycles with each region of finite connectivity n and demonstrate its usefulness in obtaining $\int_{\gamma} f\, dz$ where γ is any cycle and f is analytic in Ω.

Theorem 5.3.5
Let Ω be a region of finite connectivity $n \geq 2$. Let $A_1, A_2, \ldots A_n$ be the components of Ω^c with $\infty \in A_n$. Then we can find cycles γ_i ($i = 1, 2, \ldots n - 1$) in Ω such that $n\,(\gamma_i, w) = 1$ for $w \in A_i$ and $n\,(\gamma_i, w) = 0$ for $w \in A_j, j \neq i$.

Proof
Fix i ($i = 1, 2, \ldots n - 1$) and choose $a \in A_i$. Choose δ so that $d\,(A_i, B) \geq \delta > 0$ where $B = \bigcup_{k \neq i} A_k.$ $\Big($Here

$$d(A_i,\ B) = \inf_{\substack{x \in A_i \\ y \in B}} d(x,\ y)\Big).$$

This is possible because of the following facts:
(i) Each A_i ($1 \leq i \leq n$) is closed (A_i connected implies $\overline{A}_i$ is connected and A_i is a connected component implies $\overline{A}_i \subseteq A_i$ and so $\overline{A}_i = A_i$) and bounded for $i \neq n$ and hence A_i ($1 \leq i \leq n - 1$) is compact.
(ii) B is closed (being a finite union of closed sets). Further $B \cap A_i = \phi$.
(iii) Distance between a compact set and a closed set which are disjoint in the plane is always positive.

Cover the whole plane with a net of squares with constant length for the sides being less than $\dfrac{\delta}{\sqrt{2}}$. We are free to choose this net so that a is the centre of one particular square. We shall denote the boundary of each typical closed square Q by ∂Q. We also assume that ∂Q is oriented so that the interior of Q lies to the left of ∂Q. Consider the cycle $\gamma_i = \sum_j \partial Q_j$ where the sum ranges over all Q_j in the net which have at least one point in common with A_i and where γ_i is given the reduced representation (in the sense that cancellation have been carried out). To complete the proof we claim that

(i) the sum representing γ_i is finite.
(ii) γ_i is completely contained in Ω.
(iii) $n(\gamma_i, w) = 1$ for every $w \in A_i$.
(iv) $n(\gamma_i, w) = 0$ for every $w \in A_j$ $(j \neq i)$ and $w \notin \gamma_i$.

Proof of the claim

(i) If Q_j intersects A_i it cannot intersect B because otherwise there are two points one from A_i (say c) and another from B (say b) with c, $b \in Q_j$ so that $d(c, b) < \delta$. But this is a contradiction to $d(A_i, B) \geq \delta$. Thus Q_j must be completely contained in B^c which is bounded and the set of all such Q_j's must be finite as each Q_j has a fixed area.

(ii) From the proof of (i) it is clear that γ_i does not intersect B. On the other hand if γ_i is given its reduced representation (i.e., after cancellation of common sides) we can even prove that γ_i does not meet A_i also. (This is because each side of γ_i that intersects A_i appears as common side of exactly two squares with opposite orientation and would have cancelled each other). Thus no point of A_i or B can be in γ_i and so $\gamma_i \subseteq \Omega$.

(iii) For $w \in A_i$, $n(\gamma_i, w) = n(\gamma_i, a) = 1$ as a is the centre of exactly one Q_j.

(iv) On the other hand γ_i in its reduced form becomes a Jordan curve (Note that the interiors of the squares are disjoint) and so determines only two components one of which contains A_i and another of which contains the whole B and hence ∞. (Note that interior of γ_i contains points from Q_j intersecting A_i.) Thus $n(\gamma_i, w) = 0$ for every $w \in B$ and in particular for all $w \in A_j$, $j \neq i$.

Theorem 5.3.6
Given a region of finite connectivity $n \geq 2$ there exists cycles γ_i $(i = 1, 2, \ldots n-1)$ such that $\gamma_i \subseteq \Omega$, $\forall i$ and given any cycle γ in Ω, γ is homologous to $c_1\gamma_1 + c_2\gamma_2 + \ldots + c_{n-1}\gamma_{n-1}$ where c_i's are uniquely determined integers.

Proof
From theorem 5.3.5 we can get cycles γ_i $(1 \leq i \leq n-1)$ such that $n(\gamma_i, w) = 1$ $\forall w \in A_i$ and $n(\gamma_i, w) = 0$, $\forall w \in A_j$ $(j \neq i)$ with $w \notin \gamma_i$ where A_i

are all the components of the complement of Ω. Now let γ be any cycle in Ω. Put $c_i = n(\gamma, a_i)$ where $a_i \in A_i$ $(1 \le i \le n-1)$. Consider $\beta = \gamma - c_1\gamma_1 - c_2\gamma_2 - \ldots - c_{n-1}\gamma_{n-1}$. If $w \in \Omega^c = \bigcup_1^n A_i$, $n(\beta, w) = 0$ by the properties of γ_i. Then $\beta \sim 0$ in Ω or γ is homologous to $c_1\gamma_1 + c_2\gamma_2 + \ldots + c_{n-1}\gamma_{n-1}$ in Ω. Further this linear combination is unique because if γ is homologous to $\sum_{i=1}^{n-1} c_i\gamma_i$ and $\sum_{i=1}^{n-1} d_i\gamma_i$ then $\Gamma = \sum(c_i - d_i)\,\gamma_i \sim 0$ in Ω. But for each j and a_j in A_j, $n(\Gamma, a_j) = c_j - d_j = 0$. Thus any cycle γ is homologous in Ω to a unique integer combination of γ_i $(1 \le i \le n-1)$.

Definition 5.3.7
The cycles γ_i $(1 \le i \le n-1)$ having the properties of Theorem 5.3.6 are said to form a homology basis for the region Ω of finite connectivity $n \ge 2$.

From this definition it is clear that for any homology basis β_i $(1 \le i \le m)$, $\sum_1^m d_i\beta_i \sim 0$ in Ω implies $d_i = 0$ because $d_1\beta_1 \sim -d_2\beta_2 - \ldots -d_m\beta_m$ and the uniqueness of representation for $d_1\beta_1$ in terms of β_i implies $d_1 = 0 = d_i$ for every i.

Theorem 5.3.8
Let Ω be a region of finite connectivity $n \ge 2$. Then any homology basis for Ω consists of exactly $(n-1)$ cycles.

Proof
We already saw in Theorem 5.3.6 that γ_i, $1 \le i \le n-1$, form a homology basis and $d_1\gamma_1 + d_2\gamma_2 + \ldots + d_{n-1}\gamma_{n-1} \sim 0$ implies $d_i = 0$ for every i. Let now $\beta_1, \beta_2, \ldots \beta_m$ and $\eta_1, \eta_2, \ldots \eta_k$ be any two homology bases. We claim that $m \le k$. Suppose $m > k$. We know that each

$$\beta_j \sim \sum_{i=1}^k c_{ji}\,\eta_i \text{ where } c_{ji} \in \mathbb{Z} \text{ are unique.}$$

We now claim that there exist integers d_j $(1 \le j \le m)$ not all zero such that $\sum_{j=1}^m d_j\beta_j \sim 0$. This is equivalent to the system of simultaneous equations given by

$$\sum_{j=1}^m d_j \sum_{i=1}^k c_{ji}\,\eta_i \sim 0$$

or equivalently $\sum_{j=1}^m d_j c_{ji} = 0$ for $1 \le i \le k$.

This is a system of k equations in m unknowns where $m > k$. By a theorem on Linear algebra we can always find a non-trivial integral solution for the above equations. This contradicts the fact that β_j form a basis. Thus m is not greater than k, i.e., $m \leq k$. Interchanging the roles of these two bases we also have $k \leq m$. Since we already have a basis $\{\gamma_i\}$ consisting of $(n - 1)$ elements, the proof of the theorem is complete.

Theorem 5.3.9

Let Ω be a region of finite connectivity n. Let f be analytic in Ω. Let $\{\gamma_i\}$ $(1 \leq i \leq n - 1)$ be a homology basis for Ω; Put $P_i = \int_{\gamma_i} f(z) \, dz$ $(1 \leq i \leq n -1)$. P_i are called periods of f. Given any cycle γ, $\int_\gamma f(z) \, dz$ is a unique integral linear combination of P_i. Further a necessary and sufficient condition for the existence of a single valued analytic function F in Ω such that $F'(z) = f(z)$ is that $P_i = 0$ for every i, i.e., all the periods of f vanish.

Proof

From Theorem 5.3.6, $\gamma \sim \sum\limits_{1}^{n-1} c_i \gamma_i$ where c_i are uniquely determined integers.

Applying the general form of Cauchy's theorem we get

$$\int_\gamma f(z) \, dz = \sum_{1}^{n-1} c_i \int_{\gamma_i} f(z) \, dz = \sum_{1}^{n-1} c_i P_i.$$

We first note that a necessary and sufficient condition for the existence of indefinite integral for f is that $\int_\gamma f(z) \, dz = 0$ for every closed curve γ in Ω. (See Theorem 4.2.3 and Theorem 4.2.4).

Thus if the periods are zero then $\int_\gamma f(z) \, dz = 0$ for any closed curve and so f has a primitive. Conversely if f has a primitive we know that $\int_\gamma f(z) \, dz = 0$ for every closed curves γ in Ω and in particular for $\gamma = \gamma_i$, i.e., $P_i = 0$. The proof of our theorem is now complete.

Example 5.3.10

We shall illustrate the above conclusions with respective to a doubly connected region given by the annulus $r < |z| < R$ (we include the degenerate cases $r = 0$ or $R = \infty$). In this case the complement has the components $|z| \leq r$ and $|z| \geq R$. A homology basis for this region is evidently a circle $|z| = \rho$ where $r < \rho < R$. If this circle is denoted by C then any cycle γ in this annulus satisfies $\gamma \sim nC$ where $n = n (\gamma, 0)$. Further if f is analytic in $r < |z| < R$ then $\int_\gamma f(z) \, dz = n \int_C f(z) \, dz$. Thus $\int_\gamma f(z) \, dz$ where γ is any arbitrary cycle is just an integral multiple of $\int_C f(z)\, dz$ which is independent of γ. Thus evaluation of integrals over complicated curves inside the annulus becomes a routine exercise.

Remark 5.3.11

The detailed discussion of homology basis connected with regions of finite connectivity can be effectively put to use in the construction of single-valued analytic branches for certain multi-valued functions even in regions that are not necessarily simply connected. We shall demonstrate this by the following theorem.

Theorem 5.3.12

A single-valued analytic branch of $\sqrt{1-z^2}$ can be defined in any region for which the points ± 1 are in the same component of its complement.

Proof

Let Ω be the given region. Let γ be a curve joining -1 and 1 (including the end points) inside the component (which contains both ± 1) of the complement of Ω. Choose $|z| = R$ with R large enough so that γ is contained in $|z| < R$. Consider $\dfrac{2}{1-z^2}$ which is analytic in $G = \gamma^c \supseteq \Omega$. We observe that G is a doubly connected region (the complement of G with respect to the extended plane consists of just two disjoint components γ and $\{\infty\}$). Further a homology basis for G can be taken to be the simple closed curve $|z| = R$ (This curve has index 1 with respective to each point of γ) we now consider the period of $\dfrac{2}{1-z^2}$ say P.

$$P = \int_{|z|=R} \frac{2}{1-z^2}\, dz = 2\pi i\left[\operatorname*{Res}_{z=1} \frac{2}{1-z^2} + \operatorname*{Res}_{z=-1} \frac{2}{1-z^2} \right]$$

$$= 2\pi i\, (-1 + 1) = 0.$$

Thus $\displaystyle\int_{\Gamma} \frac{2}{1-z^2}\, dz = 0$ for all closed curves Γ in G and in particular there exists a single valued analytic function in G whose derivative is $\dfrac{2}{1-z^2}$. Call this function as $F(z)$. It follows that $e^{F(z)}\left(\dfrac{1-z}{1+z}\right)$ has a zero derivative at all points of G. Hence $e^{F(z)} = c\left(\dfrac{1+z}{1-z}\right)$ (The value of this constant c can not be 0 because for any $z \in G$, $z \neq 1$ and $e^{F(z)} \neq 0$). Taking any arbitrary value for $\log c$ say k it follows that $e^{F(z)-k} = \left(\dfrac{1+z}{1-z}\right)$. Put $F(z) - k = H(z)$ so that $H(z)$ is a single-valued analytic branch for $\log\left(\dfrac{1+z}{1-z}\right)$. We can now define a single-valued analytic branch for $\left(\dfrac{1+z}{1-z}\right)^{1/2}$ as $e^{H(z)/2} = g(z)$ (say).

Consider $(1 - z)\,g(z)$. This can be considered as a branch for $\sqrt{1-z^2}$. Indeed,

$$[(1 - z)\,g(z)]^2 = (1 - z)^2\,g^2\,(z)$$
$$= (1 - z)^2\left(\frac{1+z}{1-z}\right)$$
$$= 1 - z^2.$$

Thus a single valued analytic branch for $\sqrt{1-z^2}$ can be defined in G and therefore in $\Omega \subseteq G$.

5.4 Simply Connectedness and Equivalent Conditions

Introduction 5.4.1
In this section we shall obtain several equivalent conditions for simply connectedness of a region Ω in the plane. We have already observed that simply connectedness is a topological property. It is interesting to observe that this condition is equivalent to various analytic conditions like the existence of harmonic conjugates, the validity of Cauchy's theorem, as well as algebraic conditions like the existence of log and square roots for analytic functions. However the validity of these equivalences will be generally used only in one way (simply connectedness implies the other conditions) and thus the scope of these characterizations is limited. Nevertheless these characterizations are aesthetically most satisfying.

Theorem 5.4.2
Let Ω be a region in the finite complex plane. The following are equivalent
(1) Ω is simply connected.
(2) For any cycle γ in Ω, $\gamma \sim 0$ in Ω.
(3) The complement of Ω with respect to the extended complex plane is connected.

(4) For any closed curve γ in Ω and any analytic function f in Ω,
$$\int_\gamma f(z) \; dz = 0.$$

(5) To each f analytic in Ω there exists an analytic function F in Ω such that $F' = f$ (f has a global primitive).

(6) If f is analytic in Ω and not equal to 0 anywhere in Ω then there exists an analytic function $g \in \Omega$ such that $e^g = f$ (f has an analytic branch for its logarithm).

(7) If f is analytic in Ω and never vanishes in Ω then there exists an analytic function $\phi \in \Omega$ such that $f = \phi^2$ (f has an analytic square root).

(8) Ω is homeomorphic to the open unit disk.

Proof

Recall that by definition Ω is simply connected if and only if every closed curve in Ω is homotopic to a point. We have already encountered many of these implications and we shall freely make use of them.

(1) $\Rightarrow$ (2): Let γ be a closed curve in Ω. Then γ is homotopic to a point in Ω. By Theorem 4.2.15 taking $f(z) = \dfrac{1}{z-a}$ for $a \notin \Omega$ we get $\int_\gamma \dfrac{1}{z-a} = 0$ or that $n\,(\gamma, a) = 0$ for every $a \notin \Omega$. It follows that $n\,(\gamma, a) = 0$ for all cycles γ in Ω.

(2) $\Rightarrow$ (3): We assume $n\,(\gamma, a) = 0$ for all cycles γ in Ω. If the complement of Ω in the extended plane is not connected the complement contains at least one bounded component (say A). As in Theorem 5.3.5 we can construct a cycle $\gamma \subseteq \Omega$ such that $n(\gamma, a) = 1$ for every $a \in A$, a contradiction. This completes the proof.

(3) $\Rightarrow$ (2): Let γ be any cycle in Ω. If the complement of Ω is connected and $a \notin \Omega$, then a belongs to the only component of the complement of Ω which contains ∞. Therefore, a belongs to the unbounded component of the complement of γ (a and ∞ belong to a connected set inside the complement of γ). Therefore $n(\gamma, a) = 0$ by the property of the index.

(2) $\Rightarrow$ (4): Follows by "General form of Cauchy's theorem".

(4) $\Rightarrow$ (5): See Theorem 4.2.4 and 4.2.3.

(5) $\Rightarrow$ (6): If f is analytic in Ω and not equal to zero anywhere f'/f is analytic in Ω and has a primitive (say) F. As in the proof of Theorem 4.4.7 we can define a function $g = F + c$ (where c is a constant) such that $e^g = f$.

(6) $\Rightarrow$ (7): From (4) we get a g such that $e^g = f$. Take $\phi = e^{g/2}$ so that $\phi^2 = f$.

(7) $\Rightarrow$ (8): If Ω is the whole plane $z \to \dfrac{z}{1+|z|}$ gives the required homeomorphism. (See Solved Exercise Problem 12 of chapter I.) On the other hand if $\Omega \neq$ the whole plane we observe the following: If we look at

the proof of Riemann mapping theorem we see that the only place where we have used the simple connectivity of the region is to obtain analytic square roots for $f(z) - w$ where $w \notin \Omega$. Now this hypothesis is already given by (7). Hence the proof of the Riemann mapping theorem gives that Ω is conformally equivalent to the open unit disc and the conformal equivalence is more than a homeomorphism.

(8) $\Rightarrow$ (1): We assume that the unit disc and Ω are homeomorpic. But then the unit disc is simply connected and as we already saw, simple connectivity is a topological property. Hence Ω is simply connected.

There is another characterization of simply connectedness for regions in the finite complex plane. For technical reasons we have not included this in the previous theorem. However, we shall state and prove this now.

Theorem 5.4.3
Let Ω be a region in the finite complex plane. Ω is simply connected if and only if given u: $\Omega \to \mathbb{R}$ harmonic there exists a harmonic function v: $\Omega \to \mathbb{R}$ such that $f = u + iv$ is analytic in Ω.

Proof
We need the following:

Lemma 5.4.4
Let ϕ: $[a, b] \times [c, d] \to \mathbb{C}$ be continuous and define g: $[c, d] \to \mathbb{C}$ by $g(t) = \int_a^b \phi(s, t)\, ds$. Then g is continuous and if $\dfrac{\partial \phi}{\partial t}$ exists and is a continuous function on $[a, b] \times [c, d]$ then g is differentiable, g' is continuous and

$$g'(t) = \int_a^b \frac{\partial \phi}{\partial t}(s, t)\, ds$$

Proof of Lemma 5.4.4
ϕ is uniformly continuous on $[a, b] \times [c, d]$, since the latter is compact. Therefore given $\varepsilon > 0$ there exists a $\delta > 0$ such that for every $u, v \in [c, d]$ with $|u - v| < \delta$, $|\phi(s, u) - \phi(s, v)| < \varepsilon$. Therefore

$$|g(u) - g(v)| \leq \int_a^b |\phi(s, u) - \phi(s, v)|\, ds$$
$$\leq \varepsilon(b - a) \text{ for every } |u - v| < \delta.$$

Hence g is uniformly continuous on $[c, d]$. If we prove that g is differentiable and $g'(t) = \int_a^b \dfrac{\partial \phi}{\partial t}(s, t)\, ds$ by what we have just proved it will follow that g' is continuous. (Note that $\dfrac{\partial \phi}{\partial t}$ is also continuous by hypothesis).

Fix $t_0 \in [c, d]$ and choose $\varepsilon > 0$. Put $\dfrac{\partial \phi}{\partial t} = \psi$. Since ψ is continuous on

$[a, b] \times [c, d]$, a compact set, it is uniformly continuous there. Therefore for this $\varepsilon > 0$ there exists a $\delta > 0$ such that $(s - s')^2 + (t - t')^2 < \delta^2$ implies that

$$|\psi (s'\ t') - \psi (s, t)| < \varepsilon. \tag{1}$$

Thus for $|t - t_0| < \delta$ and $a \le s \le b$ $|\psi (s, t) - \psi (s, t_0)| < \varepsilon$. Now for a fixed $s \in [a, b]$,

$$|\phi (s, t) - \phi (s, t_0) - (t - t_0)\ \psi (s, t_0)| = |(t - t_0)\ \psi (s, t_1) - (t - t_0)\ \psi (s, t_0)|$$

$$\le |(t - t_0)|\ \varepsilon$$

where t_1 is between t and t_0 (we have used Mean value theorem for ϕ in the second variable and the inequality (1)). Now from the definition of g we have

$$\left| \frac{g(t) - g(t_0)}{t - t_0} - \int_a^b \psi(s, t_0)\, ds \right| = \left| \int_a^b \left(\frac{\phi(s, t) - \phi(s, t_0)}{t - t_0} - \psi(s, t_0) \right) ds \right|$$

$$\le \varepsilon (b - a)$$

whenever $|t - t_0| < \delta$, i.e, $g'(t_0)$ exists and is equal to $\int_a^b \psi(s, t)\, ds$. Since

t_0 is arbitrary the Lemma follows.

Lemma 5.4.5
If Ω is either the whole plane or the open unit disc and $u : \Omega \to \mathbb{R}$ is harmonic then u has a harmonic conjugate.

Proof of Lemma 5.4.5
We shall prove that if $\Omega = B (0, r) = \{z \in \mathbb{C} : |z| < r\}$ then the theorem is true. The whole plane corresponds to $r = \infty$ and the open unit disc corresponds to $r = 1$. Define

$$v (x, y) = \int_0^y \frac{\partial u}{\partial x} (x, t)\ dt + \phi(x) \tag{2}$$

where u is the given harmonic function from $B(0, r)$ to $\mathbb{R}$ and $\phi (x)$ to be determined later. From Lemma 5.4.4 we get

$$\frac{\partial v}{\partial x}(x, y) = \int_0^y \frac{\partial^2 u}{\partial x^2}(x, t)\ dt + \phi' (x)$$

$$= - \int_0^y \frac{\partial^2 u}{\partial y^2} (x, t)\ dt + \phi' (x)$$

$$= -\frac{\partial u}{\partial y}(x, y) + \frac{\partial u}{\partial y}(x, 0) + \phi'(x).$$

Thus if we choose $\phi'(x) = -\frac{\partial u}{\partial y}(x, 0)$ or $\phi(x) = -\int_0^x \frac{\partial u}{\partial y}(s, 0)\, ds$

it follows that $\quad \frac{\partial v}{\partial x} = -\frac{\partial u}{\partial y}.$ Thus

$$v(x, y) = \int_0^y \frac{\partial u}{\partial x}(x, t)\, dt - \int_0^x \frac{\partial u}{\partial y}(s, 0)\, ds$$

satisfies $\dfrac{\partial v}{\partial x} = -\dfrac{\partial u}{\partial y}$ and $\dfrac{\partial v}{\partial y} = \dfrac{\partial u}{\partial x}$. (The latter conclusion follows from the theory of Riemann integration). It also follows that v is harmonic (because u is) and it is now clear that v is the required harmonic conjugate of u.

Note that the integrations are taken over the line segments joining 0 and y or 0 and x and these line segments do actually lie in Ω at least when $\Omega = B(0, r)\ (0 < r \le \infty)$.

Theorem 5.4.6
Let f be a non-constant analytic function in a region D. Put $E = f(D)$. Let g be a real continuous function defined on E with continuous first and second order partial derivatives in E. Consider $h(z) = g(f(z))$, $z \in D$. Then

$$\Delta^2 h = |f'(z)|^2\, \Delta^2 g \text{ on } D \text{ where } \Delta^2 = \frac{\partial^2}{\partial x^2} + \frac{\partial^2}{\partial y^2}.$$

Proof of Theorem 5.4.6
We know that $E = f(D)$ is also a region by the open mapping theorem and continuity of f. Let $f = u + iv$ and $z = x + iy$. Then $h(x, y) = g(u, v)$ where $u = u(x, y)$ and $v = v(x, y)$. Using chain rule for two variables we have

$$h_x = g_u u_x + g_v v_x$$

and
$$h_{xx} = g_u u_{xx} + u_x(g_{uu} u_x + g_{uv} v_x) + g_v v_{xx} + v_x(g_{vu} u_x + g_{vv} v_x)$$
$$= g_u u_{xx} + g_{uu} u_x^2 + g_{uv} u_x v_x + g_v v_{xx} + g_{vu} v_x u_x + g_{vv} v_x^2.$$

Similarly
$$h_{yy} = g_u u_{yy} + g_{uu} u_y^2 + g_{uv} u_y v_y + g_v v_{yy} + g_{vu} v_y u_y + g_{vv} v_y^2.$$

Using the fact that $g_{uv} = g_{vu}$ and that u, v are harmonic conjugates (i.e., $u_x = v_y$ and $u_y = -v_x$) we see that

$$h_{xx} + h_{yy} = g_{uu}(u_x^2 + u_y^2) + g_{vv}(v_x^2 + v_y^2)$$

$$= (g_{uu} + g_{vv})(u_x^2 + v_x^2)$$

Thus $\Delta^2 h = |f'(z)|^2\, \Delta^2 g$ (since $f'(z) = u_x + iv_x$).

Corollary 5.4.7

If f is analytic in a region D and $E = f(D)$ and g is real harmonic on E then $h = g \circ f$ is harmonic on D.

Proof of Corollary 5.4.7

If f is constant then h is constant and so is harmonic. If f is not a constant, Theorem 5.4.6 gives

$$\Delta^2 g = 0 \text{ implies } \Delta^2 h = 0$$

Proof of Theorem 5.4.3

To prove necessity consider

Case 1 $\Omega \neq \mathbb{C}$. By Riemann mapping theorem we can find an analytic function $h : \Omega \to U$ (open unit disc) such that $h^{-1} : U \to \Omega$ is analytic and one-to-one. Consider $u_1 = u \circ h^{-1} : U \to \mathbb{R}$. u_1 is harmonic in U by Corollary 5.4.7 and by Lemma 5.4.5, u_1 has a harmonic conjugate (say) v_1 such that $v_1 : U \to \mathbb{R}$. Put $f_1 = u_1 + iv_1$ which is analytic in U and consider $f = f_1 \circ h$. Then f is analytic in Ω and

$$\text{Re } f = \text{Re } (f_1(h(z)) = u_1 (h(z)) = u(z).$$

Hence $v = \text{Im } f = v_1 \circ h$ is the harmonic conjugate of u.

Case 2   If $\Omega = \mathbb{C}$, then the theorem follows from Lemma 5.4.5.

This proves the necessary part of the theorem.

To prove the sufficiency we shall assume that every harmonic function in Ω has a harmonic conjugate and prove (6) of Theorem 5.4.2.

Let $f : \Omega \to \mathbb{C}$ be analytic and never zero in Ω. Put $f = u + iv$. Define

$$g(x, y) = \log |f(z)| = \frac{1}{2}\log (u^2 + v^2).$$

A computation shows that g is real harmonic in Ω and our hypothesis implies that g has a harmonic conjugate h (say). Thus $F(z) = (g + ih)(z)$ is analytic in Ω. Put $H(z) = e^{F(z)}$. Then H is analytic and never vanishes in Ω and

$$\left| \frac{f(z)}{H(z)} \right| = 1 \; \forall \; z \in \Omega.$$

(Since $|H(z)| = e^{\text{Re } F} = e^g = |f|$). Therefore $\dfrac{f}{H}$ is a constant function or that $f = cH$ where c is a non-zero constant. Thus $f = ce^{F(z)} = e^{F(z)+k}$ where k is any logarithm for c. Thus $F(z) + k$ is an analytic branch for $\log f$. This is (6) of Theorem 5.4.2. Thus by the very same theorem, Ω is simply connected.

Remark 5.4.8

(1) From Theorem 5.4.2 $((1) \Rightarrow (3))$ it is clear that multiply connected regions (Regions Ω whose complements have more than one component) are the ones which are not simply connected.

(2) In proving Theorem 5.4.3 we have used Theorem 5.4.6 only to the extent (Cor. 5.4.7) that f analytic and g harmonic (real) implies $g \circ f$ is harmonic. However since harmonicity is of local character we can prove this more directly by taking a simply connected neighbourhood of $f(z_0)$ and defining a harmonic conjugate h for g in that neighbourhood so that $G(z) = g(z) + ih(z)$ becomes analytic and so $G \circ f$ is also analytic in that neighbourhood. Now $(g \circ f) = \mathrm{Re}\ (G \circ f)$ and hence $g \circ f$ is locally harmonic which is what we wanted to prove. On the other hand Theorem 5.4.6 has a global character and it brings out the relationship that exists between $\Delta^2 h$ and $\Delta^2 g$ and this in fact is the reason for obtaining it (this result is of independent interest) as a separate theorem.

5.5 Analytic Automorphisms of Regions in the Complex Plane

Introduction 5.5.1

In group theory we are interested in properties of groups which are invariant under group isomorphisms and in topology we are interested in properties of topological spaces which are invariant under homeomorphisms. Similarly in function theory we are interested in properties of regions in the plane which are invariant under what are called "conformal equivalence". These are maps f from one region onto another such that f is bijective and f and f^{-1} are both analytic. Of course by Lemma 4.5.23 this condition of f is equivalent to f being bijective and analytic. For example according to Riemann Mapping Theorem, up to conformal equivalence, there are only two regions in the finite complex plane namely $\mathbb{C}$ and the open unit disc U. Thus the problem of determining the regions which are conformally equivalent to a given region becomes interesting. Another related question in this context is that if we can find one conformal equivalence f between two regions Ω_1 and Ω_2 can we find all other conformal equivalences between them? It is in this context that bijective analytic maps of a regions Ω onto itself (also called analytic automorphisms of Ω) play a significant role. Indeed if $f : \Omega_1 \to \Omega_2$ is a conformal equivalence between two regions Ω_1 and Ω_2 then any other conformal equivalence between Ω_1 and Ω_2 is given by $g = \phi \circ f$ where ϕ is an analytic automorphism of Ω_2. (Note that $\phi = g \circ f^{-1}$ is an analytic automorphism of Ω_2). Moreover as observed in Introduction 5.1.1. any analytic automorphism of Ω enables us to understand the ring theoretic structure of $H(\Omega)$, the space of all analytic functions on Ω.

For a given region Ω in $\mathbb{C}$ we shall denote by *Aut* Ω the set of all analytic automorphisms of Ω. Note that Aut Ω is a group under composition of mappings with the identity function on Ω as the multiplicative identity: Recall that in Chapter 4 (Vide Theorem 4.4.34, Theorem 4.4.38 and Theorem 4.5.24) we have characterized *Aut* Ω where $\Omega = U$ or H or $\mathbb{C}$. In this section we shall use entirely different techniques to study the group *Aut* Ω for different regions (including U, H and $\mathbb{C}$) Ω in the plane. This study will enable us to identify a few regions Ω in the plane for which *Aut* Ω reduces to the identity. i.e., regions for which there are no non-trivial analytic automorphisms (such regions are usually called rigid regions). We shall follow [12] (given at the end of this chapter) for most of our results. We shall first fix some notations and terminology.

5.5.2 Notations

The general linear group of all 2×2 non-singular matrices with complex entries will be denoted by GL $(2, \mathbb{C})$. GL $(2, \mathbb{R})$ is the general linear group of all 2×2 non-singular matrices with real entries. SL $(2, \mathbb{R})$ is the special linear group of 2×2 non-singular real matrices with determinant 1 and is a sub group of GL $(2, \mathbb{R})$. The open unit disc will be denoted by U, the upper half plane by H and the non-zero complex plane by $\mathbb{C}^*$. For any region $\Omega \subset \mathbb{C}$, *Aut* Ω denotes the group of all automorphism of Ω. Aut_c Ω denotes the subgroup of *Aut* Ω, fixing $c \in \Omega$ (also called the isometry group of $c \in \Omega$). For a subset $M \subset \Omega$ we shall use *Aut* $_M$ Ω to denote the set of all automorphisms f of Ω for which f $(M) = M$.

Definition 5.1.3

For $A = \begin{pmatrix} a & b \\ c & d \end{pmatrix} \in GL$ $(2, \mathbb{C})$ we associate the linear fractional transformation

$$h_A \ (z) \ = \ \frac{az + b}{cz + d}. \ (z \in \mathbb{C} \cup \{\infty\})$$

Theorem 5.5.4

For $A, B \in GL$ $(2, \mathbb{C})$
   (i) h_{AB} $(z) = (h_A \circ h_B)$ (z)
   (ii) h_{kA} $(z) = h_A$ (z) for any non-zero complex number k
   (iii) $h_A^{-1}(z) = h_{A^{-1}}$ (z)

   (iv) h_A $(z) = I$ $(z) \equiv z$ if and only if $A = a \begin{pmatrix} 1 & 0 \\ 0 & 1 \end{pmatrix}$

where a is any non-zero complex number.

Proof
Left to the reader.

Theorem 5.5.5

If $C = \begin{pmatrix} 1 & -i \\ 1 & i \end{pmatrix}$ and $C' = \begin{pmatrix} i & i \\ -1 & 1 \end{pmatrix}$ them h_C (z) is a conformal equivalence from the upper half plane H onto the open unit disc U with its inverse given by $h_{C'}$ (z).

Proof

Lemma 4.4.37 proves (take $z_0 = i$) the first part of our result. The second part follows from Theorem 5.5.4. (ii) and (iii) since $C^{-1} = \dfrac{1}{2i} C'$.

Theorem 5.5.6

For $\alpha \in U$ define $\phi_\alpha(z) = \dfrac{z - \alpha}{1 - \bar{\alpha}z}$. then $\phi_\alpha(z) \in Aut\ U$ and $\phi_\alpha^{-1} = \phi_{-\alpha}$.

Proof

See Theorem 4.4.35.

Theorem 5.5.7

$Aut_0\ U = \{f : U \rightarrow U/f(z) = az, |a| = 1\}$.

Proof

See Corollary 4.4.35.

Theorem 5.5.8

If $f \in Aut\ U$ and has two fixed points in U then $f(z) = z$.

Proof

Let one of the fixed points be 0. Then by Theorem 5.5.7, $f(z) = az$ with $|a| = 1$. If there is another fixed point say w with $|w| < 1$ then $f(w) = aw = w$ and so $a = 1$. Thus $f(z) = z$. On the other hand if α and β are two fixed points of $f \in Aut\ U$ (with α and β different from 0 but in U) then

$\phi_\alpha \circ f \circ \phi_\alpha^{-1}$ has 0 and $0 \neq \gamma = \phi_\alpha\ (\beta)$ as fixed points. In as much as

$\phi_\alpha \circ f \circ \phi_\alpha^{-1} \in Aut\ U$, by our previous case $\phi_\alpha \circ f \circ \phi_\alpha^{-1}$ is the identity function and so f is also the identity function, completing the proof.

Theorem 5.5.9

Let Ω and Ω' be two regions and $f : \Omega \rightarrow \Omega'$ bijective and analytic. Then f induces a group isomorphism between $Aut\ \Omega$ and $Aut\ \Omega'$ given by

$\phi_f : Aut\ \Omega \rightarrow Aut\ \Omega'$ with $\phi_f(h) = f \circ h \circ f^{-1}$.

Proof

It is clear that $\phi_f(h) \in Aut\ \Omega'$. That it is bijective (with $\phi_f^{-1}(g) = f^{-1} \circ g \circ f$, $g \in Aut\ \Omega'$) and is a group homomorphism, is easy to check.

Theorem 5.5.10

Let Ω be a region, Let J be a subgroup of $Aut\ \Omega$ such that J acts transitively on Ω (in the sense that give $z, w \in \Omega$ there exists $f \in J$ with $f(z) = w$) and $Aut_c\ \Omega \subset J$ for some $c \in \Omega$. Then $J = Aut\ \Omega$.

Proof

It suffices to prove that Aut $\Omega \subset J$. Let $f \in Aut$ Ω. Consider c and $f(c)$. By our assumption there exists $g \in J$ such that $g(f(c)) = c$. This $g \circ f \in Aut_c$ $\Omega \subset J$. Since J is a subgroup of Aut Ω and both $g \circ f$ and g belong to J we have $f = g^{-1} \circ g \circ f \in J$. Thus Aut $\Omega \subset J$ and the proof is complete.

Theorem 5.5.11

Let A be a discrete and relatively closed subset of a region Ω and $f : \Omega \backslash A \to \mathbb{C}$ be analytic and injective. Then
   (i) No point of A is an essential singularity of f.
   (ii) If $c \in A$ is a pole for f then c is of order 1.
   (iii) If every point of A is a removable singularity for f then the analytic extension $\widetilde{f} : \Omega \to \mathbb{C}$ is also injective.

Proof

Let $c \in A$. Take an open disc B around c so that $\overline{B} \subset \Omega$ and $\overline{B} \cap A = \{c\}$ (This is possible because A is discrete). If $\Omega_1 = \Omega \backslash (A \cup \overline{B})$ then Ω_1 is open (Note that A is relatively closed in Ω) and non-empty (consider a point on ∂B and a suitable neighbourhood of that point which is completely contained in Ω and which does not intersect A also. This neighbourhood obviously contains points in Ω which are not in $A \cup \overline{B}$.) and by open mapping theorem, $f(\Omega_1)$ is also non-empty and open. But $f(\Omega_1) \cap f(B \backslash \{c\}) = \phi$. (Otherwise there exists $w_1 \in \Omega_1$ and $w_2 \in B \backslash \{c\}$ such that $= f(w_1) = f(w_2)$. But then $w_1 \neq w_2$ as $\Omega_1 \cap B = \phi$ but $f(w_1) = f(w_2)$ which contradicts the injectivity of f in $\Omega \backslash A$). If c were an essential singularity, by Weierstrass theorem (Theorem 4.5.22.) $f(B \backslash \{c\})$ must be dense in $\mathbb{C}$ and so must intersect $f(\Omega_1)$. This contradiction proves that c can not be an essential singularity.

   (ii) Let $c \in A$ be a pole of order m for f. Then c is a zero of order m for $\frac{1}{f}$. i.e., there exits a neighbourhood U of c in which $g = \frac{1}{f}$ is analytic at $z = c$ and has a zero of order m there. By local correspondence theorem (Theorem 4.4.22.) g must be m-to-1 as a map in a neighbourhood of $z = c$. But then f is injective and so g is also injective, Thus $m = 1$ and c is a simple pole only.

   (iii) Let all the points of A be removable singularities for f and consider the analytic extension $\widetilde{f}$ of f to Ω. If $\widetilde{f}$ is not injective we can find a and b $(a \neq b)$ in Ω such that $p = \widetilde{f}(a) = \widetilde{f}(b)$. Choose disjoint open discs around a and b say B_1 and B_2 so that $B_1 \cap B_2 = \phi$ and $B_1 \backslash a \subseteq \Omega \backslash A$ and $B_2 \backslash b \subseteq \Omega \backslash A$ but $a \in B_1 \subseteq \Omega$, $b \in B_2 \subseteq \Omega$. Now $\widetilde{f}(B_1)$ and $\widetilde{f}(B_2)$ are also neighbourhoods of p (open connected sets containing p) and so $\tilde{f}(B_1)$ $\cap$ $\tilde{f}(B_2)$ is also a neighbourhood of p. For every $q \neq p$ in this neighbourhood we can find $a_1 \in B_1$, $b_1 \in B_2$ such that $\tilde{f}(a_1) = \tilde{f}(b_1) = q$. Now $a_1 \neq a$ and $b_1 \neq b$ as otherwise $p = q$. Thus $a_1 \in B_1 \backslash \{a\}$ and $b_1 \in B_2 \backslash \{b\}$ and $\tilde{f}(a_1) = f(a_1) = \tilde{f}(b_1) = f(b_1)$. But this contradicts the injectivity of f in $\Omega \backslash A$ and establish our claim that $\tilde{f}$ must be injective.

Theorem 5.5.12
If M discrete and is relatively closed subset of a region Ω then $Aut_M \, \Omega$ is isomorphic in a canonical way to a sub group of $Aut \, (\Omega \backslash M)$.

Proof
Define $\phi : Aut_M \, \Omega \to Aut \, (\Omega \backslash M)$ by $\phi \, (f) =$ Restriction of f to $\Omega \backslash M$. Since $f(M) = M$, ϕ is well-defined and $\phi(f) \in Aut \, (\Omega \backslash M)$ (f maps $\Omega \backslash M$ onto itself and is analytic and injective). It is easy to show that ϕ is a group homomorphism. All that is necessary is to show that ϕ is injective, Suppose $\phi(f) =$ restriction of f to $\Omega \backslash M$ is identity on $\Omega \backslash M$. We observe that all points of $\Omega \backslash M$ are interior to $\Omega \backslash M$ and so around each point a of $\Omega \backslash M$ we can get a ball $B(a; r)$ around a (and $\subset \Omega \backslash M$) in which f and the identity function agree everywhere and by principle of analytic continuation we see that $f(z) = z$ in Ω. Thus f is the identity of $Aut_M \, \Omega$. Thus the kernel of ϕ consists of just the identity element and so ϕ is injective. Thus there is an isomorphism between $Aut_M \, \Omega$ and $\phi(Aut_M \, \Omega)$ which is a subgroup of $Aut \, (\Omega \backslash M)$.

Note 5.5.13
In Theorem 5.5.12 if we merely assume Ω is open (not necessarily connected) then the theorem is still valid if we insist that $\Omega \backslash M$ intersects every connected component of Ω. On the other hand if we can just assume that the interior of M is empty (instead of the hypothesis that M is discrete which is a stronger assumption) then we can easily prove that any continuous function (not necessarily analytic) which is the identity function on $\Omega \backslash M$ must be identity throughout Ω. This is because $\Omega \backslash M$ becomes dense in Ω (in this case) and so every point of Ω is a limit point of $\Omega \backslash M$.

Theorem 5.5.14
If Ω is bounded, and has no isolated boundary points then for every discrete and relatively closed subset M of Ω the isomorphism $\phi : Aut_M \, \Omega \to Aut \, (\Omega \backslash M)$ given by $\phi(f) = f$ restricted to $\Omega \backslash M$, is bijective.

Proof
Theorem 5.5.12 guarantees that ϕ is injective. Thus it suffices to show that ϕ is surjective. Let $f \in Aut \, (\Omega \backslash M)$. Since f and $g = f^{-1}$ map $\Omega \backslash M$ into

$\Omega \backslash M$ and $\Omega \backslash M (\subset \Omega)$ is bounded no point of M can be either a pole or an essential singularity for f or g by Theorem 5.5.11 and as such f and g can be extended to Ω as analytic and injective maps (by the same theorem). Let us call these extensions as $\widetilde{f}$ and $\widetilde{g}$ respectively. Since $\widetilde{f}$ is continuous, $\widetilde{f}(\Omega) \subset \overline{\Omega}$ the closure of Ω. Suppose there were a point $p \in \Omega$ such that $\widetilde{f}(p) \notin \Omega$ but $\widetilde{f}(p) \in$ the boundary of Ω. Necessarily $p \in M$ and we can choose a disc B around p such that $B \backslash \{p\} \subset \Omega \backslash M$. (Note that M is discrete). Since $\widetilde{f}$ is open (being analytic and injective) $\widetilde{f}(B)$ must be an open set containing $\widetilde{f}(p)$. Since $\widetilde{f}$ is injective we now have

$$\widetilde{f}(B) - \widetilde{f}(p) = \widetilde{f}(B \backslash \{p\}) = f(B \backslash \{p\}) \subset \Omega$$

Hence $\widetilde{f}(p)$ becomes an isolated boundary point of Ω (Note that $\widetilde{f}(B)$ is a neighbourhood of $\widetilde{f}(p)$ and except for $\widetilde{f}(p)$ all other points of $\widetilde{f}(B)$ are points of Ω). This however contradicts our assumption on Ω. Thus $\widetilde{f}$ cannot be a boundary point of Ω and so $\widetilde{f}(p) \in \Omega$ itself. Thus $\widetilde{f}(\Omega) \subset \Omega$ and in exactly the same way we also have $\widetilde{g}(\Omega) \subset \Omega$. Hence the continuous extensions $\widetilde{f} \circ \widetilde{g}$ and $\widetilde{g} \circ \widetilde{f}$ are well defined maps of Ω into Ω and in as much as they are identity on a dense subset $\Omega \backslash M$ or Ω (Note that M is discrete and so $\Omega \backslash M$ is dense in Ω). $\widetilde{f} \circ \widetilde{g}$ and $\widetilde{g} \circ \widetilde{f}$ are also identity functions on Ω proving that the extensions $\widetilde{f}$ and $\widetilde{f}$ are members of $Aut (\Omega)$. Finally since $\widetilde{f}(\Omega \backslash M) = f(\Omega \backslash M) = \Omega \backslash M$ it follows that $\widetilde{f}(M) = M$ and so $\widetilde{f} \in Aut_M \Omega$. Hence $\phi(\widetilde{f}) = f$ and indeed we have established that ϕ is onto.

Definition 5.5.15
A region Ω in the plane is called conformally rigid if its only analytic automorphism is the identity map (i.e., $Aut \, \Omega$ reduces to the identity group)

Theorem 5.5.16.

$$Aut \, U = \left\{ \frac{az + b}{\overline{b}z + \overline{a}} : a, \, b \in \mathbb{C} \text{ with } |a|^2 - |b|^2 = 1 \right\}$$

$$= \left\{ e^{i\theta} \frac{z - w}{1 - \overline{w}z} : w \in U, \, 0 \le \theta < 2\pi \right\}$$

Proof
We shall first verify that the two descriptions of *Aut U* given above are

identical as sets. Indeed $\dfrac{az+b}{\bar{b}z+\bar{a}} = \dfrac{a}{\bar{a}}\left(\dfrac{z+\dfrac{b}{a}}{1+\dfrac{\bar{b}z}{\bar{a}}}\right)$ with $a \ne 0$ and $\left|\dfrac{a}{\bar{a}}\right| = 1$,

clearly shows that any element of the first set can be written so as to look

like an element in the second set (Take $e^{i\theta} = \dfrac{a}{\bar{a}}$ and $w = \dfrac{-b}{a}$ with $|w|^2 =$

$\dfrac{|b|^2}{|a|^2} = \dfrac{|a|^2-1}{|a|^2} < 1$). Conversely, $e^{i\theta}\dfrac{z-w}{1-\bar{w}z} = \dfrac{e^{i\theta/2}z - e^{i\theta/2}w}{-\bar{w}e^{-i\theta/2}z + e^{-i\theta/2}}$

$= \dfrac{a'z+b'}{\bar{b}'z+\bar{a}'} = \dfrac{az+b}{\bar{b}z+\bar{a}}$ where $a' = e^{i\theta/2}$, $b' = -we^{i\theta/2}$ and $a = \dfrac{a'}{1-|w|^2}$, $b =$

$\dfrac{b'}{1-|w|^2}$. Hence every element in the second set is a member of the first set.

We now denote these two identical sets as J and observer that J is a

subgroup of *Aut U* (using $h_A \circ h_B = h_{AB} \in J$, $h_I \in J$ and $(h_A)^{-1} = h_{A^{-1}} \in$

J with $A = \begin{pmatrix} a & b \\ \bar{b} & \bar{a} \end{pmatrix}$, $B = \begin{pmatrix} c & d \\ \bar{d} & \bar{c} \end{pmatrix}$, $I = \begin{pmatrix} 1 & 0 \\ 0 & 1 \end{pmatrix}$ and using Example 3.2.7).

On the other hand by Theorem 5.5.7, $Aut_0\ U \subset J$ as can be easily seen. The fact that J acts transitively on U is seen as follows. Let w and ξ in U

be given. $f(z) = \dfrac{z-w}{1-\bar{w}z} \in J$ and $f(w) = 0$. $g(z) = \dfrac{z-\zeta}{1-\bar{\zeta}z} \in J$ and $g(\zeta)$

$= 0$. Thus $g^{-1} \circ f \in J$ and takes w to ζ. Thus by Theorem 5.5.10, $J = Aut\ U$.

Theorem 5.5.17

$$Aut\ H = \left\{ h_{C'} \circ \dfrac{az+b}{\bar{b}z+\bar{a}} \circ h_C : a,\ b \in \mathbb{C},\ |a|^2 - |b|^2 = 1 \right\}$$

$$= \left\{ \dfrac{\alpha z + \beta}{\gamma z + \delta} : \begin{pmatrix} \alpha & \beta \\ \gamma & \delta \end{pmatrix} \in SL\ (2,\ \mathbb{R}), \right\}$$

Proof

Using the fact (see Theorem 5.5.5.) that h_C is a bijective analytic map of U onto H and Theorem 5.5.9 we see that $Aut\ H = \{h_C \circ f \circ h_C : f \in Aut\ U\}$ and this completes the proof of the first part of our Theorem. We now verify that this is identical with the other description given for *Aut H*.

Indeed if $A = \begin{pmatrix} a & b \\ \bar{b} & \bar{a} \end{pmatrix}$ then a matrix representing an element of *Aut H* can be written as $C'AC = \begin{pmatrix} i & i \\ -1 & 1 \end{pmatrix} \begin{pmatrix} a & b \\ \bar{b} & \bar{a} \end{pmatrix} \begin{pmatrix} 1 & -i \\ 1 & i \end{pmatrix} = \frac{1}{2i} \begin{pmatrix} \alpha & \beta \\ \gamma & \delta \end{pmatrix}$, where α

$= Re\ (a + b)$, $\beta = Im\ (a - b)$, $\gamma = -Im\ (a + b)$, and $\delta = Re\ (a - b)$.

Thus $h_{C'AC} = h_B$ with $B = \begin{pmatrix} \alpha & \beta \\ \gamma & \delta \end{pmatrix} \in SL(2, \mathbb{R})$. Conversely given $\begin{pmatrix} \alpha & \beta \\ \gamma & \delta \end{pmatrix}$

$\in SL(2, \mathbb{R})$, we can always write $B = \begin{pmatrix} \alpha & \beta \\ \gamma & \delta \end{pmatrix} = C'\ AC$ where

$A = C\begin{pmatrix} \alpha & \beta \\ \gamma & \delta \end{pmatrix}$ $C' = i\begin{pmatrix} \alpha+\delta+i(\beta-\gamma) & \alpha-\delta-i(\beta+\gamma) \\ \alpha-\delta+i(\beta+\gamma) & \alpha+\delta-i(\beta-\gamma) \end{pmatrix}$. Thus

$h_B = h_C h_A h_C = h_C\begin{pmatrix} a & b \\ \bar{b} & \bar{a} \end{pmatrix} h_C$ with $a = \frac{1}{2}(\alpha + \delta + i(\beta - \gamma))$

$b = \frac{1}{2}(\alpha - \delta) - i(\beta + \gamma))$ so that $|a|^2 - |b|^2 = a\delta - \beta\gamma = 1$. The proof of

our theorem is now complete.

Now we characterize *Aut* $\mathbb{C}$ and *Aut* $\mathbb{C}^*$ where $\mathbb{C}^* = \mathbb{C}\backslash\{0\}$.

Lemma 5.5.18
If $f : \mathbb{C} \to \mathbb{C}$ is injective and analytic then $f(z) = az + b$ for all $z \in \mathbb{C}$ and for some $a \neq 0$ in $\mathbb{C}$.

Proof

Define g $: \mathbb{C} \to \mathbb{C}$ by $g(z) = f\left(\frac{1}{z}\right)$. It is clear that g is injective and analytic

in $\mathbb{C}^*$. Hence by Theorem 5.5.11, $z = 0$ is a non-essential singularity for g. i.e., $z = \infty$ is a non-essential singularity for f. But since f is analytic in the whole plane, f reduces to a polynomial, (See Chapter 4 solved exercise 20). But f is injective implies f' never vanishes (note that by Theorem 4.4.22, $f'(z_0) = 0 \Rightarrow f$ is one-many in a neighbourhood of z_0). But any non-constant polynomial should have at least one zero by the fundamental theorem of algebra (See Corollary 4.4.9). Hence $f'(z)$ is constant or that $f(z)$ is a linear polynomial of the form $az + b$ with $a, b \in \mathbb{C}$. $a \neq 0$ follows from the injectivity of f.

Theorem 5.5.19
Aut $\mathbb{C} = \{az + b: a, b \in \mathbb{C}, a \neq 0\}$.
Since every function of the form $az + b$ $(a \neq 0) \in$ *Aut* $\mathbb{C}$ the theorem follows from Lemma 5.5.18.

Lemma 5.5.20
If $f : \mathbb{C}^* \to \mathbb{C}^*$ is injective and analytic then $f(z) = az$ for all $z \in \mathbb{C}^*$ and for some $a \neq 0$ in $\mathbb{C}$ or $f(z) = az^{-1}$ $(a \neq 0)$.

Proof

By Theorem 5.5.11, $z = 0$ is a non-essential singularity for f.

Case 1. $z = 0$ is a removable singularity for f. In this case its extension $\tilde{f} : \mathbb{C} \to \mathbb{C}$ is also injective and analytic. (again by Theorem 5.5.11). By Lemma 5.5.18, $\tilde{f}(z) = az + b(a \neq 0)$. But $f(\mathbb{C}^*) \subset \mathbb{C}^*$ and $f\left(\dfrac{-b}{a}\right) = 0$ $\Rightarrow b = 0$. Thus $f(z) = az \; \forall z \in \mathbb{C}, \; (a \neq 0)$.

Case 2 $z = 0$ is a pole for f. Define $g(z) = \dfrac{1}{f(z)}$ so that $g : \mathbb{C}^* \to \mathbb{C}^*$ is analytic and further $z = 0$ is a removable singularity for $g(z)$ (in fact $z = 0$ is also a zero for g). Thus by Case 1, $g(z) = bz \; b \neq 0$ or that $f(z) = az^{-1} \; (a = b^{-1} \neq 0)$.

Theorem 5.5.21

Aut $\mathbb{C}^* = \{f(z) = az \; a \neq 0 \text{ or } f(z) = az^{-1}, \; a \neq 0\}$.

Proof

Since every function $f(z) = az$ or $az^{-1} \; (a \neq 0))$ belongs to *Aut* C*, our theorem follows from Lemma 5.5.20.

Theorem 5.5.22

Let $U^* = U\backslash\{0\}$.

Then *Aut* $U^* = \{f : U^* \to U^* : f(z) = az; \; |a| = 1\}$.

Proof

By Theorem 5.5.14, (with $\Omega = U$ and $M^* = \{0\}$) *Aut* U^* is isomorphic to $Aut_0 U$ which inturn equals the set described here by Theorem 5.5.7.

Theorem 5.5.23

Let A be a finite non-empty subset of U^* and *Perm* $(A \cup \{0\})$ be the permutation group of the finite set $(A \cup \{0\})$. There exists a natural monomorphism $\pi : Aut \; (U^*\backslash A) \to Perm \; (A \cup \{0\})$.

Proof

Since $U^*\backslash A = U\backslash(A \cup \{0\})$ we have by Theorem 5.5.14. *Aut* $(U^*\backslash A) = Aut_{A\cup\{0\}} U$. Hence every automorphism f of $U^*\backslash A$ maps $A \cup \{0\}$ bijectively onto itself thus inducting a permutation $\pi(f)$ on $A \cup \{0\}$. It is now easy to verify that the map $f \mapsto \pi(f)$ is a group homomorphism $\pi : Aut \; (U^*\backslash A) \to Perm \; (A \cup \{0\})$. Since any non-identity automorphism of U can fix at most one point only (Theorem 5.5.8) and $A \neq \phi$ we see that π is injective. This completes the proof.

Theorem 5.5.24

For each $c \in U^*$, *Aut* $(U^*\backslash\{c\})$ is isomorphic to the cyclic group on two symbols $\{e, a\}$ with $a^2 = e$ and the map $f(z) = \dfrac{z-c}{\bar{c}z-1}$ is the only non-identity element of *Aut* $(U^*\backslash\{c\})$.

Proof

By Theorem 5.5.23, *Aut* $(U^*\backslash\{c\})$ is isomorphic to a subgroup of the permutation group on two symbols $\{0, c\}$. (Note that the permutation group on two symbols is isomorphic to the cyclic group $\{e, a\}$ with $a^2 = e$). However f defined above is a member of *Aut U*, interchanging 0 and c and its restriction to $U^*\backslash\{0\}$ is an automorphism of $U^*\backslash\{c\}$ and is different from the identity since $c \neq 0$.

Theorem 5.5.25

Suppose $a, b \in U^*$, $a \neq b$. Then *Aut* $(U^*\backslash\{a, b\}) \neq \{id\}$ if and only if atleast one of the following relations hold good. (i) $a = -b$ (ii) $2b = a + \bar{a} b^2$ (iii) $2a = b + \bar{b} a^2$ (iv) $|a| = |b|$ and $a^2 + b^2 = ab(1 + |b|^2)$.

Proof

Since *Aut* $(U^*\backslash\{a, b\}) = Aut_{\{0, a, b\}} U$, any $f \in Aut_{\{0, a, b\}} U$ which is not the identity should be of the form $f(z) = e^{i\psi}\left(\dfrac{z - w}{1 - \bar{w}z}\right)$ for a suitable $w \in U$ and

$0 \leq \psi < 2\pi$ (See Theorem 5.5.16). Further since f induces a permutaion on $\{0, a, b\}$ the following are the only possibilities.

    (i) $f(0) = 0$, $f(a) = b$, $f(b) = a$
    (ii) $f(b) = 0$, $f(a) = b$, $f(0) = a$
    (iii) $f(b) = 0$, $f(0) = b$, $f(a) = a$
    (iv) $f(a) = 0$, $f(b) = a$, $f(0) = b$
    (v) $f(a) = 0$, $f(0) = a$, $f(b) = b$

In case (i) by Theorem 5.5.7, $f(z) = cz$, $|c| = 1$ and $ca = b$, $cb = a \Rightarrow$ $c = -1$ and $a = -b$ (Note that $f(z) \neq z$). In case (ii) or (iii), $f(z) =$

$e^{i\psi}\left(\dfrac{z - b}{1 - \bar{b}z}\right)$, $f(0) = a$ (i.e., $-be^{i\psi} = a$) and $b = \dfrac{-a}{b}\left(\dfrac{a - b}{1 - \bar{b}a}\right)$ or $f(0) =$

b (i.e., $b = -be^{i\psi}$) and $a = -\left(\dfrac{a - b}{1 - \bar{b}a}\right)$. i.e., either $|a| = |b|$ and $b^2 - a|b|^2 b$

$= ab - a^2$ (ie $a^2 + b^2 = ab(1 + |b|^2)$ or $2a = b + \bar{b} a^2$. Similarly in cases (iv) or (v) we have by a simple computation, either $|a| = |b|$ and $a^2 + b^2$ $= ab(1 + |a|^2) = ab(1 + |b|^2)$ or $2b = a + \bar{a}b^2$.

Theorem 5.5.26

If $f : \mathbb{C}^* \to \mathbb{C}$ is injective and analytic then $f(z) = az + b$ $(a \neq 0)$ or $f(z)$ $= \dfrac{a}{z} + b$ $(a \neq 0)$. Hence $f(\mathbb{C}^*) = \mathbb{C}^*\backslash\{b\}$ for some $b \in \mathbb{C}$.

Proof

By Theorem 5.5.11, both $f(z)$ and $g(z) = f\left(\dfrac{1}{z}\right)$ have a non-essential singularity at $z = 0$, i.e., $z = 0$ and $z = \infty$ are both non-essential singularities for $f(z)$.

Case 1. $z = 0$ is a removable singularity for f (This includes both the cases that f has either a removable singularity or a pole at $z = \infty$). Again by Theorem 5.5.11, $\widetilde{f}$ the extension of f to $\mathbb{C}$ is entire and is also injective on $\mathbb{C}$. Hence by Lemma 5.5.18, $\widetilde{f}(z) = f(z) = az + b \ \forall z \in \mathbb{C}^*$.

Case 2. $z = 0$ is a pole of order 1 and ∞ is a removable singularity for f. In this case $g(z) = f\left(\dfrac{1}{z}\right)$ has a removable singularity at $z = 0$ and a pole of order 1 at $z = \infty$. In this case its extension $\widetilde{g} : \mathbb{C} \to \mathbb{C}$ is analytic and injective on $\mathbb{C}$. Thus by Lemma 5.5.18, $\widetilde{g}(z) = g(z) = az + b \ \forall z \in \mathbb{C}^*$ $(a \neq 0)$ or that $f(z) = \dfrac{a}{z} + b \ (a \neq 0)$.

Case 3. $z = 0$ is a pole of order 1 and ∞ is also a pole of order 1. We shall show that this case cannot happen. Indeed if f has simple poles both at $z = 0$ and $z = \infty$ then by Laurent expansion (valid for $0 < |z| < \infty$)

$$f(z) = \frac{a_1}{z} + b_1 + c_1 z + c_2 z^2 + \dots$$

and $g(z) = f\left(\dfrac{1}{z}\right) = \dfrac{a_2}{z} + b_2 + d_1 z + d_2 z^2 + \dots$ or that $g\left(\dfrac{1}{z}\right) = f(z) = a_2 z$

$+ b_2 + \dfrac{d_1}{z} + \dfrac{d_2}{z^2} + \dots$. Consequently by uniqueness of Laurent expansion we must have $a = a_1 = d_1$, $b = b_1 = b_2$ and $c_i = d_i = 0 \ (i \geq 2)$. Thus $f(z)$ $= \dfrac{a}{z} + b + cz \ (0 < |z| < \infty)$. Now $a \neq 0$ as otherwise f has no pole at $z = 0$. Also since f is injective on $\mathbb{C}^*$, $f'(z) \neq 0$ in $\mathbb{C}^*$ and this in turn implies $c = 0 \ (c \neq 0 \Rightarrow f\left(\pm\sqrt{\dfrac{a}{c}}\right) = 0$ But then $z = \infty$ is not a pole contradicting our assumption. This contradictions shows that this case can not occur. Summing up all these we see that either $f(z) = az + b \ (a \neq 0)$ or $f(z) = a/z + b \ (a \neq 0)$ as required. Further in both cases $f(\mathbb{C}^*) = \mathbb{C}^*\backslash\{b\}$. This completes the proof.

Theorem 5.5.27
The group $Aut\,(\mathbb{C}\backslash\{0, 1\})$ consists only of the following six functions.

$$f_1(z) = z, \ f_2(z) = \frac{1}{z}, \ f_3(z) = 1 - z,$$

$$f_4(z) = \frac{1}{1-z}, \ f_5(z) = \frac{z}{z-1}, \ f_6(z) = \frac{z-1}{z}$$

Proof

By Theorem 5.5.11, both $z = 0$ and $z = 1$ are non-essential singularities for f.

Case 1 Both 0 and 1 are removable. In this case as before, the extension of f, $\tilde{f} : \mathbb{C} \to \mathbb{C}$ is analytic and injective and hence $\tilde{f}(z) = az + b$ and $\tilde{f}$ fixes the set $\{0, 1\}$. Further $\tilde{f}(0) = 1$ and $\tilde{f}(1) = 0$ or $\tilde{f}(0) = 0$ and $\tilde{f}(1) = 1$. In the former case $b = 1$, $a = -1$ and $f(z) = f_3(z)$. In the latter case $b = 0$, $a = 1$ and $f(z) = z = f_1(z)$.

Case 2 $z = 1$ is a removable singularity and $z = 0$ is a pole. f gets extended to $\mathbb{C} \backslash \{0\}$ and is injective and so $\tilde{f} : \mathbb{C}^* \to \mathbb{C}$ reduces to $\tilde{f}(z) = az + b$ or $\tilde{f}(z) = \dfrac{a}{z} + b$. But in as much as f has a pole at $z = 0$, $\tilde{f}(z) = \dfrac{a}{z} + b$ only. If $\tilde{f}(1) = d$ then $d = 0$ or 1 only. (if not $\tilde{f}(c) = f(c) = d$ for some $c \in \mathbb{C} \backslash \{0,1\}$ and we can find disjoint neighbourhoods of 1 and c getting mapped onto a small neighbourhood of d under $\tilde{f}$ contradicting the injectivity of $\tilde{f}$. If $\tilde{f}(1) = 0$, $a = -b$ and $\tilde{f}(z) = a\left(\dfrac{1}{z} - 1\right)$. Further if $a + 1 \neq 0$ then at $z = \dfrac{a}{a+1} \in \mathbb{C} \backslash \{0, 1\}$ we have $\tilde{f}(z) = f(z) = 1$, a contradiction. (note that $f(\mathbb{C}^* \backslash \{0, 1\}) = \mathbb{C}^* \backslash \{0, 1\}$. Thus $a = -1$ and $f(z) = \tilde{f}(z) = f_6(z)$. On the other hand if $\tilde{f}(1) = 1$ then $a + b = 1$ and $\tilde{f}(z) = \dfrac{a}{z} + (1 - a)$. In this case if $a \neq 1$, at $z = \dfrac{a}{a-1} \neq 0$ or 1 and $f(z) = \tilde{f}(z) = 0$, again a contradiction (note that $f(\mathbb{C}^*/\{0, 1\}) = \mathbb{C}^*/\{0, 1\}$). Thus $a = 1$ and $\tilde{f}(z) = f(z) = \dfrac{1}{z} = f_2(z)$ for all $z \in \mathbb{C} \backslash \{0, 1\}$.

Case 3 $z = 1$ is a pole and $z = 0$ is a removable singularity. Consider $g(z) = f(1 - z) \in Aut\,(\mathbb{C}^* \backslash \{0, 1\})$. g has $z = 0$ as a pole and $z = 1$ as a removable singularity. By case 2, $g(z) = \dfrac{1}{z}$ or $1 - \dfrac{1}{z}$ or that $f(z) = \dfrac{1}{1-z} = f_4(z)$ or $f(z) = 1 - \dfrac{1}{1-z} = \dfrac{z}{z-1} = f_5(z)$. Theoritically there is still a last case in which $z = 0$ and $z = 1$ are both poles of order 1 for $f(z)$. We shall show that this case cannot exist. Indeed if this were the case then $g(z) = \dfrac{1}{f(z)} \in Aut\,(\mathbb{C} \backslash \{0,1\})$ and has both $z = 0$ and $z = 1$ as removable singularities (with extended values $g(1) = 0$, $g(0) = 0$). But then by case 1 $g(z) = z$ or $1 - z$. But this is impossible since $g(0) = 0$ and $g(1) = 0$.

5.6 Univalent Functions Defined on the Open Unit Disc

Introduction 5.6.1

The theory of conformal mappings on simply connected regions is of special significance from the point of view of Geometric function theory. Essentially there are three types of simply connected spaces in the extended complex plane (as a consequence of the Riemann mapping theorem).

 (i) the open unit disc
 (ii) the entire finite complex plane
 (iii) the Riemann sphere or the extended complex plane

(i) is referred to as hyperbolic case whereas (ii) and (iii) are referred to as parabolic and elliptic cases respectively. (i) and (ii) are non-compact cases and (iii) is the only compact case. Function theory on each of these cases are handled separately and investigated exhaustively in the literature and they together form a fascinating branch of Mathematics.

A simple geometric restriction namely the injectivity imposed on functions defined on the unit disc throws a wealth of information on the geometric and analytic properties of such functions. The name univalent functions or Schlicht functions is given to a set of functions defined on the open unit disc U of the complex plane which is characterized by the fact that it takes in U a value not more than once and which (as a consequence) maps U onto a Schlicht domain (a German word indicating a region which is not self overlapping and containing no branch points). The starting point in the investigation of Schlicht functions was a paper by P. Koebe in 1907 on the uniformization of algebraic curves. (P. Koebe, "Uber die Uniformisierung beliebiger analytischer kurven", Nachr. Akad. Wiss. Göttingen, Math.–Phys. Kl., 1907, 191-210). He proves in particular that there is a constant $k > 0$ such that the boundary of the image of $|z| < 1$ by any function $w = f(z)$ which is one-to-one and analytic in U with $f(0) = 0 = 1 - f'(0)$ is always at a distance not less than k from w = 0. Another related result is that there exist bounds for the modulus of the derivative of f at any point in U which depend only on $|z|$.

Koebe's result soon attracted the attention of many others. (Plemelj, Gronwall, Pick, Faber, Bieberbach etc.). Gronwall was the first to give the so called "area principle". At around that time Bieberbach proposed the so called "coefficient conjecture for Univalent functions". The research works in these directions have enriched the literature on function theory for decades later and it is still an area of current interest.

In this section we shall introduce the theory of Univalent functions in an attempt to generate interest among the readers. The choice of the unit disc as the standard frame of reference for studying function theory on simply connected regions has the added advantage of simplifying computations and leading to short and elegant formulas.

Definition 5.6.2

The class $\mathcal{F}$ denotes the class of analytic functions in U (the open unit disc) normalized by $f(0) = 0, f'(0) = 1$ which are also injective or univalent, (i.e., $f(z_1) = f(z_2) \Rightarrow z_1 = z_2$ in U).

Remark 5.6.3

If f is analytic and univalent in U, then it is always true that $f'(0) \neq 0$ (In fact f' never vanishes) and for this reason $f_1(z) = \dfrac{f(z) - f(0)}{f'(0)}$ is also univalent and analytic in U satisfying $f_1(0) = 0$ and $f_1'(0) = 1$. Thus there is no loss of generality in assuming this normalization. Further to every f in $\mathcal{F}$ we have a power series expansion $f(z) = z + \sum_{2}^{\infty} a_n z^n$ which is valid throughout U. (This is an immediate consequence of Taylor series development).

Note 5.6.4

Observe that the class $\mathcal{F}$ is not closed under either addition or multiplication as

$$(f + g)'\,(0) = 2 \text{ and } (fg)'\,(0) = 0 \text{ if } f, g \in \mathcal{F}.$$

Example 5.6.5

Let $\alpha \in \mathbb{C}$ with $|\alpha| = 1$. Let

$$f_\alpha(z) = \frac{z}{(1 - \alpha z)^2} = \sum_{1}^{\infty} n\alpha^{n-1}\, z^n.$$

then $f_\alpha(z) \in \mathcal{F}$. This follows because f_α is analytic in U as the only singularity (a pole) is at $z = (1/\alpha) \notin U$. Further $f_\alpha(0) = 0$ and $f_\alpha'(0) = 1$. Also

$$f_\alpha(z) = f_\alpha(w) \Rightarrow (z - w)\,(1 - \alpha^2 zw) = 0 \Rightarrow z - w = 0$$

as $|z| < 1$, $|w| < 1$. Note that the image of U under f_α can be found as follows: We already saw in chapter 3, (Solved Exercises, Problem 2) that

$$f(z) = \frac{z}{(1 - z)^2}$$ maps $|z| < 1$ onto the whole plane except for the slit from $-1/4$ to $-\infty$ along the negative real axis. Now

$$f_\alpha(z) = \frac{1}{\alpha}\, f(\alpha z).$$

Also since $|\alpha| = 1$, as z varies over $|z| = 1$, αz also varies over the unit circle and so $f(\alpha z)$ maps $|z| < 1$ onto the whole plane except for the same slit. Now $\dfrac{1}{\alpha}\, f(\alpha z)$ maps $|z| < 1$ onto the whole plane except for the new slit from $-1/4\alpha$ to or ∞. These functions (for various values of α) serve as "extremal functions" in various problems as we shall see later.

Theorem 5.6.6

(i) If $f \in \mathcal{F}$, $|\alpha| = 1$ and $g(z) = \overline{\alpha}f(\alpha z)$ then $g \in \mathcal{F}$.

(ii) If $f \in \mathcal{F}$ there exists $g \in \mathcal{F}$ such that $g^2(z) = f(z^2)$, $\forall z \in U$.

Proof

(i) f is analytic in U implies g is analytic in U. ($z \to \alpha z$ is analytic from U onto U as $|\alpha| = 1$). Further $g(0) = \overline{\alpha}f(0) = 0$ and $g'(0) = \overline{\alpha}f'(0)\alpha = 1$. It is clear that

$$g(z_1) = g(z_2) \Rightarrow f(\alpha z_1) = f(\alpha z_2) \Rightarrow \alpha z_1 = \alpha z_2 \Rightarrow z_1 = z_2.$$

Hence $g \in \mathcal{F}$.

(ii) Let $\phi(z) = \dfrac{f(z)}{z}$. ϕ is analytic for every $z \neq 0$ in U. Further $\lim\limits_{z \to 0} z\phi(z)$

$= \lim\limits_{z \to 0} f(z) = f(0) = 0$ and so 0 is a removable singularity for ϕ. Also the

extended value of ϕ at 0 is $\phi(0) = \lim\limits_{z \to 0} \dfrac{f(z)}{z} = \lim\limits_{z \to 0} \dfrac{f(z) - f(0)}{z - 0} = f'(0) = 1$.

Hence $\phi(0) \neq 0$ and $\phi(z) \neq 0$ for $z \neq 0$ by injectivity of f and the fact $f(0) = 0$. Thus $\phi(z)$ has an analytic square root in U (U is a simply connected region in which ϕ is analytic and never vanishes). Call $h(z) = \sqrt{\phi(z)}$ with $h(0) = 1$, i.e., $h^2(z) = \phi(z)$. Put $g(z) = zh(z^2)$. We claim that g is a required function. First $z \to z^2$ is analytic, h is analytic and multiplication by z is also analytic. Thus g is analytic in U. $g(0) = 0 \times 1 = 0$ and $g'(0) = [zh'(z^2) 2z + h(z^2)]_{z = 0} = 1$. We now prove that g is one-to-one in U. Suppose $g(z) = g(w)$, z, w in U, i.e., $g^2(z) = g^2(w)$ or that $z^2h^2(z^2) = w^2h^2(w^2)$, i.e. $z^2 \phi(z^2) = w^2 \phi(w^2)$ or that $f(z^2) = f(w^2)$. If z, w are both 0 then $z = w$. If $z = 0$ and $w \neq 0$ we get $f(w^2) = 0$ which by injectivity of f implies $w^2 = 0$ or $w = 0$ a contradiction.

Thus we can assume z, $w \neq 0$ and prove that $z = w$. $f(z^2) = f(w^2)$ implies $z^2 = w^2$ as f is injective, i.e., $z = \pm w$. We claim that $z = -w$ can not hold good. Indeed if $z = -w$, $g(z) = zh(z^2)$. Therefore $g(w) = wh(w^2) = -zh(z^2) = -g(z)$. But $g(z) = g(w)$ and so $g(z) = 0 = g(w)$. But $g(z) = zh(z^2)$ and $h(z)$ is never zero. Thus $z = 0 = w$ a contradiction. Hence $g \in \mathcal{F}$ and for

$z \neq 0$, $g^2(z) = z^2h^2(z^2) = z^2\phi(z^2) = z^2\dfrac{f(z^2)}{z^2} = f(z^2)$. However, for $z = 0$,

$g^2(0) = 0 = f(0^2)$. Therefore $g^2(z) = f(z^2) \; \forall \; z \in U$.

Theorem 5.6.7 (Area Theorem)

Let $F(z)$ be analytic in $U \backslash \{0\}$. Let F be one-to-one in U with the Laurent

expansion at the origin being $F(z) = \dfrac{1}{z} + \sum\limits_{n=0}^{\infty} b_n z^n$ then $\sum\limits_{1}^{\infty} n|b_n|^2 \leq 1$.

Proof

The name Area theorem for the above result is due to the fact that the inequality expressed by the conclusion is but an analytic expression of the geometrical fact that the complement E of $F(U\backslash\{0\})$ in the plane has an area which is non-negative.

To start the proof we shall consider the image C_r^* of C_r (C_r is the circle $|z| = r$ ($0 < r < 1$) under $F(z)$). The curve C_r^* is an analytic curve. Introducing polar coordinates R and ϕ in the $w = F(z)$ plane we observe the following.

(1) The image of $0 < |z| < r$ lies in the exterior of C_r^* (the image of this connected set lies in a single component of the complement of C_r^* and is unbounded).

(2) If C_r is positively oriented then C_r^* is negatively oriented (by Theorem 4.6.20).

(3) The area enclosed by C_r^* is given by $A_r = \dfrac{1}{2} \int\limits_{-C_r^*} R^2 d\phi$.

(Note that $-C_r^*$ is positively oriented). Now $R^2 = |F(re^{i\theta})|^2$ where r and θ are the polar coordinates of points z on C_r. Now since r is fixed using Cauchy-Riemann equations we have

$$d\phi = \frac{\partial \phi}{\partial r} dr + \frac{\partial \phi}{\partial \theta} d\theta = \frac{\partial \phi}{\partial \theta} d\theta = \frac{r}{R} \frac{\partial R}{\partial r} d\theta$$

$$A_r = \frac{r}{2} \int\limits_{-C_r} R \frac{\partial R}{\partial r} d\theta$$

$$= -\frac{r}{4} \int\limits_0^{2\pi} \frac{\partial}{\partial r}(R^2)\, d\theta.$$

(Note that $\dfrac{\partial}{\partial r}(R^2)$ denotes the partial derivative of R^2 with respect to r considered as a function of r and is evaluated at our chosen r and so becomes a function of θ only). By an argument similar to the one given in Chapter **4**, Solved exercise 42, we can conclude that

$$\int\limits_0^{2\pi} \frac{\partial}{\partial r}(R^2)\, d\theta = \frac{\partial}{\partial r} \int\limits_0^{2\pi} R^2 d\theta.$$

Hence

$$A_r = -\frac{r}{4} \frac{\partial}{\partial r} \int\limits_0^{2\pi} R^2 d\theta$$

$$= -\frac{r}{4}\frac{\partial}{\partial r}\int_0^{2\pi} |F(re^{i\theta})|^2 \ d\theta$$

$$= -\frac{r}{4}\frac{\partial}{\partial r}\int_0^{2\pi} F(re^{i\theta}) \ \overline{F(re^{i\theta})} \ d\theta$$

$$= -\frac{r}{4}\frac{\partial}{\partial r}\int_0^{2\pi} \left(\frac{1}{re^{i\theta}} + \sum_0^{\infty} b_n r^n e^{in\theta}\right)\left(\frac{1}{re^{-i\theta}} + \sum_0^{\infty} \overline{b}_m r^m e^{-im\theta}\right) d\theta .$$

Now we can multiply the infinite series and integrate it term by term since the series is uniformly convergent for $0 < r < 1$. We also note that

$$\int_0^{2\pi} e^{ik\theta} d\theta = \begin{cases} 0, & k \neq 0 \\ 2\pi, & k = 0 \end{cases}.$$

Thus

$$A_r = -\frac{r}{4}\frac{\partial}{\partial r}\int_0^{2\pi}\left(\frac{1}{r^2} + \sum_0^{\infty}|b_n|^2 r^{2n}\right) d\theta$$

$$= -\frac{r}{4}\frac{\partial}{\partial r} 2\pi\left(\frac{1}{r^2} + \sum_0^{\infty}|b_n|^2 r^{2n}\right)$$

$$= \frac{\pi}{2}\left(\frac{2}{r^2} - \sum_0^{\infty}|b_n|^2 2nr^{2n}\right)$$

$$= \pi\left(\frac{1}{r^2} - \sum_0^{\infty}n|b_n|^2 r^{2n}\right).$$

The evident geometric fact is that $A_r \geq 0$ which means

$$\frac{1}{r^2} \geq \sum_0^{\infty}n|b_n|^2 r^{2n} .$$

Since this is valid for any $0 < r < 1$, we allow $r \to 1$ and get $1 \geq \sum_0^{\infty}n|b_n|^2$.

Corollary 5.6.8

Let F be analytic in $U\setminus\{0\}$. F be one-to-one in U with Laurent expansion around origin being $F(z) = \frac{1}{z} + \sum_0^{\infty} b_n z^n$. Then $\sum_0^{\infty}n|b_n|^2 = 1$ if and only if the area of the set of points in the complement of the image of $0 < |z| < 1$ under F is precisely 0.

Proof

From the proof of the Theorem 5.6.7 it follows that the set of points enclosed by C_r^* which is the image of $|z| = r$ $(0 < r < 1)$ under F consists precisely of those points which are not in the image of $0 < |z| < r$ under F. (Since F gives a one-to-one correspondence between $0 < |z| < r$ and the exterior of C_r^*). Letting $r \to 1$ it follows that $\lim_{r \to 1} A_r$ is equal to the area of the set of points in the complement of the image of $0 < |z| < 1$ under F. But

$$\lim_{r \to 1} A_r = \pi \lim_{r \to 1} \left(\frac{1}{r^2} - \sum_1^\infty n |b_n|^2 \, r^{2n} \right)$$

$$= \pi \left(1 - \sum_1^\infty n |b_n|^2 \right) = 0$$

by our hypothesis. Hence the result.

Corollary 5.6.9
Let F be analytic in $U \backslash \{0\}$. Let F be one-to-one in U with Laurent expansion around origin being $F(z) = \frac{1}{z} + \sum_0^\infty b_n z^n$. Then $|b_1| \le 1$ with equality if and only if $F(z) = \frac{1}{z} + b_0 + b_1 z$ with $|b_1| = 1$. Further this function F for which $|b_1| = 1$ maps $U \backslash \{0\}$ bijectively and conformally onto the complement of a line segment of length 4.

Proof

We know that $\sum_0^\infty n |b_n|^2 \le 1$ which implies $|b_1|^2 \le 1$ or that $|b_1| \le 1$. On the other hand $|b_1| = 1$ if and only if $b_n = 0$ for $n \ge 2$ and hence $F(z) = \frac{1}{z} + b_0 + b_1 z$ with $|b_1| = 1$. Since b_0 is a constant, the image of $U \backslash \{0\}$ under F will be a translate of the image of $U \backslash \{0\}$ under G, where $G(z) = \frac{1}{z} + b_1 z$. Now G maps $U \backslash \{0\}$ onto the complement of $G(T)$ where T is the unit circle (see the proof Theorem 5.6.7 and Theorem 4.6.20). Putting $z = e^{i\theta}$ we see that the image of T under G is given by $\Gamma(\theta) = e^{-i\theta} + e^{2i\gamma} e^{i\theta}$ where $b_1 = e^{2i\gamma}$ (say) and θ varies in the interval $[0, 2\pi]$. Now $e^{-i\theta} + e^{2i\gamma + i\theta} = 2e^{i\gamma} \cos(\theta + \gamma)$ and $\cos(\theta + \gamma)$ assumes all the values in $[-1, 1]$ for $\theta \in [0, 2\pi]$. Thus $G(T)$ is nothing but the line joining $-2e^{i\gamma}$ and $2e^{i\gamma}$ in the plane whose length is 4. Thus G maps $U \backslash \{0\}$ onto the complement of a line segment of length 4. Since F is a translate of G, the same is true of F also.

Theorem 5.6.10

If $f(z) = z + \sum_2^\infty a_n z^n \in \mathcal{F}$ then $|a_2| \le 2$ with equality if and only if $f(z) =$

$\dfrac{z}{(1-\alpha z)^2}$ with $|\alpha| = 1$. Further $f(U)$ contains all w with $|w| < 1/4$ and

the constant $1/4$ can not be improved for all $f \in \mathcal{F}$.

Proof

By Theorem 5.6.6 there exists $g \in \mathcal{F}$ such that $g^2(z) = f(z^2)$. g is one-to-one in U with

$$G(z) = \frac{1}{g(z)} = \frac{1}{\sqrt{f(z^2)}} = \frac{1}{z}\left(1 + \sum_2^\infty a_n z^{2n-2}\right)^{-1/2}$$

$$= \frac{1}{z}\left(1 - \frac{1}{2}(a_2 z^2 + ...)\right)$$

$$= \frac{1}{z} - \frac{1}{2}a_2 z + . . .$$

Thus G satisfies the hypothesis of Corollary 5.6.9 with $b_1 = -a_2/2$. Hence $|b_1| = |a_2/2| \le 1$ with equality if and only if $G(z) = \dfrac{1}{z} - \alpha z$, $(\alpha = a_2/2)$.

Thus $f(z^2) = g^2(z) = \dfrac{1}{G^2(z)} = \dfrac{z^2}{(1-\alpha z^2)^2}$. Therefore $f(z) = \dfrac{z}{(1-\alpha z)^2}$

with $|\alpha| = 1$. To prove the concluding part of the theorem we assume that

$w \notin f(U)$ and define $h(z) = \dfrac{wf(z)}{w - f(z)}$. Now h is a composition of f and

a linear fractional transformation and so is one-to-one in U. Further h is analytic in U as $f(z) \ne w$. Further the Taylor expansion of $h(z)$ around origin is given by

$$h(z) = (z + a_2 z^2 + . . .)\left(1 - \frac{z + a_2 z^2 + ...}{w}\right)^{-1}$$

$$= z + \left(a_2 + \frac{1}{w}\right)z^2 + . . .$$

Thus $h \in \mathcal{F}$ with coefficient of z^2 being $a_2 + \dfrac{1}{w}$. By the previous Theorem

$\left|a_2 + \dfrac{1}{w}\right| \le 2$ or $\left|\dfrac{1}{w}\right| \le 2 + |a_2| \le 4$. Thus $w \notin f(U)$ implies $|w| \ge \dfrac{1}{4}$ or

that $|w| < \dfrac{1}{4}$ implies $w \in f(U)$. The example $f(z) = \dfrac{z}{(1-z)^2}$ shows that

$w = -\dfrac{1}{4}$ is not assumed by $f(z)$. (The only solution of $f(z) = -\dfrac{1}{4}$ is $z = -$

$1 \notin U$). Therefore there exists no number $a > \dfrac{1}{4}$ such that every f in $\mathcal{F}$

assumes all values in $|w| < a$.

Theorem 5.6.11

Let F be analytic in $U\backslash\{0\}$ and one-to-one in U with $F(z) = \dfrac{1}{z} + \sum_0^\infty b_n z^n$. If w_1 and w_2 are not in $F(U\backslash\{0\})$ then $|w_1 - w_2| \leq 4$. The constant 4 cannot be replaced by a smaller number.

Proof

Put $f = \dfrac{1}{F - w_1}$ so that f is analytic in $U\backslash\{0\}$ with removable singularity

at origin and the extended value at the origin being zero. (F has a pole at

origin implies $F - w_1$ has a pole at origin or that $\dfrac{1}{F - w_1}$ has a zero at origin).

Further

$$f'(0) = \lim_{z \to 0} \frac{f(z)}{z}$$

$$= \lim_{z \to 0} \frac{1}{z(F(z) - w_1)}$$

$$= \lim_{z \to 0} \frac{1}{z\left(\dfrac{1}{z} + \sum_0^\infty b_n z^n - w_1\right)} = 1.$$

Thus $f \in \mathcal{F}$ and $w_2 \notin F(U\backslash\{0\})$ implies $\dfrac{1}{w_2 - w_1} \notin f(U)$. Therefore by

Theorem 5.6.10 $\left|\dfrac{1}{w_2 - w_1}\right| \geq \dfrac{1}{4}$ or that $|w_2 - w_1| \leq 4$. The example

$F(z) = \dfrac{1}{z} + z$ (satisfying all the conditions of the theorem) shows that the

constant 4 can not be reduced. ($F(z)$ does not assume ± 2 in $U\backslash\{0\}$).

Note 5.6.12

The Bieberbach's inequality $|a_2| \leq 2$ available for the second coefficients of functions in f has other geometric implications in the theory of conformal mappings. One important consequence is to provide sharp upper and lower bounds for $|f'(z)|$ as f varies over $\mathcal{F}$. The validity of these bounds are established by the so called "Distortion theorems". Perhaps the term "distortion" arises from the geometric interpretation of $|f'(z)|$ as the infinitesimal magnification factor of arc lengths under the mapping f (Note that as $z = z(t)$ describes a curve, $w(t) = f(z(t))$ describes its image under f and

the infinitesimal arc lengths along $w(t)$ and $z(t)$ bears the ratio $\left|\dfrac{dw(t)}{dz(t)}\right| =$

$|f'(z)|$) or from the interpretation of $|f'(z)|^2$ as the infinitesimal magnification factor of the areas under f. (If $f = u + iv$ where u and v are functions of x and y then $du\ dv \doteq J\ dx\ dy$ where J is the Jacobian of the transformation u, v with respect to x, y. Thus the ratio of the elemental area in the image and in the domain is precisely given by J. Now $J =$

$$\begin{vmatrix} u_x & u_y \\ v_x & v_y \end{vmatrix} = \begin{vmatrix} u_x & -v_x \\ v_x & u_x \end{vmatrix} = u_x^2 + v_x^2 = |f'(z)|^2).$$ We now proceed to obtain the

required distortion theorems.

Theorem 5.6.13

For every $f \in \mathcal{F}$, $\left|\dfrac{zf''(z)}{f'(z)} - \dfrac{2r^2}{1-r^2}\right| \le \dfrac{4r}{1-r^2}$ $(|z| = r < 1)$ and the result is

sharp for each $r < 1$. (i.e., the bound given here is the best possible for the entire class $\mathcal{F}$).

Proof

Fix $\zeta \in U$ and consider $F(z) = \dfrac{f\left(\dfrac{z+\zeta}{1+\bar{\zeta}z}\right) - f(\zeta)}{(1-|\zeta|^2)\,f'(\zeta)}$. Since f is one-to-one,

$f'(\zeta) \ne 0$ for $\zeta \in U$. Further $z \to \dfrac{z+\zeta}{1+\bar{\zeta}z}$ is a conformal bijection from U

onto U. Hence $F(z)$ is analytic in $|z| < 1$ and $F(0) = 0$ and $F'(0) = 1$.

$$F'(z) = f'\left(\frac{z+\zeta}{1+\bar{\zeta}z}\right)\left(\frac{-\bar{\zeta}(z+\zeta)+(1+\bar{\zeta}z)}{(1+\bar{\zeta}z)^2}\right) \bigg/ (1-|\zeta|^2)\,f'(\zeta)$$

$$= \frac{f'\left(\dfrac{z+\zeta}{1+\bar{\zeta}z}\right)\left(\dfrac{1-|\zeta|^2}{(1+\bar{\zeta}z)^2}\right)}{(1-|\zeta|^2)\,f'(\zeta)}.$$

Assume $F(z) = z + A_2 z^2 + \ldots$ A quick calculation shows that

$$A_2 = \frac{F''(0)}{2!} = \frac{1}{2}\left\{(1-|\zeta|^2)\frac{f''(\zeta)}{f'(\zeta)} - 2\bar{\zeta}\right\}.$$

By Theorem 5.6.10 we get $\left|(1-|\zeta|^2)\dfrac{f''(\zeta)}{f'(\zeta)} - 2\bar{\zeta}\right| \le 4$ or that

$\left|\dfrac{\zeta f''(\zeta)}{f'(\zeta)} - \dfrac{2|\zeta|^2}{1-|\zeta|^2}\right| \le \dfrac{4|\zeta|}{1-|\zeta|^2}$. This is nothing but the required inequality

with z replaced by ζ and $|\zeta| = r$. Considering $f(z) = \dfrac{z}{(1-z)^2} \in \mathcal{F}$ we see

that $\left|\dfrac{zf''(z)}{f'(z)} - \dfrac{2r^2}{1-r^2}\right|$ is equal to $\dfrac{4r}{1-r^2}$ at least when $z = r$. Hence the

inequality is the best possible for each $|z| = r$, i.e., the result is sharp.

Theorem 5.6.14 (Distortion Theorem)

For each $f \in \mathcal{F}$ we have $\dfrac{1-r}{(1+r)^3} \le |f'(z)| \le \dfrac{1+r}{(1-r)^3}$ $(|z| = r < 1)$ with

equality if and only if f is a rotation of the Koebe function given by

$\dfrac{z}{(1-z)^2}$.

Proof

If α is a complex number with $|\alpha| \le c$ we have $-c \le \text{Re } \alpha \le c$. Thus
Theorem 5.6.13 implies

$$\frac{2r^2 - 4r}{1-r^2} \le \text{Re} \frac{zf''(z)}{f'(z)} \le \frac{2r^2 + 4r}{1-r^2}.$$

Throughout the following we keep $|z| = r$ fixed and allow θ to vary in
$[0, 2\pi]$ so that $z = re^{i\theta}$ $(0 \le \theta \le 2\pi)$. Now $f'(z)$ is never 0 and $f'(0) = 1$.
Thus we can choose a single valued analytic branch of $\log f'(z)$ in U
(simply connected) which vanishes at origin. Note that for any analytic
function $h(z)$

$$r\frac{\partial}{\partial r} (\text{Re } h(z)) = \frac{\partial}{\partial \theta} (\text{Im } h(z))$$

using C–R equations with z in polar form and $h(z)$ in Cartesian form. (See
Corollary 1.9.17 (ii)). Taking $h(z) = \log f'(z)$ we have

$$r\frac{\partial}{\partial r}(\text{Re } \log f'(z)) = \frac{\partial}{\partial \theta}(\text{Im } \log f'(z))$$

i.e., $\quad r\dfrac{\partial}{\partial r} (\log |f'(re^{i\theta})|) = \text{Im } \dfrac{\partial}{\partial \theta}(\log f' (re^{i\theta}))$

$$= \text{Im} \left(\frac{i f''(re^{i\theta})}{f'(re^{i\theta})} re^{i\theta} \right)$$

$$= \text{Re} \left(\frac{f''(re^{i\theta})}{f'(re^{i\theta})} re^{i\theta} \right)$$

$$= \text{Re} \left(\frac{z f''(z)}{f'(z)} \right) \qquad (z = re^{i\theta}).$$

Hence

$$\frac{2r^2 - 4r}{1 - r^2} \leq r \frac{\partial}{\partial r} \ (\log |f'(re^{i\theta})|) \leq \frac{2r^2 + 4r}{1 - r^2}$$

$$\frac{2r - 4}{1 - r^2} \leq \frac{\partial}{\partial r} \ (\log |f'(re^{i\theta})|) \leq \frac{2r + 4}{1 - r^2}. \qquad (1)$$

This is true for any r between 0 and 1. Now keeping θ fixed and integrating the above inequalities with respect to r from $r = 0$ to $r = \text{R} < 1$ we get

$$\int_0^R \frac{2r - 4}{1 - r^2} dr \leq \log |f' (Re^{i\theta})| \leq \int_0^R \frac{2r + 4}{1 - r^2} dr$$

$$-\log (1 - R^2) + 2\log \left(\frac{1 - R}{1 + R} \right) \leq \log |f' (Re^{i\theta})|$$

$$\leq -\log (1 - R^2) - 2\log \left(\frac{1 - R}{1 + R} \right)$$

$$\frac{1}{1 - R^2} \left(\frac{1 - R}{1 + R} \right)^2 \leq |f'(Re^{i\theta})| \leq \frac{1}{1 - R^2} \left(\frac{1 + R}{1 - R} \right)^2$$

$$\frac{1 - R}{(1 + R)^3} \leq |f'(Re^{i\theta})| \leq \frac{1 + R}{(1 - R)^3}.$$

Replacing R by r we get the required bounds. On the other hand the results are best possible because the Koebe function $f(z) = \dfrac{z}{(1 - z)^2}$ has the

derivative $f'(z) = \dfrac{1 + z}{(1 - z)^3}$ which attains these bounds at $z = \pm r$. Equalities in the above inequalities are attained if and only if equalities hold in (1) for all r between 0 and R. (Observe that $\int_0^R H(r)\ dr = 0$ for $H(r) \geq 0$ if and only if $H(r) = 0$, $0 \leq r \leq R$). But equalities hold in (1) for all r with $0 < r < 1$ if and only if

$$\text{Re}\left(\frac{zf''(z)}{f'(z)}\right) = \frac{2r^2 + 4r}{1 - r^2} \text{ or } \frac{2r^2 - 4r}{1 - r^2}$$

i.e., $$\text{Re}\left(\frac{re^{i\theta} f''(re^{i\theta})}{f'(re^{i\theta})}\right) = \frac{2r^2 + 4r}{1 - r^2} \text{ or } \frac{2r^2 - 4r}{1 - r^2}.$$

i.e., $$\text{Re}\left(\frac{e^{i\theta} f''(re^{i\theta})}{f'(re^{i\theta})}\right) = \frac{2r + 4}{1 - r^2} \text{ or } \frac{2r - 4}{1 - r^2}.$$

Letting $r \to 0$ we get

$$\text{Re}\left(e^{i\theta}\frac{f''(0)}{f'(0)}\right) = \pm 4$$

i.e., $|a_2| = 2$ where $f(z) = z + a_2 z^2 + \ldots$ Hence by Theorem 5.6.10 $f(z)$ reduces to a rotation of the Koebe function.

Theorem 5.6.15 (Growth Theorem)

For each $f \in \mathcal{F}$, $\dfrac{r}{(1+r)^2} \leq |f(z)| \leq \dfrac{r}{(1-r)^2}$ ($|z| = r < 1$). The inequalities are sharp for each r and the equalities in any one of the inequalities occur if and only if f is a rotation of the Koebe function.

Proof

Let $f \in \mathcal{F}$ and put $z = re^{i\theta}$ ($0 < r < 1$). $f(z) = \int_0^r f'(\rho e^{i\theta})\ e^{i\theta} d\rho$ (Note that $f(0) = 0 \Rightarrow f(z) = \int_0^z f'(\zeta)\ d\zeta$ where ζ varies over the line segment from 0 to z which lies in U. On this line segment $\zeta = \rho e^{i\theta}$, $0 \leq \rho \leq r$. Hence $f(z) = \int_0^r f'(\rho e^{i\theta})\ e^{i\theta}\ d\rho$).

From Theorem 5.6.14 $|f'(\rho e^{i\theta})| \leq \dfrac{1 + \rho}{(1 - \rho)^3}$. Hence

$$|f(z)| \leq \int_0^r |f'(\rho e^{i\theta})|\ d\rho \leq \int_0^r \frac{1+\rho}{(1-\rho)^3}\ d\rho = \frac{r}{(1-r)^2}.$$

This gives the required upper estimate. For the lower bound we first observe that if $|f(z)| \geq \frac{1}{4}$ trivially we have $|f(z)| \geq \frac{1}{4} \geq \frac{r}{(1+r)^2}$.

Thus we are free to assume $|f(z)| < \frac{1}{4}$. In this case by Theorem 5.6.10 the entire line segment joining 0 and $f(z)$ lies in the range of f. Let C be the pre-image of this line segment under f. Then C is a simple curve joining 0 and z and $f(z) = \int_C f'(\zeta) \, d\zeta$. Now

$$|f(z)| = \left| \int_C f'(\zeta) \, d\zeta \right| = \int_C |f'(\zeta)| \, |d\zeta|$$

(Since $\int_C |f'(\zeta)| \, |d\zeta| = \int_0^{f(z)} |dw| = |f(z)|$). Thus by Theorem 5.6.14

$$|f(z)| \geq \int_C \frac{1 - |\zeta|}{(1 + |\zeta|)^3} \, |d\zeta|.$$

Now consider the function $F(\zeta) = |\zeta|$ along a curve C with parametric equation $x = x(t)$, $y = y(t)$. Then

$$\frac{dF}{dt} = \frac{\partial F}{\partial x}\frac{dx}{dt} + \frac{\partial F}{\partial y}\frac{dy}{dt} = \frac{x}{\sqrt{x^2 + y^2}} x'(t) + \frac{y}{\sqrt{x^2 + y^2}} y'(t)$$

Now $\zeta = x(t) + iy(t)$. Therefore

$$\frac{d\zeta}{dt} = \frac{dx}{dt} + i\frac{dy}{dt} \Rightarrow \left| \frac{d\zeta}{dt} \right| = \sqrt{\left(\frac{dx}{dt}\right)^2 + \left(\frac{dy}{dt}\right)^2}$$

$$\left(\frac{dF}{dt}\right)^2 = \frac{x^2}{x^2 + y^2}\left(\frac{dx}{dt}\right)^2 + \frac{y^2}{x^2 + y^2}\left(\frac{dy}{dt}\right)^2 + \frac{2xy}{x^2 + y^2}\frac{dx}{dt}\frac{dy}{dt}$$

$$\leq \left(\frac{dx}{dt}\right)^2 + \left(\frac{dy}{dt}\right)^2$$

since $x^2\left(\dfrac{dx}{dt}\right)^2 + y^2\left(\dfrac{dy}{dt}\right)^2 + 2xy\dfrac{dx}{dt}\dfrac{dy}{dt} \leq (x^2 + y^2)\left(\left(\dfrac{dx}{dt}\right)^2 + \left(\dfrac{dy}{dt}\right)^2\right)$.

$$\left[\text{Note that the last inequality is equivalent to } 0 \leq \left(x\frac{dy}{dt} - y\frac{dx}{dt} \right)^2 \right].$$

This proves that $\left(\dfrac{d|\zeta|}{dt}\right)^2 \leq \left|\dfrac{d\zeta}{dt}\right|^2$. Using this and the fact that as ζ varies over C, $|\zeta|$ varies between 0 and r we get

$$\int_C \frac{1-|\zeta|}{(1+|\zeta|)^3}|d\zeta| \geq \int_0^r \frac{1-|\zeta|}{(1+|\zeta|)^3}d|\zeta| = \frac{r}{(1+r)^2}.$$

This completes the proof of the inequalities. The example $f(z) = \dfrac{z}{(1-z)^2}$ shows that the equalities are attained at $z = \pm r$. On the other hand if equality held in any one of the above inequalities then from the above proof we have either

(i) $\displaystyle\int_0^r |f'(\rho e^{i\theta})|\ d\rho = \int_0^r \frac{1+\rho}{(1-\rho)^3}d\rho$

or

(ii) $\displaystyle\int_0^r |f'(\zeta)|\ d|\zeta| = \int_0^r \frac{1-\rho}{(1+\rho)^3}d\rho.$

But (i) or (ii) respectively implies $|f'(\zeta)|$ must be equal to either $\dfrac{1+|\zeta|}{(1-|\zeta|)^3}$

or $\dfrac{1-|\zeta|}{(1+|\zeta|)^3}$ for every $|\zeta|$ between 0 and r. But then by Theorem 5.5.14, $f(\zeta)$ reduces to a rotation of the Koebe function.

Theorem 5.6.16 (Rotation Theorem)
If $f \in \mathcal{F}$ and $F(z) = \log f'(z)$ is the analytic branch of $\log f'(z)$ with $F(0)$

$= 1$ then $|\arg f'(z)| \leq 2 \log \left(\dfrac{1+r}{1-r}\right)$ where $|z| = r < 1$.

Proof
Since $\text{Im} \log f'(z) = \arg f'(z)$ (here $\arg f'(z)$ represents a well-defined continuous branch), which can be interpreted geometrically as the local rotation factor under the conformal mapping f, the above theorem is called a rotation theorem. (If $z_1(t)$ and $z_2(t)$ are two curves at z_0 with $z_1(t_0) = z_2(t_0) = z_0$ and $w_1(t)$ and $w_2(t)$ are the image curves under F then $w_i(t) = f(z_i(t))$ $(i = 1, 2)$ and hence

$$\lim_{t \to t_0} \arg \frac{w_1'(t) - w_2'(t)}{z_1'(t) - z_2'(t)} = \arg f'(z_0).$$

Thus arg $f'(z_0)$ represents a factor which when added to the change in the argument of $z'(t)$ gives the change in the argument of $w'(t)$ locally at z_0.

For the proof we start with $\left| \dfrac{zf''(z)}{f'(z)} - \dfrac{2r^2}{1-r^2} \right| \le \dfrac{4r}{1-r^2}$ which is valid by Theorem 5.6.13. Since $-|w| \le |\operatorname{Im} w| \le |w|$ holds for all complex numbers w, from the above inequality we get

$$\frac{-4r}{1-r^2} \le \operatorname{Im}\left(\frac{zf''(z)}{f'(z)} \right) \le \frac{4r}{1-r^2} \qquad (|z| = r < 1). \quad (1)$$

On the other hand $\operatorname{Im}\left(\dfrac{re^{i\theta} f''(re^{i\theta})}{f'(re^{i\theta})} \right) = r\dfrac{\partial}{\partial r} (\arg f'(re^{i\theta}))$.

(Recall that $\quad f'(re^{i\theta}) = \lim\limits_{\Delta r \to 0}\left(\dfrac{f((r + \Delta r)e^{i\theta}) - f(re^{i\theta})}{\Delta r\, e^{i\theta}} \right)$

$$= e^{-i\theta}\left(\lim_{\Delta r \to 0} \frac{f(r + \Delta r, \theta) - f(r, \theta)}{\Delta r} \right) = e^{-i\theta}\frac{\partial f}{\partial r}.$$

Put $F(z) = \log f'(z) = U + iV$ so that

$$\frac{f''(re^{i\theta})}{f'(re^{i\theta})} = F'(re^{i\theta}) = e^{-i\theta}\frac{\partial F}{\partial r} = e^{-i\theta}\left(\frac{\partial U}{\partial r} + i\frac{\partial V}{\partial r} \right),$$

i.e., $re^{i\theta}\dfrac{f''(re^{i\theta})}{f'(re^{i\theta})} = r\left(\dfrac{\partial U}{\partial r} + i\dfrac{\partial V}{\partial r} \right)$. Thus $\operatorname{Im}\left(\dfrac{re^{i\theta} f''(re^{i\theta})}{f'(re^{i\theta})} \right) = r\dfrac{\partial V}{\partial r}$). Changing r to ρ and integrating the inequalities (1) with respect to ρ from $\rho = 0$ to $\rho = r$, we get

$$-2 \log\left(\frac{1+r}{1-r} \right) \le \arg f'(re^{i\theta}) \le 2 \log\left(\frac{1+r}{1-r} \right)$$

whenever $z = re^{i\theta}$ with $|z| = r < 1$. Hence the result.

Note 5.6.17
The bound obtained for $|\arg f'(re^{i\theta})|$ in Theorem 5.6.16 is however not sharp. In fact the sharp bounds are given by

$$|\arg f'(re^{i\theta})| \le \begin{cases} 4 \sin^{-1} r, & r \le \dfrac{1}{\sqrt{2}} \\[2mm] \pi + \log\dfrac{r^2}{1-r^2}, & r \ge \dfrac{1}{\sqrt{2}}. \end{cases}$$

The proof of this result is however beyond the scope of this book.

Theorem 5.6.18

For each $f \in \mathcal{F}$ and $|z| = r < 1$ we have $\dfrac{1-r}{1+r} \le \left|\dfrac{zf'(z)}{f(z)}\right| \le \dfrac{1+r}{1-r}$ with

equality for $z \ne 0$ on either side if and only if $f(z)$ is a rotation of the Koebe's function.

Proof

We first note that the combination of the bounds for both f' and f given by Distortion theorem and the growth theorem do not give the bounds obtained in the above theorem. (In fact the combined upper and lower

bounds are $\left(\dfrac{1+r}{1-r}\right)^3$, $\left(\dfrac{1-r}{1+r}\right)^3$ respectively and these are less precise.) For

the proof we fix ζ with $|\zeta| < 1$ and consider

$$F(z) = \dfrac{f\left(\dfrac{z+\zeta}{1+\bar{\zeta}z}\right) - f(\zeta)}{(1-|\zeta|^2)\,f'(\zeta)}$$

$$= z + A_2(\zeta)\,z^2 + \dots$$

We already saw (Theorem 5.6.13) that $F \in \mathcal{F}$. Thus by the growth theorem (Theorem 5.6.15) we have

$$\dfrac{|\zeta|}{(1+|\zeta|)^2} \le |F(-\zeta)| \le \dfrac{|\zeta|}{(1-|\zeta|)^2}, \quad (\zeta \in U)$$

But $F(-\zeta) = \dfrac{-f(\zeta)}{(1-|\zeta|^2)\,f'(\zeta)}$ and so

$$\dfrac{1-|\zeta|}{1+|\zeta|} \le \left|\dfrac{f(\zeta)}{\zeta f'(\zeta)}\right| \le \dfrac{1+|\zeta|}{1-|\zeta|}.$$

This gives the required bounds (by replacing ζ by z). It is clear that the

Koebe's function $k(z) = \dfrac{z}{(1-z)^2}$ satisfies $\dfrac{zk'(z)}{k(z)} = \dfrac{1+z}{1-z}$ and hence the

equalities are attained for the Koebe's function $k(z)$ at $z = \pm r$. What remains to be proved is that if any one of the equalities is attained for some $f \in \mathcal{F}$ and $z \ne 0$ then f reduces to a rotation of the Koebe's function.

Since the proofs corresponding to lower and upper bounds are similar we shall assume that the lower bound is attained and prove that f is a rotation of

k. Now $\left|\dfrac{zf'(z)}{f(z)}\right| = \dfrac{1-r}{1+r}$ at $z = \zeta\,(\ne 0)$, $\zeta \in U$ implies $|F(-\zeta)| = |\zeta|/(1-|\zeta|)^2$.

By Growth theorem (equality case) we have $F(\zeta) = \zeta/(1 - e^{i\alpha}\zeta)^2$
or that $F(z) = z/(1 - e^{i\alpha}z)^2$ with $e^{i\alpha} = -\overline{\zeta}/|\zeta|$ (this is because
$|F(-\zeta)| = |\zeta|/|1 - e^{i\alpha}\zeta|^2 = |\zeta|/(1 - |\zeta|)^2$ implies the following equalities.

$$\text{Re } \zeta e^{i\alpha} = -|\zeta|, \text{ Re } \zeta e^{i\alpha} = -|\zeta e^{i\alpha}|,$$

$$\zeta e^{i\alpha} = -|\zeta|, \quad e^{i\alpha} = -|\zeta|/\zeta = -\overline{\zeta}/|\zeta|).$$

Now let $w = \dfrac{z+\zeta}{1+\overline{\zeta}z}$ or that $z = \dfrac{w-\zeta}{1-\overline{\zeta}w}$ and put $G(w) = F\left(\dfrac{w-\zeta}{1-\overline{\zeta}w}\right)$,

Our defining relation for F and f tells us that

$$F(z) = \frac{f\left(\dfrac{z+\zeta}{1+\overline{\zeta}z}\right) - f(\zeta)}{(1-|\zeta|^2)f'(\zeta)}$$

and $G(0) = F(-\zeta) = -f(\zeta)/(1 - |\zeta|^2)\,f'(\zeta).$

Hence $\qquad f(w) = F\left(\dfrac{w-\zeta}{1-\overline{\zeta}w}\right)(1 - |\zeta|^2)\,f'(\zeta) + f(\zeta)$

$$= G(w)\,(1 - |\zeta|^2)\,f'(\zeta) - G(0)\,(1 - |\zeta|^2)\,f'(\zeta).$$

Therefore $\qquad f(w) = (G(w) - G(0))\,(1 - |\zeta|^2)\,f'(\zeta).$

Now $\qquad G(w) = F\left(\dfrac{w-\zeta}{1-\overline{\zeta}w}\right) = \dfrac{\left(\dfrac{w-\zeta}{1-\overline{\zeta}w}\right)}{\left(1 - e^{i\alpha}\left(\dfrac{w-\zeta}{1-\overline{\zeta}w}\right)\right)^2}$

$$= \frac{(w-\zeta)(1-\overline{\zeta}w)}{\left((1-\overline{\zeta}w) - e^{i\alpha}(w-\zeta)\right)^2}$$

$$= \frac{(w-\zeta)(1-\overline{\zeta}w)}{\left((1-\overline{\zeta}w) + \dfrac{\overline{\zeta}}{|\zeta|}(w-\zeta)\right)^2}$$

$$= \frac{(w-\zeta)(1-\overline{\zeta}w)}{(1-|\zeta|)^2\left(1 + \dfrac{w\overline{\zeta}}{|\zeta|}\right)^2}.$$

Hence $G(w) - G(0) = \dfrac{(w - \zeta)(1 - \bar{\zeta}w)}{(1 - |\zeta|)^2 \left(1 + \dfrac{w\bar{\zeta}}{|\zeta|}\right)^2} + \dfrac{\zeta}{(1 - |\zeta|)^2}$

$$= \frac{(1 + |\zeta|)^2}{(1 - |\zeta|)^2} \frac{w}{(1 - e^{i\alpha}w)^2}.$$

Therefore $f(w) = \dfrac{(1 + |\zeta|)^3}{(1 - |\zeta|)} f'(\zeta) \dfrac{w}{(1 - e^{i\alpha}w)^2}$

$$= b \frac{w}{(1 - e^{i\alpha}w)^2} \quad \text{for some complex constant } b.$$

Since $f \in \mathcal{F}$ we have $b = 1$ and hence f reduces to a rotation of the Koebe function.

Theorem 5.6.19
Let $f \in \mathcal{F}$ and $h \in \mathcal{F}$ be defined by $h^2(z) = f(z^2)$. Put $A_r(h)$ = the area of the domain D_r which is the image of $|z| < r < 1$ under h. $L_r(f)$ = the arc length of the image of $|z| = r < 1$ under f. Then

(i) $A_r(h) \leq \pi r^2 (1 - r^2)^{-2}$

(ii) $M_1(r, f) = \dfrac{1}{2\pi} \int_0^{2\pi} |f(re^{i\theta})| \, d\theta \leq \dfrac{r}{1 - r} \quad (0 \leq r \leq 1)$

(iii) $L_r(f) \leq \dfrac{2\pi r(1 + r)}{(1 - r)^2}$

Proof

Since $f \in \mathcal{F}$, by Growth Theorem we have $|f(z)| \leq \dfrac{r}{(1 - r)^2}$ for $|z| = r$

and hence for $|z| \leq r$ by Maximum modulus theorem. Thus $|h(z)| \leq \dfrac{r}{1 - r^2}$

for $|z| \leq r$. Thus h maps $|z| < r$ conformally onto a domain which lies

entirely in a disc of radius $\dfrac{r}{1 - r^2}$ around the origin. Hence $A_r(h) \leq$

$\dfrac{\pi r^2}{(1 - r^2)^2}$. This proves (i).

Using polar coordinates and the expression for area we have

$$A_r(h) = \iint_{D_r} du \, dv$$

$$= \iint_{|z| < r} |h'(z)|^2 \, dx \, dy \qquad \text{(see Note 5.6.12)}$$

$$= \int_0^{2\pi} \int_0^r |h'\,(\rho e^{i\theta})|^2 \, \rho \, d\rho \, d\theta.$$

Let us assume $h(z) = \sum_1^\infty c_n z^n$ with $c_1 = 1$. Then by term-wise integration

$$A_r \,(h) = \pi \sum_1^\infty n \,|c_n|^2 \, r^{2n} \quad \text{(use } |h'|^2 = h' \overline{h'} \text{ and integrate)}. \text{ From (i) we have}$$

$A_r \,(h) \le \dfrac{\pi r^2}{(1 - r^2)^2}$. Therefore

$$\sum_1^\infty n|c_n|^2 \, r^{2n-1} \le \frac{r^2}{(1 - r^2)^2} \,(0 \le r < 1).$$

An integration from 0 to r gives

$$\sum_1^\infty |c_n|^2 \, r^{2n} \le \frac{r^2}{(1 - r^2)}.$$

Note that both the differentiation and the integration which are performed term wise in the above calculations are valid because the series is uniformly convergent for $r < 1$. For the same reason

$$\frac{1}{2\pi} \int_0^{2\pi} |h \,(re^{i\theta})|^2 \, d\theta = \frac{1}{2\pi} \int_0^{2\pi} h(re^{i\theta}) \, \overline{h(re^{i\theta})} \, d\theta$$

$$= \sum_1^\infty |c_n|^2 \, r^{2n}.$$

Therefore $\dfrac{1}{2\pi} \int_0^{2\pi} |h \,(re^{i\theta})|^2 \, d\theta \le \dfrac{r^2}{(1 - r^2)}.$

i.e., $\dfrac{1}{2\pi} \int_0^{2\pi} |f(r^2 e^{2i\theta})| \, d\theta \le \dfrac{r^2}{(1 - r^2)} \quad \left(\text{put } 2\theta = \phi \text{ and observe that} \right.$

$\left. \int_0^{4\pi} = 2 \int_0^{2\pi} \right).$ Thus $M_1 \,(r, f) = \dfrac{1}{2\pi} \int_0^{2\pi} |f(re^{i\theta})| \, d\theta \le \dfrac{r}{1 - r} \; 0 < r < 1$ and this

completes the proof of (ii).

Using Theorem 5.6.18 and (ii) above we have

$$L_r(f) = \int_\Gamma |\,dw| = r \int_0^{2\pi} |f'(re^{i\theta})| \, d\theta$$

where Γ is the image of $|z| = r$ under f. (Γ is given by $w = w(\theta) = f(re^{i\theta})$; $|dw| = |f'(re^{i\theta})|\, r\, d\theta$) and

$$L_r\ (f) = \int_0^{2\pi} \left| \frac{re^{i\theta} f'(re^{i\theta})}{f(re^{i\theta})} \right| |f(re^{i\theta})|\ d\theta$$

$$\leq \frac{1+r}{1-r}\ M_1\ (r,\ f)\ 2\pi$$

$$\leq 2\pi\ \frac{1+r}{1-r}\ \frac{r}{1-r}$$

$$= 2\pi \frac{r(1+r)}{(1-r)^2}$$

This proves (iii).

Theorem 5.6.20 (Littlewood's Theorem)
If $f(z) = z + a_2 z^2 + ... \in \mathcal{F}$ then $|a_n| < en$ $(n = 1, 2, ...)$.

Proof
By Cauchy's integral formula

$$f^{(n)}\ (0) = \frac{n!}{2\pi i} \int_{|z|=r} \frac{f(\zeta)\, d\zeta}{\zeta^{n+1}}$$

$$= \frac{n!}{2\pi i} \int_0^{2\pi} \frac{f(re^{i\theta})}{r^{n+1} e^{i(n+1)\theta}}\, rie^{i\theta} d\theta$$

$$= \frac{n!}{2\pi} r^{-n} \int_0^{2\pi} f(re^{i\theta})\, e^{-in\theta}\ d\theta.$$

Thus
$$|a_n| = \frac{|f^n(0)|}{n!} \leq \frac{1}{2\pi r^n} \int_0^{2\pi} |f(re^{i\theta})|\ d\theta$$

$$= \frac{M_1(r, f)}{r^n} \leq \frac{1}{r^{n-1}(1-r)}.$$

This is true for all $0 \leq r < 1$ and $\dfrac{1}{r^{n-1}(1-r)}$ attains its minimum at

$r = 1 - \dfrac{1}{n}$. $\left(\text{Indeed this function can be shown to be decreasing for } r < 1 - \dfrac{1}{n},\right.$

attains its only local minimum at $r = 1 - \dfrac{1}{n}$ and increases for $r > 1 - \dfrac{1}{n}$.
Thus this local minimum is the global minimum.) This minimum value is given by

$$n\left(1 + \frac{1}{n-1}\right)^{n-1} < en, \ \forall n.$$

This proves our theorem.

5.7 A Brief History of the Bieberbach Conjecture and Its Solution

The numbers in this section correspond to the references at the end of this section itself. In the following we shall freely use the information available in [36].

The study of geometric function theory has been given a solid foundation by Riemann around 1850 with his fundamental mapping theorem. A little over half a century later Koebe and others formulated the theory of univalent functions which are normalized (the class $\mathcal{F}$) [23, 24, 25]. Normal families of analytic functions was introduced by Montel [31] and Carathéodory [8] established his Kernel theorem on sequences of univalent functions.

Ludwig Bieberbach proposed his famous conjecture in 1916 [3, p. 946]. If $f(z) = z + a_2 z^2 + \ldots$ is analytic and univalent in the open unit disc then $|a_n| \leq n$ ($n = 2, 3, \ldots$) with equality if and only if f is a rotation of the Koebe function

$$k(z) = \frac{z}{(1-z)^2} = z + 2z^2 + 3z^3 + \ldots$$

Bieberbach [3] proved his conjecture for $n = 2$ using the area principle which had just been established by Gronwall [14]. This inequality led to distortion, growth and covering theorems.

Charles Loewner [28] in his paper during 1923 represented slit mapping in terms of a differential equation. These slit mappings are shown as dense in $\mathcal{F}$ and as an application he was able to prove $|a_3| \leq 3$ with equality if and only if f is a rotation of the Koebe function.

Goluzin [13] obtained various sharp inequalities on the values of a function $f \in \mathcal{F}$ (at prescribed points) along with rotation theorems.

Littlewood [27] considered the integral means of functions in $\mathcal{F}$ and proved the uniform bound $|a_n| < en$ (Theorem 5.6.20). A little later sharpening his method Bazilevich [2] obtained the result $|a_n| < e(n/2) + c$ for

an absolute constant c. Hayman [16, 17, 18] showed that $\alpha = \lim\limits_{n \to \infty} \left| \dfrac{a_n}{n} \right|$

exists for each $f \in \mathcal{F}$ with $f(z) = z + a_2 z^2 + \dots + a_n z^n + \dots$ and that $\alpha < 1$ unless f is a rotation of the Koebe function and also proved that $\|a_{n+1}\| - |a_n|\|$ is bounded by an absolute constant. In 1974 Baernstein completed Little-wood's methods by showing

$$\int_0^{2\pi} |f(re^{i\theta})|^p \, d\theta \le \int_0^{2\pi} |k(re^{i\theta})|^p \, d\theta$$

for all $f \in \mathcal{F}$, $0 < r < 1$, $0 < p < \infty$, $k(z) = \dfrac{z}{(1-z)^2}$. In 1936, Robertson [35] conjectured that $1 + |a_3|^2 + |a_5|^2 + \dots + |a_{2n-1}|^2 \le n$ for all odd functions $f(z) = z + a_3 z^3 + a_{2n-1} z^{2n-1} + \dots$, in $\mathcal{F}$ and also showed that his conjecture implies the Bieberbach conjecture. He also proved it for certain sub-families of univalent functions. This method along with Loewner method led to a proof of $|a_4| \le 4$ by Garabedian and Schiffer in 1955 [12].

Jenkins [19, 20, 21, 22] developed his general coefficient theorem which led to solutions of various extremal problems. This method uses the ideas of Grötzsch and Teichmüller and applies to domains slit along trajectories of quadratic differentials. In 1939. Grunsky [15] gave restrictions on the coefficients $c_{n,\,m}$ $(n,\ m \ge 1)$ of

$$\log \frac{F(z) - F(\zeta)}{z - \zeta} = \sum_{m=0}^{\infty} \sum_{n=0}^{\infty} c_{n,m} z^n \zeta^m$$

that are necessary and sufficient for the univalence of F. These inequalities were used successfully by Charzyński and Schiffer [9] to give an elementary proof of $|a_4| \le 4$. Pederson and Schiffer [34] proved $|a_5| \le 5$. Pederson [33] and Ozawa [32] independently used Grunsky's inequalities to prove that $|a_6| \le 6$. Lebedev [26] describes applications of area methods and extensions of Grunsky's inequalities.

Milin [29, 30] systematically developed a technique for exponentiating the Grunsky's inequalities to get a more direct access to the coefficients of f which led to $|a_n| < (1.243)n$. Jointly with Lebedev he also considers the logarithmic coefficients of f defined by $\log \dfrac{f(z)}{z} = 2 \sum_1^\infty \gamma_n z^n$. After studying the estimates of the growth of γ_n he conjectured that $\sum_{k=1}^\infty (n - k + 1) k \, |\gamma_k|^2 \le \sum_{k=1}^\infty (n - k + 1)\dfrac{1}{k} \ \forall \, n \in \mathbb{N}$. In 1972 FitzGerald [10] proved $|a_n| < \sqrt{7/6}\, n < 1.081 \ n$.

While detailed investigations on the general solutions of Bieberbach conjecture were taking place, elsewhere, other attempts were made to solve the conjecture. Several sub-classes of univalent functions for which the conjecture can be easily verified by other geometric means have been introduced and studied. (For example convex functions, star-like functions, close-to-convex functions, Bazilevich functions etc.) Further it was also realized that $|a_2| \le \alpha < 2$ has implications on the estimates for $|a_n|$ and the maximum value of α for which "$|a_2| \le \alpha \Rightarrow |a_n| \le n$" were also investigated. By elementary arguments we can illustrate the above remark as follows: From Theorem 5.6.7 we know that if $F(z) = \dfrac{1}{g(z)} = \dfrac{1}{z} + b_1 z$ $+ b_2 z^2 + \ldots$ for a given $f \in \mathcal{F}$, where $g^2(z) = f(z^2)$, then $\sum n|b_n|^2 \le 1$. This implies in particular that $|b_1|^2 + 3|b_3|^2 \le 1$ where $b_1 = -a_2/2$ and $b_3 = 3a_2^2/8 - a_3/2$. Note that

$$F(z) = (f(z^2))^{-1/2} = (z^2 + a_2 z^4 + a_3 z^6 + \ldots)^{-1/2}$$

$$= \frac{1}{z}(1 + a_2 z^2 + \ldots)^{-1/2}$$

$$= \frac{1}{z}\left(1 - \frac{1}{2}(a_2 z^2 + a_3 z^4 + \ldots) \right.$$

$$\left. + \frac{\left(\frac{-1}{2}\right)\left(\frac{-3}{2}\right)}{1 \times 2}(a_2 z^2 + a_3 z^4 + \ldots)^2 + \ldots \right)$$

$$= \frac{1}{z}\left(1 - \frac{a_2}{2}z^2 + \left(\frac{-a_3}{2} + \frac{3}{8}a_2^2\right)z^4 \ldots \right)$$

which implies $b_1 = -\dfrac{a_2}{2} \, b_2 = 0 \, b_3 = \dfrac{3}{8}a_2^2 - \dfrac{a_3}{2}$.

Hence $\left|\dfrac{a_2}{2}\right|^2 + 3\left|\dfrac{a_3}{2} - \dfrac{3a_2^2}{8}\right|^2 \le 1 \Rightarrow \left|\dfrac{a_3}{2} - \dfrac{3a_2^2}{8}\right| \le \dfrac{1}{\sqrt{3}}\sqrt{1 - \dfrac{|a_2|^2}{4}}$

(by triangle inequality $|a - b| \ge |a| - |b|$)

$\left|\dfrac{a_3}{2}\right| \le \dfrac{3}{8}|a_2^2| + \dfrac{1}{\sqrt{3}}\sqrt{1 - \dfrac{|a_2^2|}{4}} \le \dfrac{3}{2}$ provided $\dfrac{1}{\sqrt{3}}\sqrt{1 - \dfrac{|a_2|^2}{4}}$

$\le \dfrac{3}{2} - \dfrac{3}{8}|a_2^2| = \dfrac{3}{2}\left(1 - \dfrac{|a_2^2|}{4}\right)$ or $\dfrac{1}{\sqrt{3}} \le \dfrac{3}{2}\sqrt{1 - \dfrac{|a_2|^2}{4}}$ or

$$\frac{1}{3} \le \frac{9}{4}\left(1 - \frac{|a_2^2|}{4}\right) \text{ or } \frac{4}{27} \le \left(1 - \frac{|a_2^2|}{4}\right) \text{ or } \frac{|a_2^2|}{4} \le 1 - \frac{4}{27} = \frac{23}{27}$$

or $|a_2| \le 2\sqrt{\frac{23}{27}}$. In other words, $|a_2| \le 2\sqrt{\frac{23}{27}} \Rightarrow |a_3| \le 3$. Note that

$2\sqrt{\frac{23}{27}}$ is very near 2 and the third Bieberbach conjecture is nothing but

$|a_2| \le 2 \Rightarrow |a_3| \le 3$. Further without restrictions on $|a_2|$ we also have

$|a_3| \le \frac{3}{4}|a_2^2| + \frac{1}{\sqrt{3}}\sqrt{4 - |a_2|^2} \le \frac{28}{9} \cong 3.11$. This last inequality is got by

observing that $\frac{3}{4}x + \frac{1}{\sqrt{3}}\sqrt{4-x}$ is increasing for $x \le \frac{104}{27}$ and decreasing

for $x \ge \frac{104}{27}$ so that its absolute maximum in the interval $0 \le x \le 4$ is

attained at $x = \frac{104}{27}$ and the absolute maximum is $\frac{28}{9}$.

On the other hand, several other methods have been investigated in the literature to attack the coefficient conjecture. Though we are not able to describe these other results we encourage the reader to look into the literature. Thus the coefficient conjecture has stimulated a lot of research work on the theory of univalent functions.

Finally came the work of L. de Branges (See [7]) completing the proof of the Bieberbach conjecture in 1985. Developing some of his earlier ideas [4, 5, 6] de Branges used Loewner methods and the positivity of certain sums of Jacobi polynomials (whose proof was already contained in a paper of Askey and Gasper [1]) to complete this proof. FitzGerald and Pommerenke [11] then gave another version of the same proof.

While the completion of the proof of the conjecture removes one of the most famous unsolved problems on function theory, many other interesting problems like coefficient regions for Schilit functions (find the region of variability of $w = (a_2, a_3, \ldots a_n) \in \mathbb{C}^{n-1}$ such that $f(z) = z + a_2 z^2 + \ldots + a_n z^n + \ldots \in \mathcal{F}$) remain unsolved and the "field" continues to develop in many other directions.

References

1. Askey, R. and Gasper. G., *Positive Jacobi polynomial sums*, II, Amer. J. Math. 98, (1976), 709-737.
2. Bazilevich. I.E., *On distortion theorems in the theory of univalent functions*, Mat. Sb. 28(70), (1951), 283-292. (Russian)
3. Bieberbach. L., Über *die Koeffizienten derjenigen Potenzreihen, welche eine schlichte Abbildung des Einheitskreises vermitteln, S-B. Pruss. Akad Wiss,* (1916), 940-955.

4. Branges, L de, *Coefficient estimates*, J. Math. Anal. Appl. 82, (1981), 420-450.

5. Branges, L de, *Grunsky spaces of analytic functions*, Bull. Sci. Math. 105, (1981), 401-406.

6. Branges, L de, *Löwner expansions*, J. Math. Anal. Appl. 100 (1984), 323-337.

7. Branges, L de, A proof of the *Bieberbach conjecture*, Acta Math, 154, (1985), 137-152.

8. Caratheodory, C., *Untersuchungen Über die konformen Abbildungen von festen und veränderlichen Gebieten*, Math, Ann. 72 (1912), 107-144.

9. Charzyński, Z. and Schiffer. M., *A new proof of the Bieberbach conjecture for the fourth coefficient*, Arch. Rational Mech. Anal., 5(1960), 187-193.

10. FitzGerald, C.H., *Quadratic inequalities and coefficient estimates for schlicht functions*, Arch. Rational Mech. Anal. 46 (1972), 356-368.

11. FitzGerald, C.H. and Pommerenke. Ch., *The de Branges theorem on univalent functions*, Trans. Amer. Math. Soc., 290 (1985), 683-690.

12. Garabedian, P.R. and Schiffer. M., *A proof of the Bieberbach conjecture for the fourth coefficient*, J. Rational Mech. Anal, 4 (1955), 427-465.

13. Goluzin, G.M., *Geometric theory of functions of a complex variable*, Gosudarst. Izdat., Moscow, 1952; German transl., Deutscher Verlag, Berlin, 1957; 2nd ed., Izdat. "Nauka", Moscow, (1966); English transl., Amer. Math. Soc., Providence, R.I., 1969.

14. Gronwall, T.H., *Some remarks on conformal representation*, Ann. of Math. 16 (1914-1915), 72-76.

15. Grunsky, H., *Koeffizientenbedingungen für schlicht abbildende meromorphe Funktionen*, Math. Z., 45 (1939), 29-61.

16. Hayman. W.K., *The asymptotic behaviour of p-valent functions*, Proc. London Math. Soc., 5 (1955), 257-284.

17. Hayman. W.K., *Multivalent functions*, Cambridge University Press, London and New York, 1958.

18. Hayman. W.K., *On successive coefficients of univalent functions*, J. London Math. Soc., 38 (1963), 228-243.

19. Jenkins. J.A., *A general coefficient theorem*, Trans. Amer. Math. Soc. 77 (1954), 262-280.

20. Jenkins. J.A., *Univalent functions and conformal mapping*, Springer-Verlag, Berlin, 1958.

21. Jenkins. J.A., *"On certain coefficients of univalent functions"* in *Analytic functions*, Princeton Univ. Press Princeton, N.J., 1960, pp 159-194.

22. Jenkins. J.A., *On certain coefficients of univalent functions. II*, Trans. Amer. Math. Soc., 96 (1960), 534-535.

23. Keobe. P., *Über die Uniformisierung beliebiger analytischer Kurven*, Nachr. Akad. Wiss. Göttingen Math.-Phys. KI. (1907). 191-210.

24. Koebe. P., *Über die Uniformisierung der algebraischen Kurven, durch automorphe Funktionen mit imaginärer Substitutionsgruppe*, Nachr. Akad. Wiss. Göttingen Math.-Phys. KI. (1909), 68-76.

25. Koebe. P., *Ränderzuordnung bei konformer Abbildung*, Nachr. Königl. Ges. Wiss. Göttingen Math.-Phys. KI. (1913), 286-288.

26. Lebedev, N.A. *The area principle in the theory of univalent functions*, Izdat, "Nauka", Moscow, 1975. (Russian)

27. Littlewood. J.E., *On inequalities in the theory of functions*, Proc. London Math. Soc., 23 (1925), 481-519.

28. Loewner. C. (K. Löwner), *Untersuchungen Über schlichte Konforme Abbildungen des Einheitskreises. I*, Math. Ann. 89, (1923), 103-121.

29. Milin. I.M., *The area method in the theory of univalent functions,* Dokl. Akad. Nauk SSSR 154 (1964), 264-267; English transl. in Soviet. Math. Dokl. 5 (1964), 78-81.

30. Milin. I.M., *Univalent functions and orthonormal systems*, Izdat. "Nauka", Moscow, 1971; English transl., Amer. Math. Soc., Providence, R.I., 1977.

31. Montel. P., *Sur les suites infinies de fonctions, Ann. Sci.* École Norm. Sup. 24 (1907), 233-334.

32. Ozawa. M., *On the Bieberbach conjecture for the sixth coefficient*, Kōdai Math. Sem. Rep., 21 (1969), 97-128.

33. Pederson. R.N., *A proof of the Bieberbach conjecture for the sixth coefficient*, Arch. Rational Mech. Anal., 31 (1968), 331-351.

34. Pederson. R. and Schiffer. M., *A proof of the Bieberbach conjecture for the fifth coefficient*, Arch. Rational Mech. Anal., 45, (1972), 161-193.

35. Robertson. M.S., *A remark on the odd Schlicht functions,* Bull. Amer. Math. Soc., 42 (1936), 366-370.

36. The Bieberbach Conjecture, Proceedings of the Symposium on the Occasion of the Proof, Mathematical Surveys and Monographs, No 21, American Mathematical Society, 1986.

Solved Exercises

1. Let Ω be a proper simply connected region in the plane which is symmetric with respect to the real axis and $z_0 \in \Omega$ be real. Prove that the Riemann mapping function of Ω onto U with $f(z_0) = 0, f'(z_0) > 0$ satisfies $f(\bar{z}) = \overline{f(z)}$ for all $z \in \Omega$.

   Solution We first note that given any $z_0 \in \Omega$, there exists one and only one Riemann mapping function f satisfying the conditions that $f(z_0) = 0, f'(z_0) > 0$. By Solved Exercises 1 of Chapter 2 we also know that if Ω is symmetric with respect to the real axis then $f(z) = u + iv$ is analytic in Ω if and only if $g(z) = \overline{f(\bar{z})} = U + iV$, with $U(z) = u(\bar{z})$ and $V(z) = -v(\bar{z})$, is analytic in $\Omega^* = \Omega$ and further

$$\frac{\partial U}{\partial x}(z) = \frac{\partial u}{\partial x}(\bar{z}) \left(\text{in a similar manner } \frac{\partial V}{\partial x}(z) = -\frac{\partial v}{\partial x}(\bar{z}) \right) \forall z \in \Omega.$$

Further it is easy to see that g is also a bijection (since f is). Now since z_0 is real, $z_0 = \bar{z}_0$ and $g(z_0) = \overline{f(\bar{z}_0)} = \overline{f(z_0)} = 0$. Further

$$g'(z_0) = \frac{\partial U}{\partial x}(z_0) + i\frac{\partial V}{\partial x}(z_0) = \frac{\partial u}{\partial x}(\bar{z}_0) - i\frac{\partial v}{\partial x}(\bar{z}_0)$$

$$= \frac{\partial u}{\partial x}(z_0) - i\frac{\partial v}{\partial x}(z_0) = \frac{\partial u}{\partial x}(z_0) > 0.$$

(Note that $f'(z_0) = \frac{\partial u}{\partial x}(z_0) + i\frac{\partial v}{\partial x}(z_0)$ is real and positive implies

$\frac{\partial u}{\partial x}(z_0) > 0$ and $\frac{\partial v}{\partial x}(z_0) = 0$). Thus g is also a Riemann mapping function satisfying the same conditions as f. By uniqueness of the Riemann mapping function it now follows that $g = f$ which is what we wanted to prove.

2. Suppose $\Omega = \{z \in \mathbb{C} : -1 < \text{Re } z < 1\}$. Find the Riemann mapping function which maps Ω onto U with $f(0) = 0, f'(0) > 0$. Compute $f'(0)$.

Solution We first note down the following points which can be easily verified.

(i) $\eta(z) = iz\pi/2$ maps $|\text{Re } z| < 1$ onto $|\text{Im } \eta| < \pi/2$.

(ii) $\zeta(\eta) = e^{\eta}$ maps $|\text{Im } \eta| < \pi/2$ onto $|\arg \zeta| < \pi/2$ or $\text{Re } \zeta > 0$.

(iii) $w(\zeta) = -i \dfrac{\zeta - 1}{\zeta + 1}$ maps $\text{Re } \zeta > 0$ onto U (see Example 3.2.11).

All these maps are analytic bijections between their respective domains and ranges. Thus $w = f(z) = -i \dfrac{e^{i\pi z/2} - 1}{e^{i\pi z/2} + 1}$ maps $|\text{Re } z| < 1$ onto U and that f is an analytic bijection. By a simple computation we see that $f(0) = 0$ and $f'(0) = \pi/4 > 0$. This is the required Riemann mapping function.

3. Let $\Omega = \{z \in \mathbb{C} : |\text{Im } z| < \pi/2\}$. Fix $\alpha + i\beta \in \Omega$. Let h be a conformal mapping of Ω onto Ω which carries $\alpha + i\beta$ to 0. Prove that $|h'(\alpha + i\beta)| = 1/\cos \beta$.

Solution Note that $\alpha + i\beta \in \Omega \Rightarrow |\beta| < \pi/2$ and so $\cos \beta > 0$.

From Example 3.2.11, $f(z) = \dfrac{e^z - 1}{e^z + 1}$ maps Ω onto U with $f(0) = 0$.

Consider the following figure.

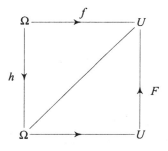

Fig. 5.S.E(i)

Define F so that the above diagram commutes, i.e., $F = f \circ h^{-1} \circ f^{-1}$. It is clear that F is an analytic automorphism of U and $F(0) = f(\alpha + i\beta) = \eta$ (say). By Theorem 4.3.34, we have $F^{-1} = \phi_{\eta}$. (Actually $F^{-1} = \lambda\phi$, with $|\lambda| = 1$ but we can assume $\lambda = 1$ without loss of generality.) From the defining relations it is easy to see that $h = f^{-1} \circ \phi_{\eta} \circ f$. Thus by chain rule and Lemma 4.5.23 we have

$$h'(\alpha + i\beta) = (f^{-1})'(0) \; \phi_{\eta}'(\eta) \; f'(\alpha + i\beta)$$

$$= (\phi_{\eta}'(\eta) \; f'(\alpha + i\beta))/f'(0).$$

By the properties of ϕ_η (see Theorem 4.3.33), $\phi'_\eta(\eta) = \dfrac{1}{1-|\eta|^2}$.

$$f'(z) = \frac{2e^z}{(1+e^z)^2} \Rightarrow f'(0) = \frac{1}{2} \text{ and } |f'(\alpha + i\beta)| =$$

$\dfrac{2e^\alpha}{1+e^{2\alpha}+2e^\alpha \cos \beta}$. Further by a simple computation we also have

$$|\eta|^2 = \frac{1+e^{2\alpha}-2e^\alpha \cos \beta}{1+e^{2\alpha}+2e^\alpha \cos \beta} \text{ so that } \frac{1}{1-|\eta|^2} = \frac{1+e^{2\alpha}-2e^\alpha \cos \beta}{4e^\alpha \cos \beta}$$

Putting these informations together we see that $|h'(\alpha + i\beta)| = 1/\cos \beta$ and this completes the solution.

4. Let f and g be analytic functions defined on the open unit disc U, mapping U onto a region Ω. Let f be one-to-one. If $f(0) = g(0)$ prove that $g(D(0, r)) \subseteq f(D(0, r))$ and $g(\overline{D(0,r)}) \subseteq f(\overline{D(0,r)})$ for all $0 < r < 1$ where $D(0, r)$ is the open disc around origin of radius r.

Solution Since f is an analytic bijection from U onto Ω, its inverse f^{-1} is well defined and analytic from Ω onto U (see Lemma 4.5.23). Consider $w(z) = (f^{-1} \circ g)(z)$. $w(z)$ is analytic in U and maps U into U. Further $w(0) = 0$ (note that $g(0) = f(0)$) and an application of Schwarz's lemma gives $|w(z)| \le |z|$. Thus $g(z) = f(w(z))$ with $|w(z)| \le |z|$. The required inclusions now follow.

5. Suppose Ω is a proper simply connected region in the plane and $z_0 \in \Omega$. Suppose that f and g are analytic bijections of Ω onto U with $f(z_0) = 0 = g(z_0)$. What relationship exists between f and g.

Solution Let $\arg f'(z_0) = \alpha$ and $\arg g'(z_0) = \beta$ with $\alpha, \beta \in [0, 2\pi]$. Then both f and $h = e^{i(\alpha - \beta)}g$ are Riemann mapping functions for the region Ω with $\arg f'(z_0) = \alpha = \arg h'(z_0)$. Thus by uniqueness of the Riemann mapping function $f \equiv h$ or that $f = e^{i(\alpha - \beta)}g$. This relationship can be described by saying that f and g are rotations of each other.

6. Define a set E in $\mathbb{C}$ to be star-like with respect to $w_0 \in E$ if the line segment joining w_0 to every other point $w \in E$ lies in E. Define E to be convex if it is star-like with respect to each of its points. A star-like function f, by definition, is a one-to-one conformal mapping of the unit disc (with $f(0) = 0$) onto a domain that is star-like with respect to the origin. A convex function, by definition, is a one-to-one conformal mapping of the unit disc (with $f(0) = 0$) onto a convex domain. Let P denote the set of all analytic functions $p(z)$ defined on U with $p(0) = 1$ and Re $p(z) > 0$ in U. Prove the following.
   (i) Let $f \in \mathcal{F}$. f is star-like if and only if $zf'(z)/f(z) \in P$.
   (ii) Let $f \in \mathcal{F}$. f is convex if and only if $1 + z(f''(z)/f'(z)) \in P$, i.e., zf' is star-like.

. (iii) $p(z) = 1 + \sum_{1}^{\infty} c_n z^n \in P \Rightarrow |c_n| \le 2.$

(iv) Every $f \in \mathcal{F}$ maps $|z| < 2 - \sqrt{3}$ onto a convex domain and that this bound $2 - \sqrt{3}$ is sharp for the family $\mathcal{F}$.

Solution We first observe certain geometrical facts.
If D is a bounded domain bounded by a simple closed curve C around the origin then D is star-like with respect to the origin if and only if there exists a single valued continuous branch for arg w which increases as w describes C in the positive direction. This is because of the following facts:

In case D is star-like with respect to the origin and $C = \partial D$ is a simple closed curve then as w describes C it will not have self intersections and since the line segment joining 0 and w ($w \in C$) lies entirely in D, arg w can not increase and decrease for w belonging to different portions of C. Conversely if arg w can be defined in an increasing way (as w describes the only boundary curve C of a domain D) then there is no possibility of the line joining 0 and w to contain a point not in D. For similar reasons a domain D whose boundary consists of a simple closed curve C is convex if and only if we can define a continuous branch for the arguments of the tangents at various points w of C in such a way that this argument function continuously increases as w describes C in the positive sense. We now start proving (i).

Let f be star-like. We first claim that f maps each sub disc $|z| < \rho < 1$ onto a domain that is star-like with respect to the origin. This is of course equivalent to proving that $g(z) = f(\rho z)$ is star-like in U. We must show that for each $0 < t < 1$ and $z \in U$ the point $tg(z)$ lies in the range of g. But since f is star-like the same condition is true for f and we see that $tf(z) = f(w(z))$ for some $w(z)$ analytic in U with $|w(z)| \le |z|$ (i.e., $f^{-1}(tf(z))$ is a function satisfying the hypothesis of Schwarz's lemma). Hence $t\,g(z) = t\,f(\rho z) = f(w(\rho z)) = g(w_1(z))$ with $|w_1(z)| = |w(\rho z)|/\rho \le |z|$. This proves that f maps $|z| = \rho$ onto a curve that bounds a domain star-like with respect to the origin. In other words we can find a continuous branch for arg $f(\rho e^{i\theta})$ as θ varies

in $[0, 2\pi]$ in such a way that $\frac{\partial}{\partial\theta}(\arg f(\rho e^{i\theta})) \ge 0$. But for $z = \rho e^{i\theta}$,

$$\frac{\partial}{\partial\theta}(\arg f(\rho e^{i\theta})) = \text{Im}\left\{\frac{\partial}{\partial\theta}\log f(\rho e^{i\theta})\right\} = \text{Im}\,\frac{iz f'(z)}{f(z)} = \text{Re}\,\frac{z f'(z)}{f(z)}.$$

Since $\text{Re}\,\dfrac{z f'(z)}{f(z)}$ is harmonic, using the minimum principle for harmonic functions we see that $\text{Re}\,\dfrac{z f'(z)}{f(z)} > 0$ for $|z| < \rho$ and as ρ is arbitrary, $\dfrac{zf'(z)}{f(z)} \in P$. Conversely suppose $f \in \mathcal{F}$ and $\dfrac{zf'(z)}{f(z)} \in P$.

Then f has a simple zero at $z = 0$ and nowhere else in U. Retracing

the above steps we see that for each $\rho < 1$, $\dfrac{\partial}{\partial\theta}(\arg f(\rho e^{i\theta}) > 0$

$(0 \le \theta \le 2\pi)$. Thus by geometric considerations described in the beginning, the image of $|z| = \rho$ under f bounds a star-like domain with respect to the origin. Since ρ is arbitrary, $f(U)$ is also a star-like domain.

We now prove (ii). Assume f is convex. As before we claim that f maps each subdisc $|z| < r < 1$ onto a convex domain. To show this let us choose z_1, z_2 in such a way that $0 < |z_1| \le |z_2| < r$ (there is no loss of generality in this assumption as the rolls of z_1 and z_2 can be interchanged) and put $w_1 = f(z_1)$, $w_2 = f(z_2)$. Let $w_0 = tw_1 + (1 - t)w_2$ for a fixed $0 < t < 1$. Since $f \in \mathcal{F}$ and is convex there exists a unique $z_0 \in U$ such that $f(z_0) = w_0$. All that we have to show is $|z_0| < r$. Consider $g(z) = tf(z_1z/z_2) + (1 - t)f(z)$. (Note that if $z_2 = 0$ then $z_1 = 0$ and $w_0 = 0 = z_0$ and hence $|z_0| < r$). Now g is analytic in U, $g(0) = 0$, $g(z_2) = w_0$. Since f is convex $h = f^{-1}(g(z))$ is well defined, analytic with $h(0) = 0$ and $|h(z)| \le 1$. By Schwarz's lemma we have $|h(z)| \le |z|$ and hence $|z_0| = |h(z_2)| \le |z_2| < r$. Thus f maps each circle $|z| = r < 1$ onto a curve which bounds a convex domain. The geometric implication of this is (as we already noted), that

$$\frac{\partial}{\partial\theta}\left(\arg\frac{\partial}{\partial\theta}f(re^{i\theta})\right) \ge 0.$$

(Note that the tangential direction of the image of $|z| = r$ under f at

the point corresponding $z = re^{i\theta}$ is given by $\arg \dfrac{\partial}{\partial\theta}f(re^{i\theta})$ which is

increasing in θ), i.e., $\dfrac{\partial}{\partial\theta}\arg[ire^{i\theta}f'(re^{i\theta})] \ge 0$

$$\text{Im}\left\{\frac{\partial}{\partial\theta}\log(f'(re^{i\theta})re^{i\theta}i)\right\} \ge 0$$

which is equivalent to $\text{Re}\left(1 + \dfrac{zf''(z)}{f'(z)}\right) \ge 0$ $(|z| = r)$.

Again as before using the minimum principle for harmonic functions

we have $\left(1 + \dfrac{zf''(z)}{f'(z)}\right) \in P$. Of course in view of (i) this condition

is equivalent to saying zf' is star-like. Conversely if this condition is satisfied retracing the above steps we see that the argument of the tangent to the image of $|z| = r$ under f is increasing or that the image of $|z| \leq r$ under f is convex. Since r is arbitrary it follows that f is convex.

To prove (iii) we first observe that every $p \in P$ has a representation called Herglotz representation formula which is given by

$$p(z) = \int_0^{2\pi} \frac{e^{it} + z}{e^{it} - z} \, d\mu(t) \tag{1}$$

where $d\mu(t)$ is a positive measure on $[0, 2\pi]$ with $\int_0^{2\pi} d\mu(t) = 1$.

(see *G* Herglotz, Über Potenzreihen mit positivem reelen Teil im Einheit-skreis, S-B Sächs. Akad. Wiss. Leipzig. Math.-Natur, Kl, 63, 1911, 501-511). Since

$$\frac{e^{it} + z}{e^{it} - z} = 1 + 2 \sum_{n=1}^{\infty} e^{-int} z^n$$

the above representation (1) gives (by comparing coefficients on both sides after integrating term by term which is permissible for each fixed z with $|z| < 1$)

$$c_n = 2 \int_0^{2\pi} e^{-int} \, d\mu(t), \qquad (n = 1, 2, \ldots)$$

Thus $|c_n| \leq 2$. Further equality holds for $p(z) = \dfrac{1 + z}{1 - z}$ as can be easily verified.

(iv) In view of the inequality

$$\left| \frac{zf''(z)}{f'(z)} - \frac{2r^2}{1 - r^2} \right| \leq \frac{4r}{1 - r^2} \quad (|z| = r < 1)$$

available for all $f \in \mathcal{F}$ (see Theorem 5.5.13). We see that (use the fact that $|w| \leq r \Rightarrow \operatorname{Re} w \geq -r$)

$$\operatorname{Re}\left(1 + \frac{zf''(z)}{f'(z)}\right) \geq \frac{1 - 4r + r^2}{1 - r^2} \tag{2}$$

Thus $\operatorname{Re}\left(1 + \dfrac{zf''(z)}{f'(z)}\right) > 0$ for $1 - 4r + r^2 > 0$. Thus f is convex for $|z| = r < 2 - \sqrt{3}$. That this bound is sharp follows by considering the

Koebe function $k(z) = z/(1 - z)^2$ for which $1 + \dfrac{zk''(z)}{k'(z)} = \dfrac{1 + 4z + z^2}{1 - z^2}$

$= 0$ for $z = -r = 2 - \sqrt{3}$ and so no constant greater than $(2 - \sqrt{3})$ can do this job for all $f \in \mathcal{F}$.

7. Denoting the set of all star-like functions in $\mathcal{F}$ by S^* and the set of all convex functions in $\mathcal{F}$ by C prove that

(i) $f(z) = z + \sum_2^\infty a_n z^n \in S^* \Rightarrow |a_n| \leq n$ for $n \geq 2$ with strict inequality

for all n unless f is a rotation of the Koebe function.

(ii) $f(z) = z + \sum_2^\infty a_n z^n \in C \Rightarrow |a_n| \leq 1$ for $n \geq 2$ with strict inequality

for all n unless f is a rotation of the function $l(z) = z/(1 - z)$.

Solution For $f \in S^*$ write

$$\phi(z) = zf'(z)/f(z) = 1 + \sum_1^\infty c_n z^n.$$

Now by the previous problem $\phi \in P$ and hence $|c_n| \leq 2$. We now have

$$zf'(z) = \phi(z)f(z).$$

Comparing the coefficients of z^n on both sides we have

$$na_n = a_n + \sum_{k=1}^{n-1} c_{n-k} a_k \quad (n = 1, 2, 3, \ldots) \tag{3}$$

with $a_1 = 1$. Since $|a_2| \leq 2$ (see Theorem 5.5.10), we shall prove our required inequality by induction. Indeed if $|a_k| \leq k$, for all $k = 1$, $2, \ldots n - 1$ (for some $n \geq 2$) then using (3) we have

$$(n - 1)|a_n| \leq \sum_{k=1}^{n-1} |c_{n-k}| |a_k| \leq 2 \sum_{k=1}^{n-1} k = n(n - 1). \tag{4}$$

Thus $|a_n| \leq n$ and the proof by induction is over. According to Theorem 5.6.10 we also have that if f is not a rotation of the Koebe function then $|a_2| < 2$ and (3) gives $|a_n| < n$ for $n > 2$. (ii) follows from (i) if we use the fact that $f \in C$ if and only if $zf' \in S^*$ by (ii) of solved exercises 6 above.

8. Prove that the range of every function $f \in C$ contains the disc $|w| < 1/2$. This constant is best possible for the entire family C.

Solution If $f \in C$ and $f(z) \neq w$ we first claim that $g(z) = (f(z) - w)^2$ is univalent. If $g(z_1) = g(z_2)$ then either $f(z_1) = f(z_2)$ or $\frac{1}{2}(f(z_1) + f(z_2)) = w$. The latter condition is impossible for a convex function which omits w (since $\frac{1}{2}(f(z_1) + f(z_2))$ is a convex combination of points in the range and so must belong to the range). Hence $g(z)$ is univalent.

Thus

$$h(z) = \frac{w^2 - g(z)}{2w} \in \mathcal{F}.$$

But $h(z) \neq w/2$ because $g(z) \neq 0$. Thus by Theorem 5.6.10 it follows that $\left|\dfrac{w}{2}\right| \geq \dfrac{1}{4}$ or that $|w| \geq \dfrac{1}{2}$. The function $l(z) = \dfrac{z}{1-z} \in C$ shows that the number $\dfrac{1}{2}$ can not be replaced by a greater constant for the entire family C.

EXERCISES

1. Find a homeomorphism of U onto U which can not be extended continuously to $\overline{U}$. Can you always extend an analytic automorphism of U continuously to $\overline{U}$?

2. Find a Schwarz-Christoffel transformation which maps the upper half plane onto the interior of an isosceles right angled triangle (try

$$w(z) = \int_0^z (t + 1)^{-3/4} (t - 1)^{-3/4} \, dt).$$

3. Prove that $f(z) = \displaystyle\int_0^z \frac{dw}{\sqrt{(1 - w^2)(1 - k^2 w^2)}}$ maps $|z| < 1$ conformally onto a rectangle with vertices ± 1, $\pm 1/k$, $(0 < k < 1)$. (These are all called "Elliptic integrals" and their inverses are doubly periodic meromorphic functions also known as "Elliptic functions").

4. Prove that the region obtained from a simply connected region by removing m points has connectivity $(m + 1)$. Find a homology basis for this region.

5. Prove that in Hurwitz theorem the limit function can be identically zero even if $f_n(z)$ never vanishes for any n. $\left(\text{Consider } f_n(z) = \dfrac{e^{z/n}}{n}\right)$

6. Suppose f is a one-to-one conformal mapping of U onto a square with centre at 0 and $f(0) = 0$. Prove that $f(iz) = if(z)$ for all $z \in U$. Deduce that if $f(z) = \displaystyle\sum_0^\infty c_n z^n$ then $c_n = 0$ unless $(n - 1)$ is a multiple of 4.

(Hint: Compare $- if(z)$ with $f(z)$ in the context of the uniqueness of the Riemann mapping function and also compare their Taylor series at origin).

7. Define f to be close-to-convex in U if f is analytic in U, $f(0) = 0 = 1 - f'(0)$ and if there exists a convex function g in U such that $f'/g' \in P$. Show that each such function is univalent and that every star-like function (and hence every convex function) is close-to-convex.

8. Prove that if f is analytic with $f'(z) \neq 0$ in U then f is close-to-convex if and only if

$$\int_{\theta_1}^{\theta_2} \text{Re}\left(1 + \frac{zf''(z)}{f'(z)}\right) d\theta > -\pi \quad (z = re^{i\theta}, \, 0 < r < 1)$$

for every pair of real numbers θ_1 and θ_2 with $\theta_1 < \theta_2$.

NOTES

As early as 1851, Riemann enunciated the basic theorem that every proper simply connected region can be mapped conformally onto the unit disc. Riemann's original proof tacitly assumed the existence of a solution to a certain extremal problem for the Dirichlet integral and was therefore incomplete. Half a century later Koebe found a proof which avoids the difficulty by posing an extremal problem over a compact normal family where the existence of an extremal function is assured. Koebe's proof has become a standard model for existential proofs in "Geometric function theory". Our presentation follows this proof and is mostly drawn from [13]. For more details on the boundary behaviour of Riemann mapping functions we refer the reader to [8, 13]. We have proved the Schwarz-Christoffel formula for the mapping of the unit disc onto the interior of a general polygon. However if the polygon happens to be either a triangle or a rectangle several interesting applications can be obtained. For instance the Legendre's modular function can be obtained by considering conformal mapping of a "triangle all of whose angles are zero" and its construction leads to a quick proof of "little Picard's theorem" [9, p 338].

By Riemann mapping theorem any two proper simply connected regions in the plane are conformally equivalent. For multiply connected regions of the same connectivity this is no longer true (see for example Theorem 5.3.2). However we can find a system of canonical regions with the property that each multiply connected region of connectivity n is conformally equivalent to one and only one canonical region. The choice of the canonical regions is to a certain extent arbitrary and there are several choices which have additional simple properties. Using the ideas of harmonic measure one can obtain these canonical regions (see for example [2]).

The following topics are suggested to the reader interested in the pursuit of knowledge in areas related to the contents of Chapter 5.

1. The Uniformization theorem: Every simply connected Riemann surface is conformally equivalent to a disc, or to the entire complex plane or to the Riemann sphere (see Chapter 10 of [1]).

2. Monodromy theorem: Define analytic continuation of a function element along arcs. State and prove the Monodromy theorem (see Chapter 16 of [13]).
3. Little Picard's theorem: Define the modular group and a modular function. Obtain its properties. Use these to prove the following: Every non- constant entire function attains each complex value with one possible exception. (That one exception can actually occur is shown by the exponential function e^z which never vanishes) (see Chapter 16 of [13]).
4. General Dirichlet's problem: Define subharmonic functions and use them in solving the problem of finding a harmonic function whose boundary values coincide with a given continuous function on the boundary. One sufficient condition could be to assume that each boundary point of the given region has a barrier. (See Chapter 6, 4.2 of [2]).

The topic on "univalent functions" is introduced in this text to motivate the reader for further studies in this area. Two of the most recent books [6, 10] can be consulted for an exhaustive treatment of this fascinating branch of Mathematics. With the help of teachers and experts in this field a large number of topics can be enlisted for writing down "Projects" at the Postgraduate/M. Phil level. A study of these topics will certainly enrich the knowledge of the reader. A few of these are given below.

1. Loewner's differential equations: Develop the theory of Loewner's differential equations and complete the understanding of the proof of the Bieberbach's conjecture given by Louis De Branges. (One can start with Chapter 3 of [6] and take up the proof of the conjecture given in [4].)
2. Polya-Schoenberg Conjecture: The convolution product of two power series $f(z) = \sum_{1}^{\infty} a_n z^n$ and $g(z) = \sum_{1}^{\infty} b_n z^n$ convergent in $|z| < 1$ is defined as $h = f * g$ where $h(z) = \sum_{1}^{\infty} a_n b_n z^n$. In   fact

$$h(r^2 e^{i\theta}) = \frac{1}{2\pi} \int_0^{2\pi} f(re^{i(\theta - t)}) \, g(re^{it}) \, dt \qquad (r < 1).$$

Develop this theory and prove the following:
"If f and g are convex functions then $f * g$ is also convex" (start from 8.3 of [6] and scan the literature).
3. Distortion of lengths under conformal mappings: Starting from [14] find out how best the length of the ray joining $z = 0$ to $z = re^{i\theta}$ is distorted under star-like functions. Understand the intricacies involved in these considerations from [3, 7, 11].

4. Grunsky inequalities: The area theorem as proved in Chapter 5 in our text has a large number of useful generalizations. They actually produce a system of inequalities called "Grunsky inequalities" which express necessary and sufficient conditions for the univalence of functions defined on the open unit disc. Together with their applications they can form the basis of a very useful topic for "Project work". One can start with Chapter 4 of [6] and go to the literature.

5. Subordination: Let U be the open unit disc. Consider two functions f and g defined and analytic in U. We say g is subordinate to f if

$$g(z) = f(w(z)) \qquad (z \in U)$$

for some analytic function $w(z)$ with $|w(z)| \le |z|$ in U. This relationship is expressed by the symbol $g \ll f$. The most interesting question in the context of subordination is as follows. "To what extent the super-ordinate function (f in $g \ll f$) dominates the behaviour of the subordinate function. A detailed study of this topic leads to many applications in the theory of univalent functions. One can start with either Chapter 2 of [10] or Chapter 6 of [6] and search the literature.

6. Analytic functions with positive real part: This class of analytic functions in the open unit disc plays a very crucial role in the theory of univalent functions. They are useful in characterizing various subclasses of the full set of univalent functions with specific properties of geometric significance. A detailed study of the properties of this class of functions together with their applications can form the basis of yet another topic. One can start with Chapter 2 of [10].

While this list is by no means exhaustive, it is suggestive of the large potential that the topic has in the development of "Geometric function theory". We shall conclude this "Notes" by quoting the translated version (see p. 7 of [12]) of the observations of R. Dedekind about the development of function theory (see [5]).

The splendid creations of this theory have exited the admiration of Mathematicians mainly because they have enriched our science in an almost unparalleled way with an abundance of new ideas and open up here-to-fore wholly unknown fields to research. The Cauchy integral formula, the Riemann mapping theorem and the Weierstrass power series calculus not only laid the ground work for a new branch of Mathematics but at the same time they furnished the first and till now the most fruitful example of the intimate connections between analysis and algebra. But it isn't just the wealth of novel ideas and discoveries which the new theory furnishes; of equal importance on the other hand are the boldness and the profundity of the methods by which the greatest of difficulties are overcome and the most recondite of truth "the mysteria functionun" are exposed to the brightest light. Complex function theory with its sheer inexhaustible abundance of beautiful and deep theorems is, as C.L. Siegel occasionally expressed it in his lectures, a one-of-a-kind gift to the Mathematician.

References

1. Ahlfors, L.V., *Conformal invariants*, McGraw-Hill (1973), New York.
2. Ahlfors, L.V., *Complex Analysis*, McGraw-Hill (3rd ed) (1979), New York.
3. Balasubramanian, R., Karunakaran, V. and Ponnusamy, S., *A proof of Hall's conjecture on star-like mappings* J. London Math. Soc., (2) 48 (1993), 278-288.
4. Branges, L.de, *A proof of the Bieberbach conjecture*, Acta Math. 154 (1985), 137-152.
5. Dedekind, R., Math. Werke 1, pp 105, 106.
6. Duren, P.L., *Univalent functions*, Springer-Verlag, (1983), New York Inc.
7. Karunakaran, V., *Length of ray images under conformal maps*, Proc. Amer. Math. Soc., 87 (1983), 289-294.
8. Lang, S., *Complex Analysis*, Springer-Verlag (2nd ed.) Graduate texts in Mathematics, 103, (1985), New York Inc.
9. Nevanlinna, R. and Paatero, V., *Introduction to Complex Analysis*, Addison-Wesley Publishing Company, (1969), Reading, Massachusetts.
10. Pommerenke, C., *Univalent functions*, (1975). Vandenhoeck and Ruprecht in Göttingen.
11. Hall, R.R., *A conformal mapping inequality for star-like functions of order 1/2, Bull. London Math. Soc.*, 12, (1980), 119-126.
12. Remmert, R., *Theory of Complex functions*, Springer-Verlag, (1991), New York Inc.
13. Rudin, W., *Real and Complex analysis*. (3rd ed.), McGraw-Hill Book Company, (1987), New York.
14. Sheil-Small, T., *Some conformal mapping inequalities for star-like and convex functions*, J. London Math. Soc., (2), 1 (1969), 577-587.

Index